Designing, Constructing, and Programming Robots for Learning

Nikleia Eteokleous
Frederick University, Cyprus

Efi Nisiforou
University of Nicosia, Cyprus

IGI Global
PUBLISHER of TIMELY KNOWLEDGE

A volume in the Advances in Educational Technologies and Instructional Design (AETID) Book Series

Published in the United States of America by
IGI Global
Information Science Reference (an imprint of IGI Global)
701 E. Chocolate Avenue
Hershey PA, USA 17033
Tel: 717-533-8845
Fax: 717-533-8661
E-mail: cust@igi-global.com
Web site: http://www.igi-global.com

Library of Congress Cataloging-in-Publication Data

Names: Eteokleous, Nikleia, 1979- editor. | Nisiforou, Efi A., 1985- editor.
Title: Designing, constructing, and programming robots for learning / Nikleia Eteokleous, and Efi A. Nisiforou.
Description: Hershey, PA : Information Science Reference, [2021] | Includes bibliographical references and index. | Summary: "This book describes and explains via theory and practice two approaches in designing, constructing and programming robots: 1) the use of educational robotics packages that are available in the market for the development of pre-fabricated robots and 2) the use of low-cost tools, materials and technologies available in our daily live activities for the development of robotics artefacts"-- Provided by publisher.
Identifiers: LCCN 2021019508 (print) | LCCN 2021019509 (ebook) | ISBN 9781799874430 (hardcover) | ISBN 9781799874447 (paperback) | ISBN 9781799874454 (ebook)
Subjects: LCSH: Robotics in education. | Robots--Design and construction. | Robots--Programming. | Educational technology.
Classification: LCC LB1028.32 .E84 2021 (print) | LCC LB1028.32 (ebook) | DDC 371.33--dc23
LC record available at https://lccn.loc.gov/2021019508
LC ebook record available at https://lccn.loc.gov/2021019509

This book is published in the IGI Global book series Advances in Educational Technologies and Instructional Design (AETID) (ISSN: 2326-8905; eISSN: 2326-8913)

British Cataloguing in Publication Data
A Cataloguing in Publication record for this book is available from the British Library.

For electronic access to this publication, please contact: eresources@igi-global.com.

Advances in Educational Technologies and Instructional Design (AETID) Book Series

Lawrence A. Tomei
Robert Morris University, USA

ISSN:2326-8905
EISSN:2326-8913

Mission

Education has undergone, and continues to undergo, immense changes in the way it is enacted and distributed to both child and adult learners. In modern education, the traditional classroom learning experience has evolved to include technological resources and to provide online classroom opportunities to students of all ages regardless of their geographical locations. From distance education, Massive-Open-Online-Courses (MOOCs), and electronic tablets in the classroom, technology is now an integral part of learning and is also affecting the way educators communicate information to students.

The **Advances in Educational Technologies & Instructional Design (AETID) Book Series** explores new research and theories for facilitating learning and improving educational performance utilizing technological processes and resources. The series examines technologies that can be integrated into K-12 classrooms to improve skills and learning abilities in all subjects including STEM education and language learning. Additionally, it studies the emergence of fully online classrooms for young and adult learners alike, and the communication and accountability challenges that can arise. Trending topics that are covered include adaptive learning, game-based learning, virtual school environments, and social media effects. School administrators, educators, academicians, researchers, and students will find this series to be an excellent resource for the effective design and implementation of learning technologies in their classes.

Coverage

- Web 2.0 and Education
- Collaboration Tools
- K-12 Educational Technologies
- Instructional Design Models
- Bring-Your-Own-Device
- Classroom Response Systems
- Virtual School Environments
- E-Learning
- Game-Based Learning
- Digital Divide in Education

IGI Global is currently accepting manuscripts for publication within this series. To submit a proposal for a volume in this series, please contact our Acquisition Editors at Acquisitions@igi-global.com or visit: http://www.igi-global.com/publish/.

The Advances in Educational Technologies and Instructional Design (AETID) Book Series (ISSN 2326-8905) is published by IGI Global, 701 E. Chocolate Avenue, Hershey, PA 17033-1240, USA, www.igi-global.com. This series is composed of titles available for purchase individually; each title is edited to be contextually exclusive from any other title within the series. For pricing and ordering information please visit http://www.igi-global.com/book-series/advances-educational-technologies-instructional-design/73678. Postmaster: Send all address changes to above address.

Titles in this Series

For a list of additional titles in this series, please visit: www.igi-global.com/book-series

Cases on Technologies in Education From Classroom 2.0 to Society 5.0
Jonathan Bishop (Congress of Researchers and Organisations for Cybercommunity, E-Learning, and Socialnomics, UK)
Information Science Reference • © 2022 • 274pp • H/C (ISBN: 9781799868781) • US $195.00

Technology-Supported Interventions for Students With Special Needs in the 21st Century
Xiongyi Liu (Cleveland State University, USA) and Patrick Wachira (Cleveland State University, USA)
Information Science Reference • © 2022 • 305pp • H/C (ISBN: 9781799888741) • US $195.00

Poverty Impacts on Literacy Education
Jill Tussey (Buena Vista University, USA) and Leslie Haas (Xavier University of Louisiana, USA)
Information Science Reference • © 2022 • 325pp • H/C (ISBN: 9781799887300) • US $195.00

TESOL Guide for Critical Praxis in Teaching, Inquiry, and Advocacy
Jenifer Crawford (University of Southern California, USA) and Robert A. Filback (University of Southern California, USA)
Information Science Reference • © 2022 • 547pp • H/C (ISBN: 9781799880936) • US $245.00

Assistive Technologies for Differently Abled Students
Sangeeta Dhamdhere-Rao (Modern College of Arts, Science and Commerce, India) and Frederic Andres (National Insitute of Informatics, Japan)
Information Science Reference • © 2022 • 305pp • H/C (ISBN: 9781799847366) • US $195.00

Education 3.0 and eLearning Across Modalities
Jeff D. Borden (Institute for Inter-Connected Education, USA)
Information Science Reference • © 2022 • 400pp • H/C (ISBN: 9781799880325) • US $195.00

Teaching Practices and Language Ideologies for Multilingual Classrooms
Ashok Bhusal (The University of Texas at El Paso, USA)
Information Science Reference • © 2021 • 210pp • H/C (ISBN: 9781799833390) • US $195.00

Practical Perspectives on Educational Theory and Game Development
Fabio Perez Marzullo (Federal University of Rio de Janeiro, Brazil) and Felipe Antonio de Oliveira (Arco Labs, Brazil)
Information Science Reference • © 2021 • 252pp • H/C (ISBN: 9781799850212) • US $185.00

IGI Global
PUBLISHER of TIMELY KNOWLEDGE

701 East Chocolate Avenue, Hershey, PA 17033, USA
Tel: 717-533-8845 x100 • Fax: 717-533-8661
E-Mail: cust@igi-global.com • www.igi-global.com

Editorial Advisory Board

Table of Contents

Detailed Table of Contents

Section 1
Educational Robotics: Teaching and Learning Practices

The first section of the book consists of four chapters that present and discuss case studies of the integration of educational robotics within classroom activities. The first chapter addresses the alignment of educational robotics (ER) tools with classroom activities. The second chapter presents a methodology for the development of activities based on educational robotics: five phases of educational robotics (5PER). The third chapter presents a method for Multi-Perspective and Multi-Scale Approach Convergence in Educational Robotics. The last one, focuses on teacher professional development training aiming to enhance the integration of educational robotics in the teaching and learning practice.

This chapter addresses the alignment of educational robotics (ER) tools with classroom activities. To this end, it first introduces a conceptualization of ER activities describing the relevant cognitive artifacts and the learning theories underlying such activities. Based on this conceptual framework, a set of design and evaluation heuristics are presented, aimed at supporting developers and educators in aligning ER tools with classroom activities and vice versa. The heuristics were elaborated in several focus groups with 35 developers and educators experienced in the domain. To evaluate the methodology in authentic contexts, two case studies with groups of developers will be presented. Moreover, to illustrate the usefulness from an educator's point of view, another example will be presented in which the devised methodology was used to guide the design of an ER classroom activity.

Section 2
Developing Computational Thinking Skills

The three chapters of the second section of the book focus on the role of educational robotics in developing student computational thinking skills. The first chapter investigates the impact of student response modality on the development of computational thinking skills in educational robotics activities. The second chapter focuses on the connection between digital storytelling and robotics. The last one discusses the relationship between educational robotics and serious games and their effect on computational thinking.

This study investigates the impact of student response modality on the development of computational thinking skills in educational robotics activities. Students of an elementary school were divided into three study groups ('Control', 'Selecting', and 'Writing') that implemented activities based on the same teacher guidance while prompted to provide responses of different modalities. The purpose was to engage students in the development of computational thinking skills, focusing on the basic skills of abstraction, generalization, algorithm, modularity, and debugging. These skills were evaluated at different phases during the activity, using different modality (selection, written, and oral) assessment tools. The results suggest that (1) prompting and eliciting thoughts in the form of written or selected answers proves to be a beneficial strategy, and (2) the two groups, 'Writing' and 'Selecting', reach the same level of CT skills, which is significantly higher than the level of the control group.

Computational thinking and educational robotics are two terms that are increasingly encountered in the literature when discussing the introduction of digital education in primary school. Computer science education is becoming increasingly important, even for young learners. Since its planned implementation in the Austrian primary school curriculum, it became necessary to find appropriate educational required concepts and methods for its implementation. Therefore, a research project was developed to investigate the use of programmable robots combined with the method of digital storytelling. To combine theoretical knowledge with pedagogical practice, the approach of educational design research was chosen. The study was conducted in five classes of primary schools. The aim was to investigate how to design a robotics-based learning environment supporting computational thinking skills. The purpose of this chapter is to give an overview of the findings of this research project.

Computational thinking (CT) is now considered an essential approach for developing critical thinking and 21st-century skills. CT as a teaching methodological approach is more connected to STEM education as it provides clearer conceptual and practical considerations to understand science, computer, and mathematical concepts. Based on the recent literature, educational robotics, applications, and serious games are the means of applying CT in teaching practice. This study examines students' needs, interests, and motivations for using a game in the context of CT. Quantitative analysis from an online questionnaire to 394 students from secondary education in different five countries (Greece, Cyprus, Italy, Poland, United Kingdom) demonstrate the students' game interests and needs that guide us to develop a game for CT's implementation in the classroom. Essential insights, considerations, and implications are providing for the design, development, and use of games for the CT in an educational environment.

The third section includes two chapters that discuss the use of robotics in promoting inclusive education. The first one focuses on using robotics to enhance social skills in children and adolescents with autism. The second chapter discusses assistive robotics integration to empower people with disabilities.

The current review focuses on the approaches that have been implemented in the development of the appropriate communication skills to children and adolescents with ASD (autism spectrum disorder) through their interaction with therapeutic robots. This study aims to emphasize the importance of the fact that robots can play a major role in enhancing social skills in students with ASD. The authors present the latest evidence on research and the most popular robots which have been used in developing social and communicational skills to children and adolescents with ASD. In particular, they focus on the use of robots in imitation, attention, eye contact, emotional regulation, initiative, and language development. It is crucial to realize that the use of robots cannot be the only educational or therapeutic procedure for ASD, but they can be used in combination with therapy and education. Robots can be a valuable tool in order to assist educators and therapists to help students with ASD developing their communication and social skills.

Science, technology, engineering, and math (STEM) education plays a vital role in order to prepare students for the twenty-first century economy. Unfortunately, people with disabilities in countries throughout the world are vastly underrepresented in scientific and technological fields. Closing the gap in the STEM is vital for improving fairness and opportunities for the people with disabilities. The STEM industry is critical in a world where there are various environmental, health, and development issues. It is essential not only for increasing productivity and competitiveness, but also for ensuring individual and societal well-being. The STEM workforce is expected to rise exponentially in the future year in today's burgeoning knowledge-intensive economy. This chapter explores a current state-of-the-art literature review on the use of assistive robots in STEM education people with disabilities.

Section 4
Different and Emergent Approaches to Educational Robotics

The final section consists of four chapters focused on different and emergent approaches to educational robotics such as online learning, making, and the use of low-cost materials to build robots. The first chapter discusses the delivery of educational robotics in online learning environments. The next one relates educational robotics, teacher training and making. The last two chapters focus on educational robotics competitions and the use of robots as tour guides.

The COVID-19 pandemic and the subsequent school closures created several challenges for teachers and students. From one day to the next, teachers had to rethink their educational activities and move to remote learning. Especially with regard to educational robotics activities, which makes large use of physical artefacts, this abrupt shift towards online learning represented a major change in how activities had to be designed and implemented. In this chapter, some experiences of online educational robotics activities carried out in compulsory schooling and teacher training are presented. The experiences are then discussed using a model for the development of educational robotics activities in order to reflect on how to design such activities that can be carried out online. The examples presented in this chapter showed there is great potential for educational robotics in online learning.

After more than 30 years of development, the designing, constructing, and programming of educational robots is still enjoying increasing popularity in formal, non-formal, and informal educational settings. Although building instructions and required technical components are easily available and accessible, the realization of own teaching projects is a special challenge and is subject to decisive influences. This

Preface

The use of robotics in education has gained global momentum due to positive effects in the teaching of science, technology, engineering, mathematics (STEM), and beyond (Andjic, Grujičić, & Markuš, 2015; Bers, & Portsmore, 2005; Bers, Seddighin, & Sullivan, 2013; Chalmers, 2013; Chalmers, 2017; Ortiz, Bos, & Smith, 2015). The introduction of robotics into the educational systems around the globe is an innovation that produces a variety of responses (positive and negative) and concerns. By introducing these technologies, medium to long-term results is sought in the teaching-learning process to keep the educational systems in line with the new era. It is also argued that cognitive skills development can be achieved when robotics and programming are integrated into developmentally appropriate ways. Existing studies have shown that through the use of robotics, students are actively engaged, collaborate, and contribute to meaningful learning outcomes (Chen, & Chang, 2008; Eteokleous & Ktoridou, 2014; Malec, 2001; Petre & Price, 2004; Rogers, 2012; Williams, Ma, Prejean, Ford, & Lai, 2007). Further, existing studies demonstrated that robotics integration could promote the development of students' higher-order thinking skills such as application, synthesis, evaluation, problem-solving, decision-making, computational thinking, and scientific investigation (Bers, Ponte, Juelich, Viera, & Schenker, 2002; Chambers & Carbonaro, 2003; Eteokleous, 2015; Khanlari, 2013; Miglino, Lund, & Cardaci, 1999; Mikropoulos & Bellou, 2013; Papert, 1980; Resnick, 2003; Talaiver & Bowen, 2010; Williams, Ma, & Prejean, 2010). Using robots for educational purposes allows the development of skills related to STEM fields and 21st-century and transversal skills (i.e., creativity, critical thinking, innovation) (Alimisis, 2013; Eteokleous, Nisiforou, & Christodoulou, 2019; Khanlari, 2013). Additionally, educational robotics allows the development of different personal abilities (Chen & Chang, 2008; Miller, Nourbakhsh & Siegwart, 2008). Previous studies, however, demonstrated that to transform and improve the educational environment, robotics activities need to be appropriately designed and henceforth implemented in the educational practice (Bauerle, & Gallagher, 2003; Benitti, 2012; Papert, 1993; Williams, Ma, & Prejean, 2010).

Jung and Won (2018) systematically and thematically reviewed the literature regarding the use of educational robotics kits, suggesting positive outcomes for teaching concepts related to the STEM areas. Along the same lines, Anwar, Bascou, Menekse, and Kardgar (2019) highlighted robotics potential as a learning and teaching tool. Henceforth, they declare robotics' impact on supporting students who do not show interest in STEM areas and its contribution to developing mental relationships to engineering, physics, and mechanistic concepts. On the other hand, Benitti (2012), after a systematic review of recently published scientific articles on the use of robotics in schools, suggests that not always educational robotics act as a parameter that promotes and enhances learning and cognitive skills. Specifically, some studies identified situations where no improvement in learning was achieved. Along the same lines, several other studies concluded that no significant gains were revealed from robotics interventions (Barker & Ansorge, 2007; Hussain, Lindh, & Shukur, 2006; Lindh & Holgersson, 2007).

Robotics education mainly follows the constructionist educational approach developed by Papert (Alimisis 2013; Eguchi, 2010; Papert 1980). Constructionist learning is based on the idea that individuals learn better when building and manipulating significant artefacts (Eguchi, 2010). In robotics, the students perceive the materials used (e.g., building blogs/bricks, sensors, gears, and motors) as toys. The educational activities designed using the aforementioned materials focus on solving a problem, identifying a solution, and explaining various phenomena by connecting theory to practice (Chambers & Carbonaro, 2003; Williams, Ma, & Prejean, 2010). There are two lines of thought around the integration of robotics in educational practice: (1) as a subject matter and (2) as a learning and cognitive tool. The first line of thought suggests that robotics should be introduced within the school curriculum as a subject matter, such as Mathematics, English, Literature, Computers (Eteokleous, Demetriou & Lambrou, 2013; Malec, 2001). However, Eteokleous and colleagues suggested that when robotics was integrated as a subject matter, as an autonomous entity (teaching about robotics), there was limited educational potential and value. The second line of thought suggests that robotics integration in the educational practice should enhance the student learning experience and support the achievement of specific learning goals (Eteokleous et al., 2013; Jonassen, 1999; Miller, Nourbakhsh, & Siegwart, 2008). Ward, Miller, Sienkiewicz, & Antonucci, 2012). Prior work indicates that when robotics is integrated as a cognitive-learning tool within a well-designed learning environment, students' full potential is exploited (Eteokleous et al., 2013). The integration of robotics as a cognitive-learning tool does not aim at helping students learn how to use a particular robotics package and its programming software (learn about robotics and programming). Rather, robotics is used as a tool within a specific educational context (learn with robotics and programming). In other words, robotics can be employed as a tool to teach and deliver concepts within various subjects at the K-12 levels such as mathematics (Whitehead, 2010), engineering (Craig, 2014), science (Vollstedt, 2005), physics, biology, psychology and beyond (Bers, Ponte, Juelich, Viera & Schenker, 2002; Craig, 2014; Eguchi, 2007; Eteokleous et al., 2013). Both lines of thought are found in the literature (i.e., Bers et al., 2002; Craig, 2014, Eguchi, 2007; Eteokleous et al., 2013; Vollstedt, 2005; Whitehead, 2010).

Teachers are a vital part of the educational process, referring to the introduction of innovations. Therefore, it is imperative to explore their attitudes to the introduction of educational robotics at school. The teachers as actors of change, which are part of the systemic change processes using technologies (Lorenz, Eickelmann, & Gerick, 2015), face new challenges based on its outcome, broadening the possibilities of integrating technologies into every aspect of the school environment (Albion, Tondeur, Forkosh-Baruch, & Peeraer, 2015).

The research in educational robotics focuses on the theoretical grounds of educational robotics use in the educational levels (primary, secondary and higher education) and settings (formal, non-formal, and informal). The various studies explain and define the concept of robotics and educational robotics, how robotics is integrated in our daily life activities and their relationship with humans, and humans' role in designing, developing, and programming robots. Research has examined and explained the educational perspective of robotics and, specifically, how they can be employed as cognitive learning tools in educational practice. It explains the two lines of thought of educational robotics as described above: as a subject matter (learning about robotics) and as a cognitive-learning tool (learning with robotics/ robotics as partners in the learning process), defining the term of educational robotics in the classroom practice. In addition, it describes the relationship of robotics to the learning theories and specifically to constructivism, constructionism, and how robotics are related to "mindtools." Finally, the knowledge constructed, and skills developed (21st-century skills, transversal skills, high-order thinking skills, transferable skills) when robotics are employed as learning tools are also presented and explained in the literature. Another significant part of the literature emphasizes the practical grounds of educational robotics and specifically, on educational robotics curricula, educational robotics clubs, the various robotics packages employed throughout all educational levels and settings, robotics artefacts and programming activities, research projects, and last but not least the various educational robotics competitions. Finally, research on educational robotics presents and explains the different approaches, case studies, and examples in developing and establishing educational robotics clubs (within school activities, within the school curriculum, and as afternoon clubs), summer camps, and competitions.

Finally, some new trends and concepts revealed in the educational research and practice of educational robotics are the following:

- The cost of the pre-prefabricated robots of the market,
- The use of low-cost material to construct robots in comparison to the pre-fabricated robots,
- The concepts of MakerSpaces, MakerEducation, making, tinkering, and fabrication and their relation to robotics,
- The concepts of STEM, STEAM, STREAM, and STREAMS and how does robotics is connected and can put into practice these three concepts,
- How robotics and programming can become an essential part of the school curriculum (formal educational settings in all levels) and not only something extra (informal and non-formal educational settings),
- Social robots and students with special needs, social robots in the classroom as a tutor and teacher assistant and how it promotes personalized teaching and learning,
- The role of educational robotics in the development of skills characterized as "other," "new," or more specifically "self-concept" (self-regulatory), as well as the development of personal abilities and skills related to learning, motivation, and student achievement.

CHALLENGES AND SEARCHING FOR SOLUTIONS

The aforementioned novel concepts and future trends can also be approached as the challenges in educational robotics that will guide and inform research, theory, and practice in the next few years. These challenges are expected to be further examined and researched for solutions to be proposed.

There is already research conducted in the field of educational robotics where it examines designing, constructing. Programming robotics using the educational robotics packages available in the market to develop pre-fabricated robots compared to the use of low-cost tools, materials, and technologies available in our daily life activities to develop robotics artefacts. However, this is an issue that will be significantly examined in future educational robotics research. The balance between the two approaches should be addressed, considering various parameters such as cost, ease of access, affordances and possibilities, relation to the school curriculum, etc. The pre-fabricated robotics package is one factor that negatively influences the systemic integration of robotics within the school curriculum.

On the other hand, using materials from our daily activities is more related to real-life situations and daily practices, and the cost is minimized. Additionally, this approach provides opportunities to students and educators to design and construct their robotics artifacts using available low-cost tools, materials, and technologies. The robotics artifact construction is expected to be further discussed in the literature concerning the following concepts: learning theories, philosophies of integration, knowledge construction, and skills developed.

Additionally, using various tools and technologies from our daily lives to construct and program robotics artifacts is directly related to the concept of MakerEducation, MakerSpaces, thinkering, and fabrication. There are lots of examples and applications where MakerSpaces are enhanced with robotics. In some cases, robotics and programming are seen as an integral part of a MakerSpace. More research is expected to be conducted to investigate the concepts of making, tinkering, and fabrication in relation to robotics and then provide guidelines and suggestions on how to design and establish a MakerSpace enhanced with educational robotics. MakerSpaces are in the process of finding their way into schools (into formal educational systems). MakerSpaces mainly operate in Museums, Libraries, Community Centers, Youth Centers as a non-formal or informal activity. Research is being conducted on how MakerSpaces can become part of a school unit in collaboration with the community and other regional organizations. Given those mentioned above, there is an alignment between the integration of educational robotics and MakerSpaces research and goals. It is believed that this alignment will further promote the expansion and development of educational robotics and MakerSpaces.

The term STEM (Science, Technology, Engineering, and Mathematics) will be directly related to educational robotics. Occasionally, the term STEM is being expanded to include more letters, thus more disciplines. The term has been developed from STEM to STEAM (A=Arts) to STREAM (R=Reading) to STREAMS (Social Sciences). The evolution of the term from STEM to STREAMS shows the direction to a more interdisciplinary integrated, interconnected approach within and across subject matters. This reveals that educational robotics can be employed within various subject topics and disciplines to promote interdisciplinarity.

In some cases, STEM, STEM Education, and STEM Clubs are synonymously used with MakerSpaces and MakerEducation. It is expected that research will focus on explaining how the concepts of STEM, STEAM, STREAM, and STREAMS are connected to educational robotics and the role of educational robotics in putting these concepts into practice. The aforementioned concepts and their connection to robotics are expected to be even more researched and examined in the literature in relation to the following concepts: learning theories, philosophies of integration, knowledge construction, and skills developed.

Another challenge that has been around for a long time, and it has been a very puzzled one, is the integration of educational robotics and programming as part of the official school curriculum. It entails a systemic approach and the involvement of various stakeholders (educators, students, instructional designers, policymakers, community members, industry representatives) to make it happened. As of now, the focus was more on robotics integration in informal and non-formal settings (i.e., afternoon clubs, summer camps, robotics competitions). Part of future research will aim to explain, compare, and contrast the integration of educational Robotics in formal, non-formal, and informal educational settings and identify and address the factors that influence robotics integration in formal educational settings.

Being part of the formal school curriculum; social and educational robots, are considered both a challenge and an opportunity for researchers, educators, and policymakers. Social robots have been related to students with disabilities and special needs. However, the potential and affordances of social robots can be seen beyond their use with students with disabilities and special needs. Research should also focus on employing social robotics as classroom tutors and/or teaching assistants and promoting personalized teaching and learning.

It seems that educational robotics influence the development of skills characterized as "other," "new," or, more specifically, "self-concept" (self-regulatory), as well as the development of personal abilities and skills related to learning, motivation, and student achievement. Literature mainly focuses on the direct impact of robotics in learning and the development of specific skills. Limited research has been conducted in investigating the relationship of educational robotics and other skills, i.e., personal abilities (Chen & Chang, 2008; Miller, Nourbakhsh & Siegwart, 2008), self-concept beliefs, and self-regulatory skills (Agatolio, Pluchino, Orso, Menegatti, & Gamberin, 2018; Eteokleous & Kolani, 2018; Master, Cheryan, Moscatelli, and Meltzoff, 2017; Schunk & Ertmer, 1999). It seems that more research studies are needed to examine the impact of robotics on the development of new skills in students (Jung & Won, 2018) since there are also studies suggesting that self-concept skills are related to learning, motivation, and student achievement (Bandura, 1993; Master, Cheryan, Moscatelli, and Meltzoff, 2017; Schunk & Ertmer, 1999).

The development of skills characterized as "other," "new," or more specifically, "self-concept" skills seemed to be essential due to the changes in the global competition and collaboration, the focus on a service economy, and the information growth. The workforce needs, job tasks, and type of work have changed, and consequently, the required skills are constantly differentiating. It is increasingly important that students, as the future citizens of the Information Society, are equipped with various cognitive, social, soft, and life skills to survive in today's globalized, rapidly changing, highly demanding interconnected world and succeed in their career aspirations. Additionally, the challenge emerges in response to how best to cultivate students' skills.

Technology plays a crucial role in the assimilation of these skills. Emerging technologies such as robotics provide challenges and opportunities for learners to develop numerous skills. What is the context and the tools through which these skills can be developed? How can technology develop students' self-concept and self-belief skills, which are essential for the Information Society?

Last but not least, the covid-19 pandemic has revealed another opportunity for educational robotics and programming to feature in online, hybrid, and hyflex learning environments. This incident entails the use of online learning platforms, technological tools, such as simulations, augmented and virtual reality applications, the design, development, and delivery of appropriate courses and educational material (i.e., handouts, exercises, activities, interactive presentations, and videos, narrated presentations, quizzes). Research should focus on further examining this newly established, and at the same time, promising approach to educational robotics and programming.

ORGANIZATION OF THE BOOK

The current book highlights the importance of integrating robotics in the educational practice arena and presents various ways to achieve it. In addition, it explains how various skills can be developed through a set of hands-on learning experiences using robotics. Besides, it describes how various 21st-century skills and problem-solving and computational thinking skills can be acquired through the hands-on experience of educational robotics. It presents and explains "the theory" based on research results and the "the practice" of educational robotics based on hands-on experiences. It presents and explains the theory behind robotics education which leads to the practical sound of educational robotics as emerged through scientific and empirical research evidence. The book is organized into thirteen chapters. A brief description of each of the chapters follows.

Chapter 1, under the title "Aligning the Design of Educational Robotics Tools With Classroom Activities," relates and discusses constructionism, social constructivism, design principles, instructional alignment, learning systems, and educational robotics. It addresses the alignment of educational robotics (ER) tools with classroom activities. To this end, it first introduces a conceptualization of ER activities describing the relevant cognitive artefacts and the learning theories underlying such activities. Based on this conceptual framework, a set of design and evaluation heuristics are presented, supporting developers and educators in aligning ER tools with classroom activities and vice versa. The heuristics were elaborated in several focus groups with 35 developers and educators experienced in the domain. To evaluate the methodology in authentic contexts, two case studies with groups of developers will be presented. Moreover, another example will be presented to illustrate the usefulness from an educator's point of view. The devised methodology was used to guide the design of an ER classroom activity.

Chapter 2 presents a methodology for the development of activities based on educational robotics. The chapter is entitled "5 Phases of educational robotics (5PER): The Scientific Method applied to Educational Robotics." First, the authors describe how technology and science have evolved to this day, recognizing their importance and impacting the schools with STEM approaches. Then, the main theories and learning approaches that have driven educational robotics in schools are reviewed, and concepts of robotics, educational robotics, and the difference between them. Finally, each of the 5 phases is explained, and two practical examples of how to implement them in learning activities are shown. The chapter discusses and explores the following concepts: educational robotics, science, technology, scientific method, STEM, pedagogy, scientist, philosopher, robotics, and system control.

Chapter 3 describes "A Method for Multi-Perspective and Multi-Scale Approaches Convergence in Educational Robotics." It connects the following concepts: educational robotics, multidisciplinarity, micro-robotics, nanorobotics, sensors, and actuators. The exponential growth of publications on educational robotics (ER) in the last ten years, undoubtedly, in many ways, is due to the introduction and the ubiquitous application of such platforms as Arduino, micro: bit, Raspberry Pi, and others. These instruments offer a variety of ways for STEM curricula introduction. Being centered on technological and engineering issues, the ER methodology offers an excellent opportunity for students of different ages. By lowering the entry-level effort, the platforms significantly simplify the first steps in the field.

Nevertheless, as every simplification presumes, a significant part of "underlying machinery" remains hidden. Indeed, the character of in-school ER projects barely holds enough space for a detailed treatment of the concepts. Nevertheless, the chapter illustrates the importance of exposing students to the currently available instruments and providing in-depth conceptual insight. To support the thesis, several examples, unifying multi-perspective and multi-scale approaches, are provided.

Chapter 4, entitled "The Impact of the Professional Development Training of Primary Education Teachers on Issues of Educational Robotics as a Learning Tool," aims to reveal the importance of professional development training in educational robotics integration as a teaching and learning tool. Specifically, studies reveal a number of factors that influence teachers' decisions to use Educational Robotics (ER)in the classroom. One of these factors is teachers' professional development training (PDF). PDF considers being an encouraging factor that obstructs the successful application of Educational Robotics (ER) among the teachers. This chapter addresses the importance of Professional Development Training (PDT) by examining its effectiveness in integrating Educational Robotics (ER) in the educational practice for teaching and learning purposes in various cognitive subjects. The sample of the study consisted of 23 elementary school teachers. The data used in this research was collected via pre and post-structured questionnaires and classroom observations. The findings reveal the importance of PDT and suggest that PDT for ER teachers was valuable in improving teachers' ER skills and knowledge.

Chapter 5 investigates the impact of students' response modality on the development of computational thinking skills in educational robotics activities. It is entitled "Efficiently Prompting Students When Developing Computational Thinking Skills: The Impact of Students' Response Modality." It aims to relate and discuss the concepts of computational thinking, guidance, prompting, response modality, and educational robotics. Students of an elementary school were divided into three study groups ('Control,' 'Selecting,' and 'Writing') that implemented activities based on teacher guidance while prompted to respond to different modalities. The purpose was to engage students in developing computational thinking skills, focusing on abstraction, generalization, algorithm, modularity, and debugging. These skills were evaluated at different phases during the activity, using different modality (selection, written, and oral) assessment tools. The results suggest that: (a) prompting and eliciting thoughts in the form of written or selected answers proves to be a beneficial strategy, (b) the two groups "Writing" and "Selecting" reach the same level of CT skills, which is significantly higher than the level of the control group.

Chapter 6 discusses the concepts of digital storytelling, educational robotics, computation thinking skills in primary education under the title "Developing Storytelling Activities Supporting Computational Thinking Using Educational Robots." Computational thinking and educational robotics are two terms that are increasingly encountered in the literature when discussing the implementation of digital education in primary school. Computer science education is becoming increasingly important, even for young learners. Since its planned implementation in the Austrian primary school curriculum, it became necessary to find appropriate didactic approaches and methods needed for its introduction. Therefore, a research project was developed to investigate the use of programmable robots combined with digital storytelling. To combine theoretical knowledge with pedagogical practice, the approach of educational design research was chosen. The study was conducted in five classes of primary schools. The aim was to investigate how to design a robotics-based learning environment that supports computational thinking skills. The purpose of this chapter is to give an overview of the findings of this research project.

Chapter 7 analyzes, compares, and contrasts concepts regarding computational thinking, serious games, robotics in secondary education, under the title "Towards the Development of a Game for Computational Thinking: Identifying the Students' Needs and Interests." Computational Thinking (CT) is now considered an essential approach for developing critical thinking and 21st-century skills. A teaching methodological approach is more connected to STEM education as it provides clearer conceptual and practical considerations to understand science, computer, and mathematical concepts. Based on the recent literature, educational robotics, applications, and games are the means of applying CT in teaching practice. This study examines students' needs, interests, and motivations for using a game in the context of CT. Quantitative analysis from an online questionnaire to 394 students from secondary education in five different countries (Greece, Cyprus, Italy, Poland, United Kingdom) demonstrate the students' game interests and needs that guide us to develop a game for CT's implementation in the classroom. Essential insights, considerations, and implications provide the design, development, and use of games for the CT in an educational environment.

Chapter 8 is entitled "The Use of Robotics in Enhancing Social Skills in School and Therapeutic Settings in Children and Adolescents With Autism Spectrum Disorder." It analyzes and discusses the concepts of robotics, social skills, communication school, and therapeutic settings. The current review focuses on the approaches that have been implemented in the development of the appropriate communication skills for children and adolescents with ASD (Autism Spectrum Disorder) through their interaction with therapeutic robots. This study emphasizes the importance that robots can play a major role in enhancing social skills in students with ASD. We will present the latest evidence on research and the most popular robots used in developing social and communicational skills for children and adolescents with ASD. In particular, we will focus on using robots in imitation, attention, eye contact, emotional regulation, initiative, and language development. It is crucial to realize that robots cannot be the only educational or therapeutic procedure for ASD, but they can be combined with therapy and education. Robots can be a valuable tool to assist educators and therapists in helping students with ASD developing their communication and social skills.

Chapter 9 discusses how students and adults with disabilities can use assistive technology and specifically assistive robotics under the title "Integrating Assistive Robotics in STEM Education to Empower People with Disabilities." Science, technology, engineering, and math (STEM) education is vital in preparing students globally more competitive in the twenty-first-century economy. Unfortunately, however, people with disabilities in countries worldwide are vastly underrepresented in scientific and technological fields. Closing the gap in STEM is vital for improving fairness and opportunities for people with disabilities. This chapter explores the current state-of-art literature review on assistive robots in STEM education for people with disabilities.

Chapter 10 is entitled "Designing Tools and Activities for Educational Robotics in Online Learning." It presents an analysis of issues in regards to educational robotics, distance and remote learning, and teacher training. The Covid-19 pandemic and the subsequent school closures created several challenges for teachers and students. From one day to the next, teachers had to rethink their educational activities and move to remote learning. Especially with respect to educational robotics activities, which make considerable use of physical artefacts, this abrupt shift towards online learning represented a major change in how activities had to be designed and implemented. This chapter presents some experiences of online educational robotics activities carried out in compulsory schooling and teacher training. The experiences are then discussed using a model for the development of educational robotics activities to reflect on how to design such activities that can be carried out online. The examples presented in this chapter showed there is great potential for educational robotics in online learning.

Chapter 11 addresses the issue of "Creating Educational Robots: Basics From Teacher Training to Making," relating and discussing the following concepts: engineering design process, design models, coding, behaving machines, construction kits, building systems, assembly, design and manufacturing tasks. After more than 30 years of development, educational robots' designing, constructing, and programming are still enjoying increasing popularity in formal, non-formal, and informal educational settings. Although building instructions and required technical components are easily available and accessible, the realization of own teaching projects is a challenge and subject to decisive influences. This includes the content-related training of teachers as well as their attitudes and ways of thinking and acting. Therefore, the first section of this chapter spans an arc from the didactic concept of the extracurricular project technikcamps related to robotics. The experiences gained from it and the consequences of teacher training to technical education's philosophical roots follow. In connection with this, the main part deals with the technological basics of creating educational robots in general. It leads from manufacturing single parts through creating a support structure and automation to the application of the engineering design process.

Chapter 12 reviews and presents how low-cost robots can be employed in schools under the title "Brushbots for Educational Robotics: Low-Cost Robots for Building and Research Activities in Schools." Due to their simple construction, brushbots are popular and widely known as craft objects. Building instructions promise tinkering fun and a promoting effect but minor as low-cost research objects. They are regarded as microbots and give opportunities for discussions. One question to answer is whether their chaotic motion can be influenced. Therefore, this contribution deals with an examination of factors influencing their movement behavior. The aim was to show experimental work and discuss the potential of brushbots in three directions: First and foremost, to propose a cheap platform to evaluate the motion capabilities of different designs. Second, offer an approach to improving through robots the experimentation results in the design process and provide a sample that can be transferred to schools as third. For this purpose, the basics of brushbots are presented, and the experiments and their results are described. The results offer inspiration for teaching applications that go beyond mere tinkering and provide approaches for further research.

Last but not least, Chapter 13 focuses on educational robotics competitions, STEM, STEAM, and interdisciplinary projects under the title "Educational Robotic Competitions: Methodology, Practical Appliance, and Experience." In this chapter, the authors present a methodology for forming and preparing teams to participate in educational robotics competitions composed of 5 stages: Recruitment, Training, Team confirmation & preparation, Prototyping, and Competition Day. The chapter begins with an introduction to educational robotics applied in schools and its use to implement Design-based Learning, followed by a description of the characteristics of the educational robotics kits and a brief description of the 5PER methodology used in the 4th stage: Prototyping, with a description of the main robotics competencies worldwide. The five stages methodology is presented in detail to be replicated by other educators around the world. The exposed methodology is based on the authors' experience in preparing equipment for participation in various robotics competitions at the Faculty of Electrical and Electronic Engineering of the Universidad Nacional de Ingenieria Lima-Peru.

Nikleia Eteokleous
Frederick University Cyprus

Efi Nisiforou
University of Nicosia, Cyprus

REFERENCES

Agatolio, P., Pluchino, P., Orso, V., Menegatti, E., & Gamberin, L. (2018). How Robots Impact Students' Beliefs about Their Learning Skills. In *HRI '18 Companion: 2018 ACM/IEEE International Conference on Human-Robot Interaction Companion*. ACM. 10.1145/3173386.3177042

Albion, P. R., Tondeur, J., Forkosh-Baruch, A., & Peeraer, J. (2015). Teachers' professional development for ICT integration: Towards a reciprocal relationship between research and practice. *Education and Information Technologies*, *20*(4), 655–673. doi:10.100710639-015-9401-9

Alimisis, D. (2013). Educational robotics: Open questions and new challenges. *Themes in Science and Technology Education*, *6*(1), 63–71.

Andjic, B., Grujičić, R., & Markuš, M. M. (2015). Robotics and Its Effects on the Educational System of Montenegro. *World Journal of Education*, *5*(4), 52. doi:10.5430/wje.v5n4p52

Anwar, S., Bascou, N. A., Menekse, M., & Kardgar, A. (2019). A Systematic Review of Studies on Educational Robotics. *Journal of Pre-College Engineering Education Research (J-PEER), 9*(2). doi:10.7771/2157-9288.1223

Bandura, A. (1993). Perceived self-efficacy in cognitive development and functioning. *Educational Psychologist, 28*(2), 117–148. doi:10.120715326985ep2802_3

Barker, B. S., & Ansorge, J. (2007). Robotics as means to increase achievement scores in an informal learning environment. *Journal of Research on Technology in Education, 39*(3), 229–243. doi:10.1080/15391523.2007.10782481

Bauerle, A., & Gallagher, M. (2003). Toying With Technology: Bridging the Gap Between Education and Engineering. In *Proceedings of Society for Information Technology and Teacher Education International Conference 2003* (pp. 3538-3541). Chesapeake, VA: AACE.

Benitti, F. B. V. (2012). Exploring the educational potential of robotics in schools: A systematic review. *Computers & Education, 58*(3), 978–988. doi:10.1016/j.compedu.2011.10.006

Bers, M., Seddighin, S., & Sullivan, A. (2013). Ready for robotics: Bringing together the T and E of STEM in early childhood teacher education. *Journal of Technology and Teacher Education, 21*(3), 355–377.

Bers, M. U., Ponte, I., Juelich, C., Viera, A., & Schenker, J. (2002). Teachers as designers: Integrating robotics in early childhood education. *Information Technology in Childhood Education Annual*, 123–145.

Bers, M. U., & Portsmore, M. (2005). Teaching partnerships: Early childhood and engineering students teaching math and science through robotics. *Journal of Science Education and Technology, 14*(1), 59–73. doi:10.100710956-005-2734-1

Chalmers, C. (2013). Learning with FIRST LEGO League. In Society for Information Technology and Teacher Education (SITE) Conference. Association for the Advancement of Computing in Education (AACE).

Chalmers, C. (2017). Preparing Teachers to Teach STEM through Robotics. *International Journal of Innovation in Science and Mathematics Education, 25*(4), 17–31.

Chambers, J., & Carbonaro, M. (2003). Designing, developing, and implementing a course on LEGO robotics for technology teacher education. *Journal of Technology and Teacher Education, 11*(2), 209–241.

Chen, G. D., & Chang, C. W. (2008, November). Using humanoid robots as instructional media in elementary language education. In *Digital Games and Intelligent Toys Based Education, 2008 Second IEEE International Conference on* (pp. 201-202). IEEE. 10.1109/DIGITEL.2008.17

Craig, D. C. (2014). Robotics Programs: Inspiring Young Women in STEM. In Girls and Women in STEM (pp. 153-174). Charlotte, NC: Information Age Publishing, Inc.

Eguchi, A. (2007). Educational Robotics for Elementary School Classroom. In *Proceedings of Society for Information Technology & Teacher Education International Conference 2007* (pp. 2542-2549). Chesapeake, VA: AACE.

Eguchi, A. (2010, March). What is educational robotics? Theories behind it and practical implementation. In Society for information technology & teacher education international conference (pp. 4006-4014). Association for the Advancement of Computing in Education (AACE).

Eteokleous, N. (2015). *Employing Educational Robotics for the Development of Problem-Based Learning Skills*. Paper presented at the 2015 Annual Meeting of the American Educational Research Association, Chicago, IL.

Eteokleous, N., Demetriou, A. Y., & Stylianou, A. (2013). The Pedagogical Framework for Integrating Robotics as an Interdisciplinary Learning – Cognitive Tool. In J. Roselli & E. Gulick (Eds.), *Information and Communications Technology: New Research* (pp. 141–158). Nova Science Publishers, Inc.

Eteokleous, N., & Kolani, E. (2018). *Building Robots, Self-Esteem, Self-Perception and Social Adequacy for Students with Learning Disabilities*. Included in the Online Paper Repository System of the American Educational Research Association Conference (AERA), New York, NY.

Eteokleous, N., & Ktoridou, D. (2014). Educational robotics as learning tools within the teaching and learning practice. In *Global Engineering Education Conference (EDUCON), 2014 IEEE* (pp. 1055-1058). IEEE. 10.1109/EDUCON.2014.6826237

Eteokleous, N., Nisiforou, E., & Christodoulou, C. (2019). Creativity Thinking Skills Promoted through Educational Robotics. In Educational Robotics in the Context of the Maker Movement. Advances in Intelligent Systems and Computing (pp. 233-237RS). Springer International Publishing.

Hussain, S., Lindh, J., & Shukur, G. (2006). The effect of LEGO training on pupils' school performance in mathematics, problem-solving ability, and attitude: Swedish data. *Journal of Educational Technology & Society*, *9*(3), 182–194.

Jonassen, D. H. (1999). *Computer as Mindtools in Schools: Engaging Critical Thinking* (2nd ed.). Prentice-Hall.

Jung, E. S., & Won, E. (2018). Systemic Review of Research Trends in Robotics Education for Your Children. *Sustainability*, *12*(905). Advance online publication. doi:10.3390u12125153

Khanlari, A. (2013). Effects of robotics on 21st-century skills. *European Scientific Journal, ESJ*, *9*(27), 27–36.

Lindh, J., & Holgersson, T. (2007). Does LEGO training stimulate pupils' ability to solve logical problems? *Computers & Education*, *49*(4), 1097–1111. doi:10.1016/j.compedu.2005.12.008

Lorenz, R., Eickelmann, B., & Gerick, J. (2015). What affects students' computer and information literacy worldwide? An analysis of school and teacher factors in high-performing countries. In *Society for Information Technology & Teacher Education International Conference* (pp. 1212-1219). Association for the Advancement of Computing in Education (AACE).

Malec, J. (2001). Some thoughts on robotics for education. *2001 AAAI Spring Symposium on Robotics and Education*.

Master, A., Cheryan, S., Moscatelli, A., & Meltzoff, A. N. (2017). Programming experience promotes higher STEM motivation among first-grade girls. *Journal of Experimental Child Psychology, 160*, 92–106. doi:10.1016/j.jecp.2017.03.013 PMID:28433822

Miglino, O., Lund, H.H. & Cardaci, M. (1999). Robotics as an Educational Tool. *Journal of Interactive Learning Research, 10*(1), 25-47.

Mikropoulos, T. A., & Bellou, I. (2013). Educational robotics as mindtools. *Themes in Science and Technology Education, 6*(1), 5–14.

Miller, D. P., Nourbakhsh, I. R., & Siegwart, R. (2008). Robots for education. In *Springer handbook of robotics* (pp. 1283–1301). Springer Berlin Heidelberg. doi:10.1007/978-3-540-30301-5_56

Ortiz, A. M. (2010). *Fifth-grade students' understanding of ratio and proportion in an engineering robotics program* (Doctoral dissertation). Tufts University.

Papert, S. (1980). *Mindstorms: Children, computers, and powerful ideas*. Basic Books.

Papert, S. (1993). *Mindstorms: Children, computers, and powerful ideas* (2nd ed.). Basic Books.

Petre, M., & Price, B. (2004). Using robotics to motivate "back door" learning. *Education and Information Technologies, 9*(2), 147–158. doi:10.1023/B:EAIT.0000027927.78380.60

Resnick, M. (2003). Playful learning and creative societies. *Educator's Update, 8*(6). https://web.media.mit.edu/wmres/papers/education-update.pdf

Rogers, C. B. (2012). Engineering in kindergarten: How schools are changing. *Journal of STEM Education: Innovations and Research, 13*(4), 4–9.

Schunk, H. D., & Ertmer, A. P. (1999). Self-regulatory processes during computer skill acquisition: Goal and self-evaluative influences. *Journal of Educational Psychology, 91*(2), 251–260. doi:10.1037/0022-0663.91.2.251

Talaiver, M., & Bowen, R. (2010). Developing 21st-Century Skills: Game Design and Robotics Exploration. In D. Gibson & B. Dodge (Eds.), *Proceedings of Society for Information Technology & Teacher Education. International Conference 2010* (pp. 2089-2090). Chesapeake, VA: AACE.

Vollstedt, A-M. (2005). *Using robotics increases student knowledge and interest in science, technology, engineering, and math* (Master's thesis). Retrieved from: ProQuest Dissertation and Theses database.

Ward, R., Miller, J. L., Sienkiewicz, R., & Antonucci, P. (2012). ITEAMS: Increasing the self-identification for girls and underserved youth in pursuing STEM careers. *Journal of Systemics, Cybernetics and Informatics, 10*(1), 95–99.

Whitehead, S. H. (2010). *Relationship of robotic implementation on changes in middle school students' beliefs and interest toward science, technology, engineering, and mathematics* (Doctoral Dissertation). Retrieved from: ProQuest Dissertation and Theses database.

Williams, D., Ma, Y., & Prejean, L. (2010). A Preliminary Study Exploring the Use of Fictional Narrative in Robotics Activities. *Journal of Computers in Mathematics and Science Teaching, Chesapeake, VA: AACE, 29*(1), 51–71.

Williams, D. C., Ma, Y., Prejean, L., Ford, M. J., & Lai, G. (2007). Acquisition of physics content knowledge and scientific inquiry skills in a robotics summer camp. *Journal of Research on Technology in Education*, *40*(2), 201–216. doi:10.1080/15391523.2007.10782505

Acknowledgment

First and foremost, the editors would like to thank all 27 authors coming from different universities and countries (Austria, Cyprus, Germany, Greece, Italy, Peru, Switzerland, Turkey, USA) around the world for their contribution and involvement in this project. The chapters developed based on the authors' knowledge, expertise, and research on the field of educational robotics made it possible. The authors' contributions provide valuable and important insights and findings in the field of educational robotics.

Second, the editors wish also to acknowledge the tremendous and valuable contribution of the Reviewers and the members of the Editorial Advisory Board that participated in the chapters review process. Their comments and feedback highly contributed to improving the quality, coherence, and content presentation of chapters. Most of the authors also served as referees; we highly appreciate their double task.

Nikleia Eteokleous
Frederick University, Cyprus

Efi Nisiforou
University of Nicosia, Cyprus

Section 1

Educational Robotics: Teaching and Learning Practices

The first section of the book consists of four chapters that present and discuss case studies of the integration of educational robotics within classroom activities. The first chapter addresses the alignment of educational robotics (ER) tools with classroom activities. The second chapter presents a methodology for the development of activities based on educational robotics: five phases of educational robotics (5PER). The third chapter presents a method for Multi-Perspective and Multi-Scale Approach Convergence in Educational Robotics. The last one, focuses on teacher professional development training aiming to enhance the integration of educational robotics in the teaching and learning practice.

Chapter 1
Aligning the Design of Educational Robotics Tools With Classroom Activities

Christian Giang
https://orcid.org/0000-0003-2034-9253
SUPSI-DFA, Switzerland

Alberto Piatti
SUPSI-DFA, Switzerland

Francesco Mondada
https://orcid.org/0000-0001-8641-8704
EPFL, Switzerland

ABSTRACT

This chapter addresses the alignment of educational robotics (ER) tools with classroom activities. To this end, it first introduces a conceptualization of ER activities describing the relevant cognitive artifacts and the learning theories underlying such activities. Based on this conceptual framework, a set of design and evaluation heuristics are presented, aimed at supporting developers and educators in aligning ER tools with classroom activities and vice versa. The heuristics were elaborated in several focus groups with 35 developers and educators experienced in the domain. To evaluate the methodology in authentic contexts, two case studies with groups of developers will be presented. Moreover, to illustrate the usefulness from an educator's point of view, another example will be presented in which the devised methodology was used to guide the design of an ER classroom activity.

INTRODUCTION

Previous research has acknowledged the potential of Educational Robotics (ER) as an innovative pedagogical approach for both technical and non-technical disciplines (Anwar, Bascou, Menekse, &

DOI: 10.4018/978-1-7998-7443-0.ch001

Kardgar, 2019; Benitti, 2021; Jung & Won, 2018; Toh, Causo, Tzuo, Chen & Yeo, 2016). For instance, ER is increasingly being used to foster the development of students' 21st century skills (Eguchi, 2014; Khanlari, 2013; Negrini & Giang, 2019) as well as computational thinking competencies (Atmatzidou & Demetriadis, 2015; Bers, Flannery, Kazakoff & Sullivan, 2014). Furthermore, ER can represent an important building block to implement progressive continuing professional development programs (CPD) in computer science for primary school teachers (El-Hamamsy et al., 2020) facilitating the adoption of computer science content by both pioneer and novice teachers (El-Hamamsy et al., 2021).

As for any kind of educational tool, the successful integration of ER in classroom education requires a proper alignment of the tools with the classroom activities. However, previous research has argued that in practice, this is often not the case for educational technology (Antonenko, Dawson & Sahay 2017). According to Antonenko et al. (2017), one possible reason is that only little conceptual research has been performed to construct useful frameworks helping educational technology developers to align the affordances of their systems with the needs of the target users (i.e., teachers and students). As a matter of fact, such conceptual frameworks are also scarce for the specific case of ER. Not surprisingly, previous research has suggested that "*future research needs to deal with the more effective design of robots to align with the educational need*" (Pachidis et al., 2018, p.7).

Proper alignment, on the other hand, is also important when conceiving classroom activities involving ER tools. Well-designed tools are meaningless if they are not appropriately used in teaching and learning activities. However, the use of ER in formal education is a relatively recent trend and it still appears that more educators need to develop the required technological pedagogical content knowledge to effectively integrate ER into classroom activities (Koehler & Mishra, 2013; Sisman & Kucuk, 2019). As Sisman and Kucuk (2019) have pointed out, many educators are still not aware of the benefits of ER, and those who are, are not prepared to leverage ER for teaching. A framework providing a comprehensive conceptualization of ER classroom activities could hence better guide educators in preparing and implementing such activities.

The technological advances of the past decades have given rise to an increased integration of educational technology in classrooms and in this context, different alignment frameworks have been developed, for instance, for the design of computer supported collaborative learning (CSCL) environments (Kirschner, Strijbos, Kreijns & Beers, 2004), e-learning tasks (Bower, 2008) and educational technologies as a whole (Antonenko et al., 2017). However, these frameworks were devised for entirely screen-based implementations of educational technology. ER activities, in contrast, often also involve objects in the physical world, suggesting that the existing frameworks may not be unreservedly applicable in these cases. Indeed, ER activities correspond to what Heersmink (2013) has denoted as "situated cognitive systems", involving three types of cognitive artifacts: Apart from the robots, they also involve programming or interaction interfaces as well as playgrounds for the learning tasks (Chevalier, Giang, Piatti & Mondada, 2020). A framework that conceptualizes ER activities thus needs to consider how these artifacts can be meaningfully integrated in the learning system as a whole in order to align with the learning goals as well as with instructional and assessment activities.

SCOPE AND MAIN FOCUS OF THE CHAPTER

The main focus of this chapter is to introduce a conceptual framework, that can support developers in the design of ER tools as well as educators in the preparation and implementation of ER classrooms activities. The next section presents the model for Educational Robotics Learning Systems (ERLS), which represents the fundamental theoretical framework of this chapter. Subsequently, a set of design heuristics will be introduced, aimed at supporting developers and educators in operationalizing the ideas of the ERLS model. To illustrate the usefulness of the proposed methodology, this chapter will finally present case studies in which it was applied to guide the development of two different ER tools as well as the design of an ER classroom activity. Specifically, the findings of this chapter aim to address the following research questions:

- How can ER activities be conceptualized?
- How can developers design ER tools that are aligned with classroom activities?
- How can educators design classroom activities that are aligned with the affordances of ER tools?

CONCEPTUALIZING ALIGNMENT IN EDUCATIONAL ROBOTICS

A well-established principle that has guided curriculum planning for many years is instructional alignment. It describes how aligning intended outcomes, instructional processes and instructional assessment can enhance teaching and learning experiences (Cohen, 1987; Lauwers, 2010; MacPhail, Tannehill, Leirhaug, & Borghouts, 2021). Indeed, it has been shown that improvements of student achievement could be observed when alignment of these three components was consciously considered (Mitchell, 1999; Wonder-McDowell et al., 2011, McCormick et al., 2020). Biggs (1996) extended the idea of instructional alignment by incorporating the learning theory of constructivism, that should guide the alignment of the components. Biggs' idea of "constructive alignment" particularly emphasizes the importance of considering teaching and learning as a system, in which ideally "*all aspects of teaching and assessment are tuned to support high-level learning*" (Biggs, 2003, p. 1). In contrast to Cohen's traditional model of alignment, integrating constructivsm as a guiding learning theory "*puts the student at the heart of the learning*" (Loughlin, Lygo-Baker & Lindberg-Sand, 2021, p.133).

An important step towards a conceptual alignment framework for ER activities was presented by Lauwers (2010). He argued that integrating educational robots in the instructional component of traditional alignment models would not be sufficient, and instead, proposed an extended alignment model that integrates a fourth component, denoted as the "tool" (Lauwers, 2010). He then presented an alignment-centered design process for the development of ER tools but did not further elaborate on the proposed alignment model. Particularly, the proposed model lacks a description of cognitive artifacts involved in ER activities as well as a discussion on how the underlying ideas can be operationalized to attain alignment. However, in order to properly discuss alignment in the context of ER activities, it is essential to specify the kind of activities that are considered as well as the cognitive artifacts that are involved. In ER activities, robots can take the roles of tutors, peers or tools (Mubin et al., 2013). When they are used as tutors or peers, they can be classified as social robots, delivering the learning experience through social interaction with the learners (Belpaeme et al., 2018). In contrast, robots taking the role of tools, do not primarily focus on social interactions with the learners. These robots, instead, represent pedagogical instruments that are mainly used as cognitive artifacts, i.e., "*human-made physical objects that functionally contribute to performing a cognitive task*" (Heersmink, 2013, p. 465). In fact, such "objects to think with", represent the key element of Papert's constructionism (Papert, 1980) which incidentally,

Figure 1. Alignment model for Educational Robotics Learning Systems (ERLS)

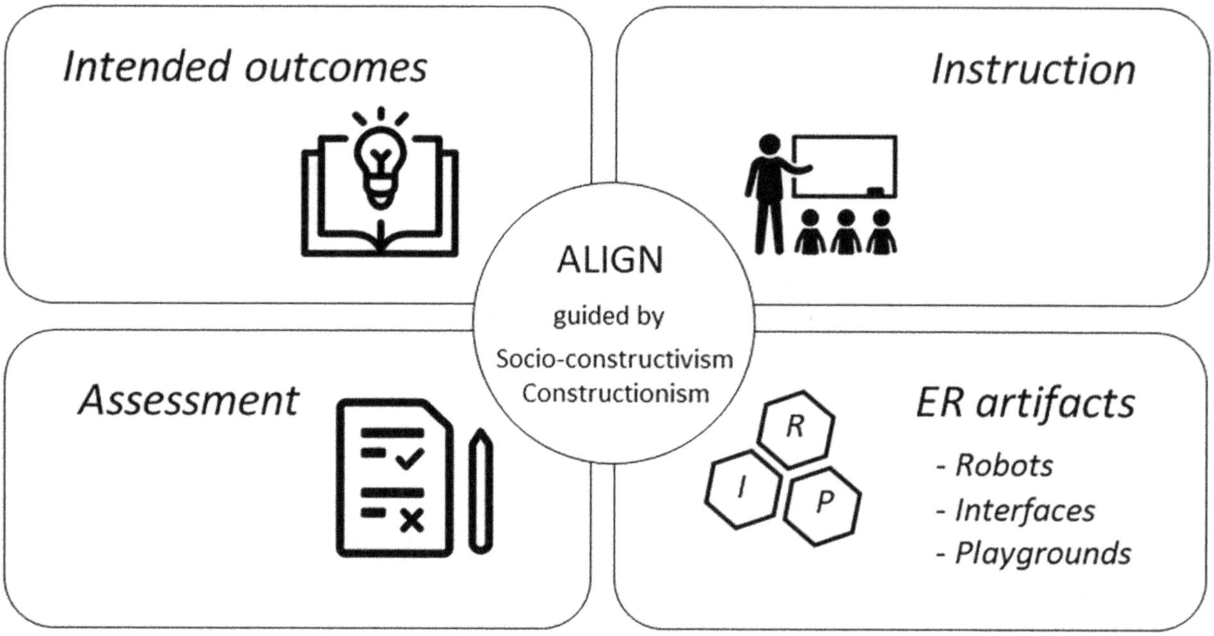

has often been referred to as the historical origin of ER. It is within this paradigm that alignment will be discussed hereafter.

Previous work has illustrated that the abovementioned kind of ER activities usually comprises a variety of different elements (Chevalier et al., 2020). Apart from the robots, they also involve programming or interaction interfaces as well as playgrounds for the learning tasks. As a whole, such activities correspond to what Heersmink (2013) has denoted as "situated cognitive systems", involving three types of cognitive artifacts. According to Heersmink's taxonomy, the robots in this system can be classified as structural ecological artifacts, while the interfaces represent symbolic representational and the playgrounds spatial ecological artifacts. Since each artifact embodies a different cognition-aiding property, this suggests an adaptation of Lauwers' alignment model. Previously denoted as the "tool", the fourth component is redefined as "ER artifacts" to account for the distinctness of the different cognitive artifacts. Moreover, in the type of ER activities considered here, students usually collaborate in small groups and construct knowledge through interactions with their peers and their teachers, who may provide guidance through scaffolding. As a consequence, the resulting model for Educational Robotics Learning Systems (ERLS, Fig. 1) explicitly incorporates two learning theories considered to be relevant: besides constructionism (Papert, 1980) which, as mentioned before, is a key learning theory underlying ER activities, it is also the theory of socio-constructivism (Vygotsky, 1978) that should guide the alignment of the different ERLS components.

The ERLS model hence provides a conceptualization of ER activities and can be used by developers and educators to align the design of ER tools and ER classroom activities, respectively. In order to attain well-aligned ERLS, it is crucial, that both developers and educators adopt a holistic perspective, always considering the entirety of the system when conceiving the individual elements. However, as discussed by Antonenko et al. (2017), one limitation of their framework is that it may not be usable by "developers that have little or no background in psychology and education". Similarly, effectively leveraging the ERLS model could be challenging for many developers and even some educators, especially if they lack experience with ER. Hence to facilitate the operationalization of the ERLS model, the next section introduces the Heuristics for ERLS. Their objective is to represent a more natural approach for developers and educators to put the ideas of the ERLS model into practice.

HEURISTICS FOR EDUCATIONAL ROBOTICS LEARNING SYSTEMS

Heuristic Evaluation

Heuristic Evaluation (HE) is a time- and cost-efficient methodology to analyze digital user interfaces, that was first introduced in the early nineties (Nielsen & Molich, 1990). HE is usually performed by experts in the domain (but sometimes also by non-experts), who evaluate a system by checking its compliance with a given set of design principles (i.e., the heuristics). The heuristics describe the characteristics of an ideal system and are often formulated as rule of thumbs such as "minimize user memory load" or "provide good error messages". By performing HE, the evaluators identify usability issues, i.e., properties of the system that do not comply with the heuristics. Representing the weak spots of the system, the issues can then be addressed by the developers in the next design iterations.

As outlined by Kirschner et al. (2004), the usefulness of a system is composed of its utility (whether it provides what is needed) and its usability (how easy and pleasant it is to use). While traditional HE is usually concerned with the latter, there have also been approaches to integrate heuristics related to utility. For instance, previous works have presented heuristics specifically compiled for the domains of educational games (Barbosa, Rêgo & de Medeiros, 2015; Mohamed & Jaafar, 2010) as well as novice programming systems (Kölling & McKay, 2016; Pane & Myers, 1996). Tailoring heuristics to a specific target domain indeed appeared to be more effective: in their study, Kölling and McKay asked half of the participants to perform HE for a novice programming system using Nielsen and Molich's set, while the other half used their tailored heuristics. Their results showed that the issues identified using the latter appeared to be more diverse and included more items that were not identified with Nielsen and Molich's set than vice versa. These findings therefore suggest that HE could also be useful in the context of ERLS, while representing a natural and intuitive approach for developers and educators to operationalize the ideas of the ERLS model. However, Kölling and McKay also emphasized that the outcome of HE heavily depends on the evaluators' findings, which in turn are strongly driven by the heuristics provided. It is, therefore, essential that heuristics are thoroughly tested for validity. Hence to develop the HERLS heuristics, focus groups were organized with developers and educators with experience in the domain of ER to ensure validity of the devised heuristics.

Devising and Evaluating the HERLS Heuristics

Previous work has suggested focus groups as a useful method in design science research, for both improving design artifacts as well as for demonstrating their usefulness in a field setting (Tremblay, Hevner & Berndt, 2010). Tremblay et al. proposed that incremental improvements can result from exploratory focus groups (EFGs), while evidence about utility can be achieved through confirmatory focus groups (CFGs). In both cases, participants of the focus groups are asked to use the design artifact under development prior to the plenary discussions. However, since EFGs and CFGs have different objectives, they may follow different experimental protocols. With respect to these notions, this work applied EFGs to devise HERLS, and CFGs to demonstrate their usefulness in authentic contexts with different groups of developers.

The development of the HERLS heuristics was based on an existing set of design heuristics for ER tools (Giang, Piatti & Mondada, 2019), which however, did not specifically consider alignment as presented in the ERLS model. To devise and evaluate HERLS, six focus groups were performed with developers and educators knowledgeable of ER (Table 1). For EFGs and CFGs different experimental protocols were applied.

Table 1. Summary of participants' details for all conducted focus groups

Focus Group	Participants	Gender (m / f)	Age (mean ± sd)	Work Experience (mean ± sd)	Evaluated ER Tools
EFG1	Four developers	3 m / 1 f	35.8 ± 3.9 y	10.5 ± 3.6 y	Two among Cozmo[a], Lego WeDo[b], mBot[c], Root[d]
EFG2	Nine teachers	1 m / 8 f	39.4 ± 6.3 y	17.6 ± 7.9 y	
EFG3	Ten teachers and two teacher trainers	7 m / 5 f	33.4 ± 6.0 y	6.1 ± 5.4 y	
EFG4	Ten teachers	6 m / 4 f	40.3 ± 4.5 y	16.1 ± 7.3 y	
CFG1	Four developers	4 m	23.3 ± 1.3 y	2.2 ± 1.4 y	Thymio PaPL[e]
CFG2	Four developers	4 m	54.3 ± 2.3 y	27.0 ± 5.4 y	Azoresbot[f]

[a]https://www.anki.com/en-us/cozmo, [b]https://education.lego.com/en-gb/product/wedo-2[c]https://www.makeblock.com/steam-kits/mbot,[d]https://root.irobot.com/, [e]Mehrotra et al. (2020), [f]Cascalho et al. (2020)

Exploratory Focus Groups to Devise HERLS

Four EFGs were performed with in total 35 participants to iteratively devise HERLS (Fig. 2). Participants were recruited with developer and educator backgrounds to allow for different perspectives on the use of ER in classrooms. For all EFGs, participants were first divided into subgroups of three to six people. Through a 15-minute presentation by one of the experimenters, the groups were then introduced to the ERLS model and the existing heuristics from Giang et al., 2019. . Each group then performed HEs for two different systems for around 30 minutes each, identifying issues for each system guided by the existing heuristics. The session concluded with a plenary discussion lasting around 30 minutes that was moderated by one of the experimenters. In the plenary discussions, the utility and the usability of the ERLS model and the HERLS heuristics was evaluated by the participants. During the discussions, the moderator was taking written notes for later analysis. To determine a questioning route, prior to the discussions the moderator presented a set of Likert-scale statements to the participants (Table 2), that were later used as references points for the discussions. However, due to organizational constraints, participants' responses to these statements were only recorded for EFG2-4.

In EFG1-4, a total of 14 HEs were performed by all participants. In total, all subgroups together identified 141 issues, with an average of 10.1±3.8 issues (mean ± std) identified per HE. The distribution of the problems across the 14 heuristics was very diverse (Fig. 3). While a large proportion of the issues were related to the Interaction heuristic (Pre-4, 34 issues), none of the 14 HEs identified issues

Figure 2. Summary of experimental protocol for EFGs

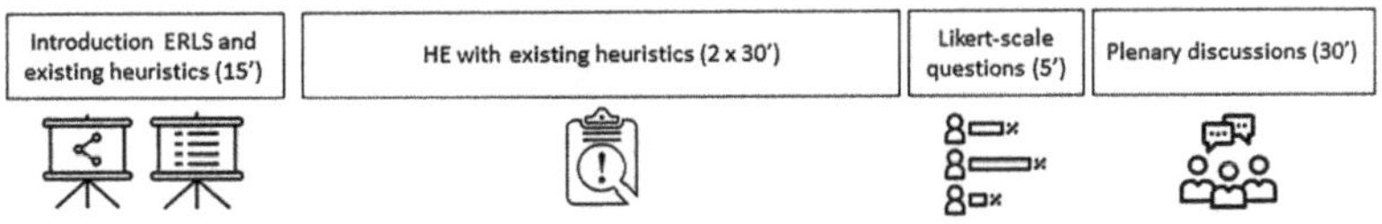

that were mapped to the Challenge (Pre-2), Level of Automation (Pre-5) and Enjoyment and Aesthetics (Pre-9) categories. Moreover, for 22 issues, participants were not able to unambiguously assign them to one heuristic and therefore indicated two or more possible categories. As a consequence, these issues were assigned to the Not Clear (NC) category by the experimenters, being the second highest proportion among all categories. No cases were observed in which issues could not be related to any heuristic.

Based on the participants' responses to the Likert-scale statements presented prior to the plenary discussions (Fig. 4), the moderator directed the discussions towards the most controversial topics. For instance, some participants believed, that the description of the existing heuristics needed to be revised, to make them clearer and more concise. Others proposed a reorganization of the existing heuristics, since they noticed thematical overlaps. Another point that was raised, was the need to better integrate the ideas of the ERLS model into the heuristics. Nevertheless, overall, the participants agreed on the importance of considering alignment as presented in the ERLS model and acknowledged that using HE would be an appropriate method to put the ideas into practice. Based on the input and feedback from the EFGs, the existing heuristics were revised by the researchers, finally resulting in HERLS (Table 3).

While some of the existing heuristics were kept in HERLS without changes, some others had to be modified. Specifically, heuristics for which only few issues were found during the HEs were either removed or merged to create new categories. Furthermore, more detailed descriptions and a different structure were implemented for each category to prevent a lack of clarity and to facilitate the use of the heuristics. To analyze the orthogonality of the new HERLS heuristics, all 141 heuristics identified during

Table 2. Questions of the surveys presented to the focus groups

Item	In Surveys Presented to EFG2-4 and CFG1-2
Q1	I think it is important to consider alignment as presented in the ERLS framework when assessing the design of ER tools.
Q2	I think that heuristic evaluation is an appropriate method to assess the design of ER tools.
Q3	I think that the overall length of the heuristics is appropriate.
Q4	I think that the description of the heuristics is sufficiently detailed and precise.
Q5	I think that the heuristics are easy to use.
Item	**Questions Additionally Presented to CFG1-2**
Q6	I would naturally (i.e., without instruction) consider alignment as presented in the ERLS framework when designing ER tools.
Q7	I think that the heuristic evaluation helps to uncover a larger amount of issues.
Q8	I think that the heuristic evaluation helps to uncover issues that cannot be found only by intuition.
Q9	I think that the organization of the heuristics is clear and well-structured.
Q10	I think that the heuristics cover all relevant aspects for the design of ER tools.

Figure 3. Issues identified by EFG1-4 mapped to the existing heuristics presented in Giang et al. 2019

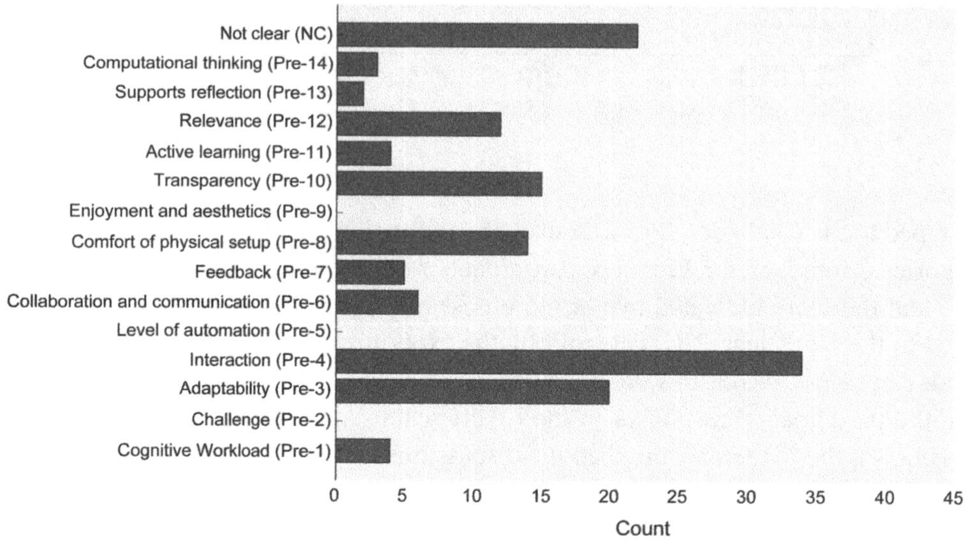

EFG1-4 were mapped to the revised HERLS heuristics by the experimenters (Fig. 5). Though also in the revised heuristics, a dominance of the Facilitating User Interaction heuristic (HERLS-4, 42 issues) was observed, the distribution across the remaining categories appeared to be more balanced than in the previous set. Particularly, in this new set, there were no categories with less than 14 issues, while for the previous set this was the case for ten categories. Moreover, the experimenters did not encounter a single instance where a problem could not unambiguously be assigned. To finally evaluate the usefulness of the revised HERLS heuristics, CFGs were performed with different groups of developers.

Confirmatory Focus Groups to Evaluate HERLS

Since the goal of the CFGs was to validate the usefulness of the heuristics, a manipulation was applied, as suggested by Tremblay et al. (2010). Therefore, after a 10-minute introduction to the ERLS model, groups were first given 30 minutes to identify issues in their systems based only on their intuition. After

Figure 4. Responses to Likert-scale questions guiding the plenary discussions in EFG2-4

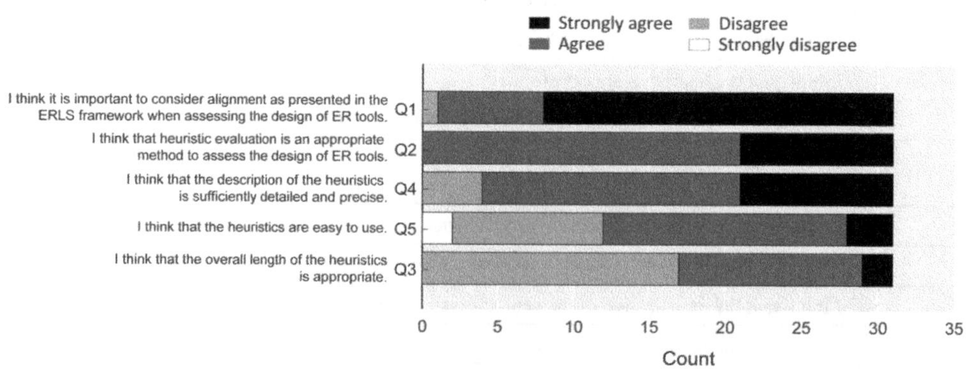

Table 3. Heuristics for Educational Robotics Learning Systems (HERLS)

Heuristic	Description
HERLS-1	**Educationally relevant** - The ERLS should embody constructionist learning approaches, involving the use of tangible objects and fostering active learning. - The ERLS should implement activities complying with desired learning goals, such as the ones listed in curricula and relevant competence models. - The ERLS should embody alignment of learning goals, instruction, ER artifacts and assessment activities.
HERLS -2	**Inviting and engaging** - The ERLS should quickly grab the learners' attention and facilitate cognitive activation and immersion in the activities. - Learners should feel comfortable exploring and using the ER artifacts, without fear of breaking them. - The ER artifacts should be aesthetically appealing.
HERLS -3	**Supporting reflection** - The ERLS should encourage learners to reflect about their performed actions. - The ERLS should highlight the learning process. - The ER artifacts should provide a transparent environment allowing the inspection of underlying mechanisms.
HERLS -4	**Facilitating user interaction** - The ERLS should be usable in an easy and intuitive way, satisfying the learners' expectations. - The ER artifacts should provide feedback to facilitate the learners' interaction with the system. - The interaction with the ER artifacts should follow the logic of the instructional activities.
HERLS -5	**Versatile** - Learners should be able to use the ER artifacts regardless of their prior knowledge. - The ER artifacts should provide a rich environment allowing to implement a large variety of different learning activities. - The ERLS should allow for personalization of the learning activities.
HERLS -6	**Promoting collaboration and communication** - The ERLS should promote small group learning (i.e., groups of 2-3). - The ERLS should provide equal opportunities for participation to all members of a learning group. - The ERLS should foster collaboration and communication within a learning group and, if appropriate, between learning groups.
HERLS -7	**Compatible with classrooms** - The ER artifacts should be easy and fast to set up, disassemble, and maintain. - The ER artifacts should support teachers in classroom orchestration. - The ER artifacts should comply with the highest quality and safety standards. - The ER artifacts should be accessible at a reasonable cost.

Figure 5. Issues identified by EFG1-4 mapped to HERLS

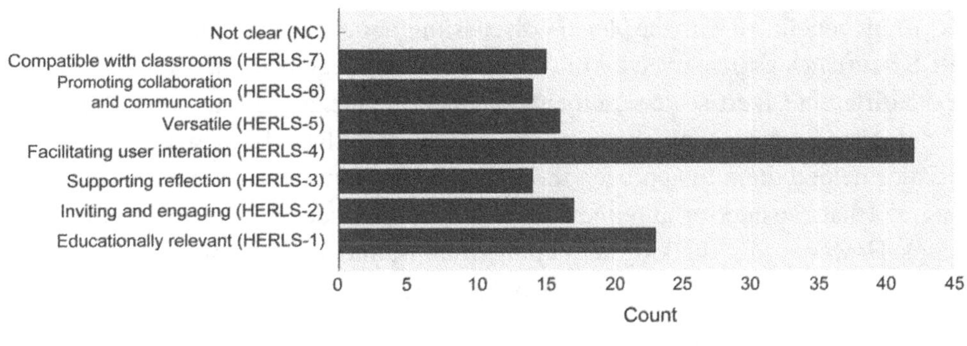

Figure 6. Summary of experimental protocol for CFGs

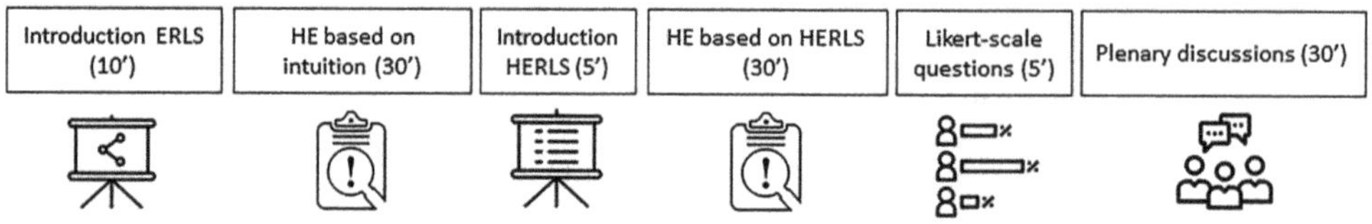

the completion of this first evaluation, the participants were then introduced to HERLS in a 5-minute presentation. The groups were then given 30 minutes to perform another assessment of their system – this time guided by HERLS. If the HE resulted in issues similar to the ones from the intuitive evaluations, the groups were asked to list them, nonetheless. Issues for which participants assigned multiple heuristics and for which no consensus was found within the group, were assigned to the "Not clear" (NC) category by the experimenters. Following the HE with HERLS, participants were also asked to map the issues identified based on intuition to the categories of HERLS.

CFGs were performed with two groups of developers to validate the usefulness of HERLS in authentic contexts. CFG1 were the developers of Thymio PaPL (Mehrotra et al., 2020) successor of the Thymio TPL project (Mussati, Giang, Piatti & Mondada, 2019). The goal of this project was to develop a tangible programming interface for the educational robot Thymio (Riedo, Chevalier, Magnenat & Mondada, 2013) based on accessible materials. CFG2 were the developers of the Azoresbot, a modular do-it-yourself educational robot based on the Arduino platform (Cascalho et al., 2020). The main objective of this project is to provide a low-cost educational robot that students and teachers can use in regional robot competitions.

For Thymio PaPL, CFG1 identified 16 issues by intuition and 14 issues by HE (Fig. 7). For the Azoresbot, CFG2 identified 9 issues by intuition and 17 issues by HE. Based on intuition, both groups identified at least two issues that were classified as Not Clear (2 for CFG1, 5 for CFG2). Excluding those in the Not Clear category, most of the issues identified by intuition (11 for CFG1 and 3 for CFG2) were mapped to the Facilitating User Interaction heuristic (HERLS-4). This result is not surprising, since it can be assumed that issues related to user interaction are generally easier to uncover. Remarkably, issues related to any of the other six categories of the heuristics appeared to be more challenging to determine (3 for CFG1, 1 for CFG2). The distribution of the issues significantly changed when the groups were guided by HERLS. With the heuristics, CFG1 found problems for all categories except for HERLS-6. CFG2 did not only find almost twice as many issues as compared to the intuitive evaluation. Guided by the heuristics, the group found at least one issue related to each category of the heuristics. As opposed to the intuitive evaluation, all issues identified by both groups were clearly assignable and none had to be associated with the Not Clear category.

The focus groups concluded with the plenary discussions about the usefulness of the ERLS framework and the HERLS heuristics as perceived by the developers. Analogous to the EFGs, the moderator used the responses to different Likert-scale statements to guide the plenary discussions (Fig. 8). Participants particularly appreciated the conceptualization provided by the ERLS framework and affirmed that the HERLS heuristics helped them to approach the evaluation in a more systematic way. Interestingly, all participants agreed that considering alignment of robot and interface was already intuitive to them before the focus groups. However, also thinking about playgrounds, intended outcomes, as well as instructional and assessment activities was natural to only two developers in the CFGs, of which one was a teacher.

Figure 7. Issues identified by CFG1-2 based on intuitive evaluation (grey) and heuristic evaluation with HERLS (black).

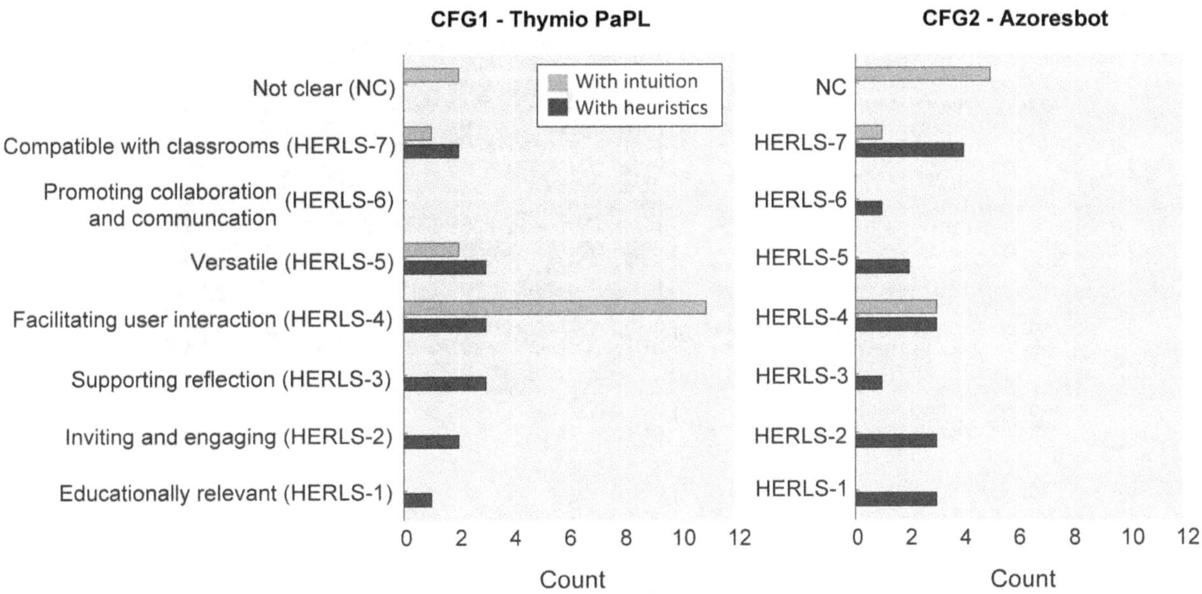

Most of the developers stated that they would first develop the robot and the interface and then think about the teaching and learning activities. As the main reason for this approach, participants indicated limited experience in the pedagogical domain. In this context, the heuristics provided them with guidance throughout the evaluation process and helped especially developers with little pedagogical background to give more attention to the alignment of the ER tool design with the other components of the ERLS model. Even those coming from a pedagogical background, namely the teacher in CFG2, acknowledged that without the heuristics, they would have missed to identify some important issues.

HERLS for the Design of Classroom Activities

The ERLS model and HERLS have been conceived to not only support developers, but also educators. To this end, the proposed methodology was also applied in a case study to guide the design of an ER classroom activity. The Thymio Escape Game (Giang et al., 2018) is an immersive learning activity based on the concept of educational escape rooms and it was developed to introduce the players to the basics of the educational robot Thymio (Riedo et al., 2013) and its programming interface VPL (Shin, Siegwart & Magnenat, 2014). In this activity, a group of up to six players has to solve a series of puzzles, eventually leading them to find three robots that have to be reprogrammed to make them autonomously drive into an energy reactor found in the room. During the activity, the players are supported by a game master and hints provided in the room. However, since the activity was not originally designed for classroom use, it had to be adapted for this purpose. To this end, the developers of the activity applied HE with HERLS to identify the necessary design changes to create a classroom version of the activity.

To illustrate how the HERLS heuristics can also be useful for educators as a support for the design of ER classroom activities, HEs were performed to identify the necessary design changes to adapt the Thymio

Figure 8. Responses to Likert-scale questions guiding the plenary discussions in the CFGs

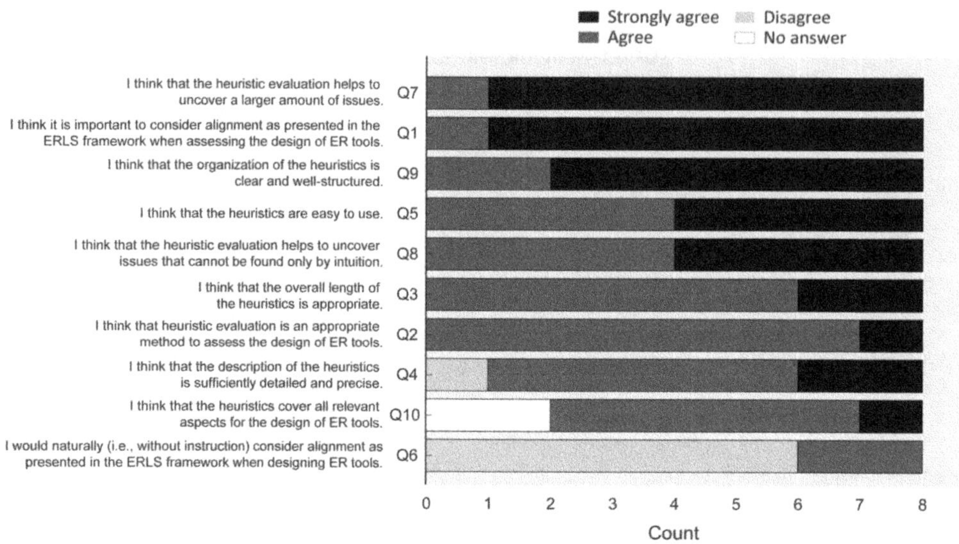

Escape Game for the use in classrooms. In the following, seven major issues that were identified through HE are presented to exemplify how the proposed methodology could be useful for educators (Table 4).

Table 4. Issues identified for the Thymio Escape Game based on HERLS

ID	Description of the Issue	Refers to Heuristic
TEG1	When players load wrong program codes on the robots, they may crash into the energy reactor causing material damage and unsettling the students. It also creates a loss of immersion in the game.	HERLS-2
TEG2	The current role of the game master causes a loss of immersion in the game when hints are given.	HERLS-2
TEG3	The incomplete preprogrammed behaviors of the robots are confusing and intimidating to the players and hamper an immediate interaction with the robots.	HERLS-2
TEG4	In the current implementation, players are encouraged to follow a blind trial-and-error approach, instead of reflecting on the performed actions.	HERLS-3
TEG5	Arranging multiple parallel games of the current version, could promote undesired competition between groups.	HERLS-6
TEG6	The activity can only be played by max. 6 players at a time.	HERLS-7
TEG7	The activity requires a significant amount of time for set up and dismantling (around 30 minutes).	HERLS-7

Based on the issues identified through HE, the following design changes were applied: first, the new activity integrated multiple identical game setups to allow the simultaneous participation of all students of a class (addressing TEG6). Moreover, the physical material needed for the activity was simplified to enable preparation and dismantling of the activity during recess (TEG7). To prevent students from blind trial-and-error approaches, a new game rule was introduced, allowing them only one trial to test their solution with the real setup (TEG4). On the other hand, test maps were provided that allowed students to extensively test and debug their solutions before trying them on the real setup (TEG1 and TEG4). Furthermore, to promote collaboration instead of competition between groups, the activity was only considered successful, if all groups presented the solution on the real setup at the same time (TEG5). Also, instead of giving hints as an external observer, the game master now assumed the role of an in-game character allowing them to give support to the students as part of the narrative (TEG2). Finally, the robots were presented to the students without any preprogrammed behavior to prevent confusion (TEG3).

Similar to the experiences with the developers, the main advantage of using HERLS was the more structured approach to identify the limitations of the activity. This allowed to determine design changes that helped to improve the activity in a more targeted and purposeful way. It can be argued that some of the issues could also be found based on intuition or observations made during the activities. However, the proposed approach with HERLS may provide a more systematic approach that allows design changes to be done more efficiently. For instance, having all the issues listed next to each other allows to identify possible design changes that can address some of the issues simultaneously, like the introduction of the test maps in the presented example (addressing both TEG1 and TEG4). Moreover, the use of HERLS allowed to identify a potential issue (TEG5), before the actual implementation and therefore helped for a more efficient design iteration. The redesigned version was finally tested with multiple school classes, illustrating its feasibility as a classroom activity (Giang, 2020). Furthermore, the game material as well as a documentation have been made publicly available and as a first measure for its acceptance, the devised activity has been independently adapted by several educators for their classroom activities

DISCUSSION

This chapter addressed the question of alignment of ER tools with classroom activities. The subject was approached based on three research questions, which in the following will be taken up to guide the final discussion and conclusion of this work.

The first research question intended to be addressed by this paper was: how can ER activities be conceptualized? To approach this question, the alignment model proposed by Lauwers (2010) was used as a starting point. By integrating the main components (i.e., intended outcomes, instructional processes and assessment) of traditional alignment frameworks (Biggs, 2003; Cohen, 1987), Lauwers' model represents an important step towards a more holistic conceptualization of ER activities. However, the present work proposes to further extend the model, in order to provide developers, educators and researchers with a more holistic view on ER activities. As it has been argued in previous work, knowledge is situated, "being in part a product of the activity, context, and culture in which it is developed and used" (Brown et al., 1989). In the context of educational technology, it has moreover been suggested that the situated perspective on learning should consider both the real and the digital environment (Osborne, 2014). The ERLS model proposed in this paper was conceived based on such a situated perspective to conceptualize ER activities. Based on the notion of situated learning systems, the relevant cognitive artifacts of ER activities (both physical and digital) were determined and classified according to Heersmink's (2013) taxonomy. Previous works have, moreover, raised the need to put more emphasis on the pedagogy and the underlying learning theories of ER (Alimisis, 2012; Jung & Won, 2018). In line with this idea, the ERLS model explicitly incorporates constructionism and socio-constructivism as guiding principles for the alignment of ERLS. As a matter of fact, previous research has discussed that both learning theories can be considered essential pillars for the type of activities discussed in this work (Alimisis, 2013; Ju, Chen & Tang, 2020; Ronsivalle et al., 2019). The focus groups with developers and educators showed that overall, participants appreciated the conceptualization provided by the ERLS model and they approved the usefulness of the framework. Nevertheless, it should be acknowledged that the findings of this study also come with the limitations that are inherent to research studies based on focus groups. Generalizability may be limited, due to the restricted number of focus groups that can be organized. The present study attempted to alleviate the effects of these limitations by a rigorous planning of the focus groups following procedures described in pertinent literature (Tremblay et al., 2010) as well as by including participants that are all knowledgeable in ER. However, it should also be noted that the ERLS model is not claimed to be exhaustive and may be subject to further revisions. Nonetheless, it may represent a reference point for developers, educators and researchers interested in studying the question of alignment in the context of ER activities.

While frameworks like the ERLS model could be useful to study alignment from a theoretical point of view, they may often be too abstract to put the ideas into practice. Indeed, Antonenko et al. (2017) highlighted that their framework may not be usable by developers with limited experience in psychology and education. To facilitate the application of the ideas from the ERLS model, this work therefore introduced HERLS, a set of design and evaluation heuristics, as its complement. The application examples presented in this chapter helped to approach the remaining two research questions, i.e., how developers and educators can design well-aligned ER tools and ER classroom activities, respectively. Previous work has suggested participatory design as part of an alignment-centered design process for ER tools (Lauwers, 2010). Other groups have followed open-source approaches to actively involve different stakeholders in the development process (Mondada et al., 2017). While such approaches can provide meaningful insights to identify design specifications for ER tools, they may not always be feasible, especially in early development stages. In this regard, HERLS may represent a time- and cost-efficient alternative for developers to identify important design changes. Naturally, a set of heuristics cannot replace a teacher's know-how and years of experience. Nevertheless, the results from the focus groups with the developers illustrate that the devised heuristics appeared to be helpful in raising awareness for pedagogical issues, that even developers with no classroom experience were able to grasp. Especially when put in contrast with an evaluation based only on their intuition, participants praised HERLS as effective and purposeful. Similar to previous studies that applied HE in the context of educational technology (Kölling & McKay, 2016), HERLS helped the developers to uncover a broader variety of issues, that were moreover, clearer defined. A precise definition of issues could provide important support to developers and make design iterations more targeted and effective. As previous works have suggested, heuristic evaluation can be considered most effective in early design stages (Clarkson & Arkin, 2007; Desurvire, Caplan & Toth, 2004). Indeed, both development projects presented in this work were still in the early design phases. For more mature systems, HERLS may therefore reach its limit as a standalone approach. However, future work could also explore whether HE could complement existing alignment-centered approaches, for instance, participatory design in which developers and educators perform the HE together. This could further leverage the advantages of the proposed methodology, as partially showcased by CFG2, which involved a teacher.

This chapter finally presented a first example with regard to the usefulness of HERLS for the design of ER classroom activities. In the context of ER, previous work has suggested the need for more teacher trainings and educational material (Chevalier, Riedo & Mondada, 2016). In this regard, the HERLS heuristics aim at providing support to educators that is generalizable, describing ideas that are applicable regardless of the specific ER tools being used. This is especially relevant with respect to teacher trainings, since such programs should ideally go beyond the idea of "just building and programming robots" (Alimisis, 2019). At the same time, the simplicity and intuitiveness of the heuristics may also allow teachers with limited ER experience to apply them. Specifically, the proposed framework could help educators in choosing appropriate technology, designing learning tasks and planning their support during the activity; three elements that have previously been considered important for engaging learning experiences with ICT (Seau Yoon, Ho & Hedberg, 2005). As a recent work has pointed out, one major challenge for the implementation of ER classroom activities, is the absence of effective pedagogical practices (Yang et al., 2020). In this regard, the proposed framework may provide educators with comprehensible guidelines, helping them in understanding how they can effectively design and implement well-aligned ER classroom activities.

CONCLUSION

In this chapter, we present a conceptual framework with the goal to support developers and educators in the design of tools and activities in ER. To this end, a theoretical model was developed describing alignment of the different cognitive artifacts involved in ER activities. Based on this model, we devised a set of heuristics aimed at supporting developers and educators in operationalizing the ideas of the model. The entire framework was evaluated in case studies in which it was applied to guide the development of new ER tools as well as the redesign of an ER classroom activity. The findings from the case studies highlighted that the proposed methodology can represent a time- and cost-efficient approach for developers to align the designs of their ER tools with classroom needs starting from early development stages. It may equally be useful to educators for the preparation and implementation of ER activities, providing them with orientation points to leverage ER tools for classroom activities in a pedagogically meaningful way. While the proposed framework is not claimed to be exhaustive, it may represent a useful conceptualization of ER activities that can provide developers, educators and researchers with hints on how to design ER tools and activities that are instructionally well aligned.

REFERENCES

Alimisis, D. (2012). Robotics in Education & Education in Robotics: Shifting Focus from Technology to Pedagogy. *Proceedings of the 3rd International Conference on Robotics in Education*, 7–14.

Alimisis, D. (2013). Educational robotics: Open questions and new challenges. *Themes in Science and Technology Education, 6*(1), 63–71. http://earthlab.uoi.gr/theste/index.php/theste/article/view/119

Alimisis, D. (2019). Teacher Training in Educational Robotics: The ROBOESL Project Paradigm. Technology. *Knowledge and Learning*, *24*(2), 279–290. doi:10.100710758-018-9357-0

Antonenko, P. D., Dawson, K., & Sahay, S. (2017). A framework for aligning needs, abilities and affordances to inform design and practice of educational technologies. *British Journal of Educational Technology*, *48*(4), 916–927. doi:10.1111/bjet.12466

Anwar, S., Bascou, N. A., Menekse, M., & Kardgar, A. (2019). A systematic review of studies on educational robotics. *Journal of Pre-College Engineering Education Research*, *9*(2), 19–42. doi:10.7771/2157-9288.1223

Atmatzidou, S., & Demetriadis, S. (2015). Advancing students' computational thinking skills through educational robotics: A study on age and gender relevant differences. *Robotics and Autonomous Systems*, *75*, 661–670. doi:10.1016/j.robot.2015.10.008

Barbosa, M. B., Rêgo, A. B., & de Medeiros, I. (2015). *HEEG: Heuristic Evaluation for Educational Games*. Academic Press.

Belpaeme, T., Kennedy, J., Ramachandran, A., Scassellati, B., & Tanaka, F. (2018). Social robots for education: A review. *Science Robotics*, *3*(21), 1–10. doi:10.1126cirobotics.aat5954 PMID:33141719

Benitti, F. B. V. (2012). Exploring the educational potential of robotics in schools: A systematic review. *Computers & Education*, *58*(3), 978–988. doi:10.1016/j.compedu.2011.10.006

Bers, M. U., Flannery, L., Kazakoff, E. R., & Sullivan, A. (2014). Computational thinking and tinkering: Exploration of an early childhood robotics curriculum. *Computers & Education*, *72*, 145–157. doi:10.1016/j.compedu.2013.10.020

Biggs, J. (1996). Enhancing teaching through constructive alignment. *Higher Education*, *32*(3), 347–364. doi:10.1007/BF00138871

Biggs, J. (2003). Aligning teaching and assessing to course objectives. *Teaching and Learning in Higher Education: New Trends and Innovations*, *2*(April), 13–17.

Bower, M. (2008). Affordance analysis – Matching learning tasks with learning technologies. *Educational Media International*, *45*(1), 3–15. doi:10.1080/09523980701847115

Brown, J. S., Collins, A., & Duguid, P. (1989). Situated cognition and the culture of learning. *Educational Researcher*, *18*(1), 32–42. doi:10.3102/0013189X018001032

Cascalho, J., Mendes, A., Ramos, A., Pedro, F., Bonito, N., Almeida, D., Augusto, P., Leite, P., Funk, M., & Garcia, A. (2020). Azoresbot. *An Arduino Based Robot for Robocup Competitions.*, *6*, 542–552. doi:10.1007/978-3-030-35990-4_44

Chen, G., Shen, J., Barth-Cohen, L., Jiang, S., Huang, X., & Eltoukhy, M. (2017). Assessing elementary students' computational thinking in everyday reasoning and robotics programming. *Computers & Education*, *109*, 162–175. doi:10.1016/j.compedu.2017.03.001

Chevalier, M., Giang, C., Piatti, A., & Mondada, F. (2020). Fostering computational thinking through educational robotics: A model for creative computational problem solving. *International Journal of STEM Education*, *7*(1), 39. doi:10.118640594-020-00238-z

Chevalier, M., Riedo, F., & Mondada, F. (2016). Pedagogical Uses of Thymio II. *IEEE Robotics & Automation Magazine*, *23*(2), 16–23. doi:10.1109/MRA.2016.2535080

Clarkson, E., & Arkin, R. C. (2007). Applying heuristic evaluation to human-robot interaction systems. *Proceedings of the Twentieth International Florida Artificial Intelligence Research Society Conference, FLAIRS 2007*, 44–49.

Cohen, S. A. (1987). Instructional Alignment: Searching for a Magic Bullet. *Educational Researcher, 16*(8), 16–20. doi:10.3102/0013189X016008016

Desurvire, H., Caplan, M., & Toth, J. A. (2004). Using heuristics to evaluate the playability of games. *Extended Abstracts of the 2004 Conference on Human Factors and Computing Systems - CHI '04*, 1509. 10.1145/985921.986102

Eguchi, A. (2014). Robotics as a Learning Tool for Educational Transformation. *International Workshop Teaching Robotics, Teaching with Robotics & International Conference Robotics in Education*, 27–34. 10.4018/978-1-4666-8363-1.ch002

El-Hamamsy, L., Bruno, B., Chessel-Lazzarotto, F., Chevalier, M., Roy, D., Zufferey, J. D., & Mondada, F. (2021). The symbiotic relationship between educational robotics and computer science in formal education. *Education and Information Technologies*, 1–31. PMID:33841027

El-Hamamsy, L., Chessel-Lazzarotto, F., Bruno, B., Roy, D., Cahlikova, T., Chevalier, M., Parriaux, G., Pellet, J.-P., Lanarès, J., & Zufferey, J. D. (2020). A computer science and robotics integration model for primary school: Evaluation of a large-scale in-service K-4 teacher-training program. *Education and Information Technologies*, 1–31. PMID:33162777

Giang, C. (2020). *Towards the alignment of educational robotics learning systems with classroom activities*. EPFL.

Giang, C., Chevalier, M., Negrini, L., Peleg, R., Bonnet, E., Piatti, A., & Mondada, F. (2018). Exploring Escape Games as a Teaching Tool in Educational Robotics. *Proceedings of Educational Robotics 2018 (EDUROBOTICS)*.

Giang, C., Piatti, A., & Mondada, F. (2019). Heuristics for the Development and Evaluation of Educational Robotics Systems. *IEEE Transactions on Education, 62*(4), 1–10. doi:10.1109/TE.2019.2912351

Heersmink, R. (2013). A Taxonomy of Cognitive Artifacts: Function, Information, and Categories. *Review of Philosophy and Psychology, 4*(3), 465–481. doi:10.100713164-013-0148-1

Ju, A., Chen, D., & Tang, J. (2020). Design of Educational Robot Platform Based on Graphic Programming. In M. Atiquzzaman, N. Yen, & Z. Xu (Eds.), *Big Data Analytics for Cyber-Physical System in Smart City* (pp. 382–391). Springer Singapore. doi:10.1007/978-981-15-2568-1_53

Jung, S. E., & Won, E. S. (2018). Systematic review of research trends in robotics education for young children. *Sustainability (Switzerland), 10*(4), 1–24. doi:10.3390u10040905

Khanlari, A. (2013). Effects of robotics on 21st century skills. *European Scientific Journal, ESJ, 9*(27), 26–36.

Kirschner, P., Strijbos, J. W., Kreijns, K., & Beers, P. J. (2004). Designing electronic collaborative learning environments. *Educational Technology Research and Development, 52*(3), 47–66. doi:10.1007/BF02504675

Koehler, M. J., Mishra, P., & Cain, W. (2013). What is Technological Pedagogical Content Knowledge (TPACK)? *Journal of Education*, *193*(3), 13–19. doi:10.1177/002205741319300303

Kölling, M., & McKay, F. (2016). Heuristic Evaluation for Novice Programming Systems. *Trans. Comput. Educ.*, *16*(3), 12:1--12:30. doi:10.1145/2872521

Lauwers, T. (2010). *Aligning capabilities of interactive educational tools to learner goals*. ProQuest Dissertations and Theses, 3470163, 366. http://ezproxy.net.ucf.edu/login?url=http://search.proquest.com/docview/742526809?accountid=10003%5Cnhttp://sfx.fcla.edu/ucf?url_ver=Z39.88-2004&rft_val_fmt=info:ofi/fmt:kev:mtx:dissertation&genre=dissertations+&+theses&sid=ProQ:ProQuest+Dissertations+&+T

Loughlin, C., Lygo-Baker, S., & Lindberg-Sand, Å. (2021). Reclaiming constructive alignment. *European Journal of Higher Education*, *11*(2), 119–136. doi:10.1080/21568235.2020.1816197

MacPhail, A., Tannehill, D., Leirhaug, P. E., & Borghouts, L. (2021). Promoting instructional alignment in physical education teacher education. *Physical Education and Sport Pedagogy*, 1–12. doi:10.1080/17408989.2021.1958177

McCormick, M. P., Weiland, C., Hsueh, J., Maier, M., Hagos, R., Snow, C., Leacock, N., & Schick, L. (2020). Promoting content-enriched alignment across the early grades: A study of policies & practices in the Boston Public Schools. *Early Childhood Research Quarterly*, *52*, 57–73. doi:10.1016/j.ecresq.2019.06.012

Mehrotra, A., Giang, C., Duruz, N., Dedelley, J., Mussati, A., Skweres, M., & Mondada, F. (2020). Introducing a Paper-Based Programming Language for Computing Education in Classrooms. *Proceedings of the 2020 ACM Conference on Innovation and Technology in Computer Science Education*, 180–186. 10.1145/3341525.3387402

Mitchell, F. M. (1999). *All Students Can Learn: Effects of Curriculum Alignment on the Mathematics Achievement of Third-Grade Students*. Academic Press.

Mohamed, H., & Jaafar, A. (2010). Development and potential analysis of Heuristic Evaluation for Educational Computer Game (PHEG). *Computer Sciences and Convergence Information Technology (ICCIT), 2010 5th International Conference On*, 222–227. doi:10.1109/ICCIT.2010.5711061

Mondada, F., Bonani, M., Riedo, F., Briod, M., Pereyre, L., Retornaz, P., & Magnenat, S. (2017). Bringing Robotics to Formal Education: The Thymio Open-Source Hardware Robot. *IEEE Robotics & Automation Magazine*, *24*(1), 77–85. doi:10.1109/MRA.2016.2636372

Mubin, O., Stevens, C. J., Shahid, S., Al Mahmud, A., & Dong, J.-J. (2013). a Review of the Applicability of Robots in Education. *Technology for Education and Learning*, *1*(1). Advance online publication. doi:10.2316/Journal.209.2013.1.209-0015

Mussati, A., Giang, C., Piatti, A., & Mondada, F. (2019). A Tangible Programming Language for the Educational Robot Thymio. *2019 10th International Conference on Information, Intelligence, Systems and Applications (IISA)*, 1–4.

Negrini, L., & Giang, C. (2019). How do pupils perceive educational robotics as a tool to improve their 21st century skills? *Journal of E-Learning and Knowledge Society*, *15*(2).

Nielsen, J., & Molich, R. (1990). Heuristic Evaluation of User Interfaces. *Proceedings of the SIGCHI Conference on Human Factors in Computing Systems*, 249–256. 10.1145/97243.97281

Osborne, R. (2014). *An ecological approach to educational technology : affordance as a design tool for aligning pedagogy and technology*. Academic Press.

Pachidis, T. P., Macedonia, E., Vrochidou, E., Kaburlasos, V., & Macedonia, E. (2018). *Social Robotics in Education : State-of-the-Art and Directions Social Robotics in Education : State-of-the-Art and Directions In recent years, robots attract considerable attention and become ever more popular in*. Academic Press.

Pane, J. F., & Myers, B. a. (1996). Usability Issues in the Design of Novice Programming Systems. *Education*, *78*(August). http://repository.cmu.edu/isr/820/

Papert, S. (1980). Mindstorms: Children, computers and powerful ideas. In New Ideas in Psychology (Vol. 1). doi:10.1016/0732-118X(83)90034-X

Riedo, F., Chevalier, M., Magnenat, S., & Mondada, F. (2013). Thymio II, a robot that grows wiser with children. *Proceedings of IEEE Workshop on Advanced Robotics and Its Social Impacts, ARSO*, 187–193. 10.1109/ARSO.2013.6705527

Ronsivalle, G. B., Boldi, A., Gusella, V., Inama, C., & Carta, S. (2019). How to Implement Educational Robotics' Programs in Italian Schools: A Brief Guideline According to an Instructional Design Point of View. Technology. *Knowledge and Learning*, *24*(2), 227–245. doi:10.100710758-018-9389-5

Seau Yoon, F., Ho, J., & Hedberg, J. G. (2005). Teacher understandings of technology affordances and their impact on the design of engaging learning experiences. *Educational Media International*, *42*(4), 297–316. doi:10.1080/09523980500237716

Shin, J., Siegwart, R., & Magnenat, S. (2014). *Visual Programming Language for Thymio II Robot. Interaction Design and Children*. IDC. doi:10.3929/ethz-a-010144554

Sisman, B., & Kucuk, S. (2019). An Educational Robotics Course: Examination of Educational Potentials and Pre-Service Teachers' Experiences. *International Journal of Research in Education and Science*, *5*(2), 510–531.

Toh, L. P. E., Causo, A., Tzuo, P.-W., Chen, I.-M., & Yeo, S. H. (2016). A Review on the Use of Robots in Education and Young Children. *Journal of Educational Technology & Society*, *19*(2), 148–163. http://www.jstor.org/stable/jeductechsoci.19.2.148

Tremblay, M. C., Hevner, A. R., & Berndt, D. J. (2010). The Use of Focus Groups in Design Science Research. In Design Research in Information Systems: Theory and Practice (pp. 121–143). Springer US. doi:10.1007/978-1-4419-5653-8_10

Vygotsky, L. S. (1978). Mind in Society: The Development of Higher Psychological Processes. doi:10.1007/978-3-540-92784-6

Wonder-McDowell, C., Reutzel, D. R., & Smith, J. A. (2011). Does instructional alignment matter? Effects on struggling second graders' reading achievement. *The Elementary School Journal, 112*(2), 259–279. doi:10.1086/661524

Yang, Y., Long, Y., Sun, D., Van Aalst, J., & Cheng, S. (2020). Fostering students' creativity via educational robotics: An investigation of teachers' pedagogical practices based on teacher interviews. *British Journal of Educational Technology, 51*(5), 1826–1842. doi:10.1111/bjet.12985

ADDITIONAL READING

Antonenko, P. D., Dawson, K., & Sahay, S. (2017). A framework for aligning needs, abilities and affordances to inform design and practice of educational technologies. *British Journal of Educational Technology, 48*(4), 916–927. doi:10.1111/bjet.12466

Biggs, J. (2003). Aligning teaching and assessing to course objectives. *Teaching and Learning in Higher Education: New Trends and Innovations, 2*(April), 13–17.

Cohen, S. A. (1987). Instructional Alignment: Searching for a Magic Bullet. *Educational Researcher, 16*(8), 16–20. doi:10.3102/0013189X016008016

Giang, C. (2020). *Towards the alignment of educational robotics learning systems with classroom activities*. EPFL.

Heersmink, R. (2013). A Taxonomy of Cognitive Artifacts: Function, Information, and Categories. *Review of Philosophy and Psychology, 4*(3), 465–481. doi:10.100713164-013-0148-1

Lauwers, T. (2010). Aligning capabilities of interactive educational tools to learner goals. *ProQuest Dissertations and Theses, 3470163*, 366.

Nielsen, J., & Molich, R. (1990). Heuristic Evaluation of User Interfaces. *Proceedings of the SIGCHI Conference on Human Factors in Computing Systems*, 249–256. 10.1145/97243.97281

Tremblay, M. C., Hevner, A. R., & Berndt, D. J. (2010). The Use of Focus Groups in Design Science Research. In Design Research in Information Systems: Theory and Practice (pp. 121–143). Springer US. doi:10.1007/978-1-4419-5653-8_10

KEY TERMS AND DEFINITIONS

Educational Robotics: The use of robots as cognitive artifacts in formal and informal learning activities.

Educational Robotics Learning System: The entirety of elements involved in educational robotics activities, i.e., intended objectives, instruction, assessment, and educational robotics artifacts.

Focus Groups: A research method that aims at gaining new insights through moderated discussions in small groups of participants.

Heuristic Evaluation: An evaluation method based on a list of specifications for an ideal system (i.e., the heuristics).

Identified Issues: Weak points of a system identified by evaluators using the list of heuristics provided to them.

Instructional Alignment: The idea of aligning intended outcomes, instruction and assessment to enhance the teaching and learning experience.

Tangible Programming: The concept of using physical tiles/blocks to program educational robots and/or virtual agents.

Chapter 2
5 Phases of Educational Robotics (5PER):
The Scientific Method Applied to Educational Robotics

Mauricio P. Galvez-Legua
https://orcid.org/0000-0002-4845-4218
Universidad Nacional de Ingeniería, Peru

Diego E. Galvez-Aranda
Texas A&M University, USA

ABSTRACT

This chapter presents a methodology for the development of activities based on educational robotics: five phases of educational robotics (5PER). First, the authors describe how technology and science have evolved to this day, recognizing their importance, and how they have impacted the schools with approaches such as STEM. Then, the main theories and learning approaches that have driven to the use of educational robotics in schools are reviewed, as well as concepts of robotics, educational robotics, and the difference between them. Finally, each of the five phases are explained, and two practical examples of how to implement them in the development of learning activities are shown.

INTRODUCTION

In recent years, the use of Information and Communication Technologies (ICT) in education has intensified, framed in what we currently call the knowledge society. (Välimaa & Hoffman, 2008) Initially, many of these applications were aimed to provide better tools for teachers in their classrooms: computers, multimedia projectors, electronic whiteboards, etc. Subsequently, the ICT began to be used as means of learning: the internet generated information and interpretation skills and the simulators developed capacities in the management of virtual environments.(Boud & Prosser, 2002) The problem with these types of technological applications is that relativized the real world. Theories such as Seymurt Papert's

DOI: 10.4018/978-1-7998-7443-0.ch002

constructionism,(Harel & Papert, 1991) gave us a more active and natural approach to learning, his Learning by doing theory promoted the creation of Educational Robotics (ER) as a means of learning, which privileges creativity and experimentation through the design and construction of prototypes.(S. Papert, 1987, 1993; S. A. Papert, 1973, 2020), allowing students to experiment in the real world. Later, with the appearance of STEM as a new interdisciplinary educational approach, different disciplines such as science, technology, engineering, and mathematics, were encouraged in the classroom. Nowadays, the teaching of these disciplines are currently considered essential for the integral development of people.(Bybee, 2013; Xie, Fang, & Shauman, 2015) In this context, ER allows a more natural approach of students with these areas, initially as a game and then through experimentation, it allows the student to interact with physical phenomena and later propose possible solutions to observed problems and formulate conclusions in a way very similar to the scientific method.(Benitti, 2012; Mikropoulos & Bellou, 2013; S. A. Papert, 2020)

ER has great potential in education, and it is necessary to define a framework that standardizes its use and its relationship with research through experimentation. In the present work, the main objective of the authors is to propose an ER model based on the traditional scientific method called "5 Phases of educational robotics (5PER)" and an example of its implementation in educational environments.

BACKGROUND

Technology and Science

Technology appears as a result of understanding the world in which we, humans, live. Experimentation was the primary tool that man used, initially to survive, and later to dominate his environment.(Fan, Yu, & Hunter, 2004) The method called: trial-and-error, was refined as the human race evolved. An example of the application of the trial-and-error method could be a primitive man using a hard stone to carve a soft stone as shown in Figure 1. Another example could be the opening scene: The dawn of man, from the 2001 movie: Space Odyssey from Stanley Kubrick

Figure 1. Primitive man learning to carve a stone.

During the stone age, man learned how to build tools. The first craftsmen appeared and began to modify their environment.(Henshilwood & Dubreuil, 2009) During the Metal Age, technology continued to evolve and did not require science to do so. Authors can cite two examples, the first is the construction of ships that the Egyptians developed before Archimedes discovered the physical law, principle, that governed.(Pickover, 2008) The second example is the steam engine that was created before the appearance of thermodynamics, which later explained its working principle. These advances were developed by artisans, not scientists, it was in general terms, practices or techniques that arose from experimentation.

"The craftsman applies, does not explain."

Science arises in part, in response to religion; faith-based beliefs versus evidence-based beliefs. It is the search of the human being to understand his environment based on logical reasoning. For a long time, these two ways of understanding the world around us have been antagonistic, however today there are many scientists who have expressed their belief in a god. According to Saint John Paul II, in his encyclical Fides et ratio, he says: faith and reason (fides et ratio) are like two wings on which the human spirit rises to the contemplation of truth.

In this context, the first philosophers emerged, trying to understand and explain the phenomena of the world around us. Philosopher's tools were observation and reasoning, they did not require experimentation. Authors can cite as an example of that time, what Aristotle said about the fall of two bodies: a body with greater mass falls faster than another with less mass. What Aristole claimed would have been quickly refuted with experimentation. This science, called "Ancient Science", lacked a method of replication and verification, it only required the theories to be logically valid.(Lloyd, 2012)

The ancient science ends when Galileo shows that when two unequal stones are dropped simultaneously, both stones hit the ground at the same time.(Adler & Coulter, 1978) Galileo used observation and experimentation as a way to test theoretical assumptions generating a more rational relationship with nature.

The incorporation of experimentation generated what is called: classical science. From that moment on, science begins to have greater relevance than philosophy, in part because science proves to be a more reliable source of knowledge because it uses experimentation and has Karl Popper's falsificationism, as a way of declaring his theories wrong. The falsificationism proposes that science is not able to verify if a hypothesis is true, but it can demonstrate if the hypothesis is false.(Popper, 1999) All this boom led to the Industrial Revolution of the 19th century.(Ashton, 1997) Science continued evolving and the so-called modern science appears. Authors can cite as an example of modern science the contributions of Michael Faraday taken by James Clerk Maxwell for the discovery of electromagnetic waves,(Garratt, 1994; Sengupta & Sarkar, 2003) which drove the discovery of the X-rays by Wilhelm Conrad Röntgen,(Glasser, 1993) and the radio waves by Heinrich Rudolf Hertz among others.

The termn scientist appears, which refers to people who seek new knowledge using observation, measurements, hypothesis formulation, experimentation, and interpretation of results. This set of steps forms what is called the scientific method. Both the philosopher and the scientist seek the truth, but the philosopher does so through pure reasoning and the scientist through experimentation. There is currently a great debate about whether science requires philosophy or not. It is a subject that is beyond the scope of this article (Figure 2).

Figure 2. Evolution of human understanding about the universe

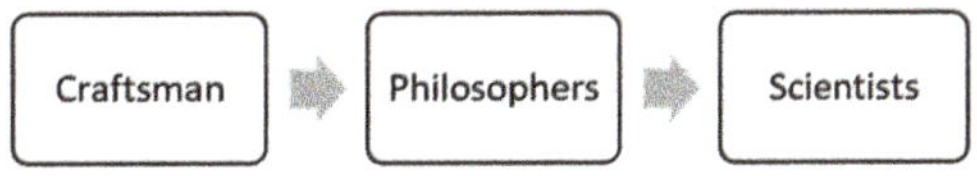

Scientific Method

Science is the search and generation of knowledge through experimental methods, for the understanding of nature and the prediction of its behavior, following a methodology based on research.(Heidegger, 1977; Scheffler, 1982; Tauber, 2016) Science uses what is called the scientific method, which was born from the trial-and-error with which technology began to evolve and led to the transformation of artisans into scientists.(Gauch Jr & Gauch, 2003) The central concept of the traditional scientific method is a refutable hypothesis about some phenomenon of interest.(Harris, 2014) This hypothesis must be tested experimentally. The test results can support or refute the hypothesis, generating a new round of hypothesis formulation and testing. This is what is called "Traditional scientific method: Hypothesis-based deduction."(Voit, 2019).

An interesting approach to the model of the Traditional scientific method: Hypothesis-based deduction, was proposed by Dewey in his book, How We Think, where he proposes that the investigation is divided in five steps: (i) a felt difficulty, (ii) its location and definition, (iii) suggestion of a possible solution, (iv) development by reasoning of the bearing of the suggestions, (v) further observation and experiment leading to its acceptance or rejection.(Dewey, 1997)

The scientific method is a set of steps established in advance by a discipline in order to achieve valid knowledge through reliable instruments. The scientific method tries to avoid subjectivity in knowledge.(Letherby, 2013) The scientific method is a sequence to formulate and answer a question. In the present work, the authors propose a model based on the traditional scientific method (Figure 3).

Figure 3. Diagram of the model based on the traditional scientific method.

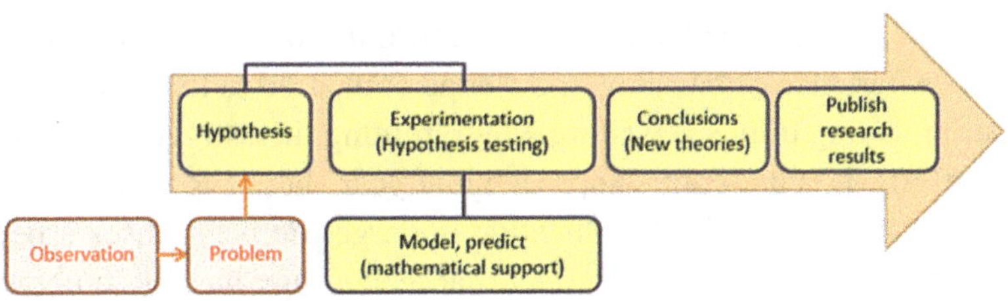

- Observation: It is a carefully application of the senses over an object or a phenomenon, to study it as it appears.
- Problem: It defines, delimits, what is the problem or what is trying to explain.
- Hypothesis: Proposal of an explanation or solution to the problem.
- Experimentation: Test the hypothesis through the use of experiments.
- Conclusions: If the hypothesis is verified, a new theory is developed.

- Publication of results: Disseminate the new theory so that others can replicate it and eventually can refute it: falsificationism.

Learning Theories

According to Merriam-Webster, pedagogy is the art, science, or profession of teaching.(Erlam, Smythe, & Clair, 2017) Pedagogy is the study of how knowledge is imparted in an educational context, making use of theories, and learning approaches. Traditionally, pedagogy in education has been centered on the teacher as the possessor of knowledge and the student as the receiver of knowledge, (Freire, 2018) however, these learning theories and approaches increasingly identify students as an active participant in their own learning process and the teacher as a facilitator.

Learning theories are conceptual frameworks which describe how knowledge is absorbed, processed, and retained during the learning process. Cognitive, emotional, and environmental influences, as well as prior experience, all play a part in the learning process.(Knud, 2004) Many learning theories have been developed, which in turn have generated diverse learning approaches. The first learning theories were formulated mostly by psychologists and/or philosophers who had an emphasis on the internal processes that occur in the individual when he learns.

One of the first learning theories is the behaviorism, advocated by Iván P. Pávlov, in which the subject is passive, and his learning is measured through observable and measurable behavior changes.(Clark, 2018) Subsequently, **constructivism** appears, formulated by Jean Fritz Piaget who stated that people learn by building new ideas based on previous experiences or knowledge from their interaction with the world.(Wadsworth, 1996) In constructivism, the subject is active in the learning process. With the Neobehaviorism of Burrhus Frederic Skinner, and his proposal for programmed or linear teaching raised in 1958, technology began to be used as a learning tool.(Fromm, 1973)

Then Seymour Papert, who worked with Piaget at the University of Geneva, developed a vision of learning called **constructionism**, which is based on Piaget's constructivism and advocates learning by doing.(Harel & Papert, 1991) Papert argues that learning is best when children are engaged in building a meaningful product. When children build things in the external world, they are simultaneously building knowledge within their minds, generateing more knowledge in a self-reinforcing cycle. Papert argues that children actively create their knowledge, and that education should provide them with tools to carry out activities that promote this knowledge creation. According to Papert, learning is intricately linked to doing, it is an active experience of building knowledge compared to the usual experiences of passive reception of information. Doing things is writing, diagramming, investigating, trying, making mistakes and above all being curious. It is under this approach that years later ER was born at MIT.

As an example of the impact of ICT in education, George Siemens raises a pedagogical perspective called **connectivism**, which explains the effect that technology has had on how we currently live, communicate and learn.(Siemens, 2017) Siemens argues that the ability to learn for the future is more important than what is known now. Students need to generate new skills in a changing context, where science and technology are defining new paradigms such as the internet, virtual reality, artificial intelligence, computational thinking, etc. ICT entail transformations and restructuring that give rise to the creation and exchange of knowledge as well as new ways of acquiring, addressing, and organizing the training process.

Technologies in Education

Technology has impacted education in various ways. The first case of the use of technology in education dates back to the 1960s in the United States, where Skinner launched his theory of Programmed Teaching, which consisted of using question-response-stimulus linear chain teaching machines.(B. Skinner, 1986; B. F. Skinner, 1960) Thus, a progressive linear sequence (always the same) of the key ideas was presented, which the student was supposed to acquire and internalize (Behaviorism).(Clark, 2018; Watson, 2017)

Seymour Papert, who was a disciple of Piaget, experimented at MIT with technologies that provide new ways of learning and teaching mathematics in children. He was one of the first to recognize the revolutionary potential of computers in education. Papert suggested that the computer could also serve as an environment to think in thinking, as opposed to the trend of computer serving as a transmission system to present content. Papert argues that children actively create their knowledge, and that education should provide tools to carry out activities that promote this activity. According to Papert, learning is closely related to doing, it is an active experience of building knowledge compared to the usual experiences of passive reception of information.(S. Papert, 1993) In 1968 he invented the Logo programming language, which functions as a learning medium that allows children to build their knowledge through the development of logical-mathematical thought processes.(Solomon et al., 2020). Logo was the first computer language designed specifically for kids. It is worth to mention that at that time personal computers did not exist yet, they appeared in 1981.(Thacker, McCreight, Lampson, Sproull, & Boggs, 1981)

The children used Logo to program the movements of a "turtle", either in the form of a small mechanical robot or a graphic object on the computer screen.(F. Martin, 1988) In his book, Mindstorms: Children, computers and powerful ideas, Papert presented an approach in which "the child programs the computer and, in doing so, both acquires a sense of mastery over a piece of the most modern and powerful technology and establishes an intimate contact with some of the deepest ideas from science, from mathematics, and from the art of intellectual model building".(S. A. Papert, 2020) A partnership agreement is subsequently signed between Lego and MIT whereby Lego would fund the research from the MIT epistemology and learning group on how children learn and, in return, obtain new ideas for its products. One result of this collaboration was the development of the MIT Programmable Brick, which was a programmable controller brick, which was combined with construction parts, sensors, and actuators, allowing from static structures to dynamic structures.(F. G. Martin, 1994) This gave birth to the Lego Mindstorms following Papert's 'learning by doing' paradigm. The MIT original idea was leaving the building possibilities open to children, however, Lego opted for a more instructive approach by using building guides. Mitchel Resnick (1956), a disciple of Papert and creator of the Scratch programming language, promotes learning by designing, and proposes to leave the possibilities of construction more open. In the words of Mitchel Resnick the goal is: to design things that will enable kids to design things. (Resnick & Silverman, 2005)

Nowadays, the term educational technology is becoming popular, and it refers to the incorporation of ICT in education, in order to support learning processes in different contexts, both formal education and non-formal education. New paradigms in the Education such as STEM (science, technology, engineering, and mathematics) and its variants STEAM and STREAM are promoting an education closely related to science and technology to reduce the education gap in the face of the ever-changing requirements of our society. Within the framework of the STEM educational approach, ER allows a more natural approach to the science, technology, engineering, and mathematics.

Robotics and Educational Robotics

Robotics is an interdisciplinary area (Birk, 2011) that deals with the study, development, and applications of robots, and is the result of the interaction of various engineerings (mechanics, electronics, and computer science) therefore it has a direct relationship with science and mathematics that are the base of the study of any engineering. Robotics summarizes the incessant search of the human to build intelligence, from the implementation of logical functions with hydraulic and pneumatic systems, which gave rise to mechanical automation, through the use of electrical panels based on relays and contactors in the electric automation, up to the integrated circuits like the microprocessors and microcontrollers in the electronic automation (Figure 4).

Figure 4. Robotics

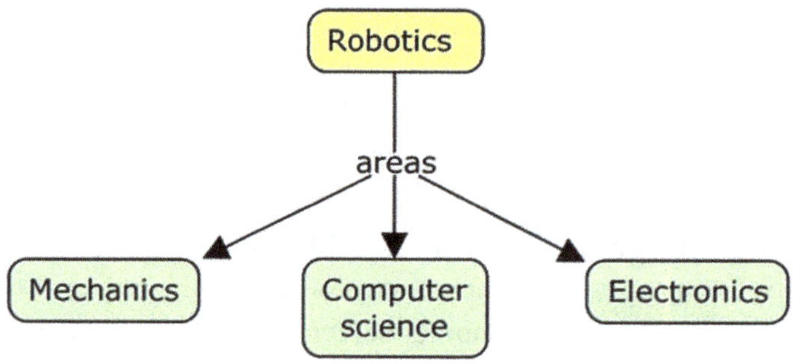

Robotics is the highest level of industrial evolution, currently called Industry 4G. Robotics combined with artificial intelligence, internet of things, virtualization, and cloud computing, generate what is called smart and ubiquitous factories. Figure 5 shows the industrial evolution.

Figure 5. Impact of technology on the evolution of the industry

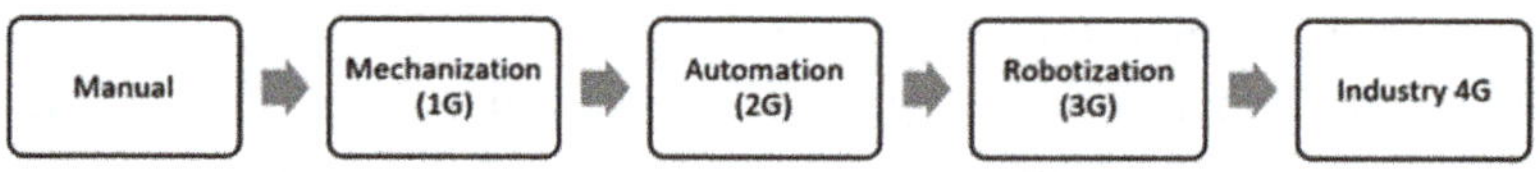

A robot is a programmable intelligent machine,(Holland, 2004) however this definition is relatively ambiguous since a computer can meet this definition. Therefore, a set of features must be included in the definition of a robot. The authors define robot as follows: A robot is a mechanical electronic device, with capacity of movement and action, with a certain degree of autonomy, that performs tasks automatically exhibiting computational intelligence and it must be programmable. Authors understand by capacity of action, the interaction of the robot with the nature (environment), through the use of sensors and actuators. Autonomy refers to the fact that it does not require an operator to function. Computational intelligence refers to having a CPU internally. The operation of a traditional robot is summarized in three stages (Perceive, process, act); however, a fourth stage (learn) is being considered, which is an ability that certain robots have to use machine learning algorithms based on data analysis. These robots use what is known as machine learning, which is an application of artificial intelligence to learn. Each of these stages are described in Table 1.

Table 1. How a robot works

Stages	Description
Perceive	The robot becomes aware of its position and the environment that surrounds it through the use of sensors (input data).
Process	Based on the input data and the programmed algorithm, the robot processes the information and makes decisions.
Act	The decisions are executed by activating actuators that interact with the environment.
Learn	Using machine learning algorithms, the robot analyzes data and detects patterns that are incorporated as the robot response or behave.

Robotics is related to programming, which is a highly required skill nowadays, largely because human beings are increasingly surrounded by machines and have the need to interact with them in a more automated world. Technologies such as Internet of Things (IoT), cloud computing, Artificial Intelligence (AI), Software Defined Networking (SDN) are a demonstration of what has been said. Programming is the ability of people to implement algorithms that are executed by a CPU performing a certain number of tasks. Programming is performed in devices from cell phones to trains, planes, or spaceships. Programming languages are the language of communication between humans and machines. Programming is essentially a creative act, as is drawing, painting, composing a song.

"A program is the implementation of an algorithm. An algorithm is the implementation of an idea, which has been refined by logic".

ER is a learning environment that allows the understanding of abstract concepts such as: mathematics, geometry, programming, etc., as well as physical concepts from our environment such as: structures, forces, speed, center of gravity, etc., through the construction of prototypes. ER stimulates the imagination, creativity, motivation, concentration, and manual skills in student by solving challenges. ER has the main virtue of serving as an interface between the logical world and the physical world and it does so in a playful way. In many countries it is becoming part of the curriculum due to its ability to integrate into areas such as: science, technology, mathematics, etc. ER is an example of the application of technology in education. Unlike educational programs or simulators that guide students' work in virtual environments, ER brings the student closer to a direct interaction with the real world.

The main difference between robotics and ER is pedagogy (Figure 6). While robotics is an "end", the second is a "process" applied to the understanding of our environment and the laws that govern our universe. In addition, ER allows us to approach critical pedagogy (Paulo Freire) that advocates the development of critical and reflective skills in students (critical thinking) as well as student-centered learning. In Table 2, authors summarize what ER provides to each type of user who uses it.

Figure 6. Educational Robotics (ER)

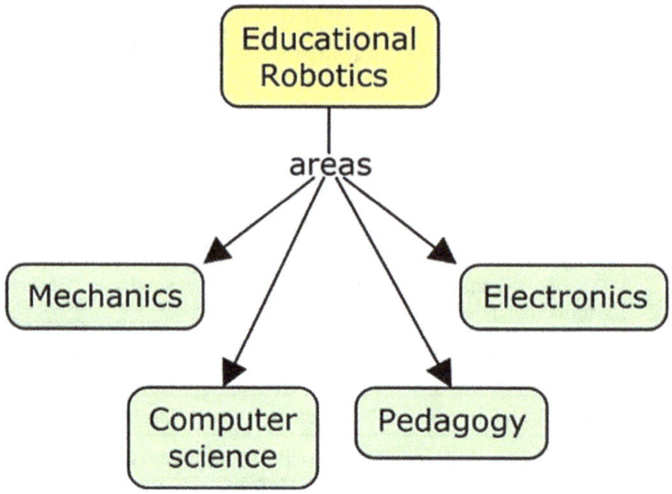

"It is the teacher who turns robotics into educational robotics with his/her pedagogical contribution."

Table 2. Contribution of ER to each type of user

Kid	Teenager	Teacher
An environment to play	An environment to compete	A learning environment
Improve your psychomotor skills	Build your ideas	Show the interaction of the physical world with the virtual world
Get closer to math	Approach to science	Develop multidisciplinary activities
Understand how things work	Understand the laws that rules our universe	Encourage design and creativity in students

ER allows us to develop various levels of learning activities: from explaining how things work, then how our world and the universe work through understanding its laws and finally through design: the application of knowledge and creativity.

Educational Robotics Kit

An ER kit allows the design, construction, programming, and testing of prototypes. We use the word prototype to refer to any construction that may or may not have a CPU. Example: the construction of a hammock or a house are only mechanical constructions without the use of CPU.

In general, every robotics kit is composed of the following components (Figure 7):

1. Electronic components:
    a. Control System: It is the intelligent part of the kit, normally called CPU, which is implemented by a microcontroller and has input ports where the sensors are connected, output ports where

the actuators are connected, and communication ports such as USB or bluetooth. Some robotics kits do not come with a CPU, instead they use the CPU of the computer to which they are connected.

   b. Sensors: used to enter information from the environment to the Control System for processing. Example: temperature, inclination, acceleration, speed sensor, etc.
   c. Actuators: they are used to activate something in the environment, for example to turn on a motor or a light.

2. Building elements: consisting of beams, wheels, gears, rods, endless screw, conctors, etc. They can be interconnected to create complex machines.
3. Programming software: it is a programming environment that allows the Control System to be programmed through the use of a programming language. There are two types of programming languages: textual (commands that must be written) and iconographic (it uses icons that represent the commands and have a Drag and Drop feature).
4. Design software: it is used to document the sequence of steps to assemble the prototype. Useful when designing new constructions.

Figure 7. Components of an ER kit

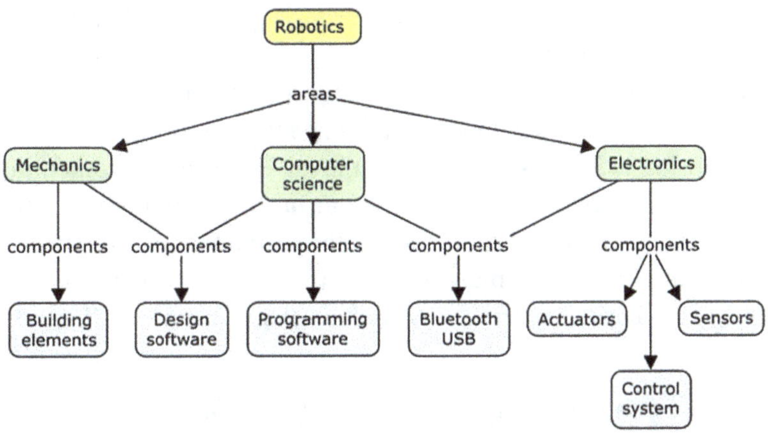

Problems in the Implementation of the Educational Robotics in Schools

The main problem with ER is that in many educational institutions it is not well implemented. ER is used in most cases as activities to build previously designed robotic models, leaving aside its main characteristic: being a powerful learning environment that encourages students to approach science and mathematics under a learning by doing and a learn by design approach. This missuse of the ER is due to various factors, but one of the most important is that many teachers are not clear about the scope of ER and how to create activities that can involve its use. Therefore, it is necessary to define a framework that standardizes its use and its relationship with learning.

ER Evolution

The applications of ER are evolving, from the guided construction that is developed by the student, their psychomotor abilities, and the understanding of our world through simple constructions. Then, as a learning medium that allows the student to approach science and mathematics in a playful way, experiencing concepts such as strength, speed, and logical thinking through programming. Later as a learning environment, where skills for research, design and implementation are promoted.

ER is currently being promoted as a form of implementation of the so-called "Design-based learning" (DBL), which is found in many disciplines such as engineering, architecture, business, etc.(Burdick, 2009)

DBL serves to promote the skills required today and, in the future, as well as a deeper learning, because it is based on the development of projects, which require the understanding and application of knowledge. The complexity of a design often requires collaboration and role assignment, giving students the opportunity to begin to specialize in a particular area and thereby discover their "areas of interest" (vocation). Design projects require students to set goals and constraints, come up with ideas, build prototypes, and test them until requirements are met.(Doppelt, 2009)

An example of an ER-based DBL are robotic competitions, which are design-based learning activities, where each member of a participating team is assigned a role to: design, build, program, test and document your solutions to the challenge. The design process is an iterative process that has a variety of steps: investigate context, identify needs, develop criteria, generate alternatives, select alternative, prototype / test, produce and evaluate.(Mehalik, Doppelt, & Schuun, 2008) A similar approach is the ADDI model, which is used to design learning experiences for training. It consists of five phases: analysis, design, development, implementation, and valuation.(Morrison, Ross, Morrison, & Kalman, 2019)

Understanding that ER-based DBL is a process that involves designing something (creativity), these models serve to understand the steps that are carried out in the development of that activity.(Galvez-Aranda & Galvez Legua, 2021) The scientific method is the search for the truth through experimentation, a hypothesis is raised (creativity), which must be tested (trial and error) and finally draw conclusions. ER-based DBL has many similarities with the traditional scientific method (Figure 8).

Figure 8. Similarities between scientific method and ER-based DBL

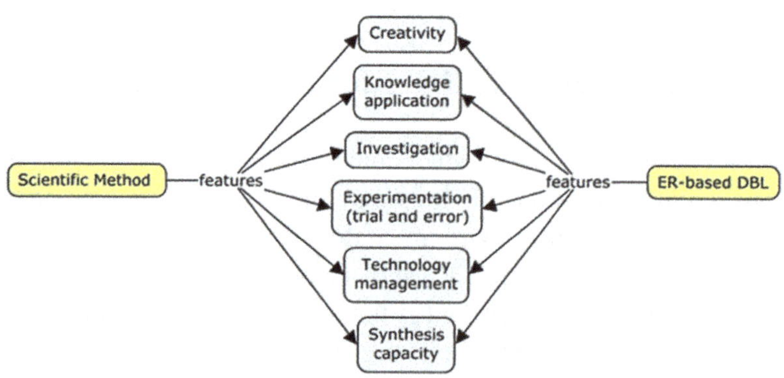

An analysis of the various models presented allows us to observe that they are remarkably similar. Based on these referential models, the authors have proposed the 5PER (Table 3).

Table 3. Contribution of ER to each type of user

Scientific Method	5PER	Design Process	ADDI Model
Observation		Investigate context	Analysis
Problem		Identify needs	
Hypothesis	Design	Develop criteria Generate alternatives Select alternative	Design
Experimentation	Build	Prototype	Develop
	Program		
	Test	Test	
		Produce	Implement
Conclusions	Document & Share	Evaluate	Evaluate
Publication of results			

5 PHASES OF EDUCATIONAL ROBOTICS (5PER)

In the present work authors propose an implementation methodology called: The Five Phases of ER (5PER), which is based on the traditional scientific method. The methodology present sequential and relatively independent activities that define a manual and/or intellectual action in the use of ER (Figure 9).

Figure 9. The Five Phases of ER (5PER)

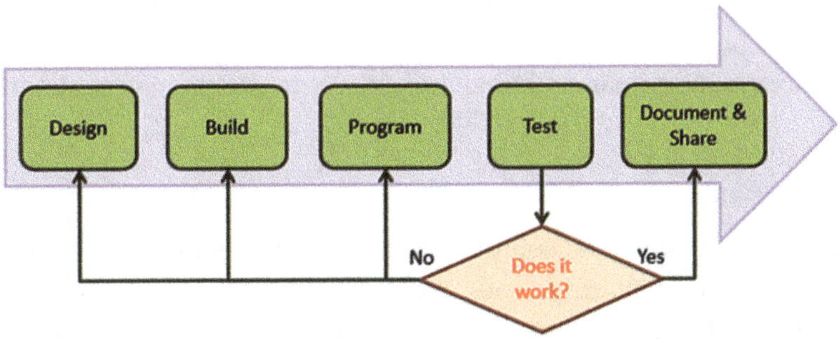

Educational Robotics Levels

The use of ER as a learning environment can be divided into three levels, according to the level of complexity we want to reach. In the first two levels, it is the teacher who must develop the activity. The third level is what Mitchel Resnink calls learning by designing, encouraging a personal vision of things, new points of view, new concepts, and new theories.

- Level I: it is used to develop the psychomotor capacities of the students, the construction of prototypes previously made using psychomotor skills and with the use of iconographic programming (cognitive skills), "giving life" to the constructions made.
- Level II: it is used to integrate and articulate ER within courses such as mathematics, physics, natural sciences, etc. In this way, students begin to approach science and mathematics under a learning by doing approach.
- Level III: it is used as a "small experimentation laboratory", where the student can unleash his curiosity to investigate, design (creativity and imagination) and build "new ways of understanding things". At this level, students begin to transform themselves into little scientists, who through experimentation continue to learn but autonomously.

All these three levels are presented in each of the phased of the 5PER methodology.

Phase I: Design

This phase starts from the observation of a problem or the establishment of a challenge. It requires the use of imagination and creativity to mentally conceive an idea that could be a potential solution to the problem or challenge. Then, through logical analysis, convert the idea into a design (set of specifications that can be realized in our space-time). It is highly recommendable that the design idead can be sketched in a physical medium, such as drawing it on paper, to extract the necessary requirements to implement the prototype. We can implement this phase in the following way for each of the proposed levels of ER:

- Level I and II: In these first two levels, students begin by copying examples from their environment: imitating reality. It is a way of expressing the student conception of the different object shapes.
- Level III: Students use their imagination and creativity to design new prototypes.

Phase II: Build

In this phase, a prototype of the potential solution to the problem begins to be built. Building parts such as beams, bricks, gears, wheels, connecting rods, etc. are used. Additionally, sensors such as proximity, inclination, sound, among others, and actuators such as motors and LEDs can be used. This phase can be implemented in the following way:

- Level I: As a game to build basic prototypes, which are representations of things in the everyday environment such as a house, a bridge, etc.
- Level II: Build intermediate prototypes which can have movement and/or displacement. For example, machines such as fan, mixer, etc., or animals such as lion, crocodile, etc.
- Level III: Build advanced prototypes that are representations of machines such as a drill, a sewing machine, among others, and especially your own creations.

Phase III: Program

In this phase, the prototype is brought to life through the use of a programming environment called Integrated Development Environment (IDE), which includes a programming language that is used to give orders to the prototype. It should be noted that the element that understands and executes these commands is the CPU or controller, which can be located in the same prototype, or the prototype can use the CPU from the computer. This phase can be implemented in the following way:

- Level I: Learn basic instructions of a programming language, read flowcharts of an algorithm, and implement them through an iconographic programming language.
- Level II: Learn more programming instructions and develop programs that interact with sensors and actuators to perform more complex movements and displacement.
- Level III: Use creativity and logic to create and code algorithms. With this, learn a new way of expressing yourself.

Phase IV: Test

In this phase, your prototype is evaluated and verifies that meets the requirements initially set out during the design phase. All your skills must be used to detect errors, which may be due to flaws in programming, construction, or design. If faults are found, correct them, and test the prototype again: trial-and-error methodology. This phase can be implemented in the following way:

- Level I and II: The design is contrasted with the implementation to visually verify that the implemented prototype works as designed.
- Level III: Checks that the prototype's operation complies with a previously defined set of specifications and/or standards. During testing, find areas for potential improvement for your prototype's performance.

Phase V: Document and Share

Once the prototype has been tested and verified that it works according to the design requirements, it is then documented and shared with other students for verification and future improvements. Sharing encourages the exchange of ideas, the argumentation of the design and the recognition of our creations. This phase can be implemented in the following way:

- Level I: Document through photos, videos, freehand drawing, etc.
- Level II: Prepare a document using a word processor or web page where various types of media such as graphics and videos are combined.
- Level III: Prepare a presentation and present the prototype design criteria. This generates a debate where the points of view must be argued. A publication on the internet to share the knowledge generated can be also a good practical method.

Figure 10. Equivalence between the traditional scientific method and the five-phase methodology.

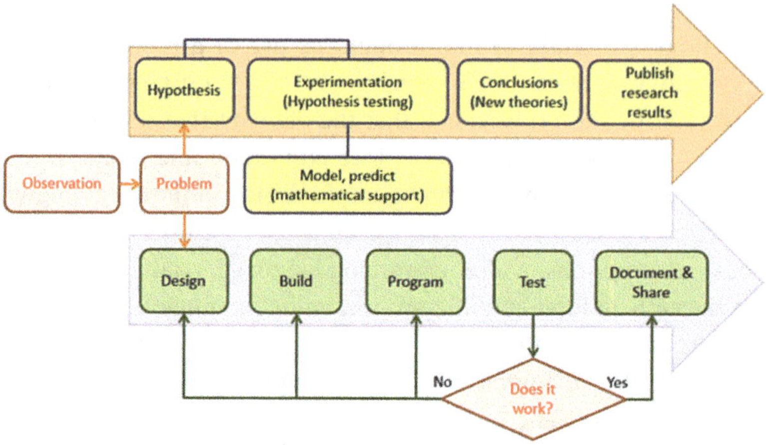

In Figure 10 an equivalence between the traditional scientific method and 5PER methodology is described. Both frameworks start from the observation of a problem and have the same purpose: to propose a solution. In the case of the scientific method, it is through experimentation and in ERit is through the construction, programming, and testing of a prototype.

In the scientific method, if the proposed solution does not work, then the hypothesis is rejected. In the case of ER, if the solution does not work or does not work well, the previous phases of programming, construction or design are reviewed to find where it is failing and correct it.

In the scientific method, once the hypothesis is validated, there is a new theory. In the case of ER, if the prototype passes the tests, you have the solution to the problem.

In both cases, the work is documented and published. In the case of the scientific method, the publication has the purpose of spreading the new theory so other scientists can reproduce it. In the case of ER, the purpose is to share experience and knowledge, so the students contribute to society.

In the next section some examples of how the 5PER methodology is applied in an activity are described. These examples assume that the teacher and students are using an ER kit.

5PER APPLICATION: MEASURING OBJECTS LENGHTS

Introduction and Problem

Introduction: The teacher comments on how the lengths of objects were measured in antiquity. Examples:

- The Egyptians used the elbow.
- The Greeks and Romans used the foot.

Problem: Students want to measure the length of an object, for example the length of a shoes box.

Phase I Design

The idea is to design a rover that moves along the entire length of the object through discrete steps using a proximity sensor to detects the object. A counter which starts with the value of zero incrementes by one once a step is performed by the rover. The rover will perform another step until the proximity sensor does not detect the object anymore, then the rover stops, and the counter value will be proportional to the length of the object (Figure 11a)

What problems can the students face with this idea?

- When the distance sensor detects the object, the counter starts to increase. How fast will it count? Will the distance sensor have enoughtime to know if there is an object? Keep in mind that most remote sensors work by emitting light or ultrasound and receive the rebound of the emitted light or ultrasound on the object. If the rover moves too fast, the sensor could miss the bounced light and thus falsely detect that there is no object and stop the measurement. A good to avoid the missed

Figure 11. (a) Rover moving with discrete steps. (b) Big discretes steps generates a big error in the measurement. (c) Small discretes steps generate a small error in the measurement.

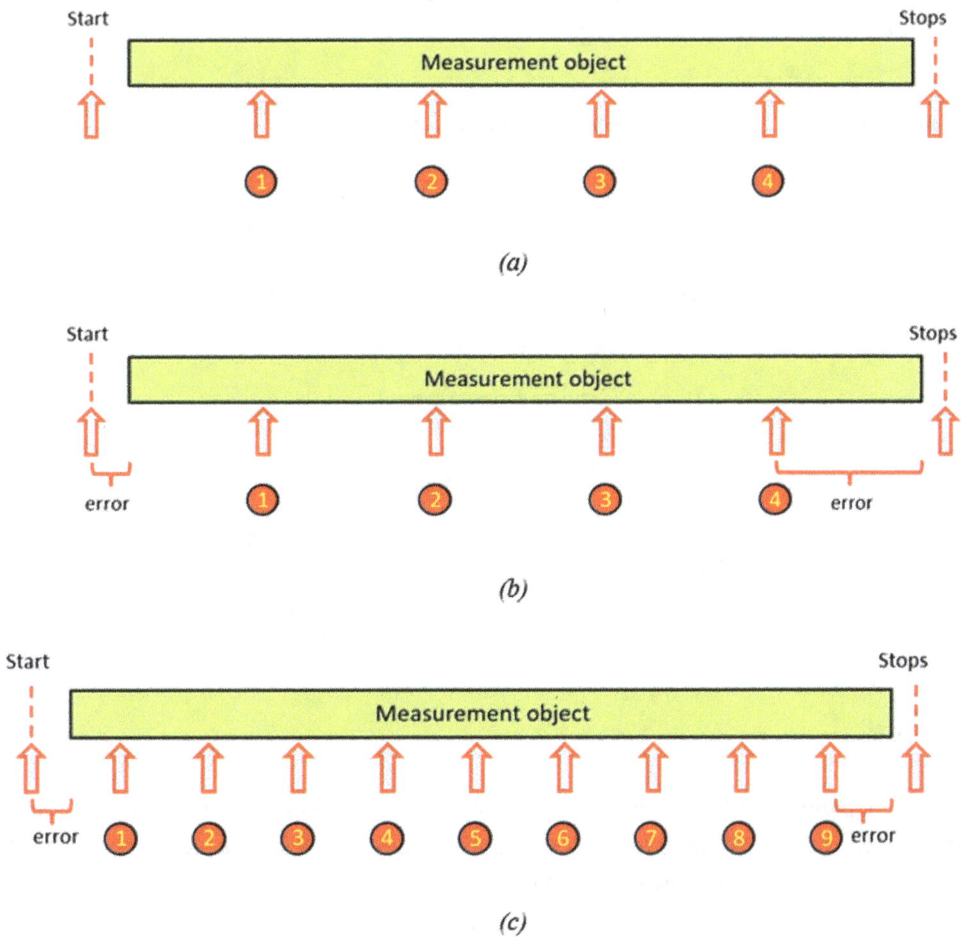

measurement if to make the rover move through discrtete steps, meaning that every steps the rover stops and reads the proximity sensor.

How accurate will be the measurements?

- If the rover moves in big discrete steps, this will lead to a big error (Figure 7b).
- If the rover moves in small discrete steps, this will lead to a small error (Figure 7c).

Questions to have in mind during design phase:

- Which method measures the object faster? Using small discrete steps or big discrete steps? Why?
- Does rhe surface of the object to be measured influence the sensor detection?

Before going to the next phase, a design sketch of the how the rover would looke like must be done (Figure 12). Creativity is put into practice. This will help to have a better idea of how the final rover should look like. Some of the requirements in the designing of the rover are:

- Rover must move in a linear way. There is no need to build a rover that can rotate.
- Rover must have a proximity sensor mounted in such a way that the sensor can detect the presence of an object while the rover is moving.

Figure 12. Rover design to measure the length of an object.

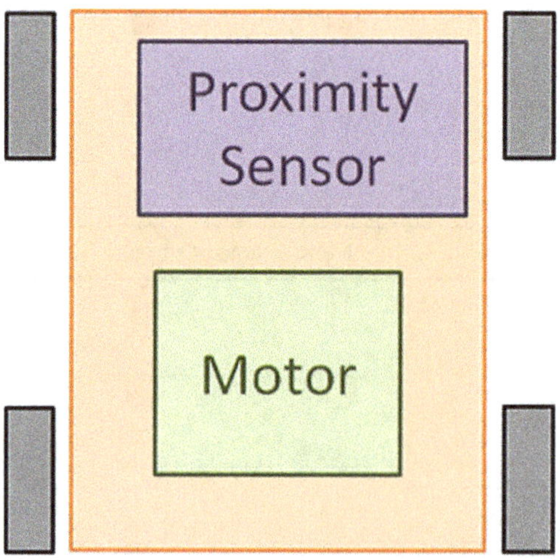

Phase II: Build

The construction depends on the type and quantity of building parts that the ER kit has. There are some building parts that are necessary for a rover design such as wheels. A motor is also necessary to make the rover move the wheels. There are several mechanisms that can be used to transmit the motion generated by the motor to the rover's wheels such as gear train, gear worm or pulleys.

Phase III: Program

Once the rover has been built, the next step is to program the rover to perform as described in the Design Phase. Rover must move discreetly and measure the length of a certain object, such as a shoe box. Before coding the program, the elaboration of a flowchart might be helpful. Example of a flowchart:

1. Start Counter = 0
2. Configure the direction of rotation of the motor (clockwise or counterclockwise).
3. Configure the motor power (this depends on the type of robotics kit that is being used).
4. If the proximity sensor does not detect anything, go forward one step, and return to step 4.
5. If the proximity sensor detects an object, increment the Counter by one and return to step 5.
6. The motor stops and the counter value at this moment is proportional to the length of the object.

Phase IV: Test

Once programmed, it is time to test the operation of the rover. Several king of errors could occur during the testing phase. Preferably the teacher should ask some questions that indirectly lead the students to detect their error.

1. The rover does not move.
    a. It may be that the programmed power for the motor is very small and must be increased to break the initial inertia of the rover.
2. The rover is moving too fast, and the proximity sensor has measurement errors.
    b. Create a slower rover motion by using a worm gear mechanism to transmit the rotation from the motor to the wheels.
3. The distance sensor was located pointing in the same direction of the rover movement. This is a design error. The sensor must be pointing a perpendicar direction to the rover movement.
4. Does the value of the counter is equal to the length of the object?
    c. Students should realize that the value of the counter is proportional to the length. When the cart moves it does so at a constant speed, therefore between one stop and another, the distance traveled is the same, it is a uniform rectilinear movement.
5. What should students do to obtain the real value of the length? There are several alternatives:
    d. Measure the length that the rover travelled in one step, and then enter that value into the program and multiplied by the value of the counter when the proximity sensor no longer detects the object.
    e. Measure the length of the object with a ruler and carry out the following operation: constant = Length of the object / counter. Report that value.

Phase V: Document and Share

After all the functional tests have been performed, and having corrected all the errors, a working prototype can be finally reported. The documentation process starts, so others can replicate the design, modify it and improve it.

- All the problems found, the solutions that were planted and which ones worked should be recorded.
- Document design and building diagrams using freehand drawings or by creating a building guide using photos, videos, or a 3D/CAD software.
- Indicate possible improvements that can be made to the design.

5PER APPLICATION: SOLIDS OF REVOLUTION

Introduction and Problem

Introduction: The teacher begins by talking about flat geometric figures such as the triangle, rectangle, and trapezoid. Ask students to complete Table 4.

Table 4. Characteristics of plane geometric figures

2D Figure	Number of Sides	Perimeter Formula	Area Formula
Triangle			
Rectangle			
Trapeze			

- The sides of a triangle can have any value or is there a relationship between them?
- What are the types of triangles?
- What are the types of trapezoids?

Then the teacher explains what the solids of revolution are and from which 2D figure they are generated. Ask your students to complete Table 5.

Table 5. Characteristics of solids of revolution

Solid of Revolution	2D Figure That Generates It	Perimeter Formula	Area Formula
Cone			
Cylinder			
Cone trunk			

Problem: How to generate the three solids of revolution seen in the previous table, by using 2D figures.

Phase I Design

Students must design a prototype that, using 2D figures, generates a solid of revolution. What characteristics should the prototype have?

- How does a plane geometric figure become a solid of revolution?
- Do you need wheels?
- Is a motor required? For what?

Remind students that the prototype must allow the 2D figures to be exchanged to generate a new solid of revolution. This means that they require some kind of mechanism to interchange the 2D figures. Students use their creativity to design the prototype and sketch it on paper.

Phase II Build

Using the building blocks from the ER kit, students build their prototype. Students must be careful to not use all the pieces until having that plane geometric figures.

- Verify that the plane geometric figure must be easily interchanged.

Phase III Program

Once the prototype has been built, does your prototype require a program?

- What should the program do? Does the program should read the value of a sensor or activate an actuator?
- If students are using a motor. Is it important to determine the direction of rotation?

Phase IV Test

The prototype must be tested.

- Is it relatively easy to change the 2D figures?
- When students rotate it, does the 2D figures deattach easily from the rotation platform? Why?
  - Maybe the structure design of the plane geometric figure is not stable.
  - Maybe how the 2D figures are connected to the platform is not strong.
  - Maybe the motor is rotation too fast. If so, how can you slow down the motor speed?
- Do the 2D figures rotate with difficulty or vibrate?
  - Maybe the mechanism to attach the 2D figure to the rotation platform of the prototype is not the most appropriate.

Phase V Document and Share

Once the testing of the prototype has been completed and all the errors have been corrected, student finally have a working prototype. Now, the student must document everything that they have done on the previous phases.

- All the problems found, the solutions that were planted and the ones that worked must be reported.
- Document the design and construction using freehand drawings or by creating a building guide using a 3D/CAD software. Students can also use photos and videos.
- Indicate possible improvements that can be made to the design.

Finally share all the information, so that other students can repeat it, modify it, or even improve it.

CONCLUSION

It is important to understand that robotics becomes ER due to the pedagogical contribution that the teacher adds. Pedagogy is about analyzing, planning, learning processes and developing teaching. Pedagogy gives an educational purpose to an activity and thereby generate learning in the students.

The use of the 5PER allows the teacher to develop learning activities using ER, where the development of each phase involves the student, through their knowledge prior to solving a specific problem. The teacher approaches the theory in a playful way and through challenges, awakening in the student their innate curiosity to solve them, making them the center of learning and always maintaining their interest.

5PER provides a methodology to use RD as a driver of science and mathematics, provides a frame of reference to direct students' innate curiosity towards research, and promotes teamwork and knowledge sharing.

RECOMMENDATIONS

Educational institutions should promote ER as a way to bring mathematics, science, and technology closer to students. The ER allow, for example, that children learn programming in a playful way and thereby train people to use programming in a natural way, without the fears that it generates today, due to the complexities of the syntax of textual programming languages. which are being largely outmatched by iconographic programming languages.

Teachers dhould be trained to use ER as learning environment and to be able to carry out activities contextualized to their courses. The teacher can design interdisciplinary activities that combine knowledge from several courses, bringing the student closer to real problems. This presupposes, on the part of the teacher, knowledge in basic robotics and programming.

The development of competition-type activities should be encouraged among teachers, taking advantage of the design concept, immersed in ER. It is an activity that should be used with greater emphasis. ER-based DBL allows the creation of multidisciplinary learning environments, which through challenges, encourages student participation by applying their knowledge and skills in a context remarkably close to reality.

Authors have focused on this article the importance of mathematics, science, and technology in the formation of students; however, it is also necessary to open spaces for art.

ACKNOWLEDGMENT

This research received no specific grant from any funding agency in the public, commercial, or not-for-profit sectors.

REFERENCES

Adler, C. G., & Coulter, B. L. (1978). Galileo and the Tower of Pisa experiment. *American Journal of Physics*, *46*(3), 199–201. doi:10.1119/1.11165

Ashton, T. S. (1997). *The industrial revolution 1760-1830*. OUP Catalogue.

Benitti, F. B. V. (2012). Exploring the educational potential of robotics in schools: A systematic review. *Computers & Education*, *58*(3), 978–988. doi:10.1016/j.compedu.2011.10.006

Birk, A. (2011). What is robotics? an interdisciplinary field is getting even more diverse. *IEEE Robotics & Automation Magazine*, *18*(4), 94–95. doi:10.1109/MRA.2011.943235

Boud, D., & Prosser, M. (2002). Appraising new technologies for learning: A framework for development. *Educational Media International*, *39*(3-4), 237–245. doi:10.1080/09523980210166026

Burdick, A. (2009). *Design without designers*. Paper presented at the conference on the future of art and design education in the 21st century, Brighton, UK.

Bybee, R. W. (2013). *The case for STEM education: Challenges and opportunities*. NSTA press.

Clark, K. R. (2018). Learning theories: Behaviorism. *Am Soc Radiol Tech.*

Dewey, J. (1997). *How we think*. Courier Corporation.

Doppelt, Y. (2009). Assessing creative thinking in design-based learning. *International Journal of Technology and Design Education*, *19*(1), 55–65. doi:10.100710798-006-9008-y

Erlam, G. D., Smythe, L., & Clair, W.-S. (2017). Simulation is not a pedagogy. *Open Journal of Nursing*, *7*, 779–787. doi:10.4236/ojn.2017.77059

Fan, J., Yu, W., & Hunter, L. (2004). *Clothing appearance and fit: Science and technology*. Woodhead Publishing.

Freire, P. (2018). *Pedagogy of the oppressed*. Bloomsbury Publishing USA.

Fromm, E. (1973). *Review of Skinner's theory*. Revista de Psicoanálisis Psiquiatrîa y Psicologîa.

Galvez-Aranda, D., & Galvez Legua, M. (2021). What to Know Before You Start? In D. Galvez-Aranda & M. Galvez Legua (Eds.), *Robotics Models Using LEGO WeDo 2.0: Design, Build, Program, Test, Document and Share* (pp. 1–25). Apress. doi:10.1007/978-1-4842-6846-9_1

Garratt, G. R. M. (1994). *The early history of radio: from Faraday to Marconi*. Iet.

Gauch, H. G. Jr, & Gauch, H. G. (2003). *Scientific method in practice*. Cambridge University Press.

Glasser, O. (1993). *Wilhelm Conrad Röntgen and the early history of the Roentgen rays*. Norman Publishing.

Harel, I. E., & Papert, S. E. (1991). *Constructionism*. Ablex Publishing.

Harris, E. E. (2014). *Hypothesis and perception: The roots of scientific method*. Routledge. doi:10.4324/9781315830018

Heidegger, M. (1977). Science and reflection. *The question concerning technology and other essays, 173*.

Henshilwood, C. S., & Dubreuil, B. (2009). Reading the artefacts: gleaning language skills from the Middle Stone Age in southern Africa. *The Cradle of Language, 2*, 61-92.

Holland, J. M. (2004). *Designing autonomous mobile robots: Inside the mind of an intelligent machine*. Elsevier.

Knud, I. (2004). *The three dimensions of learning*. Krieger Pub. Co.

Letherby, G. (2013). Theorised subjectivity. *Objectivity and subjectivity in social research*, 59-78.

Lloyd, G. E. R. (2012). *Early greek science: Thales to Aristotle*. Random House.

Martin, F. (1988). *Children, cybernetics, and programmable turtles*. Massachusetts Institute of Technology.

Martin, F. G. (1994). *Circuits to control—learning engineering by designing LEGO robots*. Massachusetts Institute of Technology.

Mehalik, M. M., Doppelt, Y., & Schuun, C. D. (2008). Middle-school science through design-based learning versus scripted inquiry: Better overall science concept learning and equity gap reduction. *Journal of Engineering Education, 97*(1), 71–85. doi:10.1002/j.2168-9830.2008.tb00955.x

Mikropoulos, T. A., & Bellou, I. (2013). Educational robotics as mindtools. *Themes in Science and Technology Education, 6*(1), 5–14.

Morrison, G. R., Ross, S. J., Morrison, J. R., & Kalman, H. K. (2019). *Designing effective instruction*. John Wiley & Sons.

Papert, S. (1987). Information technology and education: Computer criticism vs. technocentric thinking. *Educational Researcher, 16*(1), 22–30. doi:10.3102/0013189X016001022

Papert, S. (1993). *The children's machine: Rethinking school in the age of the computer*. ERIC.

Papert, S. A. (1973). *Uses of technology to enhance education*. Academic Press.

Papert, S. A. (2020). *Mindstorms: Children, computers, and powerful ideas*. Basic books.

Pickover, C. (2008). *Archimedes to Hawking: Laws of science and the great minds behind them*. Oxford University Press.

Popper, S. K. (1999). *Falsificationism*. Academic Press.

Resnick, M., & Silverman, B. (2005). Some reflections on designing construction kits for kids. *Proceedings of the 2005 conference on Interaction design and children*. 10.1145/1109540.1109556

Scheffler, I. (1982). *Science and subjectivity*. Hackett Publishing.

Sengupta, D. L., & Sarkar, T. K. (2003). Maxwell, Hertz, the Maxwellians, and the early history of electromagnetic waves. *IEEE Antennas & Propagation Magazine*, *45*(2), 13–19. doi:10.1109/MAP.2003.1203114

Siemens, G. (2017). Connectivism. *Foundations of Learning and Instructional Design Technology.*

Skinner, B. (1986). Programmed instruction revisited. *Phi Delta Kappan*, *68*(2), 103–110.

Skinner, B. F. (1960). Teaching Machines. *The Review of Economics and Statistics*, *42*(3), 189–191. doi:10.2307/1926170

Solomon, C., Harvey, B., Kahn, K., Lieberman, H., Miller, M. L., Minsky, M., . . . Silverman, B. (2020). History of Logo. *Proc. ACM Program. Lang.*, *4*(HOPL), Article 79. 10.1145/3386329

Tauber, A. I. (2016). *Science and the Quest for Reality*. Springer.

Thacker, C., McCreight, E., Lampson, B., Sproull, R., & Boggs, D. (1981). *Alto: A personal computer*. Academic Press.

Välimaa, J., & Hoffman, D. (2008). Knowledge society discourse and higher education. *Higher Education*, *56*(3), 265–285. doi:10.100710734-008-9123-7

Voit, E. O. (2019). Perspective: Dimensions of the scientific method. *PLoS Computational Biology*, *15*(9), e1007279. doi:10.1371/journal.pcbi.1007279 PMID:31513575

Wadsworth, B. J. (1996). *Piaget's theory of cognitive and affective development: Foundations of constructivism*. Longman Publishing.

Watson, J. B. (2017). *Behaviorism*. Routledge. doi:10.4324/9781351314329

Xie, Y., Fang, M., & Shauman, K. (2015). STEM education. *Annual Review of Sociology*, *41*(1), 331–357. doi:10.1146/annurev-soc-071312-145659 PMID:26778893

ADDITIONAL READING

Papert, S. (1987). Information technology and education: Computer criticism vs. technocentric thinking. *Educational Researcher*, *16*(1), 22–30. doi:10.3102/0013189X016001022

Scheffler, I. (1982). *Science and subjectivity*. Hackett Publishing.

Sengupta, D. L., & Sarkar, T. K. (2003). Maxwell, Hertz, the Maxwellians, and the early history of electromagnetic waves. *IEEE Antennas & Propagation Magazine*, *45*(2), 13–19. doi:10.1109/MAP.2003.1203114

Siemens, G. (2017). Connectivism. *Foundations of Learning and Instructional Design Technology.*

Välimaa, J., & Hoffman, D. (2008). Knowledge society discourse and higher education. *Higher Education*, *56*(3), 265–285. doi:10.100710734-008-9123-7

Voit, E. O. (2019). Perspective: Dimensions of the scientific method. *PLoS Computational Biology*, *15*(9), e1007279. doi:10.1371/journal.pcbi.1007279 PMID:31513575

Watson, J. B. (2017). *Behaviorism*. Routledge. doi:10.4324/9781351314329

Xie, Y., Fang, M., & Shauman, K. (2015). STEM education. *Annual Review of Sociology*, *41*(1), 331–357. doi:10.1146/annurev-soc-071312-145659 PMID:26778893

KEY TERMS AND DEFINITIONS

Algorithm: Algorithm is a finite and ordered sequence of tasks to follow in order to solve a problem.

Electrical Motor: An actuator that transform electrical energy into mechanical energy.

Electronics: Electronics allows the addition of senses to the prototype, making it able to detect and recognized its environment.

LED: It is an actuator that allows to generate visible light when it is activated. It is an electronic device of the digital type; this means that it has two states: on (ON) and off (OFF).

Microcontroller: It is an intelligent integrated circuit that is programmable. It has implemented a set of instructions that allows it to process information according to a program previously stored in its ROM or RAM memory.

Pedagogy: it is a social science that studies the methods and techniques applied to teaching.

Program: A program is a set of instructions that a computer understands and execute.

Robotics: Is the design and construction of machines with a certain degree of intelligence, capable of replacing human beings in certain activities.

Teaching: It is a communicative process, whose purpose is to transmit information to generate knowledge and/or transmit procedures to generate skills.

Technology: Is the set of skills, methods or processes used for the design and construction of machines or services to satisfy human needs.

Chapter 3
A Method for Multi-Perspective and Multi-Scale Approach Convergence in Educational Robotics

Alden Meirzhanovich Dochshanov
European Training and Research Association for a Cooperation Key to Business, Italy

Michela Tramonti
European Training and Research Association for a Cooperation Key to Business, Italy

ABSTRACT

The exponential growth of publications on educational robotics (ER) in the last 10 years, undoubtedly, in many ways, is due to the introduction and the ubiquitous application of such platforms as Arduino, micro:bit, Raspberry Pi, and others. These instruments offer a variety of ways for STEM curricula introduction. Being centered on technological and engineering issues, the ER methodology offers an excellent opportunity for students of different ages. By lowering the entry-level effort, the platforms significantly simplify the first steps in the field. Nevertheless, as every simplification presumes, a significant part of "underlying machinery" remains hidden. Indeed, the character of in-school ER projects barely holds enough space for a detailed treatment of the concepts. Nevertheless, the chapter illustrates the importance of exposing students to the currently available instruments and providing in-depth conceptual insight. To support the thesis, several examples, unifying multi-perspective and multi-scale approaches, are provided.

INTRODUCTION

The modern advancement of educational robotics (ER) has received a decisive impact by introducing such platforms as Arduino, Raspberry PI, micro:bit and others with user-friendly Integrated Development Environments (IDEs) and simplified treatment of coding and electronics. Indeed, the primary enthusiasm in co-involvement such systems provide is an outstanding achievement by itself because it

DOI: 10.4018/978-1-7998-7443-0.ch003

enables all interested in unveiling a deeper level of technology passing from users to co-creators. At the same time, the teaching community must be aware that the future evolution of robotics, rooted deeply in science, is linked both to the latest ICT achievements and the progress in material science based on our current understanding of the nanoscale in particular.

As a natural bridge between the atomic and macroscopic worlds, nanosized objects are under the close attention of a large and diverse army of scientists. The results of this collaborative effort are manifested in the new types of materials for sensors, actuators and, in a longer perspective, create the foundations for a new element base of the future electronics, so-called functional electronics, and robotics (Kramer, 2015).

Currently, the robotics research community faces a variety of problems, e.g. new materials (Tan, Susanto, Anwar Ali, & Tee, 2021), energetic efficiency (Boscariol & Richiedei, 2019; Hobart, Mazumdar, Spencer, Quigley, Smith, Smith, Bertrand, Pratt, Kuehl, & Buerger, 2020), cognitive and artificial intelligence issues (Taylor, Berrueta, & Murphey, 2021). To be solved, the tasks require deep professional knowledge and such skills as lateral and creative thinking and problem-solving (Durak, Yilmaz, & Yilmaz, 2019), i.e. the ones considered ER and STEM (Science, Technology, Engineering, Mathematics) methodology pillars. These refer to all skills, named transversal, which are those determining how a person relates within a working or social context. According to the ISFOL (italian abbreviation of the Institute for Vocational Training of Workers) definition, transversal is "a set of skills of wide thickness that are involved in numerous types of tasks, from the most elementary to the most complex, and that are expressed in situations between them different and therefore widely generalisable" (D'Agostino, 2013). Unlike the technical skills, which relate to specific knowledge acquired through study or professional experience, crosscutting skills are an added bonus that can make a difference during job recruitment. Among those considered most important there were time management, motivation and flexibility. However, the world of work is a scenario in continuous evolution. Therefore, today's most requested skills are different from those sought in the past and especially from those that will be asked to workers in the coming years. According to the report "The Future of Jobs" of the World Economic Forum (Forum, 2020; Rainie & Anderson, 2017), in the next five years, the possession of certain transversal or soft skills will become crucial for almost every profession. In particular, it is expected that the most in demand will be:

- Ability to analyse and be innovative;
- Learning skills and strategies;
- Problem solving;
- Critical thinking;
- Creativity, originality and initiative;
- Leadership and the ability to influence others.

In this regard, present chapter illustrates that due to its introductory character, the last should be able to provide students with a coherent exposure of the related concepts and solid contextualised problem solving skills. In particular, two approaches are touched upon here - multi-perspective and multi-scale. While the former concerns the variety of perspectives the ER projects are generally characterised by, the latter instead aims at assisting the acquisition of a deeper view of the robotic artefacts at different scales and levels of consideration. STEM methodology is considered as the unifying framework for both approaches.

Thus, with the view of the abovementioned approaches, the chapter provides several examples of how they can be combined in practice. In addition, a particular emphasis on the nanoscale considerations and their relation to future robotics is given.

BACKGROUND

The inherent multidisciplinary character of robotics is manifested in the heterogeneity of technologies the robotic engineer should master and integrate consistently and effectively (Schlegel, Steck, Brugali, & Knoll, 2010). The actual complexity of robotics could be roughly estimated from the range of the topics offered by the Handbook of Robotics (Siciliano & Khatib, 2016), humanoids evolution (Cingolani & Metta, 2015), the rise of artificial intelligence technology (Chassignol, Khoroshavin, Klimova, & Bilyatdinova, 2018).

On the other hand, the continuous actual and future challenges the society is facing and foreseeing imposes a natural imperative for adequate preparation of future generations of scientists, engineers and active citizens capable of dealing with the problems of increasing complexity (Eguchi, 2014). In this regard, ER represents a suitable framework favouring the acquisition of cross-disciplinary skills, knowledge and attitudes towards dealing with the issues in a systemic and contextualised way (Eguchi, 2010). The last is naturally caused by the fact that the interdisciplinarity of robotics is dictated by the need to involve multiple sectoral knowledge within the process of making a robot. In fact, engineering, computer programming, psychology, automation, mechanics and biology are just some of the aspects that are interconnected during the design.

In this context, the ER becomes an effective teaching method, which allows students to learn from creating a robot, starting from scratch, through its programming and development, passing from all stages of the process. In addition, the diversity of instruments at the ER teaching community disposal permits to reach a wide range of students, guiding them throughout diverse complexity projects and introducing the robotic design starting from primary schools (Khanlari & Mansourkiaie, 2015), and even earlier (Liu, Lin, Liou, Feng, & Hou, 2013). At the same time, the degree of detail for the related disciplines may enable different strategies and perspectives for their exposition. Thus, for example, ER enables not only an engaging and contextualised introduction to coding and electronics but many concepts of physics (Williams, Ma, Prejean, Ford, & Lai, 2007), biology (Crowder & Zauner, 2013; Garofalo, Sandler, & Seth, 2020), chemistry (Verner & Revzin, 2017).

The use of educational robots allows students to develop essential skills not only to enter the world of work in the future, but also to solve problems and seize the opportunities that life puts before them every day. In fact, thanks to this type of teaching, the children manage to develop important skills, such as problem solving, computational thinking, creativity, teamwork, critical thinking, ability to analyse the relationship between cause and effect, observation of micro events to place them in a wider context.

Moreover, the educational robotics can be defined as the method in which STEM subjects, the scientific subjects, such as programming, are learnt in a practical and entertaining way: one learns to use logic, to solve problems with increasing difficulty, increasing the ability to form what in the sector is called "computational thinking", which is also part of coding.

In particular, STEM, in its turn, defined as the purposeful integration of the individual disciplines (Breiner, Harkness, Johnson, & Koehler, 2012), represents a synthetic whole that is greater than the sum of its parts (Takeuchi, Sengupta, Shanahan, Adams, & Hachem, 2020). The last presumes that following the approach in an integrated manner is different and richer than simply learning multiple disciplines (Shanahan, Burke, & Francis, 2016). Therefore, instead of just approaching a problem from various and separate perspectives, STEM methodology, according to its very definition, may provide a multi-perspective vision. In addition, teaching STEM in a more connected manner, especially contextualising it in real-world issues, enhances the disciplinary practices within the individual STEM disciplines, making them more relevant to students and teachers (Honey, Pearson, & Schweingruber, 2014).

Students, with the help of their teachers, are facilitated to move from "using" to "doing", e.g. learning to program in general and realising projects of educational robotics in particular, by developing new and different skills for understanding reality, such as:

- Design sequences of actions;
- Devise loops to run the same sequence several times;
- Use parallelism to make things happen at the same time;
- Identify the events and causes that determine the effects;
- Managing the conditions under which decisions are taken;
- Understand the importance of data.

In addition to the specific knowledge and skills, the most important attitude, stimulated by ER, is curiosity, defined as the ability to understand the changes in progress, in order to guide the students indicating the most suitable opportunities, training them on updated topics, sharing information with passion and interconnecting their own experiences with those of teachers in other subjects.

In this regard, authors of the chapter claim that the multidisciplinary aspect of STEM and robotics integrated practices may be related both to the selection of micro- and macro-level thinking tools, as described in (Chalmers & Nason, 2017), and in a narrower and exclusively to a robotic artefact related sense. In particular, depending on the age of the audience, a robot may be seen at different scales. The macroscale, in this case, is the robot as an integral combination of hardware, software – "general plan" of the system - and its interaction with the surrounding. With scale reduction, the focus gradually shifts to constituent parts down to basic components and phenomena. For example, speaking about motors can be naturally developed and reduced to the demonstration and discussion of the copper wire with the current and its interaction with magnet (providing references to the Ohm's law and Lorentz force correspondingly). Whereas discussing the algorithm, one can demonstrate the elementary conditionals, loops, and functions again contextualised. In this way, following the continuous evolution of technology and the complexity of instruments, student obtains a clear and connected view of interdependence between the basic building blocks and their role within more complex systems. In providing this perspective, the lower scale limit can be determined by the age of the students and the familiar scientific notions from STEM curricula. In other words, the combination of methods to employ is intended to provide a coherent treatment of the arguments while covering them in breadth and depth.

As one can see, in this sense, STEM methodology obtains direct pathways for a contextualised subject delivery. And from this multi-scale perspective, students may perceive the scientific background not as an obstacle or something abstract to treat in the copybook only but as solid conceptual support for going further.

In this respect, ER provides a fertile ground and contextual framework for STEM integration. Furthermore, the exponential growth of research literature related to ER during the last decade presents numerous examples of how beneficial the auxiliary role of robotics is when it comes to scientific and mathematical concepts and skills acquisition (Altin & Pedaste, 2013; Dochshanov & Tramonti, 2020; Lammer, Lepuschitz, Kynigos, Giuliano, & Girvan, 2017). In general, studies support the notion that when students can observe a problem realised in robotics behaviour, they gain the opportunity for experimenting where ideas, scientific theories, and computer coding merge with the real world (Anwar, Bascou, Menekse, & Kardgar, 2019). In other words, while stemming deeply from science, robotics provides the perspectives in their dynamic, real and tangible interaction.

MOTIVATION: WHY THE INTEGRATION OF ROBOTICS INTO STEM CURRICULA IS IMPORTANT?

Maintaining a mutually beneficial relationship between Education and Robotics is critical for many different reasons. First, the overall participation of women in STEM has been marginal. The main reason is driven by the dominant masculine composition of technology-related career culture. Moreover, lack of female role models in the STEM fields, negative stereotypes regarding female abilities, low self-efficacy and lack of positive early experiences discourage women from pursuing education in STEM as well (Beede, Julian, Langdon, McKittrick, Khan, & Doms, 2011; Cheryan, Ziegler, Montoya, & Jiang, 2017). At the same time, the overall dropout rates in the STEM courses is also rising at an alarming rate, such that between 2003 and 2009, 69% of the students in associate degree programs and 48% of the students pursuing bachelor's degree left their respective degree programs (Balch, Summet, Blank, Kumar, Guzdial, O'hara, Walker, Sweat, Gupta, & Tansley, 2008; Chen, 2013). Among others, literature reports the following factors contributing to the rising dropout rates: lack of faculty involvement, unavailability of support from peers, and widening gap in expectations regarding quality of work between students and professors (Ahmed & La, 2019; Vieyra, Gilmore, & Timmerman, 2011).

From the other side, based on the US Bureau of Statistics (Ebrahimipour & Suzuki, 2005), for the coming decade (2016-2026), careers related to STEM are expected to experience the highest growth, resulting in increased demand for a skilled workforce shortly. The related dilemma in regard connects the two poles: the overall interest and participation towards pursuing STEM-related careers are declining while rising innovation and technological revolution are expected to increase the overall demand for skilled labour with STEM-related qualifications and expertise. Robotics education is seen as one way to address the difficulties mentioned above, bridging the STEM-related employment gap. (Balch et al., 2008). The efforts of educational institutions at various levels of education (early childhood education, middle school, high school, colleges, and undergraduate education) can increase the attractiveness and stimulate interest in STEM-related subjects among students (mostly female students) at a young age (Ahmed & La, 2019).

Another critical aspect of Robotics-Education symbiosis concerns utilising technological advancements towards developing innovative practices to meet evolving educational needs of students. Robotics and the associated innovative educational tools have demonstrated to possess the potential of incorporating multidisciplinary concepts related to Mathematics, Physics, Electronics, Computer Science, Programming, Engineering, Problem Solving and Troubleshooting (Vieyra et al., 2011). Simultaneously, researchers have found that using AI-based applications, social robots, and robotic kits in educational settings has a positive impact on problem-solving abilities, skill development, social and cognitive development, and motivation to pursue STEM-related courses and careers (Ebrahimipour & Suzuki, 2005; Nugent, Barker, & Grandgenett, 2014). Meanwhile, a variety of obstacles obstruct the capacity to realise the promise of the Education-Robotics Symbiosis fully. For example, high expenses of robotic platforms, a lack of teacher training and confidence, a perception of incapacity to teach technology courses, and substantial gaps between the age of students and the technical knowledge needs of learning activities (Nugent et al., 2014).

Being complex, modern robots are comprised of an assortment of integrated elements treated within multiple disciplines. Consequently, the development of robotic systems has inevitably shifted from isolated single components designs to fully cross-functional in team development. In this regard, multidisciplinary vision and the capacity to interact with peers inside a team are essential (Weinberg, White, Karacal, Engel, & Hu, 2005). Moreover, while the curriculum in any specific area of study tends to focus on that area narrowly, real-world systems integrate electrical, mechanical and computing components (Weinberg, Engel, Gu, Karacal, Smith, White, & Yu, 2001), presuming a continuous flow and interchange of ideas and professional collaboration. Therefore, to provide formal guidance, given the need for cross-disciplinary knowledge, instructors need to have sufficient background knowledge to formulate efficacious learning exercises to form a basic understanding of mechanical, electronic, algorithmic and multitasking issues. In this regard, the experience of Multidisciplinary Project Action Group (MPAG) composed of faculty members from Computer Science, Electrical and Computer Engineering, Industrial Engineering and Mechanical engineering represents a successful example of the methodological and pedagogical solution (Weinberg et al., 2001). A main pedagogical goal pursued by the authors was to teach group dynamics and the skills necessary for interpersonal communication in different disciplines and multidisciplinary teams. A similar approach has been outlined in (Moulton & Johnson, 2010), where authors claim that courses might attempt to ensure graduates work effectively in cross-discipline teams rather than produce robotics graduates skilled in every sub-discipline.

As recent research and development in the humanoid robots (Cingolani & Metta, 2015), industry 4.0 insights demonstrate (Lu, 2017), the further progress in robotics technology is intimately related to the continuous advancements in ICT, microelectronics and materials science. In this regard, ER methodology fits the multidisciplinary imperative while stimulating the open and flexible mindset indispensable in this context. At the same time, one must understand that current tools actively used in the ER practice represent an upper, or, adopting the ICT term, “high-level” of deeply rooted technologies. Whereas the low-level hardware basis and its relation to the basic fundamental physical concepts remains frequently unveiled to a student. The fact is mainly due to the evident complexity of underlying hardware for non-professional from one side and the other – due to the availability of simplified alternatives, enabling the “less painful” entry to the coding and electronics. This situation can be regarded as natural at the professional level, given the complexity of tasks a modern robot performs (Cingolani & Metta, 2015). However, as concerns the introductory level, the approach may be considered unreasonably one-sided, hiding much of underlying fundamentals and is frequently reduced simply to assembling ready-to-use pieces and subsequent coding, without any particular insights even into the basic electronics. In most cases, the central core methodology is constrained to assembling and coding a pre-defined structure, which is acceptable in the case of children but is by no means sufficient at the secondary school and especially engineering high-school level.

Undoubtedly, the coherent and detailed coverage of all the relevant topics standing at the base of, e.g. Arduino, is absolutely out of the scope of any ER class and by no means was the intention of its creators (Banzi & Shiloh, 2014). Nevertheless, the authors believe it is the introductory level of robotics classes that possesses enough methodological room to help the students figure out inter-and multidisciplinary interaction of the robotics domain areas and acquire a deeper insight into its constituents. Thus, to build a wider conceptual reference mind-map at the earlier stage of a learning path, the example presented further intends to help a student figure out the evolutionary development and inter-relationship of the underlying ideas.

As can be seen, the problems of robotics education are faced from different perspectives. Indeed, the multidisciplinary character of the discipline is basically at the root of the methodological issues. Although authors (Filippov, Ten, Fradkov, & Shirokolobov, 2017) underline that the use of kits in most cases meets educational needs, the best results can be achieved when using task-oriented materials. In addition, the variability of the ER instruments, kits and methodologies expose the numerous facets of electronics, permitting one to estimate and obtain an insight into the work of discrete components, different integrated circuits (ICs), direct current (DC) and step motors, and, as a central "brain" of the system – microcontroller. What is vital in this regard – is to provide a sense of a gradual evolution of the instruments, particularly electronics, while accentuating that what is available today serves as the basement for future technological development.

"BLACK" AND "WHITE" BOX PARADIGMS WITHIN MULTI-PERSPECTIVE AND MULTI-SCALE APPROACHES

The concept of multidisciplinarity itself presumes a uniform coverage of disciplines arranged around a context, i.e. at a level adequate to understand the issues lying behind the subject (Weinberg et al., 2001; Weinberg et al., 2005). At the same time, without going too deeply into the electronic part, but possessing an adequate coding environment and a variety of plug-and-play sensors, one can develop much more complex robotic projects, never thought before, see e.g. (Phal, Phal, & Jacob, 2014).

Based on the constructivism theory (Piaget, 1964, 1973a, 1973b), robotics education presumes the active involvement in a process when learning occurs due to a mental construction of a learner (Khanlari & Mansourkiaie, 2015). In such an approach, "a learner is actively constructing new understandings, rather than passively receiving and absorbing 'facts'" (Jacobson & Wilensky, 2006). Right from Piaget's teaching, Seymour Papert (Harel & Papert, 1991; Papert & Harel, 1991) developed a new theory: constructionism. If constructivism argues that learning is based on building mental models that serve to understand the world, constructionism argues that learning is more efficient if learners use tangible objects to build these models, e.g. the production of a mechanical turtle (Papert & Harel, 1991).

The elements that make the educational robotics laboratory an incredibly powerful educational tool are three: learning by discovery, problem solving and the recognition of the positive role of error.

Learning by discovery: teacher gives some initial indications on the use of the robot and challenges the students, explaining what the goal is and what the rules are, but all the rest is up to the students. They need to work independently or in groups to find out how to solve the problem (Hammer, 1997).

Problem solving: students have a problem to solve and have to find a possible solution. And each group working on the same problem will come up with a different solution. This process will provide students with an opportunity for comparison and analysis, bringing their problem-solving skills to a higher level (Baker, 2017).

The keyword of educational robotics is undoubtedly an *attempt*: there is no proper solution to the first blow; it is not a problem to make mistakes: students learn by trial and error (Steen-Utheim & Wittek, 2017). If the *attempt* fails, students can start again until they are satisfied with the solution found. The errors made by the students allow them to understand how to move forward and how to improve their work.

As a result, ER emerged as a unique learning tool, can offer hands-on, fun activities in an attractive environment feeding interest and curiosity (Eguchi, 2010). On the other hand, constructivism and constructionism methods require the transition to the design of "transparent" ("white-box") robots. In this case, learners can deal with robots from scratch and gain a deeper insight into the primary artefacts rather than just consume pre-engineered pieces (Alimisis, 2013). Also, this approach can generate a lot of creative thinking and involvement in learners (Resnick, Berg, & Eisenberg, 2000). Furthermore, the emerging issue is how deeply one is supposed to go into a specific discipline topics' coverage while preserving at the same time the multidisciplinary point of view onto the robotic major or, in other words, to combine micro-and macro-level thinking tools (Chalmers & Nason, 2017). With this in mind, nowadays, the robotics educational community, nevertheless, is facing a dilemma outlined in (Alimisis, 2013) as "Shifting from "black box" to "white box" paradigm: learners as "makers" rather than just consumers".

Also, it is worth noting that the diversity of available platforms and different additional auxiliary pre-engineered sensors and shields substantially eliminates many difficulties the professional robot developers are supposed to face. Therefore, all these pieces are becoming "black boxes" ready to be connected and programmed. Alimisis (2013) argues that the main reason behind these solutions is the perception that the robot's construction and programming is a highly demanding task, especially for children. However, the difficulties of robotics tasks perceived have been attributed to poor design rather than cognitive problems of learners (Blikstein & Krannich, 2013).

Of course, depending on the purpose sought, the "white box"-only approach might be considered, in a certain sense, as partially retrogressive too, especially if one takes into account the continuous introduction of technological innovations. However, when providing the pupils with a "black box" perspective alone, there is a risk of narrowing down not their worldview only but a dire need for underlying/hidden or basic, fundamental issues. Needless to say, that by using only the Robotic kits available, despite their enhanced functionality, pupils become the recipients of pre-determined paths in building a robot. Without any particular insights into their functioning basics, work's whole experience reduces to digital/analogue read or control. Moreover, by the time today's children will have grown, the design, development, installation, and maintenance of robotic systems might be completely automated. Therefore, for a robotics course to remain both modern and essentially appealing to the issues demanded in the future, adequate and contextualised coverage of fundamental topics should not be overlooked.

With the general aim to connect the challenges of the forefront research and ER methodology, authors demonstrate that a balanced exposition of these within the context of ER classes may be beneficial in content delivery and may serve as an updated reference point for primary and secondary school teachers. In addition, it is claimed that ER offers sufficient space to encourage curiosity from multi-perspective and multi-scale points of view (Dochshanov & Lapina, 2019). Furthermore, the authors believe that a proper combination of both may provide a more diversified framework to arouse the interest of students in a broader range of aspects and eventual choice of a future professional career.

AN EXAMPLE OF A COMBINATION OF MULTI-SCALE AND MULTI-PERSPECTIVE APPROACHES IN EDUCATIONAL ROBOTICS

Thus, as one of the possible strategies combining "black" and "white" box approach, or multi-scale, with multi-perspective approaches in robotics classes, here below the combination of different variants of a widely known line-follower robot model and concept, with a particular accent to "in-hardware" realisation, is given and discussed.

The choice is due to a widely-known principle and diversity of both "in-hardware"/micro-controller free (Baharuddin, Abidin, Mohideen, Siah, & Chuan, 2005) and Arduino-based platform (Chia, Siew, & Chee, 2014) (Hymavathi & Kumar, 2011) realisations. The detailed description and strategies to build a line follower robot are beyond the chapter's scope and available elsewhere (Hymavathi & Kumar, 2011; Pakdaman, Sanaatiyan, & Rezaei, 2010).

Arduino Line Follower

The core elements of slightly different Arduino-based line-follower robot realisations remain the same (see Fig.1), namely IR sensors or an array of IR sensors, responsible for the crossing-the-line moment detection with the subsequent reaction of the guiding motors (Pakdaman et al., 2010). Consequently, the primary difficulty is related to an appropriate sensors' calibration and the corresponding feedback control of the DC-motors (Hymavathi & Kumar, 2011).

In addition to the attractiveness and relative simplicity of realisation (Chia et al., 2014), the approach, depending on the age group of students, may represent an insufficient basis for fully grasping the problem and understanding what happens at a lower level. Multi-perspective, but constrained to the hardware assembly and code development, the method in many ways can be likened to "the tip of an iceberg". Because in this way, instead of starting from the bare physical phenomena, the main challenge shifts towards the adjusting robot's reaction to a digital replica of the latter. Of course, the project may be acceptable if the purpose presumes nothing but to attract or intrigue pupils with the simplicity and showiness of the final result.

As one may see, the version holds more multi-perspective features, mainly composed of "black box" interconnected components controlled through a high-level programming language. Therefore, to provide an audience with a deeper understanding of the phenomena behind it, it would be advisable to consider another variant of the line-follower robot. In that way, a class is furnished with an enhanced multi-perspective and multi-scale views of the issue. The advantages of this additional approach involvement are outlined in the next section.

Microcontroller-Free Line Follower

The principal scheme of in-hardware commercial line-follower robot realisation is presented in Fig. 2. A well-known structure can be found as a commercially available DIY-kit as well (educ8s.tv, 2015). As opposed to the previous case, the basic principle is translated into the terms of voltages, feedbacks, signal amplification and comparison (Pakdaman et al., 2010). As the essential elements of every electronics engineer's alphabet, all this becomes "tangible" and evident. Without an add-on microcontroller, the version may bring the following topics for an eventual in-depth study:

- A deeper level of the overall function understanding due to a system view at a low level;
- The contextualised introduction of the hardware feedback control;
- Work with discrete sensors (LDR - light-dependent resistors);
- Role of the potentiometer;
- Setting the gain factor for an operational amplifier;

Figure 1. Arduino line follower.

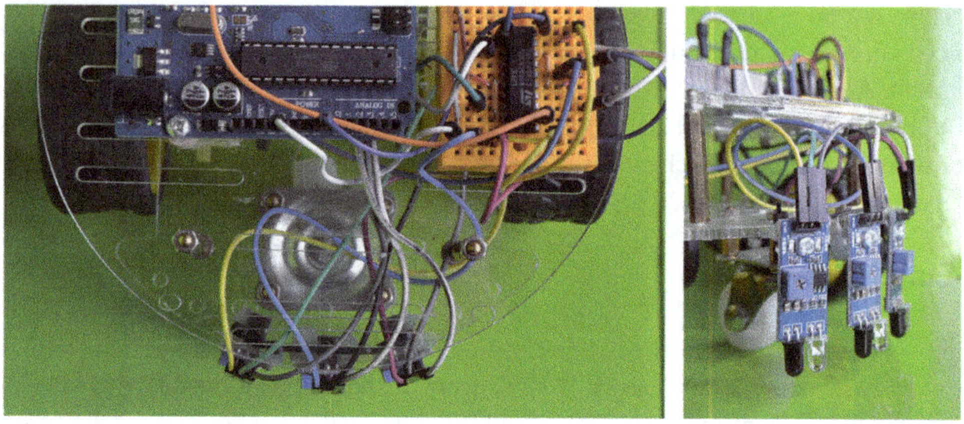

- Comparator working principle as the introduction to the ADC;
- LEDs;
- Amplifying characteristics of transistors;
- Motor coil issues (e.g. Lorentz' force, stall current).

The list presented is by no means exhaustive. However, as can be seen, the variety of additional themes for in-depth elaboration may be of essential support to be used alone and as an auxiliary tool for the Arduino-based realisation. In such a way, students are provided with a more comprehensive perspective that advances their reasoning through the new combination of points of view.

Comparison of the Arduino and Microcontroller-Free Variants

Both approaches described above possess both advantages and disadvantages. In the Arduino case example, one directly approaches the task through the code development without going into particulars of the underlying hardware organisation. Thus, leaving a great deal backstage, pupils obtain an immediate but relatively superficial tackling of the problem despite all the possible difficulties at the software level. On the other hand, whereas leaving the add-on microcontroller issues outside, the second variant provides an essentially enriched and integrated perspective, widening the pupils' perception and understanding of the pure electronics issues.

The methods may be considered as the "black" and "white" box complementary manifestations if taken together. Both variants offer a multi-perspective insight to a greater or lesser extent as examples of ER exercises. However, as far as the multi-scale view is concerned, only the combined application can provide the whole panorama. In particular, as shown in Figure 3, the same integrated circuit responsible for comparing the light levels from sensors is used in both variants (LM393). At the same time, whereas in the microcontroller-free version, there is only one single piece (see element C on Fig. 1), in the Arduino-based variant, it becomes a part of an integrated module with the light-sensor as well (Fig. 3). So, while seen in the microcontroller-free variant, the function of the IC is naturally more evident, being deprived of the Arduino shell and single-board integration. But once another perspective is considered, the same

Figure 2. Circuit diagram of a microcontroller-free line-follower based on LM393 dual differential comparator. A – the complete circuit diagram of hardware-line follower; B1, B2 – light sensor and LED C – dual differential comparator; D – driving transistors for DC motors.

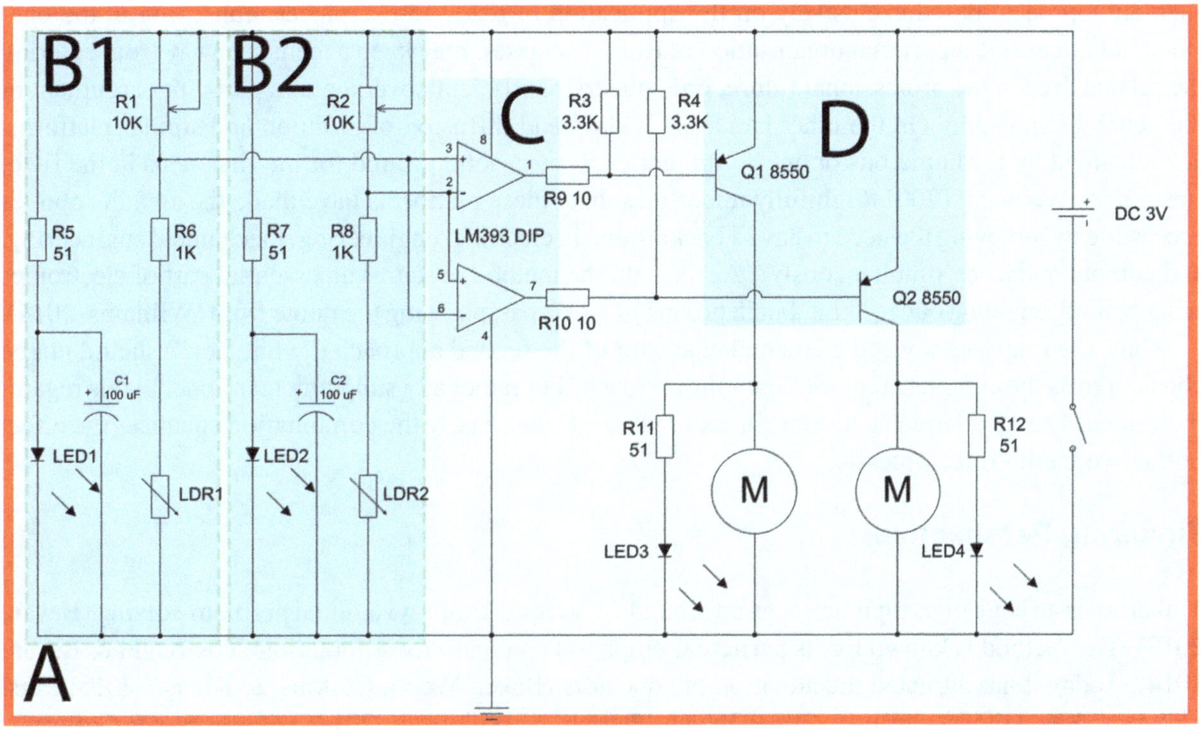

component emerges in a new form. Therefore, combining both variants provides a coherent panorama of technological evolution from one side and functional development from the other.

Indeed, the choice of a strategy and instruments to apply is defined by the goals and interests of students and a teacher. Indeed, the microcontroller-free option does not provide all the flexibility as compared to the Arduino variant. For example, adding the ultrasonic sensor and extending the initial design with the obstacle avoidance function will not take much. This approach is consistent with continuous technological progress. However, from the other side, to the authors' opinion, the approach may weaken the perception of core fundamentals. Thus, to compensate for both shortcomings and avoid one-sidedness, one should opt for a combined path.

Figure 3. The multi-scale aspect of shifting between microcontroller-free and Arduino line follower (A, C, B are referencing the corresponding elements of Figure 2).

From a teenager's perspective, who is generally younger than the Arduino platform, the instrument and auxiliary pieces are naturally perceived as integral, ground zero all-in-ones. As a consequence, a significant part of the underlying fundamental issues remains inevitably hidden or missed. Furthermore, such an approach, based exclusively on the application of ready-to-use pieces, consisting in the one-sided, identical and superficial organisation of Robotic classes, may keep a teenager away from entering the related area in the professional future, only due to insufficient coverage of a particular argument or the general exposition. On the other hand, the widespread diffusion of Arduino and similar platforms accompanied by the numerous online communities forms a solid ground for the first steps in the field. In support, Weinberg (2001) rightfully underlines that "these platforms have made the area of robotics accessible by removing the need to have a background in electrical engineering, mechanical engineering, and computer science simultaneously". As a result, the method obscures an essential part of electronics lying behind, enabling one to "knit, but in boxing gloves", paraphrasing the quote from (Williams, 2014).

Thus, the transparency and elemental character of the second approach ("white box") should not be considered as the complete replacement for the "black" but rather as a supplementary one. In this regard, there should be no dilemma between "black or white" because only the combination enhances the effect of the two methods reciprocally.

Tinkering Perspective

Tinkering is an educative approach centred around creative and improvisational problem-solving (Bevan, 2017). The method is known for its particular emphasis on creativity stimulation (Vossoughi & Bevan, 2014). Today it has attracted the attention of educators (Baker, Wood, Corkins, & Krause, 2015; Bers, Flannery, Kazakoff, & Sullivan, 2014; Bevan, Petrich, & Wilkinson, 2014; DiGiacomo & Gutiérrez, 2014; Gutwill, Hido, & Sindorf, 2015; Tramonti & Dochshanov, 2018) and policymakers (Richmond, Bartell, & Dunn, 2016). In addition, it is argued to be a vehicle to gain knowledge and skills profitable for engineering, design and science (Mader & Dertien, 2016).

On an example of a counter-driven clapper fan, below, tinkering is considered a primer of activities held within the robotics classes. It is shown how a combination of the approaches discussed above may be introduced through "thoughtful" tinkering (Dochshanov & Tramonti, 2018). The results show that the latter is preferable compared to iterative, mechanistic, and therefore less efficient making-based tinkering. The technique's core is to provide a student with a complex vision of different aspects of modern electronics combined in a single project .

The project (see Figure 4) combines the online tutorial (Afrotechmodes, 2010) with the Arduino platform. The constituent components are discrete elements (resistors, potentiometer, capacitors, MOSFET - Metal-Oxide-Semiconductor Field-Effect Transistor), integrated circuits, brushless fan (taken from an old PC power supply). In this regard, the authors believe that integrating the pieces corresponding to different electronics evolution levels (discrete elements, ICs, Arduino) may serve as an effective supplementary tool. Thus, the perception of students forms a more comprehensive, coherent, and consistent image of related concepts.

As one can see, the variety of the subjects covered in a single example represents a rich fusion of the concepts and topics. In particular, they include:

- Operational amplifiers (as a stabilised amplifier and comparator);
- Diodes as components for peak detectors;

Figure 4. Clapper counter-driven fan.

- Analogue-to-digital (ADC) conversion and vice versa (DAC);
- MOSFET application;
- Pulse-width modulation (PWM);
- Microcontrollers programming.

As a result, a student receives a concentrated piece of different topics combined in a contextualised fashion. Furthermore, the core function - to transform the number of claps detected (input) into the fan's speed (output) – serves as a fruitful framework to trigger an intuitive understanding of the different electronic components and the system as a whole. In addition, the combination of discrete elements and integrated circuits permits one to gain an unconscious insight into the evolving continuity of engineering ideas and their interplay. This point is of utmost importance, especially when dealing with nanotechnologies - a recent achievement forming the core basis for tomorrow's hardware.

Finally, the range of the objects and topics covered in a single project enables one to become more flexible in terms of their combination and, as a result, to acquire if not a purely multidisciplinary vision, but a predisposition to it, which, in turn, is of fundamental importance for the future nanotechnologist.

Taken together, approaches described above in the given context compose a basis of what the authors of the chapter would call "thoughtful tinkering", an approach aimed at diversifying the concept of tinkering as a maker activity only.

The authors of the present chapter believe that to deal with such an endeavour, tinkering and thoughtful tinkering, in particular, may represent a limitless field of the enormous potential for didactical exploration as an auxiliary methodological tool.

Figure 5. Schematic representation of the Educational Robotics components.

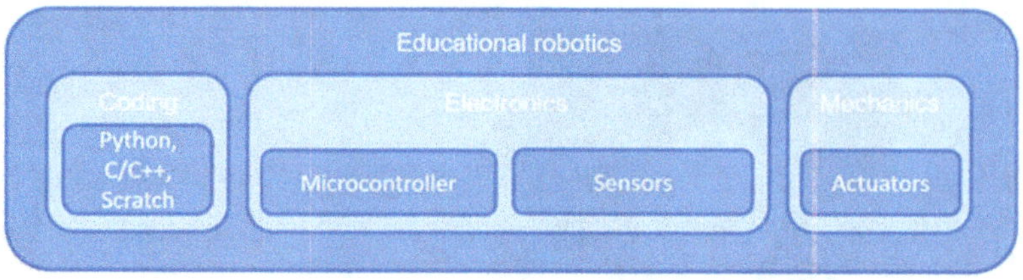

Multi-Scale Approach as Seen From Nano-Level

A gradual evolution of semiconductor device fabrication technology during the last 50 years through 10 μm (1971) down to 5 nm (2020) is manifested in the number of transistors per cm2 of the silicon wafer and microprocessor's clock rate (operations performed by a per second). Ultimately this has affected the complexity of the tasks the robotic systems today are capable of. The evolution has also significantly impacted ER due to the advent of so-called computers-on-chip, i.e. microcontrollers, and simplified means for their programming. At the same time, one should admit that an essential part of robotic coding literacy today in many ways can be considered as ones' capacity to adapt already existing projects to his/her own needs to avoid the reinvention of the wheel. Moreover, the maturity of current commercial technology furnishes significant initial entry support in starter kits, tutorials, and code examples, ready for use and adaptation.

Given the above, one may ask whether the ER context bears enough methodological space to help students figure out how a robotic artefact/system may behave once scaled down to micro and nanoscale? Or, to put it the other way, could the basic principles and ideas standing behind the micro-and nanorobotics be conveyed to the students taking the first steps in robotics? Considering that fundamental complication – the changes of the matter behaviour with the scale – as recent studies demonstrate (Mandrikas, Michailidi, & Stavrou, 2020), does not seem to cause any particular difficulty, the answer to the question is somewhat optimistic. Indeed, considering the steady and incremental development in the field, it is better to approach it at the earliest, especially by those who attend primary school today and are destined to contribute to the subject in the future.

To support the introduction of a multi-scale perspective within nano-context, one may adopt the following reasoning. As can be seen from the diagram (Figure 5) ER domain may be subdivided into the following core elements: coding, mechanics and electronics. While coding may be regarded as the principal component the students are dealing with due to a wide range of available IDEs, mechanics may be seen as the primary provider of feedback, immediate and intuitive, e.g. motors. Finally, electronics, being, for the most part, represented by the microcontroller board and the set of relevant sensors, because of its supporting role, is given a minor priority. Therefore a shift in focus to sensors and actuators may be a starting point for introducing micro-and nanoelectromechanical systems (MEMS and NEMS).

The continual increase of MEMS devices and NEMS applications and technology have emerged since the 2000s. Today, the micro-and nanorobotics field deals with the robotic manipulation of objects with dimensions in the millimetre to the sub-micron range and the design and fabrication of autonomous robotic agents falling in this size range (Siciliano & Khatib, 2016).

Sensors and actuators, the principal constituent elements of MEMS and NEMS, are antagonistic between them. Whereas the former converts the external physical stimuli (e.g. heat, light, pressure, touch, magnetic field, acceleration) into electrical energy, the latter transforms the electrical energy into the external physical response (chiefly movement). The fundamental idea to deliver is that the relation between input and output varies with a scale change. Yet quantitatively and qualitatively. From one side, the dominating physics at these levels is different as opposed to the macroscale. In particular, the surface and intermolecular forces, such as adhesion forces originating from surface tension forces, van der Waals forces, and electrostatic forces when confronted with gravitational forces, become more significant for objects with sizes well below 1000 μm (Fearing, 1995). On the other hand, given every transducing principle (e.g. electrostatic, piezoelectric) is characterised by a set of parameters, e.g. range of motion, force, dynamic range, speed, sensitivity and others, the selection of the most adapted sensor/actuator is in striking a balance between the last and scaling laws' constraints.

Certainly, artificially manufactured nanosized manipulators, and robotics systems in general, remain a rich research field with many open problems (Siciliano & Khatib, 2016). For example, the most traditional MEMS sensor systems do not scale down well to the nanoscale due to noise, sensitivity and fabrication limits (Cullinan, Panas, Dibiasio, & Culpepper, 2012). As an alternative, new types of nanomaterials, such as carbon nanotubes, nanowires, nanobelts, and graphene, may enable the detection and actuation of ultra-small quantities of objects with unprecedented accuracy. Moreover, the so-called "molecular robotic systems", motivated by dramatic progress in chemical sciences, resulted in various molecular sensors, nanoscale machines, and molecular computing devices over recent years. In the last case, the DNA molecule is considered the primary building block. It is already nominated as one of the most promising molecular materials for the control and organisation of matter at the nanoscale to date (Seeman, 2005).

Thus, further progress in this vibrant field is possible thanks to a convergence of nanoscale materials, biotechnology and molecular biology. Currently, nanorobotics mainly deals with fabricating systems for nanoworld exploration (Fukuda, Nakajima, Liu, & ElShimy, 2009). While in a longer perspective, the effort will lead to the basic building blocks for future nanorobots (Dong, Subramanian, & Nelson, 2007; Requicha, 2003).

Besides, when seen at the nanoscale, the field naturally enters into a close relationship with other adjacent disciplines' levels due to the prevalence of surface effects. To be more precise, the interaction becomes much more pronounced and interdependent than in conventional macro-scale robotics case. For example, where, e.g. photosensors, even if governed by the quantum mechanics laws, are nevertheless managed quite distantly through the code line.

As a platform for future advancements in science and technology, nanotechnologies are intimately tied with the current technologies (i.e. micro- nanofabricated tools (Cecchi, Verotti, Capata, Dochshanov, Broggiato, Crescenzi, Balucani, Natali, Razzano, Lucchese, Bagolini, Bellutti, Sciubba, & Belfiore, 2015; Dochshanov, Verotti, & Belfiore, 2017; Verotti, Dochshanov, & Belfiore, 2017a, 2017b). Therefore, for the future nanotechnology engineering specialist, it is essential to be aware of the principles behind the current state of the art and acquire a clear vision of the evolutionary continuity of ideas that lead to it.

CONCLUSION

Given all the current IT dynamism, perceptible through the widespread presence of mobile technologies, the attention of future potential nanotechnologist (today's student) is shifted towards readily available software and hardware solutions. However, the last, being enriched through the vast majority of low-entry-level platforms, enables but superficial interaction with technology despite all the appeal. Which in turn, leads to an inevitable shift of students' attention and effort towards coding. At the same time, the knowledgeable contact with the fundamental principles and ideas gets inevitably attenuated. Within the context of ER, the situation may be regarded as reasonable, given the great variety of perspectives in play. Nevertheless, the value of fundamental physical basics cannot be overestimated when it comes to future robotics perspectives and nanotechnological issues in particular. Otherwise, a potential risk may arise, which may be condensed in the formula: "the simpler now, the harder after". Fortunately, the nanotechnological issues, particularly nanoscale phenomena, have already found their way to classes (Mandrikas et al., 2020). In this regard, to provide the topic with an applied context, ER framework may also be of use.

Aiming at remediating the contact with basics, the approaches described in the present chapter, according to the authors, respond to the need for updating the ER curricula. In this regard, Honey, 2014 underlines that contextualising STEM in real-world issues can make the disciplinary procedures more relevant to students and teachers (Honey et al., 2014). Finally, authors hope to have demonstrated that robots' design, construction, and programming for learning offers enough methodological space for such exposition.

Thus, the chapter introduced practical examples of combining multi-perspective and multi-scale approaches within the context of ER. Based on the examples provided, the authors illustrated the methodological paths for teachers of primary and secondary schools for the approaches combination. Undoubtedly, the choice of the concrete robotic artefacts is left to the teacher, provided that the approach embraces both perspectives.

REFERENCES

Afrotechmodes. (2010). *Comparator tutorial & clapper circuit* [Videotutorial]. YouTube. https://www.youtube.com/watch?v=y0Q0ERSP24A

Ahmed, H., & La, H. M. (2019). *Education-Robotics Symbiosis: An Evaluation of Challenges and Proposed Recommendations*. Academic Press.

Alimisis, D. (2013). Educational robotics: Open questions and new challenges. *Themes in Science and Technology Education*, *6*(1), 63–71.

Altin, H., & Pedaste, M. (2013). Learning approaches to applying robotics in science education. *Journal of Baltic Science Education*, *12*(3), 365.

Anwar, S., Bascou, N. A., Menekse, M., & Kardgar, A. (2019). A systematic review of studies on educational robotics [Review]. *Journal of Pre-College Engineering Education Research, 9*(2), 19-42. Advance online publication. doi:10.7771/2157-9288.1223

Baharuddin, M. Z., Abidin, I. Z., Mohideen, S. S. K., Siah, Y. K., & Chuan, J. T. T. (2005). *Analysis of line sensor configuration for the advanced line follower robot*. University Tenaga Nasional.

Baker, D. R., Wood, L., Corkins, J., & Krause, S. (2015). Tinkering and Technical Self-Efficacy of Engineering Students at the Community College. *Community College Journal of Research and Practice*, *39*(6), 555–567. doi:10.1080/10668926.2014.902780

Baker, R. (2017). Problem-solving. In *Agile UX Storytelling* (pp. 107–116). Springer. doi:10.1007/978-1-4842-2997-2_13

Balch, T., Summet, J., Blank, D., Kumar, D., Guzdial, M., O'hara, K., ... Tansley, S. (2008). Designing personal robots for education: Hardware, software, and curriculum. *IEEE Pervasive Computing*, *7*(2), 5–9. doi:10.1109/MPRV.2008.29

Banzi, M., & Shiloh, M. (2014). *Getting started with Arduino: the open source electronics prototyping platform*. Maker Media, Inc.

Beede, D. N., Julian, T. A., Langdon, D., McKittrick, G., Khan, B., & Doms, M. E. (2011). *Women in STEM: A gender gap to innovation*. Economics and Statistics Administration Issue Brief (04-11).

Bers, M. U., Flannery, L., Kazakoff, E. R., & Sullivan, A. (2014). Computational thinking and tinkering: Exploration of an early childhood robotics curriculum. *Computers & Education*, *72*, 145–157. doi:10.1016/j.compedu.2013.10.020

Bevan, B. (2017). The promise and the promises of Making in science education. *Studies in Science Education*, *53*(1), 75–103. doi:10.1080/03057267.2016.1275380

Bevan, B., Petrich, M., & Wilkinson, K. (2014). Tinkering is serious play. *Educational Leadership*, *72*(4), 28–33.

Blikstein, P., & Krannich, D. (2013). The Makers' Movement and FabLabs in Education: Experiences, Technologies, and Research. *IDC '13 Proceedings of the 12th International Conference on Interaction Design and Children*. 10.1145/2485760.2485884

Boscariol, P., & Richiedei, D. (2019). Trajectory design for energy savings in redundant robotic cells. *Robotics*, *8*(1), 15. Advance online publication. doi:10.3390/robotics8010015

Breiner, J. M., Harkness, S. S., Johnson, C. C., & Koehler, C. M. (2012). What is STEM? A discussion about conceptions of STEM in education and partnerships. *School Science and Mathematics*, *112*(1), 3–11. doi:10.1111/j.1949-8594.2011.00109.x

Cecchi, R., Verotti, M., Capata, R., Dochshanov, A. M., Broggiato, G. B., Crescenzi, R., Balucani, M., Natali, S., Razzano, G., Lucchese, F., Bagolini, A., Bellutti, P., Sciubba, E., & Belfiore, N. P. (2015). Development of micro-grippers for tissue and cell manipulation with direct morphological comparison. *Micromachines*, *6*(11), 1710–1728. doi:10.3390/mi6111451

Chalmers, C., & Nason, R. (2017). Systems thinking approach to robotics curriculum in schools. In *Robotics in STEM Education* (pp. 33–57). Springer. doi:10.1007/978-3-319-57786-9_2

Chassignol, M., Khoroshavin, A., Klimova, A., & Bilyatdinova, A. (2018). Artificial Intelligence trends in education: A narrative overview. *Procedia Computer Science*, *136*, 16–24. doi:10.1016/j.procs.2018.08.233

Chen, X. (2013). *STEM Attrition: College Students' Paths into and out of STEM Fields. Statistical Analysis Report. NCES 2014-001*. National Center for Education Statistics.

Cheryan, S., Ziegler, S. A., Montoya, A. K., & Jiang, L. (2017). Why are some STEM fields more gender balanced than others? *Psychological Bulletin*, *143*(1), 1–35. doi:10.1037/bul0000052 PMID:27732018

Chia, K. S., Siew, K., & Chee, E. N. G. (2014). *Build A Line Follower Robot: A User-Friendly Guide*. Academic Press.

Cingolani, R., & Metta, G. (2015). *Nanotechnology for Humans and Humanoids A vision of the use of nanotechnology in future robotics*. Academic Press.

Crowder, R. M., & Zauner, K. P. (2013). A project-based biologically-inspired robotics module. *IEEE Transactions on Education, 56*(1), 82-87. Advance online publication. doi:10.1109/TE.2012.2215862

Cullinan, M. A., Panas, R. M., Dibiasio, C. M., & Culpepper, M. L. (2012). Scaling electromechanical sensors down to the nanoscale [Review]. *Sensors and Actuators. A, Physical, 187*, 162–173. doi:10.1016/j.sna.2012.08.035

D'Agostino, L. (2013). *Dalle competenze agite al "Campo delle competenze trasversali"*. Academic Press.

DiGiacomo, D. K., & Gutiérrez, K. D. (2014). *Learning and becoming in an after school program: The relationship as a tool for equity within the practices of making and tinkering* [Conference Paper]. Academic Press.

Dochshanov, A., & Lapina, M. (2019). Robotics in STEM education: A multiperspective strategy case study. *CEUR Workshop Proceedings*.

Dochshanov, A., & Tramonti, M. (2018). Re-making Classroom Borders with Tinkering Approach. *Conference Proceedings*.

Dochshanov, A., & Tramonti, M. (2020). Computational Design Thinking and Physical Computing: Preliminary Observations of a Pilot Study. *Robotics, 9*(3), 71. https://www.mdpi.com/2218-6581/9/3/71

Dochshanov, A., Verotti, M., & Belfiore, N. P. (2017). A Comprehensive Survey on Microgrippers Design: Operational Strategy [Review]. *Journal of Mechanical Design, 139*(7), 070801. Advance online publication. doi:10.1115/1.4036352

Dong, L., Subramanian, A., & Nelson, B. J. (2007). Carbon nanotubes for nanorobotics. *Nano Today, 2*(6), 12–21. doi:10.1016/S1748-0132(07)70169-X

Durak, H. Y., Yilmaz, F. G. K., & Yilmaz, R. (2019). Computational thinking, programming self-efficacy, problem solving and experiences in the programming process conducted with robotic activities. *Contemporary Educational Technology, 10*(2), 173–197. doi:10.30935/cet.554493

Ebrahimipour, V., & Suzuki, K. (2005). US Department of Labor Bureau of Labor Statistics; Report 948 US Department of Labor Bureau of Labor Statistics; Report 948, 2000. *Journal of Chemical Engineering of Japan, 38*(6), 436–445. doi:10.1252/jcej.38.436

educ8s.tv. (2015, 26 September). *Line Follower Robot without Arduino or Raspberry Pi - A DIY project* [Video]. YouTube. https://www.youtube.com/watch?v=BcUCDADzndw

Eguchi, A. (2010, 03). What is Educational Robotics? Theories behind it and practical implementation.

Eguchi, A. (2014). Educational robotics for promoting 21st century skills. *Journal of Automation, Mobile Robotics and Intelligent Systems*, 5-11.

Fearing, R. S. (1995). Survey of sticking effects for micro parts handling. *Proceedings 1995 IEEE/RSJ International Conference on Intelligent Robots and Systems. Human Robot Interaction and Cooperative Robots.*

Filippov, S., Ten, N., Fradkov, A., & Shirokolobov, I. (2017). Robotics education in Saint Petersburg secondary school. *International Conference on Robotics and Education RiE 2017.*

Forum, W. E. (2020). *The Future of Jobs Report 2020.* Academic Press.

Fukuda, T., Nakajima, M., Liu, P., & ElShimy, H. (2009). Nanofabrication, nanoinstrumentation and nanoassembly by nanorobotic manipulation. *The International Journal of Robotics Research, 28*(4), 537–547.

Garofalo, L., Sandler, S., & Seth, D. (2020). *Evaluation of a Snake Jaw Robot to Teach Integrated Biology*. Mathematics, and Engineering.

Gutwill, J. P., Hido, N., & Sindorf, L. (2015). Research to Practice: Observing Learning in Tinkering Activities. *Curator, 58*(2), 151–168. https://doi.org/10.1111/cura.12105

Hammer, D. (1997). Discovery learning and discovery teaching. *Cognition and Instruction, 15*(4), 485–529.

Harel, I. E., & Papert, S. E. (1991). *Constructionism.* Ablex Publishing.

Hobart, C. G., Mazumdar, A., Spencer, S. J., Quigley, M., Smith, J. P., Smith, J. P., . . . Buerger, S. P. (2020). Achieving Versatile Energy Efficiency with the WANDERER Biped Robot. *IEEE Transactions on Robotics, 36*(3), 959-966. doi:10.1109/TRO.2020.2969017

Honey, M., Pearson, G., & Schweingruber, H. A. (2014). *STEM integration in K-12 education: Status, prospects, and an agenda for research* (Vol. 500). National Academies Press Washington.

Hymavathi, V., & Kumar, V. (2011). Design and implementation of double line follower robot. *International Journal of Engineering Science and Technology*, 3.

Jacobson, M. J., & Wilensky, U. (2006). Complex Systems in Education: Scientific and Educational Importance and Implications for the Learning Sciences. *Journal of the Learning Sciences, 15*(1), 11–34. https://doi.org/10.1207/s15327809jls1501_4

Khanlari, A., & Mansourkiaie, F. (2015, July). Using robotics for STEM education in primary/elementary schools: Teachers' perceptions. *2015 10th International Conference on Computer Science Education (ICCSE).*

Kramer, R. K. (2015). Soft electronics for soft robotics. *Micro-and Nanotechnology Sensors, Systems, and Applications VII.*

Lammer, L., Lepuschitz, W., Kynigos, C., Giuliano, A., & Girvan, C. (2017). ER4STEM educational robotics for science, technology, engineering and mathematics. In *Robotics in Education* (pp. 95–101). Springer.

Liu, E. Z.-F., Lin, C.-H., Liou, P.-Y., Feng, H.-C., & Hou, H.-T. (2013). An Analysis of Teacher-Student Interaction Patterns in a Robotics Course for Kindergarten Children: A Pilot Study. *Turkish Online Journal of Educational Technology-TOJET, 12*(1), 9–18.

Lu, Y. (2017). Industry 4.0: A survey on technologies, applications and open research issues. *Journal of Industrial Information Integration, 6*, 1-10.

Mader, A., & Dertien, E. (2016). Tinkering as method in academic teaching. *DS 83: Proceedings of the 18th International Conference on Engineering and Product Design Education (E&PDE16), Design Education: Collaboration and Cross-Disciplinarity.*

Mandrikas, A., Michailidi, E., & Stavrou, D. (2020). Teaching nanotechnology in primary education. *Research in Science & Technological Education, 38*(4), 377–395.

Moulton, B., & Johnson, D. (2010). Robotics education: A review of graduate profiles and research pathways. *World Transactions on Engineering and Technology Education, 8*(1), 26–31.

Nugent, G. C., Barker, B. S., & Grandgenett, N. (2014). The impact of educational robotics on student STEM learning, attitudes, and workplace skills. In *Robotics: Concepts, methodologies, tools, and applications* (pp. 1442–1459). IGI Global.

Pakdaman, M., Sanaatiyan, M. M., & Rezaei, M. (2010). *A line follower robot from design to implementation: Technical issues and problems.* Academic Press.

Papert, S., & Harel, I. (1991). Situating constructionism. *Constructionism, 36*(2), 1–11.

Phal, D. D., Phal, K. D., & Jacob, S. (2014). Design, implementation and reliability estimation of speech-controlled mobile robot. *IEEE, 2014 2nd International Conference on Emerging Technology Trends in Electronics, Communication and Networking.*

Piaget, J. (1964). Part I: Cognitive development in children: Piaget development and learning. *Journal of Research in Science Teaching, 2*(3), 176–186.

Piaget, J. (1973a). *The child and reality: Problems of genetic psychology* (A. Rosin, Trans.). Grossman.

Piaget, J. (1973b). *To understand is to invent: The future of education.* Academic Press.

Rainie, L., & Anderson, J. (2017). *The Future of Jobs and Jobs Training*. Pew Research Center.

Requicha, A. A. (2003). Nanorobots, NEMS, and nanoassembly. *Proceedings of the IEEE, 91*(11), 1922–1933.

Resnick, M., Berg, R., & Eisenberg, M. (2000). Beyond Black Boxes: Bringing Transparency and Aesthetics Back to Scientific Investigation. *Journal of the Learning Sciences, 9*(1), 7–30. https://doi.org/10.1207/s15327809jls0901_3

Richmond, G., Bartell, T., & Dunn, A. H. (2016). Beyond "Tinkering": Enacting the Imperative for Change in Teacher Education in a Climate of Standards and Accountability [Editorial]. *Journal of Teacher Education, 67*(2), 102–104. https://doi.org/10.1177/0022487116628837

Schlegel, C., Steck, A., Brugali, D., & Knoll, A. (2010). Design abstraction and processes in robotics: From code-driven to model-driven engineering. *International Conference on Simulation, Modeling, and Programming for Autonomous Robots.*

Seeman, N. C. (2005). Structural DNA nanotechnology. *NanoBiotechnology Protocols*, 143-166.

Shanahan, M.-C., Burke, L. E. C.-A., & Francis, K. (2016). Using a boundary object perspective to reconsider the meaning of STEM in a Canadian context. *Canadian Journal of Science, Mathematics and Technology Education, 16*(2), 129–139.

Siciliano, B., & Khatib, O. (2016). Springer handbook of robotics. Springer.

Steen-Utheim, A., & Wittek, A. L. (2017). Dialogic feedback and potentialities for student learning. *Learning, Culture and Social Interaction, 15*, 18–30.

Takeuchi, M. A., Sengupta, P., Shanahan, M. C., Adams, J. D., & Hachem, M. (2020). Transdisciplinarity in STEM education: A critical review [Review]. *Studies in Science Education, 56*(2), 213–253. https://doi.org/10.1080/03057267.2020.1755802

Tan, Y. J., Susanto, G. J., Anwar Ali, H. P., & Tee, B. C. (2021). Progress and Roadmap for Intelligent Self-Healing Materials in Autonomous Robotics. *Advanced Materials, 33*(19), 2002800.

Taylor, A. T., Berrueta, T. A., & Murphey, T. D. (2021). Active learning in robotics: A review of control principles. *Mechatronics, 77*, 102576.

Tramonti, M., & Dochshanov, A. (2018). *Students' engagement through computational thinking and robotics*. Digital Presentation and Preservation of Cultural and Scientific Heritage.

Verner, I. M., & Revzin, L. B. (2017). *Robotics in school chemistry laboratories*. Academic Press.

Verotti, M., Dochshanov, A., & Belfiore, N. P. (2017a). Compliance Synthesis of CSFH MEMS-Based Microgrippers. *Journal of Mechanical Design, 139*(2), 022301. https://doi.org/10.1115/1.4035053

Verotti, M., Dochshanov, A., & Belfiore, N. P. (2017b). A Comprehensive Survey on Microgrippers Design: Mechanical Structure [Review]. *Journal of Mechanical Design, 139*(6), 060801. https://doi.org/10.1115/1.4036351

Vieyra, M., Gilmore, J., & Timmerman, B. (2011). Requiring research may improve retention in STEM fields for underrepresented women. *Council on Undergraduate Research Quarterly, 32*(1), 13–20.

Vossoughi, S., & Bevan, B. (2014). Making and tinkering: A review of the literature. *National Research Council Committee on Out of School Time STEM, 67*, 1–55.

Weinberg, J. B., Engel, G. L., Gu, K., Karacal, C. S., Smith, S. R., White, W. W., & Yu, X. W. (2001). A multidisciplinary model for using robotics in engineering education. *Proceedings of the 2001 ASEE annual conference and exposition.*

Weinberg, J. B., White, W. W., Karacal, C., Engel, G., & Hu, A.-P. (2005). Multidisciplinary teamwork in a robotics course. *ACM SIGCSE Bulletin.*

Williams, D. C., Ma, Y., Prejean, L., Ford, M. J., & Lai, G. (2007). Acquisition of physics content knowledge and scientific inquiry skills in a robotics summer camp. *Journal of Research on Technology in Education, 40*(2), 201–216.

Williams, E. (2014). *AVR Programming: Learning to Write Software for Hardware*. Make Community, LLC.

KEY TERMS AND DEFINITIONS

Line-Follower: A robot that follows a particular path controlled by a feedback mechanism, consisting of continuous monitoring of the reflected light ratio incident on at least a couple of the sensors and subsequent motor activation.

Microelectromechanical System (MEMS): A miniature machine composed of mechanical and electronic components sized between 1 and 100 micrometres.

Multi-Perspective Approach: The use of multiple perspectives when approaching a robotic artefact.

Multi-Scale Approach: The method consisting of a combined consideration of a robotic artefact at different scales: from basic physical phenomena lying behind single electronic components up to the combinations of the latter.

Nanoelectromechanical System (NEMS): The technological evolution of MEMS with typical device dimensions in the nanometer range. The systems manifest qualitatively new behaviour and therefore extended potential functionality.

Nanotechnology: A branch of modern science and technology dealing with objects of dimensions less than 100 nm.

Tinkering: An educative approach, emphasising hands-on, creative problem-solving for unintentional gaining the knowledge and skills profitable for engineering, design and science.

Chapter 4
The Impact of the Professional Development Training of Primary Education Teachers on Issues of Educational Robotics as a Learning Tool

Raphaela Neophytou
Frederick University, Cyprus

Nikleia Eteokleous
https://orcid.org/0000-0003-4364-9558
Frederick University, Cyprus

ABSTRACT

Studies reveal a number of factors which influence teachers' decisions to use educational robotics (ER) in the classroom. One of these factors is teachers' professional development training (PDF). PDF is considered to be an encouraging factor that obstructs the successful application of educational robotics (ER) among the teachers. This chapter addresses the importance of professional development training (PDT) by examining its effectiveness in integrating educational robotics (ER) in the educational practice for teaching and learning purposes in various cognitive subjects. The sample of the study consisted of 23 elementary school teachers. The data used in this research was collected via pre and post structured questionnaires and classroom observations. The findings reveal the importance of PDT and suggest that PDT for teachers in ER were valuable in contributing to improve teachers ER skills and knowledge.

INTRODUCTION

The rapid development of Information and Communication Technology (ICT) has brought significant changes in the 21st century as it has affected the demands of modern societies (Buabeng, 2012; Schina,

DOI: 10.4018/978-1-7998-7443-0.ch004

Esteve-González & Usart, 2021). ICT have become part of our lives so they could not be missing from the field of education. When refering to technology in education, it means the digital tools that are used in the learning process, in order to enhance the teaching practice. One of these technologies is Educational Robotics (ER). ER has been already employed in formal educational settings, becoming a powerful tool in the hands of the teachers. ER when appropriately integrated it can facilitate and contribute to the achievement of various learning goals. However, ER seems to be used by a relatively small number of primary school teachers in the Cypriot educational system, as there are various factors that affect this integration.

Over the years, several studies (Baek, Jung & Kim 2008; Tondeur, 2008; Buabeng, 2012; Player-Koro, 2012; Agyei & Voogt 2014; Ames, 2016; Khasawneh, 2018) have focused on investigating the factors that influence the integration of ICT into education. However, little research has been found to examine the factors that influence teachers' integration of ER in the teaching and learning practice (Papadakis, Vaiopoulou, Sifaki, Stamovlasis, Kalogiannakis & Vassilakis, 2021).

Professional development training (PDT) is able to positively influence teachers' attitudes towards the integration of ER as an educational tool as well as the various internal and external factors that discourage teachers in integrating ER in school settings. Such factors may be the lack of familiarity of teachers with ICT (Haydn & Barton 2007; Sang, Valcke, Braak & Tondeur, 2009), the time required to prepare a technology-supported course (Askan & Erylimaz, 2011; Hadjithoma & Eteokleous, 2007; Eteokleous, 2008; Vrasidas, et al., 2010), the technical problems that arise (Preston, Cox & Cox, 2000), the work environment (Cuban, 2001; Fullan, 2007) and teachers' attitudes (Cox, et al., 2000).

Research has shown the need for PDT. Specifically, according to Abuhmaid (2011), training courses usually aim at the development of two levels of skills: 1) digital literacy skills and 2) the pedagogical use of these skills. The role of the training regarding the smooth introduction of ICT is considered to be particularly important in education. The professional development programs that involve teachers in lessons' collaborative design, are considered to be promising in regards to the development of teachers' knowledge and skills, as well as achieving increased level of sustainability over time (Agyei & Voogt, 2012; Agyei and Voogt 2014; Alayyar, Fisser & Voogt, 2011; Jimoyiannis, 2010). Some of the skills that teachers develop are the following: thinking skills, social skills, problem solving skills, design skills in terms of content, appearance, accuracy, testing, processing and testing of data, metacognitive skills (Rogers, 2003; Sullivan, 2008), activation of observation and creative thinking skills (Bers, 2018).

BACKGROUND

Educational Robotics (ER)

The evolution of ER and its application in the classroom have increased its presence in recent years (Arís & Orcos, 2019). In recent years both teachers and researchers have shown great interest in robotics, as a useful tool in the hands of the teacher during his/her teaching. ER is identified as an educational tool through which digital skills can be accomplished (Schina, Estev-Gonzalez & Usart, 2021). Robotics also helps to develop both the cognitive and social skills of students of different ages. In other words, robotics responds to a wide range of ages, ranging from preschool to university (Alimisis, 2013; Chan et al., 2013). ER has the potential to take place in classrooms as it can support learning in many disciplines (Taylor, 2016) such as science, mathematics, technology and computer science (Alimisis, 2009; Chan et al., 2013; Demetriou, Lambrou, Eteokleous, & Sisamos, 2013). According to Demetriou et al. (2013), ER is an interdisciplinary approach that consists of a mix of knowledge from various disciplines such as engineering, electrical engineering, computer science, communication and psychology.

Through ER, high level of mental skills can be developed as it is an innovative teaching tool. In addition, ER can cultivate problem-solving skills (Blanchard, Freiman & Lirrete-Pitre, 2010; Petre & Price, 2004). Furthermore, the integration of robotics in teaching promotes the development of 21st century skills such as critical thinking, creativity, communication, collaboration, metacognition and problem solving. (Karakoyun & Lindberg, 2020; Liesa-Orús, Latorre-Cosculluela, Vázquez-Toledo & Sierra-Sánchez, 2020; Maryuningsih, Hidayat, Riandi & Rustaman, 2020; Talaiver & Bowen, 2010). In this way, students will be able to meet the needs of a technological demanding society (Kalogiannakis &Papadakis, 2017; Mertala, 2019). First of all, it should be mentioned that the acquisition of knowledge is developed by the students themselves through the design of the structures they use in authentic examples (Castledine & Chalmers, 2011). In addition, children learn through their actions (learning by doing) in the virtual (software part) but also in the real (hardware part) world. Finally, they learn to collaborate, discuss and argue (Mikropoulos & Bellou, 2013) while acquiring skills that meet the demands of the 21st century (Ma & Williams, 2013). Several studies indicate that if students receive instructional material using the robot as an educational tool, then they enhance collaboration with their peers, cultivate critical thinking, discover the potential of research (Williams et al., 2007) and learn new (visual) programming languages (Nourbakhsh, Crowley, Bhave, Hammer & Perez Bergquist, 2005).

The Importance of ER

Although the majority of robotic activities are observed at university level, in recent years the educational use of robots in primary and secondary education has developed a very strong presence (Alimisis, 2013; Bers, Ponte, Juelich, Viera & Schenker, 2002; Duncan, Bell & Atlas, 2017; Kazakoff & Bers, 2012; Van Lith & Caska, 2007).

Students already encounter robotics in their daily lives. It would be good for students to become familiar with it and understand its operations and functions (Leonard, Mitchell, Barnes-Johnson, Unertl, Outka-Hill, Robinson & Hester-Croff, 2018; Piedade, 2021) and become digital literate. Digital literate student in relation to robotics is the one who is able to design, build and program a robot in order to perform certain functions (Slangen et al., 2010). Robotics also requires utilization of calculation, estimation, manipulation and observation. Students involved in robotics come into contact with research and programming elements (Sullivan, 2008). By engaging in ER activities, students change roles in the educational process. From passive receivers, who may receive sterile knowledge from the teacher, they turn into active learners who experiment and discover knowledge on their own, but there is always the supportive guidance of the teacher. In this way a variety of skills are developed. According to Sisman and Kucuk (2019), teachers which are unaware of benefits of ER, they are remain unprepared to use ER as an educational tool. This lack of teachers' readiness of teachers may be due to the limited or non-attendance of PDT programs about ER.

Philosophy of ER Integrated

There are two philosophies about how robots can be used in education. One view supports the use of robots to teach robotics as a subject. The main goal of this approach is to develop knowledge and skills in building and programming a robot (Eteokleous, Demetriou & Lambrou, 2013) through various training activities that focus on building robots (Rosenblatt & Choset, 2000).

The second approach embraces the idea that robots can be used as educational tools to teach various concepts. The activities in this case are integrated around an interdisciplinary approach and allow students to participate in collaborative and creative activities (Eteokleous, Demetriou & Lambrou, 2013; Jonassen & Hung, 2015). The main courses in which educational robotics is applied are mathematics, natural sciences and computer science (Barker, Grandgenett & Adamchuck, 2010; Bers, 2002; Eteokleous & Ktoridou, 2014). According to this approach, robotics is used within a specific learning context in order to achieve learning objectives. Throughout the literature (Burke & Kafai, 2012; Kafai & Burke, 2010) there are significant efforts to expand the applications of robotics in education, not only as a learning object, but also as a learning tool for other scientific fields, such as language and the arts.

Factors That Influence the Integration of ER in the Educational Practice

According to the literature (Conti, Commodari & Buono, 2017; Eteokleous, 2008; Hadjithoma & Eteokleous, 2007) Kalogiannakis, Ampartzaki, Papadakis & Skaraki, 2018; Shin, 2020; Van Braak & Valcke, 2008) it seems that the integration of new technologies (and robotics) in education is influenced by many factors. Research in the use of technology by teachers in the classroom, recognizes a complex pattern of interrelated factors that are considered determinants of the successful implementation of ICT. in education. Overall, the research generally classifies the factors that facilitate or act as barriers to the use of ICT in schools by teachers, whether they come from the external environment or from the personal characteristics of teachers, including beliefs, values and attitudes (Player-Koro, 2012).

There are several categories of factors that influence the attitudes of teachers towards the integration of robotics in the learning process. In the present study the factors are divided into personal (demographic characteristics), internal and external. Teachers' demographics are classified as personal factors (Tondeur et al., 2008).

Internal factors include the lack of confidence in technological knowledge and skills that teachers have (Anderson & Maninger, 2007, Bas, Kubiatko, & Murat, 2016), technophobia and stress that may encounter them during integration robotics in the learning process (Agyei & Voogt 2014).

External factors include lack of technological knowledge and skills (Dawson & Cavanaugh, 2012), lack of technological equipment, preparation time, various forms of support (Bingimlas, 2009) and PDT (Gil-Flores, et al., 2017). PDT can positively influence many factors that hinder the use of robotics in the teaching and learning process.

The Importance of PDT

The PDT should be based primarily on constructive theory in order to learn in a practical, constructive, collaborative and active environment (Agatolio et al. 2017; Alimisis et al. 2009; Alimisis 2014; Alimisis 2019; Castro et al. 2018; Negrini 2019; Scaradozzi et al. 2019; Sisman and Kucuk 2019).

Undoubtedly, teachers need to use their technical knowledge to practice with robotic packages, the construction and programming part (Kim et al. 2015) through solving exercises (Agatolio et al. 2017). Adequate time should also be given to teachers to build and program a robot during training and to familiarize themselves with the hardware and software (Agatolio et al. 2017; Hamner et al. 2016). However, PDT on ER should not be limited in learning how to use the robotics package, but also to how to pedagogically use it in any educationa setting (formal, informal and non-formal) (Castro et al. 2018; Kucuk & Sisman, 2018; Scaradozzi et al. 2019).

It is very important in a PDT for teachers to learn to handle robotic packages but mainly to get acquainted with the pedagogical approaches of ER integration in the learning process (Esteve-González, & Usart, 2021). In the context of PTDs for robotics, participants can be asked to become creative themselves. They may have to create a lesson plan, using robotics as a learning tool, and apply it in their classroom (Castro, Cecchi, Salvini, Valente, Buselli, Menichetti & Calvani, 2018). They can also prepare worksheets and material for various classes to have them for future use in conjunction with other courses (Agatolio, Pivetti, Battista, Menegatti & Moro, 2017). Another good practice to be employed within a PDT is to create a database that includes the material created by all participants, thus creating a material exchange community on the use of ER in the educational process (Negrini, 2019). This practice facilitates the implementation of ER activities in the classroom (Castro et al. 2018; Hamner et al. 2016; Negrini 2019) as teachers do not have enough free time due to the material they have to cover. So, it would be good idea during a PDT to develop material by the teachers themselves (Hamner et al. 2016) and share it with everyone through a database (Negrini, 2019). The material will be there and will be enriched with new material (Kucuk and Sisman 2018; Sisman and Kucuk 2019) and after the end of the PTD.

Professional development training on ER is becoming increasingly important in educational research (Schina, Esteve-González & Usart, 2021). According to the bibliographic data of the scientific community, the biggest problem that teachers point out is their lack of PDF about ICT (one of ICT example is ER) (Jimoyiannis 2010; Leask & Younie, 2013) and this has the effect of alienating them from the integration of robotics, since they don't have the skills needed. At the same time, they don't feel comfortable to teach a lesson using new technologies (robotics) because they of the lack of self-confidence (Mathevula, & Uwizeyimana, 2014). Many teachers who attended PDT programs, showed higher order knowledge and skills, more positive attitudes towards technology integration, greater understanding about the value and usefulness of technology as well as increased use of technology in various aspects of their daily lives (Sergis & Koutromanos, 2013). Longer-term PDT develops digital skills to a greater extent (Mathevula, & Uwizeyimana, 2014). On the other hand, teachers who have not taken part in any PDT show more stress during their teaching process in contrast to teachers who attended PDTs. Therefore, the role of PDT is particularly important for the application of technologies and innovative tools in the educational practice (Bingimlas, 2009). According to numerous researchers (Bingimlas, 2009; Drent & Meelissen, 2008; Koh & Chai, 2014; Lee & Lee, 2014; Valtonen, Kukkonen, Kontkanen, Sormunen, Dillon & Sointu, 2015), teacher training should to provide effective integration of technologies as learning tools in the educational process (Hutchison, 2012; Gil-Flores, et al., 2017). ER is one of the many digital tools that teacher can integrate in the teaching and learning process. The difference between ER and other technologies is the fact that ER consists of 2 dimensions: construction and programming. Therefore, the training focused on both dimensions in regards to the technical part as well as on how to integrate the ER as an educational learning tool within the educational practice (the pedagogical part). This unique characteristic of ER increase the time of the PDT needed to acquire sills,

knowlegde and experiences. The increased time and the need for equipment (hardware and software – robotics construction and programming platform), another unique characteristic of ER, may discourage teachers to attend PDTs and therefore to experiment and integrate ER as tools in their educational practice.

MAIN FOCUS OF THE CHAPTER

The current chapter aims to examine the effectiveness of elementary school teachers' professional development training in integrating ER in the educational practice for teaching and learning purposes in various cognitive subjects. The research objectives of the study focuses on 1) exploring the effect of the PDT to teachers' views, attitudes and perceptions in regards to ER, and 2) the development of teachers' knowledge and skills in regards to ER and specifically the development of knowledge and skills in building and programming robots and in designing and delivering learning environments enhanced with ER.

METHODOLOGY

Research Design

Teachers' training was delivered in 5 three-hour workshops by one of the authors. The purpose of the workshops was twofold: to develop teachers' technical as well as pedagogical skills. Teachers' technical skills on robotics and programming of using various educational robotic packages were developed as the foundation of pedagogical skills acquisition on how to integrate ER as a tool within the teaching and learning process to achieve specific learning objectives. The educational material provided to teachers included: robotics packages and programming platforms presentations, handouts, interactive and collaborative activities, exercises, video tutorials, unplugged activities as well as programming activities. The professional development training covered the following areas: 1) Introduction to Educational Robotics, Bee Bot and Robot Mouse, 2) Pro-robot, 3) Lego Wedo 2.0, 4) and 5) Lego Mindstorms EV3 (level I & II). Besides learning how to use the robotic package, the training focused on how to integrate ER as an educational tool in the teaching and learning process.

In each workshop, the teachers came in contact with different robotics packages. The process of the meetings followed a spiral approach with the degree of difficulty being gradually increased. Thus, in this way they came in contact with ways and examples of integrating various educational robotics packages in order to address the needs of elementary school students.

The various training packages were selected according to different criteria, i.e. to correspond to elementary students' age range. The variety of packages employed throughout the PDT addressed the needs of all elementary grades (from 1st to 6th grade). Also, all the robotics packages chosen were easily accessible by the teachers. Finally, after informal discussion with the teachers, the packages selected considered as the most popular ones and the ones that teachers aimed to work with.

Data Collection Instruments

In order to address the main aim of the study quantitative and qualitative data were collected. Quantitative data collected via pre- and post- survey instruments and qualitative data collected via participatory observation. The data collection instruments (survey instrument and participatory observation) were developed based on the grounds of other studies (Heinssen, Glass & Knight, 1987; Nickell & Pinto, 1986) and the study's authors.

The questionnaire consists of four sections. The first section deals with demographic characteristics. The second section refers to internal factors which influence teachers regarding the use of ER in educational process. The third section deals with the external factors that may influence the integrated of ER in education. The last section deals with teachers' attitudes regarding the integration of educational robotics in education.

Participatory observations were also employed for data collection purposes. Two other elementary school teachers who had experience in the field of ER participated at the data collection process, as instructor-assistants (observers). Specifically, the instructor needed their contribution during the participatory observations. sThe observation protocol was divided into 5 categories which included relevant questions: a) construction part, b) programming part, c) cooperation, d) support needed and e) interest about the activities. The observation was based on a Likert scale from 1 to 4 (1: Not valid, 2: Valid to a small extent, 3: Valid, 4: Valid to a large extent). Observers (instructor assistants) also had the opportunity, at the end of the protocol, to provide any other comments or to comment on some of their observations. The participation of the 2 observers in addition to the researcher, enhance the reliability of the research, since it provided the opportunity for comparison and discussion of the data collected as well as to ensure data collected unanimity. The quantitative data from the survey instrument questions were analysed using descriptive and inferential statistics (SPPS 23 package was used), and for the qualitative data obtained from the observations, a qualitative thematic analysis approach was followed (Braun, & Clarke, 2006; Guest, MacQueen, & Namey, 2012).

Participants

For this purpose, 23 teachers participated in 15 hours PDT. The teachers who took part in the research had to meet two specific criteria: a) teach in Primary Education and b) don't have pre-existing knowledge/ experience regards in ER. The participants voluntarily took part in PDT following a relevant announcement issued mainly via social media and in collaboration with non-profit educational organizations.

Robotics Packages

Throughout the ER development process, a wide range of robotic packages is available to the educational community, thus facilitating the implementation of ER activities (Schina, Esteve-González & Usart, 2021). In the present study, 5 robotics packages were used addressing all 6 elementary school level grades.

The first robot used at the PDT was the Bee-Bot. The bee robot is one of the most common programmable floor robots. This robot has the ability to accept simple orientation commands (front, back, right, left). This means that the student can't modify the movement distance of the robot since it is predetermined (Janka, 2008). It has no code writing and it does not necessarily require reading, writing and computer use.

The second robot that was used for the purposes of the PDT, was the Pro-bot, which can be considered as the evolution of the robot bee (Bee-Bot). The main difference between the two is that this package also has sensors (Sullivan, Kazakoff & Bers, 2013). Pro-Bot is a floor robot in the shape of a car which accept commands via keys (arrows and numbers) that can be found at the top. With this tool students design paths, shapes, numbers, etc.

The Code & Go Robot Mouse robot is another of the robots that belong to the category of floor robots. The buttons on the top of the mouse are color-coded and each color corresponds to an instruction. In addition, the robot is accompanied by maze instructions, which can be followed and created by children.

The fourth robot package used was Lego Wedo 2.0 which is a robotic platform that consists of hardware (brings and sensors for construction) and software parts (visual programming platform) and is designed for elementary school children. The package contains 150 pieces, including a motor, a tilt sensor, a motion sensor, lego bricks, straps and gears (Kazakoff, Sullivan & Bers, 2013). In the Lego Wedo 2.0 there are more models that can be built with complete or partial instructions.

The last robotics package is the Lego Mindstorms EV3. Eventhough EV3 is more appropriate for secondary and higher education students, it is common to be used at the upper level of elementary school grades and mainly with children that have already gained experience in ER through various other robotics packages.

EV3 robots are characterized as optical programming robots that have sensors. They consist of a wide range of bricks, components, motors and sensors. The programming environment is child friendly as it is controlled by visual programming. However, it can be programmed with other programming languages (C, C ++, Java, etc). In visual programming the process is quite simple, without meaning that the programming is not complex.

All packages employed throughout the training embrace the view that the student must independently build knowledge through a playful way (learning by playing) (Hussain at al., 2006). The whole process is a motivation for children to learn (Jarvinen & Hiltunen, 2000).

FINDINGS AND RESULTS

Quantitative Analysis

The data for this study were collected from 23 teachers from different schools without any experience in ER. The current chapter aims to examine the effectiveness of elementary school teachers' professional development training in integrating ER in the educational practice for teaching and learning purposes in various cognitive subjects. The research objectives of the study focuses on 1) identifying the factors that influence teachers, 2) exploring the effect of the PDT to teachers' views, attitudes and perceptions in regards to ER, and the development of teachers' knowledge and skills in regards to ER and specifically the development of knowledge and skills in building and programming robots and in designing and delivering. The main statistical analysis was a test for dependent samples which showed a difference in teachers' ratings for internal (F(49,245) = 4.917, p=.000), external factors (F(5,253) = 26,861, p=.000) and their attitudes (t (22) = -7.430. p= .000) before and after intervention (PDT).

This study identified two internal factors: a) the perceptions of teachers' digital skills and b) technophobia/ anxious. In regards to the first internal factor, statistical significant differences revealed before and after the training (F(49,245) = 4.917, p=.000). Before the PDT teachers believed that they didn't have enough skills and knowledge to introduce ER (Negrini, 2020) in their teaching practice. After the engagement with ER via the PDT they felt confident with their digital skills and appeared to be very positive and receptive in integrating ER as a learning tool (Zhou, Yuen, Popescu, Guillen & Daviset, 2015). On the other hand, the second internal factor was negatively influenced after the PDT. According to data findings, the PDT decreased the level of teachers' anxiety in using ER. Thus, being more likely for teachers to integrating ER in their teaching and learning practice.

The external factors identified as important by the study were the following: a) the lack of using specific robotics packages (t (22) = -12.733. p= .000), b) technical support about ER (t (22) = 3.224.

p= .004), c) lesson preparation time required when ER is integrated as a tool (t (22) = -5.075. p= .000), d) the importance of PDT on ER (t (22) = -1.155. p= .001). According to the results, the PDT helped to enrich the knowledge and skills of teachers regarding ER as well as increased the time of teachers' involvement with ER by searching for information from various sources. Another important finding was that after the PDT the perceived need for technological support was decreased. This was due to the fact that the acquired knowledege and skills on ER, causing the decrease of the anxiety and stress level. The PDT had no effect on the deficiencies that teachers faced regarding ER, the preparation time to develop a lesson using ER and the importance of PDT. This was expected since the deficiencies in the schools did not change within 2 months. Also, the preparation time of a lesson using the ER is necessary even if a teacher is trained.

Teachers before taking part in the training, appeared to have negative attitude towards the integration of ER in the educational process. This was evident from the various factors mentioned above. Their attitude has changed after the training (became more positive and receptive), revealing its importance.

According to the analysis, it has been revealed that the PDT had positive effect on teachers' attitudes (r=.189, p=.000) towards the integration of educational robotics in the teaching and learning practice. Finally, the value of professional development training and the need to provide related ER training were revealed (F(11,252) = 3,735 p=.000).

Qualitative Analysis

During the first meeting, the participants dealt with the bee robot (Bee-Bot) and the mouse robot (Code & Go Robot Mouse). They were very excited (positive attitudes). The participants were divided into 8 stations and spent 10 minutes at each station. They really liked those activities since they were experiential and required collaboration among the teachers. the fact that in order to solve the activities they had to work together. They seemed to be having fun with the whole process and discussing ways in which they could integrate these activities to their classroom. They commented that it's not so difficult to learn how to program a robot (built new knowledge) and discussed many possible integration ideas. They took pictures of the material in order to prepare something similar for their students and asked to borrow the robots. They said that they would ask the students, during Art lesson, to prepare floor mats for various topics that they would use in other subject matters. In general, the first meeting left participants with a pleasant sense of what ER is.

During the second meeting, teachers worked with the educational robotic package "Pro-Bot" which is also available in schools. As they commented, it's regrettable that while they have this particular robot in their school, they don't know how to use it as a robot and as a learning tool. They reported not knowing how to use the robotics package, its uses and functions (programming buttons)) and not knowing how to integrate it within classroom activities. The first activities seemed simple and very useful, since as they said they could definitely integrate the robot in the subject of Mathematics and specifically in the section of geometry. They were excited and didn't give up until they were able to complete the missions and scenarios given. At some point, the teams also competed each with other for which team would achieve the final goal. Then the activities became more complex. In the end, after a lot of practical training, they managed to understand how programming works and all together solved the activities. The same thing happened when they worked with the sensors. Some teachers asked the researcher if she could visit their school for co-teaching purposes, incorporating this package into their classroom. They commented on the fact that it develops many skills, transmits knowledge while at the same time the students consider that they are engaged in a game (gamification).

In the third meeting the teachers worked with the Lego Wedo II robotic package.

According to observations, the teams immediately began to build the robot (race car) following the steps in order. In all groups the members worked in the same way. They took one step at a time, but they all helped if someone was having difficulty at one point or could not find the right bricks and/ or sensors they were looking for. At some points during the robot construction process, the educators were in a hurry to finish because they considered them too simple and chose the wrong parts, but in the process, they realized that they had placed the wrong part and accordingly replaced them. They were looking forward to finishing the construction part to move on. They wondered if the color of the pieces mattered and if they could replace one piece, they could not find in the box with another that was similar. In reality, they thought that they could not find the correct piece, and after closer and more thorough search they were able to locate it. They forgot to measure the size of the axes, but in general they did not have much difficulty. They dud not give attention to detail. They worked in groups and asked the instructor and the assistants when they had difficulties. They had more questions than the previous robots they worked with. All teams completed their construction in about 30 minutes. Then they started programming the robots they assembled. The majority of team members remembered the basic commands. They looked excited and tried to solve the exercises. They only had difficulty using the sensors and combining them with other commands. In the end, they managed to address all the missions and scenarios given. They found it more difficult than the bee bot and the robot mouse. However, they did not get bored and were engaged to a large extent with both the construction and the programming. They wanted to learn more on how to employ them in their classrooms.

In the fourth meeting, participants had to build the "Castor Bot" model of the Lego Mindstorms EV3 robotic package. Most discussed the fact that they did not have the opportunity to work with this robot before and were looking forward to dealing with the package. At the same time, they were worried that it would be too difficult and that they might not be able to build the robot and finish it within the time frame they had. They were impatient and excited, because they thought they would deal with a robot they might not have the opportunity to see again. Teams began building robots. During the robot construction process, they faced lots of difficulties in comparison to the previous constructions with Lego WeDo 2. They were in a hurry to finish because they were worried that they would not be able to finish all the steps, so they left steps behind (some of them unconsciously and some of them consciously) and therefore made several mistakes. Unlike the previous construction, in this construction all teams chose to work as a whole and not individually, taking 1 step each in turn. Specifically, in each group they searched for the bricks all together in each step and each one completed a step in construction,, but overall everyone helped in all the steps. They collaborated in a great degree. They mainly had difficulty in measuring the axes, since they forgot to measure the rods, as a result of which they were placed incorrectly. The main mistakes they made were similar in all groups (incorrect axis measurement, misalignment of parts and incorrect centimeter measurement of cables). They generally had lots of questions and needed more guidance in comparison to Lego WeDo 2 constructions. They had difficulty going back to previous steps to find the point where they made a mistake in placing a piece. Although initially, all the teams proceeded almost simultaneously, some teams completed their construction in 1.5 hours while others took 2 hours. They were discussing whether they could use this robot in the subjects they teach, what ages of students it is suitable for, and they were worried about the time students would need to build. They wanted to program it to see how it worked and were discussing how anxious they were about the next lesson.

During the last meeting, participants were asked to program the robot they assembled in the 4th meeting (Castor Bot - Lego Mindstorms EV3). First, the programming environment was explained and demonstrated. During the explanation, everyone listened carefully and took notes, because it seemed more difficult to them than everything they had done so far. They had questions mainly about sensors, degrees and time. They confused the categories in which the commands were located, and thus referred to the notes they kept when explaining the software. They were interested in learning more and asked about the more complex commands they would not use at this stage, but wanted to know their operations and functions. The participants were divided into groups. In the first activity they had to program the robot to make a square on the floor. All participants began to perform the exercise reluctantly and found it difficult to measure the degrees. They made several attempts to program the robot and worked as a team doing experiments. They did not give up until they achieved their goal. Much more guidance was needed from the instructor and the assistants. They were generally hesitant throughout the meeting but worked together and supported each other to solve the exercise. In another activity they were requested to program the robot to move an obstacle. In this challenge they mainly had difficulty calculating the distance needed to move the robot back and forth. In subsequent activities they had to activate sensors. Although they had many questions about the sensors, they nevertheless stated that they liked these exercises much more than the previous ones. They felt it was a challenge for them.

Part of the PTD was to provide and discuss examples on how to integrate the robotics packages within classroom activities. The educators also presented and discussed their ideas. Very important was the fact that the educators began to think ways of integrating the robotics packages experienced during the PDT into their educational practices.

SOLUTIONS AND RECOMMENDATIONS

Teachers' perceptions can influence the integration of ER (Inan et al., 2010; Tondeur et al., 2008). A teacher may accept or reject the integration of a new technology such as robotics into the learning process, depending on his/hers perceptions. One of the reasons that teachers do not use technology as an educational tool is because they are not possessed by positive perceptions and that is why they feel uncomfortable (Setiawan & Koimah, 2019). Based on the results of the research, technophobia (which arises from the lack of knowledge) influences teachers' attitudes towards the ER integration in the teaching and learning practice. In other words, it is revealed that the higher the levels of anxiety and technophobia teachers have, the more negative attitudes they possess. This is because technophobia brings about many negative emotions (Martinez - Corcoles, Teichmann & Murdvee, 2017) with predominant stress (Khasawneh, 2018) which is an inhibitory factor. According to the results of this research, the knowledge that teachers have about the construction and programming of robots has a significant effect on their attitudes towards ER. Lack of knowledge and skills is a key factor that affects teachers as they feel digitally illiterate and uncomfortable in relation to the demands and developments of the modern. The aforementioned attitude in addition to the lack of experience prevents teachers from realizing the benefits of using technological tools (Kim, Kim, Lee, Spector & DeMeester, 2013; Kim & Santiago, 2005). All this has an impact on the integration of ERs since the less knowledge and skills the teachers possess, the more negative their attitudes on this issue are. The greatest parameter that teachers emphasize is their lack of training on the use of modern technologies (Jimoyiannis, 2010; Leask & Younie, 2013). It is widely recognized that a strong relationship exists between educators' attitudes towards technology and their actions in regards to technology (in this case ER) (Kalogiannakis, Ampartzaki, Papadakis & Skaraki, 2018; Vidal-Hall, Flewitt, & Wyse 2020). All this results in the alienation of teachers from new technologies due to lack of training and thus the lack of necessary self-confidence to teach a technologically designed course (Ertmer, 2006; Mathevula & Uwizeyimana, 2014). According to the research findings, the more trained a teacher is, the more positive attitudes he/she has about the integration of ER. This is why various researches emphasize the importance of vocational training (Abuhmaid, 2011). The results of this research highlight the importance and necessity of training and support towards the teachers. Additionally, it reveals the significance in gaining knowledge and understanding of the benefits and educational value of ER. When we manage to achieve the aforementioned then, there is a greater possibility in integrating ER in the classroom practice. Of course, there are some other factors that play significant role such as the equipment, the appropriate educational material and the continuous support and guidance needed. Consequently, actions and decisions are needed to be made on multiple levels such as policy and school level. On policy level there is a need to take decision for a more holistic and systematic approach of the integration of new and emerged technologies where the teachers will be appropriately trained, the schools will be equipped, the appropriate material will be developed by a group of experts in collaboration with the teachers (the base) and the Ministry), and the support will be provided (human resource to technically and pedagogically support the teachers). All the aforementioned policy decisions will be implemented on school level. In regards to the appropriate material, it is suggested that ER study guides, lesson plans and activities could be prepared for teachers and students. This would greatly contribute to the successful implementation of the ER in the classroom practice. The material to be developed should correspond to the thematic units of each subject matter Curriculum. It would be much easier for teachers to be able to integrate ER as learning tool. We would like also to suggest the design, development and implementation of an intervention ER program. Specifically, we suggest the use of a mobile unit that supports the teachers' effort to integrate ER into their classroom. The mobile unit will have the appropriate human resources (i.e. ER experts and technicians) as well as the appropriate equipment. The ER experts will be able to co-teach with teachers that are interested in integrating ER in their classroom practices. The mobile unit can also offer additional or missing equipment. Finally, ER communities of practice should be initiated by the Ministry in collaboration with other educational institutions to support and guide teachers. Communities of practices can be developed online.

FUTURE RESEARCH DIRECTIONS

Initially, as in the present study, PDT can be conducted on how to integrate ER into educational practice.

The PDT is expected to consist of 3 stages. The first stage of training should focus on learning how to use the robotics package, its operations and functions. The second stage, should go one step further and focus on explaining how ER can be integrated in the classroom practices and employed as a learning tool that provides added value and educational significance. This can be achieved through authentic classroom learning activities, which is stage 3. One aspect of the seminar should request teachers to apply what they have learned during the two first stages in authentic learning contexts, i.e. in the classrooms. When implementing ER by teachers in their classroom, it would be good for other teachers, researchers, academics to take the role of the observer in the classroom in order to record targeted observations. It is suggested that a professional development ER training to have integrated the implementation and application phase (stage 3), meaning that within the training the teachers will be requested to integrate robotics in their daily classroom practices. They would be requested to develop lesson plans and supporting educational material, as well as to complete reflective journals in order to be able to document and reflect on their practices. Participatory observations and interviews should take place throughout the professional development training in order to collect data that would be useful to improve training processes and procedures and identify additional teacher needs.

Further research should be designed in regards to examining students' knowledge and skills development on a particular subject matter. Experimental and control groups are expected to participate in the research studies, where the students will be requested to complete pre- and post- surveys. The goal would be to examine and identify differences between the control and experimental groups (integrating ER as a learning tool) in regards to skills and knowledge development. Additionally, the differences among various subject matters will be identified. Finally, the study guides, lesson plans and materials to be developed (as suggested above), should be pilot tested to a group of teachers and students in order to be able to make improvements and adjustments based on their feedback.

CONCLUSION

There are several factors that may affect the integration of ER into the educational process. Some of the factors are internal (i.e. lack of knowledge) while others are considered external (i.e. lack of knowledge).

Teachers' attitudes and perceptions can influence the integration of ER (Inan et al., 2010; Tondeur et al., 2008). A teacher may accept or reject the integration of a new technology into the learning process, depending on his or her perception towards technology (Becker & Riel, 2000; Eteokleous, 2008; Overbay, Ppatterson, Vasu & Grable, 2010). According to the results of the research, knowledge that teachers have about the construction and programming of robots has a significant effect on teachers' attitudes towards ER. Lack of experience, knowledge and skills are key factors that affect teachers as they feel digitally illiterate with the demands and developments of the modern era. Consequently, they cannot realize the benefits and educational value of using digital tools (Kim, Kim, Lee, Spector & DeMeester, 2013; Kim & Santiago, 2005). The aforementioned negatively influence the integration of ER since lack of knoelwde, skills and experience contribute to the development of negative attitudes towards ER.

If the teacher does not have the necessary experience, knowledge and skills to be able to integrate an educational tool in his/her teaching, it is very difficult to be receptive to use it. Also, due to the fact that technology is being highly integrated in our daily lives and activities, teachers must be properly trained to be able to address the needs and demands of their students, which are considered to be digital natives.

The greatest challenge that the teachers have to face and keep emphasizing on, is the lack of training on the use of modern technologies (Jimoyiannis, 2010; Leask & Younie, 2013), which results in the alienation of teachers from new technologies. Lack of training does not promote teachers' confidence in designing, developing and delivering a digital designed course (Mathevula & Uwizeyimana, 2014). According to the research findings, the more trained a teacher is, the more positive attitudes he/she would develop towards ER integration. That is why various researches give emphasis on the importance of vocational training (Abuhmaid, 2011). It is important to organize PDTs that allow teachers to integrate ER into their teaching to make it more effective. Teachers who have attended PDTs have higher knowledge and skills, more positive attitudes, increased confidence, and are more comfortable towards the use of ICT. Thus, these results highlight the necessity and usefulness of PDT and lead to the conclusion that training positively affects the integration of ER. On the other hand, teachers who have not received any training experience more stress and anxiety especially when it comes to new trends and developments. In conclusion, the role of the raining is particularly important for the application of new and emergent technologies and innovative tools in the educational practice (Bingimlas, 2009; Van Braak, 2001; Vanderlinde, Aesaert & van Braak, 2014).

ACKNOWLEDGMENT

This research was supported by the Cyprus State Scholarship Foundation.

REFERENCES

Abuhmaid, A. (2011). ICT training courses for teacher professional development in Jordan. *The Turkish Online Journal of Educational Technology, 10*(4).

Agatolio, F., Pivetti, M., Di Battista, S., Menegatti, E., & Moro, M. (2017). A training course in educational robotics for learning support teachers. In *International Conference EduRobotics 2016* (pp. 43-57). Springer. 10.1007/978-3-319-55553-9_4

Agyei, D., & Voogt, J. (2014). Examining factors affecting beginning teachers' transfer of learning of ICT-enhanced learning activities in their teaching practice. *Australasian Journal of Educational Technology, 30*(1). Advance online publication. doi:10.14742/ajet.499

Agyei, D. D., & Voogt, J. (2012). Developing technological pedagogical content knowledge in pre-service mathematics teachers through collaborative design. *Australasian Journal of Educational Technology, 28*(4). Advance online publication. doi:10.14742/ajet.827

Alayyar, G. M., Fisser, P., & Voogt, J. M. (2011). ICT integration through design teams in science teacher preparation. *International Journal of Learning Technology, 6*(2), 125–145. doi:10.1504/IJLT.2011.042645

Alimisis, D. (2009). *Teacher education on robotics-enhanced constructivist pedagogical methods.* Athens: School of Pedagogical and Technological Education.

Alimisis, D. (2013). Educational robotics: Open questions and new challenges. *Themes in Science and Technology Education, 6*(1), 63–71.

Alimisis, D. (2014). Educational robotics in teacher education: An innovative tool for promoting quality education. *The Teacher of the 21st Century: Quality Education for Quality Teaching*, 14.

Alimisis, D. (2019). Teacher training in educational robotics: The ROBOESL project paradigm. Technology. *Knowledge and Learning, 24*(2), 279–290. doi:10.100710758-018-9357-0

Ames, C. W. (2016). *Teacher perceptions of factors influencing technology integration in k-12 schools.* Utah State University.

Arís, N., & Orcos, L. (2019). Educational robotics in the stage of secondary education: Empirical study on motivation and STEM skills. *Education Sciences, 9*(2), 73. doi:10.3390/educsci9020073

Baek, Y., Jung, J., & Kim, B. (2008). What makes teachers use technology in the classroom? Exploring the factors affecting facilitation of technology with a Korean sample. *Computers & Education*, *50*(1), 224–234. doi:10.1016/j.compedu.2006.05.002

Bers, M. U. (2018). Coding, playgrounds and literacy in early childhood education: The development of KIBO robotics and ScratchJr. In *2018 IEEE Global Engineering Education Conference (EDUCON)* (pp. 2094-2102). IEEE. 10.1109/EDUCON.2018.8363498

Bingimlas, K. A. (2009). Barriers to the successful integration of ICT in teaching and learning environments: A review of the literature. *Eurasia Journal of Mathematics, Science and Technology Education*, *5*(3). Advance online publication. doi:10.12973/ejmste/75275

Blanchard, S., Freiman, V., & Lirrete-Pitre, N. (2010). Strategies used by elementary schoolchildren solving robotics-based complex tasks: Innovative potential of technology. *Procedia: Social and Behavioral Sciences*, *2*(2), 2851–2857. doi:10.1016/j.sbspro.2010.03.427

Braun, V., & Clarke, V. (2006). Using thematic analysis in psychology. *Qualitative Research in Psychology*, *3*(2), 77–101. doi:10.1191/1478088706qp063oa

Buabeng-Andoh, C. (2012). Factors influencing teachers' adoption and integration of information and communication technology into teaching: A review of the literature. *International Journal of Education and Development Using Information and Communication Technology*, *8*(1), 136.

Burke, Q., & Kafai, Y. B. (2010). Programming & storytelling: opportunities for learning about coding & composition. In *Proceedings of the 9th international conference on interaction design and children* (pp. 348-351). 10.1145/1810543.1810611

Burke, Q., & Kafai, Y. B. (2012). The writers' workshop for youth programmers: digital storytelling with scratch in middle school classrooms. In *Proceedings of the 43rd ACM technical symposium on Computer Science Education* (pp. 433-438). Academic Press.

Castledine, A. R., & Chalmers, C. (2011). LEGO Robotics: An authentic problem solving tool? *Design and Technology Education: An International Journal, 16*(3).

Castro, E., Cecchi, F., Salvini, P., Valente, M., Buselli, E., Menichetti, L., Calvani, A., & Dario, P. (2018). Design and impact of a teacher training course, and attitude change concerning educational robotics. *International Journal of Social Robotics*, *10*(5), 669–685. doi:10.100712369-018-0475-6

Chan, N., Wai, K., & Chiou, A. (2013). *Framework for educational robotics : a multiphase approach to enhance user learning in a competitive arena*. Academic Press.

Conti, D., Commodari, E., & Buono, S. (2017). Personality factors and acceptability of socially assistive robotics in teachers with and without specialized training for children with disability. *Life Span and Disability*, *20*(2), 251–272.

Cox, M. J., Cox, K., & Preston, C. (2000). *What factors support or prevent teachers from using ICT in their classrooms?* Academic Press.

Cuban, L. (2001). *Oversold and Underused: Reforming Schools Through Technology, 1980- 2000.* Harvard University Press.

Demetriou, G. A., Lambrou, A., Eteokleous, N., & Sisamos, C. (2013). The Engino Robotics Platform (ERP) controller for education. In *Control & Automation (MED), 2013 21st Mediterranean Conference on* (pp. 567-572). IEEE.

Drent, M., & Meelissen, M. (2008). Which factors obstruct or stimulate teacher educators to use ICT innovatively? *Computers & Education, 51*(1), 187–199. doi:10.1016/j.compedu.2007.05.001

Duncan, C., Bell, T., & Atlas, J. (2017). What Do the Teachers Think?: Introducing Computational Thinking in the Primary School Curriculum. In *Proceedings of the Nineteenth Australasian Computing Education Conference* (pp. 65-74). ACM. 10.1145/3013499.3013506

Eteokleous, N. (2008). Evaluating computer technology integration in a centralized educational system. *Computers in Education Journal, 51*(2), 669–686. doi:10.1016/j.compedu.2007.07.004

Eteokleous, N., & Ktoridou, D. (2014). Educational robotics as learning tools within the teaching and learning practice. In *2014 IEEE Global Engineering Education Conference (EDUCON)* (pp. 1055-1058). IEEE. 10.1109/EDUCON.2014.6826237

Fullan, M. (2007). *The new meaning of educational change*. Routledge.

Gil-Flores, J., Rodríguez-Santero, J., & Torres-Gordillo, J. J. (2017). Factors that explain the use of ICT in secondary-education classrooms: The role of teacher characteristics and school infrastructure. *Computers in Human Behavior, 68*, 441–449. doi:10.1016/j.chb.2016.11.057

Guest, G., MacQueen, K. M., & Namey, E. E. (2012). *Validity and reliability (credibility and dependability) in qualitative research and data analysis. In Applied thematic analysis*. Sage Publications.

Hadjithoma, C., & Eteokleous, N. (2007). ICT in primary schools: Explaining the integration in relation to the context. *Mediterranean Journal of Educational Studies, 12*(1), 1–25.

Hamner, E., Cross, J., & Zito, L. (2016). *Training teachers to integrate engineering into non-technical middle school curriculum. In 2016 IEEE Frontiers in education conference*. FIE.

Haydn, T. A., & Barton, R. (2007). Common needs and different agendas: How trainee teachers make progress in their ability to use ICT in subject teaching. Some lessons from the UK. *Computers & Education, 49*(4), 1018–1036. doi:10.1016/j.compedu.2005.12.006

Heinssen, R. K. Jr, Glass, C. R., & Knight, L. A. (1987). Assessing computer anxiety: Development and validation of the computer anxiety rating scale. *Computers in Human Behavior, 3*(1), 49–59. doi:10.1016/0747-5632(87)90010-0

Hiltunen, J., & Jarvinen, E. M. (2000). Automation Technology in Elementary Technology Education. *Journal of Industrial Teacher Education, 37*(4), 51–76.

Hutchison, A. (2012). Literacy teachers' perceptions of professional development that increases integration of technology into literacy instruction. *Technology, Pedagogy and Education, 21*(1), 37–56. doi:10.1080/1475939X.2012.659894

Janka, P. (2008). Using a programmable toy at preschool age: Why and how. In *Teaching with robotics: didactic approaches and experiences. Workshop of International Conference on Simulation, Modeling and Programming Autonomous Robots* (pp. 112-121). Academic Press.

Jimoyiannis, A. (2010). Designing and implementing an integrated technological pedagogical science knowledge framework for science teachers professional development. *Computers & Education*, *55*(3), 1259–1269. doi:10.1016/j.compedu.2010.05.022

Jonassen, D. H., & Hung, W. (2015). All problems are not equal: Implications for problem-based learning. *Essential readings in problem-based learning*, 7-41.

Kalogiannakis, M., Nirgianaki, G. M., & Papadakis, S. (2018). Teaching magnetism to preschool children: The effectiveness of picture story reading. *Early Childhood Education Journal*, *46*(5), 535–546. doi:10.100710643-017-0884-4

Kalogiannakis, M., & Papadakis, S. (2017). Combining mobile technologies in environmental education: A Greek case study. *International Journal of Mobile Learning and Organisation*, *11*(2), 108–130. doi:10.1504/IJMLO.2017.084272

Karakoyun, F., & Lindberg, O. J. (2020). Preservice teachers' views about the twenty-first century skills: A qualitative survey study in Turkey and Sweden. *Education and Information Technologies*, *25*(4), 2353–2369. doi:10.100710639-020-10148-w

Kazakoff, E., & Bers, M. (2012). Programming in a robotics context in the kindergarten classroom: The impact on sequencing skills. *Journal of Educational Multimedia and Hypermedia*, *21*(4), 371–391.

Kazakoff, E. R., Sullivan, A., & Bers, M. U. (2013). The effect of a classroom-based intensive robotics and programming workshop on sequencing ability in early childhood. *Early Childhood Education Journal*, *41*(4), 245–255. doi:10.100710643-012-0554-5

Khasawneh, O. Y. (2018). Technophobia without boarders: The influence of technophobia and emotional intelligence on technology acceptance and the moderating influence of organizational climate. *Computers in Human Behavior*, *88*, 210–218. doi:10.1016/j.chb.2018.07.007

Kim, C., Kim, D., Yuan, J., Hill, R. B., Doshi, P., & Thai, C. N. (2015). Robotics to promote elementary education pre-service teachers' STEM engagement, learning, and teaching. *Computers & Education*, *91*, 14–31. doi:10.1016/j.compedu.2015.08.005

Kim, C., Kim, M. K., Lee, C., Spector, J. M., & DeMeester, K. (2013). Teacher beliefs and technology integration. *Teaching and Teacher Education*, *29*, 76–85. doi:10.1016/j.tate.2012.08.005

Kim, C. J., & Santiago, R. (2005). Construction of e-learning environments in Korea. *Educational Technology Research and Development*, *53*(4), 108–114. doi:10.1007/BF02504690

Koh, J. H. L., & Chai, C. S. (2014). Teacher clusters and their perceptions of technological pedagogical content knowledge (TPACK) development through ICT lesson design. *Computers & Education*, *70*, 222–232. doi:10.1016/j.compedu.2013.08.017

Kucuk, S., & Sisman, B. (2018). Pre-service teachers' experiences in learning robotics design and programming. *Informatics in Education*, *17*(2), 301–320. doi:10.15388/infedu.2018.16

Lee, Y., & Lee, J. (2014). Enhancing pre-service teachers' self-efficacy beliefs for technology integration through lesson planning practice. *Computers & Education*, *73*, 121–128. doi:10.1016/j.compedu.2014.01.001

Leonard, J., Mitchell, M., Barnes-Johnson, J., Unertl, A., Outka-Hill, J., Robinson, R., & Hester-Croff, C. (2018). Preparing teachers to engage rural students in computational thinking through robotics, game design, and culturally responsive teaching. *Journal of Teacher Education*, *69*(4), 386–407. doi:10.1177/0022487117732317

Liesa-Orús, M., Latorre-Cosculluela, C., Vázquez-Toledo, S., & Sierra-Sánchez, V. (2020). The technological challenge facing higher education professors: Perceptions of ICT tools for developing 21st century skills. *Sustainability*, *12*(13), 5339. doi:10.3390u12135339

Ma, Y., & Williams, D. C. (2013). The Potential of a First LEGO League Robotics Program in Teaching 21 st Century Skills. *An Exploratory Study*, *6*, 13–28.

Martínez-Córcoles, M., Teichmann, M., & Murdvee, M. (2017). Assessing technophobia and technophilia: Development and validation of a questionnaire. *Technology in Society*, *51*, 183–188. doi:10.1016/j.techsoc.2017.09.007

Maryuningsih, Y., Hidayat, T., Riandi, R., & Rustaman, N. Y. (2020). Profile of information and communication technologies (ICT) skills of prospective teachers. *Journal of Physics: Conference Series*, *1521*(4), 042009. doi:10.1088/1742-6596/1521/4/042009

Mathevula, M. D., & Uwizeyimana, D. E. (2014). The challenges facing the integration of ICT in teaching and learning activities in South African rural secondary schools. *Mediterranean Journal of Social Sciences*, *5*(20), 1087. doi:10.5901/mjss.2014.v5n20p1087

Mertala, P. (2019). Teachers' beliefs about technology integration in early childhood education: A meta-ethnographical synthesis of qualitative research. *Computers in Human Behavior*, *101*, 334–349. doi:10.1016/j.chb.2019.08.003

Mikropoulos, T. A., & Bellou, I. (2013). Educational robotics as mindtools. *Themes in Science and Technology Education*, *6*(1), 5–14.

Negrini, L. (2019). Teacher training in educational robotics. In *International Conference on Robotics and Education RiE 2017* (pp. 92-97). Springer.

Negrini, L. (2020). Teachers' attitudes towards educational robotics in compulsory school. *Italian Journal of Educational Technology*, *28*(1), 77–90.

Nickell, G. S., & Pinto, J. N. (1986). The computer attitude scale. *Computers in Human Behavior*, *2*(4), 301–306. doi:10.1016/0747-5632(86)90010-5

Nourbakhsh, I. R., Crowley, K., Bhave, A., Hsium, T., Hammer, E., & Perez Bergquist, A. (2005). The robotic autonomy mobile robotics course: Robot design, curriculum design and educational assessment. *Autonomous Robots*, *18*(1), 103–127. doi:10.1023/B:AURO.0000047303.20624.02

Nugent, G., Barker, B., Grandgenett, N., & Adamchuk, V. I. (2010). Impact of robotics and geospatial technology interventions on youth STEM learning and attitudes. *Journal of Research on Technology in Education, 42*(4), 391–408. doi:10.1080/15391523.2010.10782557

Papadakis, S., Vaiopoulou, J., Sifaki, E., Stamovlasis, D., Kalogiannakis, M., & Vassilakis, K. (2021). Factors That Hinder in-Service Teachers from Incorporating Educational Robotics into Their Daily or Future Teaching Practice. In CSEDU (2) (pp. 55-63). Academic Press.

Petre, M., & Price, B. (2004). Using robotics to motivate "back door" learning. *Education and Information Technologies, 9*(2), 147–158. doi:10.1023/B:EAIT.0000027927.78380.60

Piedade, J. M. N. (2021). Pre-service and in-service teachers' interest, knowledge, and self-confidence in using educational robotics in learning activities. *The Educational Forum, 6*(1), 2.

Player-Koro, C. (2012). Factors influencing teachers' use of ICT in education. *Education Inquiry, 3*(1), 93–108. doi:10.3402/edui.v3i1.22015

Rogers, E. M. (2003). *Diffusion of innovations* (5th ed.). Free Press.

Rosenblatt, M., & Choset, H. (2000). Designing and implementing hands-on robotics labs. *IEEE Intelligent Systems & their Applications, 15*(6), 32–39. doi:10.1109/5254.895856

Sang, G., Valcke, M., van Braak, J., & Tondeur, J. (2009). Investigating teachers' educational beliefs in Chinese primary schools: Socioeconomic and geographical perspectives. *Asia-Pacific Journal of Teacher Education, 37*(4), 363–377. doi:10.1080/13598660903250399

Scaradozzi, D., Screpanti, L., Cesaretti, L., Storti, M., & Mazzieri, E. (2019). Implementation and assessment methodologies of Teachers' training courses for STEM activities. Technology. *Knowledge and Learning, 24*(2), 247–268. doi:10.100710758-018-9356-1

Schina, D., Esteve-González, V., & Usart, M. (2021). An overview of teacher training programs in educational robotics: Characteristics, best practices and recommendations. *Education and Information Technologies, 26*(3), 2831–2852. doi:10.100710639-020-10377-z

Sergis & Koutromanos. (2013). Effects of an INSET in ICT in education to the educators. *Journal of Issues of Science and Technology in Education, 6*, 1–2.

Setiawan, A. R., & Koimah, S. (2019). *Effective learning and teaching*. Academic Press.

Shin, W. S. (2020). An Analysis of Teachers' TPACK on Robotics in Education. *Journal of the Korea Academia-Industrial Cooperation Society, 21*(6), 196–203.

Sisman, B., & Kucuk, S. (2019). Educational robotics course: Examination of educational potentials and preservice teachers' experiences. *International Journal of Research in Education and Science, 5*(2), 510–531.

Slangen, L., Keulen, H., & Gravemeijer, K. (2010). What pupils can learn from working with robotic direct manipulation environments. *International Journal of Technology and Design Education, 21*(4), 449–469. doi:10.100710798-010-9130-8

Slangen, L., Keulen, H., & Gravemeijer, K. (2011). Preparing teachers to teach robotics in primary schools. In Professional Development for Primary Teachers in Science and Technology (Vol. 9, pp. 181-198). doi:10.1007/978-94-6091-713-4_14

Sullivan, F. R. (2008). Robotics and Science Literacy : Thinking Skills. *Science Process Skills and Systems Understanding*, *45*(3), 373–394.

Talaiver, M., & Bowen, R. (2010). Developing 21st-Century Skills: Game Design and Robotics Exploration. In *Society for Information Technology & Teacher Education International Conference* (pp. 2089-2090). Association for the Advancement of Computing in Education (AACE).

Taylor, P., & C. (2016). Why is a STEAM curriculum perspective crucial to the 21st century? In *14th Annual conference of the Australian Council for Educational Research.* Brisbane: Murdoch University Research Repository.

Tondeur, J., Valcke, M., & Van Braak, J. (2008). A multidimensional approach to determinants of computer use in primary education: Teacher and school characteristics. *Journal of Computer Assisted Learning*, *24*(6), 494–506. doi:10.1111/j.1365-2729.2008.00285.x

Valtonen, T., Kukkonen, J., Kontkanen, S., Sormunen, K., Dillon, P., & Sointu, E. (2015). The impact of authentic learning experiences with ICT on pre-service teachers' intentions to use ICT for teaching and learning. *Computers & Education*, *81*, 49–58. doi:10.1016/j.compedu.2014.09.008

Van Lith, P., & Caska, J. (2007). *RoboDidactics Manual.* Academic Press.

Vidal-Hall, C., Flewitt, R., & Wyse, D. (2020). Early childhood practitioner beliefs about digital media: Integrating technology into a child-centred classroom environment. *European Early Childhood Education Research Journal*, *28*(2), 167–181. doi:10.1080/1350293X.2020.1735727

Vrasidas, C., Pattis, I., Panaou, P., Antonaki, M., Aravi, C., Avraamidou, L., & Zembylas, M. (2010). Teacher use of ICT: Challenges and opportunities. *Proceedings of the 7th International Conference on Networked Learning*.

Williams, M. D., & Williams, J. (2007). A change management approach to evaluating ICT investment initiatives. *Journal of Enterprise Information Management*, *20*(1), 32–50. doi:10.1108/17410390710717129

Younie, S., & Leask, M. (2013). *Teaching With Technologies: The Essential Guide: The Essential Guide*. McGraw-Hill Education.

Zhou, H., Yuen, T. T., Popescu, C., Guillen, A., & Daviset, D. G. (2015). *Designing teacher professional development workshops for robotics integration across elementary and secondary school curriculum.* Academic Press.

ADDITIONAL READING

Alimisis, D. (2019). Teacher training in educational robotics: The ROBOESL project paradigm. Technology. *Knowledge and Learning*, *24*(2), 279–290. doi:10.100710758-018-9357-0

Jaipal-Jamani, K., & Angeli, C. (2017). Effect of robotics on elementary Preservice teachers' self-efficacy, science learning, and computational thinking. *Journal of Science Education and Technology*, *26*(2), 175–192. doi:10.100710956-016-9663-z

(2016). Robotics-based learning interventions for preventing school failure & early school leaving. InAlimisi, R. (Ed.), *ROBOESL conference 2016 proceedings*. EDUMOTIVA.

Scaradozzi, D., Screpanti, L., Cesaretti, L., Storti, M., & Mazzieri, E. (2019). Implementation and assessment methodologies of Teachers' training courses for STEM activities. Technology. *Knowledge and Learning*, *24*(2), 247–268. doi:10.100710758-018-9356-1

Schina, D., Esteve-González, V., & Usart, M. (2021). An overview of teacher training programs in educational robotics: Characteristics, best practices and recommendations. *Education and Information Technologies*, *26*(3), 2831–2852. doi:10.100710639-020-10377-z

Sisman, B., & Kucuk, S. (2019). Educational robotics course: Examination of educational potentials and preservice teachers' experiences. *International Journal of Research in Education and Science*, *5*(2), 510–531.

KEY TERMS AND DEFINITIONS

Digital Literate: Refers to the ability to find, evaluate, and clearly communicate digital information through various devices.

Educational Robotics (ER): An alloy of sciences that deals with the creation and programming of simple and complex constructions.

Educational Technology: Refers to Information and Communication Technologies (ICT) in Education that have been created to support the educational process and learning.

ER Robotics Skills: The ability to use robotics as a learning tool that adds value to the learning process.

Professional Development Training: Refers to all training, certification, and education that a worker needs to succeed in his or her career.

Robotics Skills: The ability to assemble and operate robotic systems.

Technological Learning Tool: A way for getting students working together or competing against each other. It is also a tool for achieving learning objectives via a playful approach.

Section 2
Developing Computational Thinking Skills

The three chapters of the second section of the book focus on the role of educational robotics in developing student computational thinking skills. The first chapter investigates the impact of student response modality on the development of computational thinking skills in educational robotics activities. The second chapter investigates the impact of student response modality on the development of computational thinking skills in educational robotics activities. The last one discusses the relationship between educational robotics and serious games and their effect on computational thinking.

Chapter 5
Efficiently Prompting Students When Developing Computational Thinking Skills:
The Impact of Students' Response Modality

Soumela K. Atmatzidou
https://orcid.org/0000-0002-6640-9066
Aristotle University of Thessaloniki, Greece

Chrysanthi N. Bekiari
Aristotle University of Thessaloniki, Greece

Stavros N. Demetriadis
Aristotle University of Thessaloniki, Greece

ABSTRACT

This study investigates the impact of student response modality on the development of computational thinking skills in educational robotics activities. Students of an elementary school were divided into three study groups ('Control', 'Selecting', and 'Writing') that implemented activities based on the same teacher guidance while prompted to provide responses of different modalities. The purpose was to engage students in the development of computational thinking skills, focusing on the basic skills of abstraction, generalization, algorithm, modularity, and debugging. These skills were evaluated at different phases during the activity, using different modality (selection, written, and oral) assessment tools. The results suggest that (1) prompting and eliciting thoughts in the form of written or selected answers proves to be a beneficial strategy, and (2) the two groups, 'Writing' and 'Selecting', reach the same level of CT skills, which is significantly higher than the level of the control group.

DOI: 10.4018/978-1-7998-7443-0.ch005

INTRODUCTION

Educational Robotics (ER) is a constructivist teaching and learning tool which promotes the development of students' computational thinking (CT) and social skills. Through ER activities, students are motivated and guided to solve authentic problems by programming the behaviour of a physical object (the robot) and getting immediate feedback on their proposed solutions. Researchers argue that ER is a problem-based learning tool which facilitates collaboration, enhances conceptual understanding and critical thinking, and promotes higher-order learning in all scientific fields (e.g. Atmatzidou & Demetriadis, 2016; Atmatzidou, Demetriadis, & Nika, 2018; Eteokleous-Grigoriou & Psomas, 2013; Eteokleous, Neophytou, Kolani & Christodoulou, 2020; Giang, Piatti & Mondada, 2019; Ponticorvo, Rubinacci, Marocco, Truglio, & Miglino, 2020; Sapounidis, Alimisis, 2021; Stewart, 2021). Especially increased is the interest for the contribution of robotics in the development of CT skills, which have been recognized as fundamental for all students. CT, as described by Wing (2011, 2017), "*is the thought processes involved in formulating a problem and expressing its solution(s) in such a way that a computer—human or machine—can effectively carry out*" and is highly important for controlling and managing cognitive activities in all disciplines (Wing, 2008).

Regarding the guidance provided by the teacher, researchers have focused on the impact that prompting strategies may have on students' effective engagement in the deeper processing of the learning material. Certainly, it is not always clear what type of prompting, and what conditions, can maximize the benefits of this process. Many researchers agree that an effective prompting strategy is providing written answers. However, this type of student response ('writing') is also reported to be tiresome and boring for students, causing heavy workload (Anewalt, 2002; Atmatzidou et al., 2018; Papadopoulos, Demetriadis, Stamelos, & Tsoukalas, 2011).

In the light of above, this study investigates the effectiveness of prompting strategies that trigger students' cognitive processing in ER activities. The focus is on the impact of different students' response modality and the study explores how to achieve improved learning outcomes without the negative impact of imposing on students a heavy workload, as students are not always willing to write down and explicitly report their thoughts during problem solving activity.

Analytically, the study presents the implementation of ER activities conducted with the participation of 56 elementary school students. Students worked in small teams, guided by worksheets to solve authentic complex problems and were guided to develop CT skills based on a CT model that includes: abstraction, generalization, algorithm building, modularity and debugging. The students were distributed in three conditions employing different student response modality as follows: a) *'Control condition'* -the baseline condition teams were prompted during problem solving without being mandatory for them to explicitly answer any of the questions during the activity. b) *'Writing condition'* teams were asked to provide explicitly written answers, in order to describe, document, justify and advocate their suggested solution to the prompts, and c) *'Selecting condition'* teams were asked to select the correct answer in closed-type questions relevant to the type of prompts presented to the other conditions. In the following, we present: a) the theoretical background of our work, and b) a study conducted in an elementary school that focuses on the development of computational thinking skills, and contrasts the impact of three different prompting strategies on response. The results provide encouraging evidence regarding the positive impact on the development of students' computational thinking skills.

BACKGROUND

The term Computational thinking has received intense attention over the past several years as a fundamental skill, which promotes new ways of thinking to the students across all disciplines of science. Wing (2006) described Computational Thinking (CT) as a way of "*solving problems, designing systems, and understanding human behavior by drawing on the concepts fundamental to computer science.*" She argues that CT is a type of analytical thinking that shares many similarities with mathematical, engineering and scientific thinking (Wing, 2006, 2008). CT roots go back, to Papert's work on Logo programming language and the idea of the computer being the children's machine that would allow them to develop procedural thinking through programming (Papert, 1993). Nowadays, researchers report that CT is an essential skill for all individuals that needs to be cultivated from early education to higher education, as part of the student's analytical thinking and as a principal component of Science-Technology-Engineering-Mathematics (STEM) education (Ioannou & Makridou, 2018; Fessakis, Komis, Mavroudi, & Prantsoudi, 2018; Shanmugam, Yassin, & Khalid, 2019).

According to Wing (2008) the keys of CT are abstraction, decomposition, separation of concerns and modularity. The elements that are widely accepted as components of CT skills and form the basis of activities that focus on supporting and developing these skills, are: computation, communication, coordination, recollection, automation, evaluation, design, algorithm building, conditional logic, debugging, simulation, working effectively in teams and analyzing problems (Astrachan et al., 2011; Settle & Perkovic, 2010; Csizmadia et al., 2015; Fagerlund, Häkkinen, Vesisenaho, & Viiri, 2021). Lu and Fletcher argue that, teaching CT should focus on establishing languages that can be used to annotate and describe concepts of CT and provide notation around which mental models of processes can be built (Lu & Fletcher, 2009). Saidin, Khalid, Martin, Kuppusamy and Nalini (2021) report that CT skills can be fostered through game-based learning. Deng, Pi, Lei, Zhou, & Zhang (2019) refer that the development of students' CT skills can be improved using programming environments. Kazimoglu, Kiernan, Bacon and Mackinnon (2012) report that programming a robot enables students to practice working with introductory programming constructs, within an environment that explicitly supports the acquisition of CT skills (such as algorithm building, debugging and simulation). Bottino and Chioccariello (2015) advocate that "*learners should practice computational thinking in playful contexts where they can develop personal projects, for example building videogames and/or robots, share and discuss their construction with others.*" Researchers believe that educational robotics encompasses main cognitive skills of CT, as students: 1) build algorithms to implement the activities by designing their own solutions 2) often use modules (my block) in order to create reusable solutions, and 3) can debug their solution in order to detect errors in their logic (Kazimoglu et al., 2012).

Robotics is usually seen as an interdisciplinary activity drawing mostly on Science, Maths, Informatics and Technology and offering major new benefits to education in general at all levels (Alimisis, 2009; Rogers & Portsmore, 2004). Drawing on the theoretical perspective of Piaget's constructivism, Papert's constructionism and Vygotsky's collaborative learning, ER activities help students transform from passive to active learners, developing many mental skills as researchers and creating new knowledge. Many studies indicate that ER activities have a positive effect on the development of critical thinking, problem solving and metacognitive skills (Grover & Pea, 2013; Atmatzidou & Demetriadis, 2017; Atmatzidou et al., 2018), and also on the learning of a programming language (Alimisis, 2009). Researchers report that students who learn by programming robots are actively involved in the problem solving process and display skills related to the basic components of computational thinking throughout this process

(Atmatzidou & Demetriadis, 2016; Bers, Flannery, Kazakoff, & Sullivan, 2014; Chen et al., 2017; Durak & Saritepeci, 2018; García-Valcárcel-Muñoz-Repiso & Caballero-González, 2019). In addition, a great advantage of using robots is that abstract concepts can be turned into real-world problems and solutions (Yadav, Zhou, Mayfield, Hambrusch, & Korb, 2011).

Studies have focused on the environment of robots, as an appropriate tool for the development of CT. In 2011, a research by the National Science Foundation, examined how abstraction, automation and analysis in problem-solving take shape for middle and high school youth. In a robotics project, student programmers needed to think about how the robotic agent would interact within its world and the results indicated that the students were able to use abstraction, automation, and analysis to create original products. Still the field requires systematic assessment procedures. Prior research demonstrates that children as young as four–six years old can build and program simple robotics projects as well as learn powerful ideas of engineering, technology, and computer programming while also building their computational thinking skills (Bers, Flannery, Kazakoff, & Crouser, 2010; Bers et al., 2014). Studies showed a significant increase in participants' computational thinking skills while programming robots (Atmatzidou & Demetriadis, 2016; Baek et al., 2019; Chen et al., 2017).

Although CT is a concept that has received considerable attention in recent years and the literature on implementing CT in a K-12 setting is rapidly increasing, with studies showing numerous benefits in this field, more research is needed. Many studies report the need to explore an effective and efficient pedagogical framework for the development of CT and its utility for learning of CT (Constantinou & Ioannou, 2018; Kotsopoulos et al., 2017; Ioannou & Makridou, 2018) and more empirical studies should be conducted to investigate the classroom realization of CT (Fessakis et al., 2018). Moreover, many researchers have signaled a gap in the research literature regarding the validity and reliability of CT assessments tools that should comprehensively assess CT and its relation to learning, using scientifically substantiated approaches, for all age groups (Cutumisu, Adams, & Lu, 2019; Rodriguez et al., 2020; Poulakis & Politis, 2021)

Guidance - Prompting Strategies

In literature, the dispute concerning the role and the impact of guidance on the students' learning process received considerable attention in the last half century. The relative effectiveness of guidance as compared to no or minimal guidance instruction is ambiguous. Researchers indicate that minimally guided instruction is less effective and less efficient than instructional approaches that place a strong emphasis on guidance of the student learning process (Anewalt, 2002; Papadopoulos et al., 2011).

In particular, Kirschner, Sweller and Richard (2006) argue that, although unguided or minimally guided instructional approaches are very popular and intuitively appealing, they are less effective and efficient for novices than guided instructional approaches, because they ignore the structures that constitute human cognitive architecture (Schmidt, Loyens, Van Gog, & Paas, 2007). In addition, as Clark, Kirschner and Sweller (2012) have mentioned, students learn best when provided with full, explicit instructional guidance. However, other researchers argue that strong assignments are expected to induce increased course workload, which might intimidate students and educators alike (Anewalt, 2002). Others still report that problem-based learning is itself an instructional approach that allows for flexible adaptation of guidance and therefore no further guidance is required (Schmidt et al., 2007), and also that students learn best when instructed with no or minimal guidance, because they must discover or construct some or all of the essential information for themselves, rather than having all the essential information presented to them and being asked to practice using it. Many studies point out that the learning process must be guided and monitored with various strategies, including question prompts, in order to maximize the learning benefits (Ge & Land, 2003).

Embedding prompts in the learning material is one of the most effective ways to achieve meaningful learning that triggers and guides students' learning processes, and makes the classrooms dynamic and meaningful learning places that are student-centered (Wilhelm, 2007). Prompting students to follow various strategies and to respond to questions, we motivate them to investigate, analyze, and demonstrate knowledge and procedures which are required to be learnt. Question prompts may guide students to implement the different stages of a task, by directing them to important features of the problem (Rosenshine, Meister, & Chapman, 1996) and helping them identify goals and analyze factors and constraints. Moreover, they can help students to activate prior knowledge, to monitor and reflect on the solution process and compare and evaluate alternatives, to elicit relevant explanations, and thereby to engage in high level procedures during the process of the solution. Especially, specific prompting questions may support students to develop high order skills or skills such as: problem solving, metacognition, computational thinking, collaboration etc. In this case, the steps comprising this particular skill must be determined, and with appropriate prompting questions, during the task, which will incite students to think and to assimilate these skills. Moreover, question prompts are one important method to direct the efforts of a group more productively, and to support group members in order to avoid problems that can arise.

Despite the well-founded support for the value of question prompts as a guidance tool, some other studies mention that they were often insufficient as a scaffold (Ge & Land, 2003), and also, that the continuous and extensive guidance may have negative consequences in learning. One other important point is, as Greene and Land (2000) reported that students sometimes ignored question prompts or answered superficially, because of a lack of attention to important aspects of a problem, thereby failing to engage in deeper processing.

In view of the above, the impact of various response modalities in prompting questions needs to be investigated, with the aim of defining which type of prompting and what conditions can maximize students' substantive engagement. Available studies highlight the effectiveness of prompting strategies when students provide written answers (Greene & Land 2000; Menary 2007; Stewart, Myers, & Culley, 2010; Papadopoulos et al., 2011; Çavdar & Doe, 2012; Kim et al., 2021). And they also claim that these strategies trigger an additional cognitive activity that leads to better learning outcomes and help to avoid superficial engagement. Although this response strategy proved to be effective, researchers reported that consistently writing answers create heavy workload, and is boring and tiring for students (Anewalt, 2002). Researchers (Collins, 2006; Nückles, Hübner, Dümer, & Renkl, 2010; Wecker & Fischer, 2011) believe that teachers should provide explicit instruction when introducing a new topic, but gradually fade it out as knowledge and skill increase. They argue that a gradual and adaptive fading of the prompts might moderate the over-prompting, giving the opportunity to the students to increase individual competence in applying their cognitive strategies. These differences encourage deeper research that will determine the impact of different students' response modalities on achieving improved learning outcomes through the systematic design of instructional experiments.

Research Motivation and Key Research Questions

Considering the above background, the current study investigated the use of prompting strategies focusing on the response modality, aiming to maximize the learning benefits and minimize the negative impact of a heavy workload.

Thus, the research questions set by the study are:

1. Do the prompting strategies to respond with varied modality -oral, written or selection- have the same impact on the development of CT skills in ER activities? Which of these strategies leads to better learning benefits?
2. What are the advantages and disadvantages of each different modality response?

METHOD

Participants and Conditions

This study was conducted with the participation of 56 Elementary School students (27 boys and 29 girls of 6th grade (12 years old) in Kozani, Greece. The students took part in ER activities which were implemented in 10 sessions each of which lasted two hours per week, 10 weeks totally. The sessions were conducted during the typical school time schedule and the class teachers remained in the classroom during the activity simply helping maintain the flow of each lesson.

Specifically, in the seminars the students were engaged in three response conditions:

- 'Control condition' group (C): 18 students, 7 boys and 11 girls
- 'Writing condition' group (WR): 19 students, 11 boys and 8 girls
- 'Selecting condition' group (SR): 19 students, 9 boys and 10 girls

In this study we used the Lego Mindstorms NXT 2.0 educational tool.

For the children's participation in the ER activities the parents gave their consent after being informed by the school head-teacher. Moreover, data processing was anonymized for compliance with ethical considerations.

Operationalization of CT Skills

In order to operationalize our approach for the development of CT skills, we follow a model that we also used in our previous studies (Atmatzidou & Demetriadis, 2016, 2017), focusing on specific CT components, namely: abstraction, generalization, algorithm, modularity, and debugging. Table 1 presents the model in detail providing definitions for each CT component and the suggested student guidance for developing each specific CT skill.

Table 1. A model for the development of CT skills

CT Skills	Definitions	Guidance for Development CT Skills
Abstraction	Abstraction is the process of creating something simple from something complicated by leaving out the irrelevant details, by finding the relevant patterns, and by separating ideas from tangible details.	1. Separate the important from the redundant information. 2. Analyze and specify common behaviors or programming structures between different scripts. 3. Identify abstractions between different programming environments.
Generalization	Generalization is transferring a problem-solving process to a wide variety of problems.	1. Expand an existing solution in a given problem in order to cover more possibilities/cases. 2. Use variables in solution
Algorithm	Algorithm is a practice of writing step-by-step specific and unambiguous instructions for carrying out a process.	1. Explicit wording of the steps of the algorithm. 2. Possibility of different algorithms for the same problem. 3. Effort to find the most effective algorithm.
Modularity	Modularity is the development of autonomous processes, which encapsulate a set of often used commands that perform a specific function and might be used in the same or different problems.	Develop autonomous sections of code to be used for the same or different problems.
Debugging	Debugging is the process of finding and resolving defects that prevent correct operation of a system.	1. Prompts to think about which part of the code might be causing the bug. Designing and running test plans and interpreting the results (testing); 2. Help break down the task by trying to get one code segment to work before moving onto another. 3. Stepping through algorithms step by step to work out what they do (dry run/tracing);

Implementation Procedure

Students of three classes of the 6th grade participated in the study. In order to distribute students to three condition groups, we applied convenient sampling, whereby the students of each class were assigned to one specific condition. All condition groups implemented Educational Robotics activities based on the same trainer-teacher guidance, while prompted to provide different modalities answers to a set of knowledge integration prompts. The specific differences in guidance between the groups are described below:

1. **'Control condition' group (C):** This group was the baseline condition in which students were prompted to implement the activities following the worksheet guidance without explicitly answering any of the questions during the activity. The instructor as facilitator provided support and prompted the students, from time to time during the activities, but did not prompt students to give an explicit answer.
2. **'Writing condition' group (WR):** the students were prompted to provide explicit written answers, in order to describe, document, justify and advocate their suggested solution to the prompts.
3. **'Selecting condition' group (SR):** the students were prompted to respond to Closed-type questions relevant to the type of prompts presented to the other conditions. The instructor encouraged the students to read carefully each suggested answer before selecting the correct one, and also underlined the importance of investigating and understanding the reasons why each suggested answer is correct or false.

Table 2. The responsibilities of each role that every member of the group assumed.

Role	Responsibilities/Guidance for Developing CT Skills	CT Skills
Analyst	1. Reads the problem to the whole team. 2. Analyses the problem for the team, specifying the data and the objects of the problem. 3. Separates the important from the redundant information. Reduces complexity by removing unnecessary detail. 4. Refers to similar problems that the team faced in the past. 5. Analyses and specifies common behaviors between different scripts	Abstraction
Programmer	1. Records in detail and step by step the actions to be done by the robot. (Algorithm) 2. Investigates the possibility of different algorithms for the same problem. (Algorithm) 3. Attempts to find the most effective algorithm. (Algorithm) 4. Develops autonomous sections of code to be used for the same or different problems. (Modularity) 5. Creates and operates the program.	Algorithm Modularity
Debugger/ Evaluator	1. Checks whether the program operated correctly and spots possible mistakes. 2. Suggests corrections when needed and checks whether these are the proper corrections. 3. Repeats the two previous processes until the program operates correctly. 4. Prompts to think about which part of the code might be causing the bug. 5. Helps break down the task by trying to get one code segment to work before moving onto another. 6. Designs and runs test plans and interprets the results (testing); 7. Evaluates the final solution and suggests improvements if he thinks they are necessary. 8. Expands an existing solution in a given problem in order to cover more cases.	Debugging Generalization

In total, we conducted 10 sessions (2 hours each, conducted once a week) that were organised and supervised by the main researchers (authors of this study). Also, trained postgraduate students ("trainers") assisted with the practicalities of the activity (e.g. handing out worksheets, encouraging and scaffolding teams, administering questionnaires, etc.). In the activities, the students collaborated in small teams consisting of three or two members. Each team was formed by the students themselves and if necessary with the guidance of their teacher. The teams were guided by collaboration scripts aiming to trigger peer interaction and to engage students in fruitful and productive learning. The script organized the activity into a sequence of phases, assigned student roles and tasks, and explicated the specificities of role playing and task development. Each student assumed roles such as analyst, programmer or debugger/evaluator, with the roles rotating in each activity. This rotation meant that all the students would familiarize with the responsibilities of each role, developing at the same time the relevant skills (as presented in Table 2).

During each session one member of the team is assigned the "moderator" role. The moderator was guided to: (a) expresses an opinion first in the team, regarding each specific prompt, (b) moderate the discussion between the team members, and (c) decide on the final answer to the prompt to be recorded in the worksheet (for the "writing" and "selecting" conditions). Students were also offered step-by-step instructions through worksheets to engage them in the development of basic CT skills. As already mentioned, the CT model included abstraction, generalization, algorithm building, modularity and debugging. In the worksheets, there were specific instructions that helped students develop those skills. For example, to support abstraction, the students were instructed to: (1) separate the important from the redundant information, (2) analyse and specify common behaviors or programming structures between different problems, and (3) identify abstractions between different programming environments.

The implemented ER activities were divided into three phases: the "Introduction", "Trainings" and the "Challenge". In the beginning, an introduction to robotics in general, the Lego Mindstorms NXT robot and the Lego NXT-G programming environment took place. Then, nine "training" sessions followed. In the first three sessions we handed out worksheets to students for familiarizing with robotics and basic programming concepts. At the core of each worksheet is the understanding and assimilation of the basic CT skills. In the next sessions the learning activities integrated more complex problems with gradually increased difficulty and addressing more CT skills. In the robotic activities, the students programmed the robots to implement authentic scenarios such as an alarm, a car that follows the rules of traffic, a recycler, a theme-park, a doll game, etc. In the last activity of each session after the 5th, in order to moderate the workload induced by the continual prompting, the explicit guidance faded out, as the worksheet did not provide any detailed guidelines, but only reflection questions which incited the students to integrate the skills that they acquired in their suggested solution. In this way, students were further given the opportunity to expand on the problem by using their imagination. Finally, in the Challenge phase, a demanding (yet within their reach) problem was given to the groups, and clear rules were set for grading and appointing the winner.

During the ER activity, the trainers employed the facilitator role prompting students with appropriate questions and explaining and analysing the CT skills. Also, she observed and checked whether the students internalized the role or skill and if the circumstances require it, she prompts students to assume the relevant role and triggers students' reflection and development of CT skills with appropriate prompting questions.

DATA COLLECTION INSTRUMENTS AND ANALYSIS

In the present study the level of CT skills that student developed was evaluated at different phases during the activity, using qualitative and quantitative methods with instruments of varied modality (written and oral). In particular, the evaluation tools we used were:

Profile Questionnaire (PQ): An individual questionnaire was administered in the beginning of seminar. The questionnaire recorded some simple demographic data, the students' background on computer use (for example, frequency of computer use, computer experience, etc.) and experience with robotics (such as previous knowledge on constructing and programming robots).

Two intermediate questionnaires (CT-pre & CT-post): CT-pre was handed out in the Introduction phase and CT-post after the Challenge phase. The questionnaires were developed for the purposes of our study, they consisted of fourteen multiple choice questions, each question had a form of a small scenario problem and students had to identify and select the answer closely related to the CT components "behind" each problem. For example, one of fourteen questions that focuses on abstraction is:

"Maria installed in her car an automatic light system which is activated when it gets dark. Paul installed in his house an alarm system, which activates the alarm when the front door of his house opens. What is the common behavior in these two systems?

1. There isn't any common behavior. Their actions are completely different.
2. Both of them check a condition and if the condition is satisfied then the system is activated.
3. Both of them often repeat the same action.
4. The system is activated in both cases."

Two intermediate activities (CT-3 & CT-9): CT-3 was handed out after the 3rd and CT-9 after the 9th session. Students were asked to solve programming problems and cite the CT components that they use during the solution process, e.g. identify the common behaviour between two or more tasks (abstraction), suggest a more general solution (generalisation), describe step-by-step the solution process (algorithm), etc. The assessment of students' answers was based on a graded criterion instrument (rubric) using a 4-point Likert scale (1='unsatisfactory', 2='quite satisfactory', 3='satisfactory', 4='excellent'). There were specific criteria for each component of the CT model (Abstraction, Generalisation, Algorithm, Modularity, Debugging) and so each student was assigned a grade for each CT construct after answering the CT-3 and CT-9 questionnaires.

Think-aloud protocol: After the training, students individually were given a certain robot programming task and were asked to describe aloud the solution process and cite the CT components that they integrated in it. Simultaneously, the researcher, when and if clarification was required, prompted students to reflect on CT components relevant to their solution. The assessment of students' answers in Think-aloud protocol was based also on the same graded criterion instrument (rubric) as in CT-3 & CT-9. The main difference is that the 'think-aloud' method allows students to express their thinking more freely (orally) as opposed to the highly structured form of the questionnaire instruments (written).

Semi-structured Interview: After the think-aloud activity, semi-structured individual interviews were conducted, with a fairly open framework, to elicit students' subjective views and opinions regarding the following key aspects of the activity: (1) the advantages and disadvantages of the different response modalities, (2) the contribution of the guidance to the educational process, (3) benefits or possible drawbacks in group collaboration, and (4) likes and dislikes relevant to the robotics activity. We used content analysis on collected data to identify the views and opinions that dominated in the statements of the students.

Observation: Systematic monitoring of the students' work was applied by taking notes on a structured observation sheets focused on key aspects of the activity (the four aspects described above in the "Semi-structured Interview" section). Both the supervising researcher and trainers filled in the sheets and after each session extensively discussed their observations to reach consent and decide on their importance.

RESULTS

Statistical Analysis

Profile Questionnaire (PQ): Data revealed that none of the participating students had any previous experience with robotics.

After the data collection, to analyze the results we used a) a paired-samples t-test to control for pre/post differences within students, b) an independent-samples t-test to compare pre-questionnaire between teams, and c) a one-way between subjects ANCOVA to compare the effect of post conditions on pre as covariate, in teams. The statistical processing was as follows:

CT-pre & CT-post: Table 3 presents statistical controls applied on the students' CT-pre and CT-post scores in three groups and as a total for all groups.

Table 3 showed that there are no significant differences between the groups, as regards the students' level of CT skills in the Introduction phase (CT-pre). However, when focusing on the results of the ANCOVA analysis after the sessions have been completed, it was observed the existence of statistically significant differences between the groups ($p=0.012$), without being able to specify though between which groups these differences exist. Therefore, a series of Tukey post-hoc tests were performed on the

Figure 1. The structure of sessions and the data collection instruments

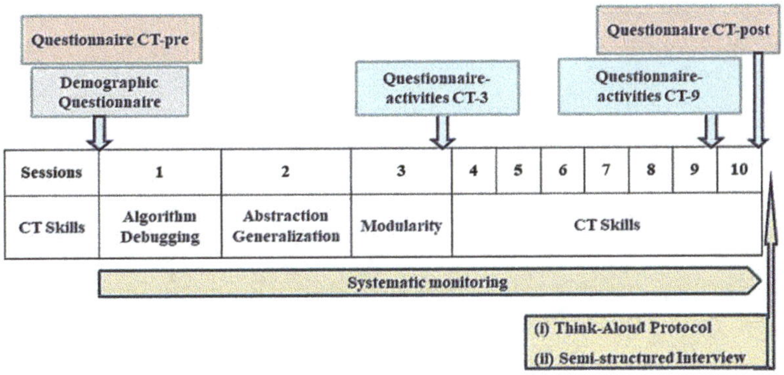

dependent variables that show statistically significant differences. From this analysis it emerged that we have statistically significant differences between ‘SR’ and ‘C’ groups (p=.049), while there are no statistically significant differences between ‘ SR’ and ‘WR’ (p=.828) and between ‘WR’ and ‘C’ groups (p=.167).

CT-3 & CT-9: Table 4 presents statistical controls applied on students’ CT-3 and CT-9 scores in groups.

Table 4 showed that there was no statistically significant difference in CT skills between groups in the 3rd session (CT-3), but at the end of the sessions (CT-9) statistically significant difference were observed between the groups. A Tukey post-hoc test revealed that the CT-9 score of ‘WR’ group was statistically significantly higher (2.83 ± .50, p=.012) as compared to that of group ‘C’ (2.35 ± .46), while there are no statistically significant differences between the students of ‘WR’ and ‘SR’ groups (p=.464) and between ‘C’ and ‘SR’ groups (p=.176).

Table 3. Comparing students’ CT-pre and CT-post scores between groups

Group	N	CT-pre M *(SD)*	CT-post M *(SD)*	Paired t-Test CT-Post Compared to CT-Pre (Same Student Group)	ANCOVA Comparing CT-Post Across Student Groups (CT-pre as Covariate)	
C	18	6.89 *(1.13)*	8.94 (1.77)	t(17)=9.994, p<.001*		
SR	19	7.11 *(1.20)*	10.26 (1.85)	t(18)=9.409, p<.001*		
WR	19	6.95 *(1.18)*	9.95 (1.31)	t(18)=12.406, p<.001*	F(2,53)=4.831 p=.012 η^2 =.154	
ANOVA		F(2,53)=0.171 p=0.843 η^2 =.006				
Total	56	6.98 *(1.15)*	9.73 (1.72)	t(55)=16.603, p<.001*		

* Significant difference at the 0.05 level

Figure 2. Comparing students' CT-pre and CT-post scores between groups

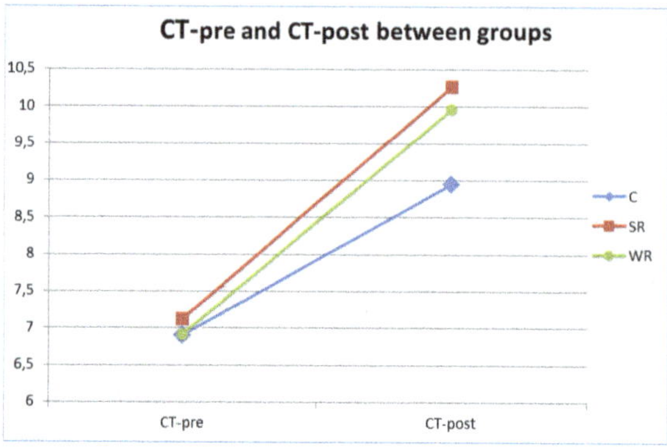

Table 4. Comparing CT-9 and CT-3 scores between groups

Group	N	CT-3 M (SD)	CT-9 M (SD)	Paired t-Test CT-9 Compared to CT-3 (Same Student Group)	ANCOVA Comparing CT-9 Across Student Groups (CT-3 as Covariate)
C	8	2.14 (.41)	2.35 (.46)	t(17)=-3.948, p=.001	F(2,52)=9.368 p<.001 η^2=.265
SR	19	1.92 (.33)	2.64 (.51)	t(18)=-6.391, p<.001	
WR	19	2.09 (.26)	2.83 (.50)	t(18)=-7.496, p<.001	
ANOVA		F(2,53)=2.365 p=0.104 η^2=.082			
Total	56	2.05 (.35)	2.613 (.52)	t(55)=-9.038, p<.001	

* Significant difference at the 0.05 level

Figure 3. Comparing CT-9 and CT-3 scores between groups

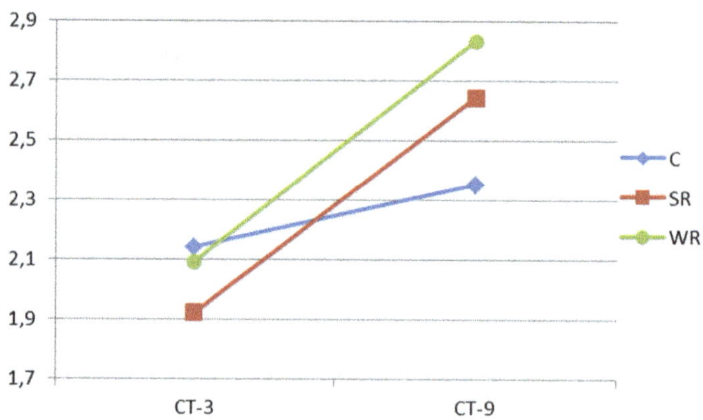

In the following, we compare CT-9 with CT-3 as covariate to examine possible statistical differences in students' CT skills among the three student groups at the end of the training. The ANCOVA control

Table 5. Comparison of CT-9 scores among different group (CT-3 as covariate)

CT Skills	C		SR		WR		F(2,52)	p	η^2
	M	*SD*	M	*SD*	M	*SD*			
Abstraction	2.36	*.61*	2.47	.49	2.632	.74	1.590	.214	.058
Generalization	2.22	*.71*	2.24	*1.06*	2.50	*.93*	.599	.553	.023
Modularity	1.81	*.86*	2.61	*1.19*	2.658	*.83*	3.440	**.040**	.117
Algorithm	2.53	*.61*	2.74	*.56*	3.263	*.65*	9.367	**<.001**	.265
Debugging	2.83	*.79*	3.16	*.76*	3.105	*.79*	.965	.388	.036
Total	2.35	*.46*	2.64	.51	2.83	.498	9.368	**<.001**	.265

was implemented analytically in all CT skills (Table 5).

In Table 5 we identified statistically significant differences between the groups in the total CT and also in Modularity and Algorithm skills. Specifically, a Tukey post-hoc test revealed the following:

1. **Modularity skill:** the students of the 'C' group develop this skill to a statistically significantly lower degree (1.81 ± .86) than the 'SR' group (2.61 ±1.19, p=.041) and the 'WR' group (2.66 ± .83, p=.028), while between the students of 'SR' and 'WR' there was no difference (p=.985).
2. **Algorithm skill**: the students of the 'WR' group showed a statistically significantly higher score (3.26 ± .65) compared to the 'SR' group (2.74 ± .56, p=.027) and the 'C' group (2.53 ± .61, p<.002), while between the students of 'SR' and 'C' groups there were no significant differences (p=.552).

CT-TA: The three groups' scores regarding CT-TA measure were analyzed implementing ANOVA to examine possible statistical differences in students' CT skills scores between groups at the end of the training. The ANOVA control was implemented analytically in all CT skills (Table 6).

Table 6. Comparison of CT-TA scores between groups

CT Skills	C		SR		WR		F	p	η^2
	M	*SD*	M	*SD*	M	*SD*			
Abstraction	2.44	*.78*	2.93	*.92*	2.711	*1.19*	1.135	.329	.041
Generalization	2.39	*.98*	2.84	*.90*	2.95	*1.13*	1.583	.215	.056
Modularity	2.00	*.77*	2.50	*1.42*	2.74	*1.11*	2.002	.145	.070
Algorithm	2.78	*.55*	3.06	*.85*	3.06	*.84*	.876	.422	.032
Debugging	2.44	*.73*	2.42	*.56*	2.74	*1.05*	.904	.411	.033
Total	2.41	.53	2.75	.71	2.84	.94	1.655	.201	.059

Table 7 presents statistical controls applied on students' CT-TA and CT-3 scores in groups. Finally, the CT scores that emerged from all assessment instruments are summarized in Table 8.

Table 7. Comparing CT-TA and CT-3 scores between groups

Group	N	CT-3 M *(SD)*	CT-TA M *(SD)*	Paired t-Test CT-TA Compared to CT-3 (Same Student Group)	ANCOVA Comparing CT-TA Across Student Groups (CT-3 as Covariate)
C	18	2.14 *(.41)*	2.41 *(.53)*	t(17)=-3.759, p=.002*	F(2,52)=3,494 p=.038* η^2=.118
SR	19	1.92 *(.33)*	2.75 *(.71)*	t(18)=-5.669, p<.001*	
WR	19	2.09 *(.26)*	2.84 *(.94)*	t(18)=-3.555, p=.002*	
Total	56	2.05 *(.35)*	2.67 *(.76)*	t(55)=-6.596, p<.001*	

* Significant difference at the 0.05 level

Figure 4. Comparing CT-TA and CT-3 scores between groups

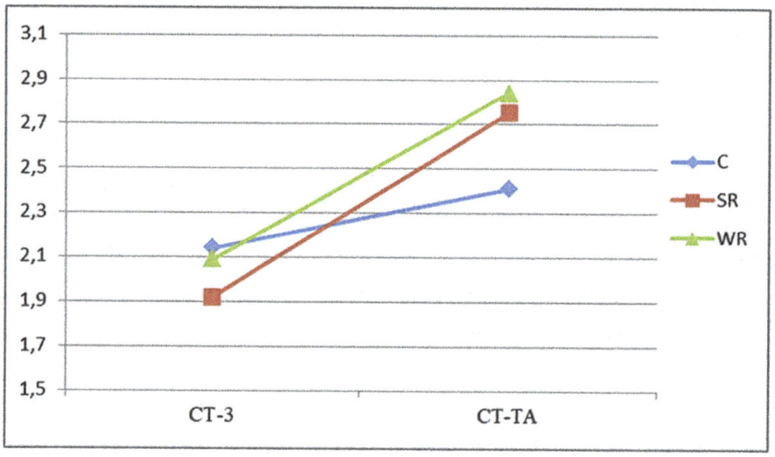

Table 8. Summary table: CT scores of all assessment instruments

		Pre 1st Session	Post 3rd Session	Post 9th Session	Post10th Session	Post 10th Session
Assessment Instrument Modality		Multiple Choice Questions	Writing	Writing	Think-Aloud	Multiple Choice Questions
Group	N	CT-pre M *(SD)* Scale 1-14	CT-3 M *(SD)* Scale 1-4	CT-9 M *(SD)* Scale 1-4	CT-TA M *(SD)* Scale 1-4	CT-post M *(SD)* Scale 1-14
C	18	6.89 *(1.13)*	2.14 *(.41)*	2.35 *(.46)*	2.41 *(.53)*	8.94 *(1.77)*
SR	19	7.11 *(1.20)*	1.92 *(.33)*	2.64 *(.51)*	2.75 *(.71)*	10.26 *(1.85)*
WR	19	6.95 *(1.18)*	2.09 *(.26)*	2.83 *(.50)*	2.84 *(.94)*	9.95 *(1.31)*
Total	56	6.98 *(1.15)*	2.05 *(.35)*	2.61 *(.52)*	2.67 *(.76)*	9.73 *(1.72)*

Semi-Structured Interviews and Researchers' Observations Analysis

Reflecting further on the students' semi-structured interviews and the researchers' observations we report the following, regarding the key aspects of the activity:

Different response modalities: Most students reported that the answers to the prompting questions helped them better understand the process they would have to follow in order to complete the activities. However, they said that this process was something new for them and they needed time to get familiar with it and also to find the way to respond to the prompting questions. The answers of each group varied, as was expected due to the different modality of the answers. Specifically:

- The students of the **'WR' group** stated that at first it seemed difficult for them to express their thoughts in writing but soon learned how to respond. Specifically, students reported:

"..having to write my answer was an additional difficulty for me, because I had to think how to formulate it correctly", "There is a significant difference between oral and written speech."

Some students reported that they became more aware that written answers are an important factor in learning, but most found it tedious and sometimes discouraging to complete the activity. As students reported:

"..It seemed tedious to us, so sometimes we answered hurriedly so we could get on with the task and see how the robot works"

"..It would have been good if we did not have to always answer in writing .."

- The students of the **'SR' group** stated that at first it seemed simple to select the correct answer, as they made a very superficial choice, devoting minimal time. But they soon realized that this process required more time and special attention. Specifically, the students mentioned the following:

".. Selecting the correct answer is not simple, as the answers are so similar one to the other and thus great attention is required."

"..As we hurried to the next step, without having answered the previous one or having responded superficially, we had difficulty solving the problem. In this case our trainer did not provide answers but advised us to go back to the previous steps and try carefully to give appropriate answers... gradually we understood and followed the instructions."

The students considered that the trainer's help was important for their learning how to think when selecting the right answer. In addition, several of them stated that they learned from suggested answers which were incorrect, as they identified points that they either had formulated in their minds mistakenly or had not thought through.

- The students of the **'C' group** stated that at first they spent little time to discuss and usually answered superficially. But then, with the guidance of the trainers, the discussion between them became more substantial. They underlined that the trainers' encouragement was decisive and quite supportive for the completion of their activities.

"..In each step, we reflected quickly and moved on to the next step... when we could not complete the activity, the trainer did not give us the solution but advised and prompted us to address each step carefully… so we realized that we had to devote time to each step, although this caused a delay".

Guidance: Several students reported that the guidance and prompting questions helped them understand and apply the concepts of CT, such as identifying commonalities or suggesting a generalization, and moreover that their role was crucial to the implementation of activities. As students reported:

"..the step-by-step instructions in the worksheets were helpful, as some of these steps come to our mind when we are solving problems in other lessons such as maths and physics."

"For example, I learned to pay attention to the wording of a problem, to highlight the important points by underlining them and to delete the unnecessary ones.."

In particular, the students of the groups 'WR' and 'SR' found that it was very important to apply the fade-out method, in the last activity from the 4th session onwards. As they characteristically stated,

"..it was important for us to be more free and have the opportunity to experiment and 'play' with the robot without the need to respond -writing or select- to the prompting questions"

They also said that they would have preferred it if their guidance had gradually declined and had not existed in the last sessions, as, according to the statements,

"..we have now learned the steps we have to follow".

Some students suggested that each time they completed their activities correctly, they should be rewarded with a fade-out of the guidance in the last activity.

Collaboration: Regarding the collaboration in the team, the roles and the responsibilities they took, the students mentioned that it took some time and the trainer's guidance for them to be able to get into the roles. They also said that their collaboration script helped them to work together.

"..The collaboration rules helped us to work together because it was clear what each member had to do"

"..Without the collaboration rules I think we wouldn't have been able to cooperate effectively, in the other courses when it is needed.. we usually end up arguing.."

Robotics activity: In relation to the whole robotics activity, the students expressed positive views. Most of the participants stated that the activity with the robots was an important learning experience for them, which they would like to repeat. They especially liked the fact that they learned to program and control their robot.

"..It was something new, innovative... it is a great satisfaction to see your robot do immediately what you designed and programmed it for.."

"..We would like to deal with this activity longer.... in school we would learn more if we had such lessons in our curriculum..".

DISCUSSION

The present study investigated the impact of students' response modality on the development of computational thinking skills in the context of Educational Robotics activities. The different modalities in which the students were prompted to respond during problem-solving were as follows: a) 'Control condition' - no explicit answer documenting their suggested solution is mandatory, b) 'Writing condition'- providing explicitly written answers, in order to document and advocate their suggested solution, and c) 'Selecting condition'- selecting the correct answer in closed-type questions.

The study provides evidence from evaluation instruments administered at various times during the activity, thus offering a picture of how CT skills develop in three conditions, each with a different response modality. Students' CT skills are also evaluated using different modalities assessment tools, namely questionnaires consisting of multiple choice questions, questionnaires-activities in written and problem-solving think-aloud protocols. Finally, researchers' observations and qualitative data from students' semi-structured interview help triangulate data and more deeply understand their meaning.

A first observation shows that students develop CT skills at the end of their training irrespective of their response modality. More specifically, CT skills are statistically significantly improved as the training proceeds, as is shown by the CT scores of all evaluation instruments in Table 3, 4, 7 and in the summary Table 8. Thus, one key conclusion is that guiding students through their learning process and encouraging them to solve authentic problems in ER activities based on Collaboration, Problem, Game-competition leads to the development of students' CT skills. This conclusion is in line with studies emphasising that ER promotes CT skills (Atmatzidou & Demetriadis, 2016; Bers et al., 2014; Clark et al., 2015).

Reflecting further on Table 3, we see that there were no significant differences between the groups, regarding the students' level of CT skills in the introduction phase (CT-pre) (p=.843). However, at the end of the training CT skills were significantly improved in all groups, although this improvement was not the same for all groups. We observe that the 'SR' group reached statistically significant higher levels than the 'C' group, while no significant differences were observed between 'SR' and 'WR', and between 'C' and 'WR' groups (Figure 2).

Moving on to Table 4, we observe that the students reached the same CT skills level after the 3rd session. This shows that the satisfactory development of CT skills needs a considerable number of training sessions -independently of student's response modality- and is not simply a matter of a few training sessions. This conclusion is in line with studies reporting that skill development in general requires adequate amount of training time (Atmatzidou & Demetriadis, 2016, 2017; Bers et al., 2014; Clark et al., 2015).

Focusing on Table 4, we see that significant differences between CT-9 and CT-3 measures were identified independently of the students' response modality. However, the 'WR' group reached statistically significant higher levels than the 'C' group, while no significant differences were observed between 'SR' and 'WR', and between 'SR' and 'WR' groups (Figure 3).

Reflecting further on Table 5, where the results of CT skills are presented analytically, we observe that significant differences between CT-9 and CT-3 measures were identified, in certain cases and also in the total CT skills score, independently of students' response modality. This is clear for the development of the Abstraction, Generalization, and Debugging skills, which is at the same level for all groups, with students in the 'WR' group reaching better scores. By contrast, in Modularity and Algorithm skills significant differences are observed between groups. Regarding Modularity, the students in the 'WR' and 'SR' groups reached the same level, as opposed to the students of 'C' group who showed significantly lower levels of this skill (p=.40). As regards the Algorithm skill, the 'WR' group showed significantly higher levels than either of the other groups.

One possible explanation for the results regarding: (a) the abstraction and generalization skills is that in the first sessions it was observed that the students gave superficial answers and avoided identifying the abstract concept or proposing a more general solution, as they stated "..it is something new for us and it seems difficult for us..". Therefore, special attention was paid by the trainers-teachers to the concepts of subtraction and generalization in order for the students to become familiar and integrate them in their activities, (b) the improvement of the Debugging skill is that Educational robotics by its nature significantly motivates students to debug, regardless of their guidance or their response modality. The students' motive is to see their robot "work properly," as they stated. The "check-find error-correct-test" process is repeated by the students until the robot implements "its challenge", without the need of guidance from their trainers and this is reflected in the students' CT scores, (c) Modularity is that since the NXT-G programming language involves the processing of module (My Block), and consequently allows the students to easily develop this skill through the activities, we expected that there would be no differences between the groups in the improvement of this skill. However, it was observed that the control group does not develop this ability to the same degree compared to the other groups. Since all groups followed the same guidance to complete their activities, this difference may be due to the lack of prompting to the 'C' group to submit their solution in writing or by selecting in close-type questions, and this may be the reason why they sometimes avoided incorporating the modularity in activities. Finally, (d) Algorithm skill, we identify the highest statistically significant differences between the groups in CT-9, with the students in the WR group reaching the highest scores. We believe that this is explained by the fact that the students in WR group, following the worksheet guidelines and expressing their thinking by capturing in writing the algorithmic steps that their robot will follow, became more familiarized with the algorithm skill.

Moving on to Table 6 (CT-TA scores), we see that when evaluating students' CT skills orally (Think-Aloud protocol) no between-group differences are identified in CT skills. Additionally, comparing CT-TA and CT-3 scores in Table 7, a key conclusion here is that all groups significantly improved their CT skills level as the training proceeded. However, some interesting evidence emerges, that no significant differences in CT skill scores appear between 'WR' and 'SR' group, as students of both groups reached the same level (p=.695) and in fact a statistically significantly higher level than 'C' group (Figure 4).

By reflecting on researchers' observations and students' Semi-structured Interview, we report the most important of them as follows:

1. The effectiveness of guidance and prompting strategies in the development of CT skills is reflected in the qualitative and quantitative scores of the students in all groups. More specifically, the students followed the guidance instructions in the worksheets, became familiar with and incorporated the CT skills into the process of implementing the activities. This finding is in line with studies argu-

ing that the prompting question strategy has a positive impact on learning (e.g. Rosenshine et al., 1996; Ge, 2001).

2. The modality of responses to prompting questions has a crucial impact on the development of the students' CT skills. Students who answer in writing or select the answer in close-type questions develop skills on a higher level than those who are not prompted to answer explicitly. In the beginning, the students faced difficulties as to how they had to respond to prompting questions, so they needed time to become familiar with the response modality, and also the trainer's guidance and encouragement to follow the prompts consistently. It was observed that each group faced different difficulties: (i) Regarding 'WR' group: In the beginning, most students preferred to describe the solution orally and faced difficulty formulating it in writing. For example, in the case of the algorithm, they described a process in general rather than analysing it step by step with clarity and accuracy. Soon they became familiar with this, and also they understood that written answers were an important factor in learning, as many students reported. However, they reported that they would prefer not to respond always in writing, as they found it tedious and time-consuming, and it sometimes discouraged them from implementing the activity. (ii) Regarding 'SR' group: In the beginning, it was observed that most students faced the stage of response in close-type questions superficially as they chose an answer quickly and proceeded to the next step. However, they soon realized that this needed more time and attention, as they had to go back to the previous step and re-examine the gaps that they had left in order to be able to complete their activity. In this case, the role of the trainer was crucial, as she supported and motivated the students to follow this strategy, and in addition, helped them to understand that we can also learn from the wrong answers. Finally, (iii) As regards the 'C' group, the students followed the instructions on the worksheet superficially and proceeded quickly to the implementation of the program, after which if the robot did not work properly, there might follow some disagreement. They needed constant encouragement to follow the guidance and share their thoughts and opinions within their group so that they could have a productive dialogue between them.
3. The role of guidance was essential for the implementation of the activities and the development of CT skills. As several students mentioned, the guidance and the prompting strategies helped them to understand and incorporate the components of CT in their activities, such as identifying the common behaviour, suggesting a generalization, etc. However, sustained guidance increased the students' workload, and it had a discouraging effect. This problem was settled with the encouragement given by the trainers as well as the possibility, only in the last activity of each session after the 5th, to fade out the prompting to the students to describe their answer in detail. The students' need to apply this fade-out method, even in these few activities was obvious, as they wanted to feel more 'freedom' in carrying out their activity. Additionally, this method kept their interest undiminished and strengthened their self-confidence, as it was due to their own skills that their robot completed the activity. From the trainer's point of view, this process was important as it helped her to observe if the students followed the steps and strategies of the guidance, to identify the points of the process to which they do or do not respond satisfactorily, to point out if the students assimilated the CT skills, and finally to intervene accordingly in the next steps.

CONCLUSION

In the present research, we studied the effect of different response modality on the development of Computational Thinking skills in Educational Robotics activities. Overall, this study provides encouraging evidence regarding the positive impact on the development of students' CT skills. Specifically, the results indicate that: (a) Prompting students without asking them to explicitly answer any of the questions during the activity is not an effective strategy, as opposed to prompting and eliciting thoughts in the form of written or selected answers, which proves to be a beneficial strategy, (b) The students in the two treatment conditions ("Writing" and "Selecting") reached the same level of CT skills, and also both of these conditions show a statistically important improvement of the CT skills in comparison with the control group, (c) In the "Selecting" condition it was recorded that, when students were required to select the correct answers, they went through the resolution stage superficially, selected an answer quickly and moved on to the next step. Soon, however, the need arose to turn back and review the gap steps in order to be able to complete the rest of the activity. To avoid this problem, the instructor should make sure and motivate them to apply this strategy on their own, d) Students reported that the step-by-step instructions in the worksheets, modeling the CT skills during the problem-solving process, were very helpful, and that steps emerged to their mind when they were solving problems in other domains such as mathematics, physics etc.

We believe that a blending of response modalities would lead to effective guidance, motivate students to learn, and keep their interests unabated. Particularly, a type of response which combines selecting the answer in closed-type questions with the writing response can be a useful tool for students to improve CT skills and it can also engage them in a higher-level learning process.

REFERENCES

Alimisis, D. (Ed.). (2009). *Teacher Education on Robotics-Enhanced Constructivist Pedagogical Methods. School of Pedagogical and Technological Education.* ASPETE.

Anewalt, K. (2002). Experiences teaching writing in a computer science course for the first time. *Journal of Computing Sciences in Colleges, 18*(2), 346–355.

Astrachan, O., Barnes, T., Garcia, D., Paul, J., Simon, B., & Snyder, L. (2011). CS principles: piloting a new course at national scale. In *Proceedings of the 42nd ACM technical symposium on Computer science education* (pp. 397-398). 10.1145/1953163.1953281

Atmatzidou, S., & Demetriadis, S. (2012). Evaluating the role of collaboration scripts as group guiding tools in activities of educational robotics. *IEEE International Conference on Advanced Learning Technologies (ICALT 2012).* 10.1109/ICALT.2012.111

Atmatzidou, S., & Demetriadis, S. (2016). Advancing students' computational thinking skills through educational robotics: A study on age and gender relevant differences. *Robotics and Autonomous Systems, 75*, 661–670. doi:10.1016/j.robot.2015.10.008

Atmatzidou, S., & Demetriadis, S. (2017). A Didactical Model for Educational Robotics Activities: A Study on Improving Skills through Strong or Minimal Guidance. In D. Alimisis, M. Moro, & E. Menegatti (Eds.), *Educational Robotics in the Makers Era. Edurobotics 2016. Advances in Intelligent Systems and Computing* (Vol. 560, pp. 58–72). Springer. doi:10.1007/978-3-319-55553-9_5

Atmatzidou, S., Demetriadis, S., & Nika, P. (2018). How Does the Degree of Guidance Support Students' Metacognitive and Problem Solving Skills in Educational Robotics? *Journal of Science Education and Technology*, *27*(1), 70–85. doi:10.100710956-017-9709-x

Baek, Y., Wang, S., Yang, D., Ching, Y.-H., Swanson, S., & Chittoori, B. (2019). Revisiting second graders' robotics with an Understand/Use-Modify-Create (U2MC) strategy. *European Journal of STEM Education*, *4*(1), 1–12. doi:10.20897/ejsteme/5772

Bers, M., Flannery, L., Kazakoff, E., & Crouser, J. (2010). *A Curriculum Unit on Programming and Robotics. Computer Science*. Faculty Publications, Smith College. https://scholarworks.smith.edu/csc_facpubs/9

Bers, M., Flannery, L., Kazakoff, R., & Sullivan, A. (2014). Computational thinking and tinkering: Exploration of an early childhood robotics curriculum. *Computers & Education*, *72*, 145–157. doi:10.1016/j.compedu.2013.10.020

Bottino, R., & Chioccariello, A. (2015). Computational Thinking: videogames, educational robotics, and other powerful ideas to think with. *KEYCIT 2014: Key competencies in informatics and ICT, 7,* 301.

Çavdar, G., & Doe, S. (2012). Learning through writing: Teaching critical thinking skills in writing assignments. *PS, Political Science & Politics*, *45*(2), 298–306. doi:10.1017/S1049096511002137

Chen, G., Shen, J., Barth-Cohen, L., Jiang, S., Huang, X., & Eltoukhy, M. (2017). Assessing elementary students' computational thinking in everyday reasoning and robotics programming. *Computers and Education, 109*, 162–175. http:// doi:10.1016/j.compedu.2017.03.001

Chevalier, M., Giang, C., Piatti, A., & Mondada, F. (2020). Fostering computational thinking through educational robotics: A model for creative computational problem solving. *International Journal of STEM Education*, *7*(1), 39. Advance online publication. doi:10.118640594-020-00238-z

Chookaew, S., Howimanporn, S., Pratumsuwan, P., Hutamarn, S., Sootkaneung, W., & Wongwatkit, C. (2018). Enhancing High-School Students' Computational Thinking with Educational Robotics Learning. In *Proceedings - 2018 7th International Congress on Advanced Applied Informatics, IIAI-AAI 2018* (pp. 204–208). Institute of Electrical and Electronics Engineers Inc. 10.1109/IIAI-AAI.2018.00047

Clark, R., Kirschner, P., & Sweller, J. (2012). Putting students on the path to learning: The case for fully guided instruction. *American Educator*, *36*(1), 6–11.

Collins, A. (2006). Cognitive apprenticeship. In R. K. Sawyer (Ed.), *The Cambridge handbook of the learning sciences* (pp. 47–60). Cambridge University Press.

Constantinou, V., & Ioannou, A. (2018). *Development of computational thinking skills through educational robotics* [Paper presentation]. *13th European Conference On Technology Enhanced Learning*, Leeds, UK.

Csizmadia, A., Curzon, P., Dorling, M., Humphreys, S., Ng, T., Selby, C., & Woollard, J. (2015). *Computational thinking-A guide for teachers*. Computing at School. https://eprints.soton.ac.uk/424545/1/150818_Computational_Thinking_1_.pdf

Cutumisu, M., Adams, C., & Lu, C. A. (2019). Scoping Review of Empirical Research on Recent Computational Thinking Assessments. *Journal of Science Education and Technology*, *28*(6), 651–676. doi:10.100710956-019-09799-3

Deng, W., Pi, Z., Lei, W., Zhou, Q., & Zhang, W. (2019). Pencil Code improves learners' computational thinking and computer learning attitude. *Computer Applications in Engineering Education*, *2020*(28), 90–104. doi:10.1002/cae.22177

Durak, H. Y., & Saritepeci, M. (2018). Analysis of the relation between computational thinking skills and various variables with the structural equation model. *Computers & Education*, *116*, 191–202. doi:10.1016/j.compedu.2017.09.004

Eteokleous, N., Neophytou, R., Kolani, E., & Christodoulou, C. (2020). The case of the Robotics Academy@ Frederick University: 21st Century Skills Developed through a Non-formal Educational Setting. *International Conference in Open & Distance Learning*, *10*(1B), 162-170. 10.12681/icodl.2434

Eteokleous-Grigoriou, N., & Psomas, C. (2013). Integrating Robotics as an Interdiscipli-nary-Educational Tool in Primary Education. In R. McBride & M. Searson (Eds.), *Proceedings of Society for Information Technology & Teacher Education International Conference 2013* (pp. 3877-3881). Chesapeake, VA: AACE.

Fagerlund, J., Häkkinen, P., Vesisenaho, M., & Viiri, J. (2021). Computational thinking in programming with scratch in primary schools: A systematic review. *Computer Applications in Engineering Education*, *29*(1), 12–28. doi:10.1002/cae.22255

Fessakis, G., Komis, V., Mavroudi, E., & Prantsoudi, S. (2018). Exploring the scope and the conceptualization of computational thinking at the K-12 classroom level curriculum. In M. S. Khine (Ed.), *Computational thinking in the STEM disciplines: Foundations and research highlights* (pp. 181–212). Springer. doi:10.1007/978-3-319-93566-9_10

García-Valcárcel-Muñoz-Repiso, A., & Caballero-González, Y. (2019). Robotics to develop computational thinking in early Childhood Education. *Comunicar. Media Education Research Journal*, *27*(1).

Ge, X., & Land, S. (2003). Scaffolding students' problem-solving processes in an ill-structured task using question prompts and peer interactions. *ETR&D*, *51*(1), 21–38. doi:10.1007/BF02504515

Giang, C., Piatti, A., & Mondada, F. (2019). Heuristics for the development and evaluation of educational robotics systems. *IEEE Transactions on Education*, *62*(4), 278–287. doi:10.1109/TE.2019.2912351

Greene, B., & Land, S. (2000). A qualitative analysis of scaffolding use in a resource-based learning environment involving the World Wide Web. *Journal of Educational Computing Research*, *23*(2), 151–179. doi:10.2190/1GUB-8UE9-NW80-CQAD

Grover, S., & Pea, R. (2013). Computational Thinking in K–12 A Review of the State of the Field. *Educational Researcher*, *42*(1), 38–43. doi:10.3102/0013189X12463051

Ioannou, A., & Makridou, E. (2018). Exploring the potentials of educational robotics in the development of computational thinking: A summary of current research and practical proposal for future work. *Education and Information Technologies*, *23*(6), 2531–2544. doi:10.100710639-018-9729-z

Kazimoglu, C., Kiernan, M., Bacon, L., & Mackinnon, L. (2012). A Serious Game for Developing Computational Thinking and Learning Introductory Computer Programming. *Procedia: Social and Behavioral Sciences*, *47*, 1991–1999. doi:10.1016/j.sbspro.2012.06.938

Kim, S., Yang, J. W., Lim, J., Lee, S., Ihm, J., & Park, J. (2021). The impact of writing on academic performance for medical students. *BMC Medical Education*, *21*(1), 1–8. doi:10.118612909-021-02485-2 PMID:33468118

Kirschner, P., Sweller, J., & Richard, E. C. (2006). Why minimal guidance during instruction does not work: An analysis of the failure of constructivist, discovery, problem-based, experiential and inquiry-based teaching. *Educational Psychologist*, *41*(2), 75–86. doi:10.120715326985ep4102_1

Kotsopoulos, D., Floyd, L., Khan, S., Namukasa, I. K., Somanath, S., Weber, J., & Yiu, C. (2017). A pedagogical framework for computational thinking. *Digital Experiences in Mathematics Education*, *3*(2), 154–171. doi:10.100740751-017-0031-2

Lu, J., & Fletcher, G. (2009). Thinking about computational thinking. *ACM SIGCSE Bulletin*, *41*(1), 260–264. doi:10.1145/1539024.1508959

McNeill, K., Lizotte, D., Krajcik, J., & Marx, R. (2006). Supporting students' construction of scientific explanations by fading scaffolds in instructional materials. *Journal of the Learning Sciences*, *15*(2), 153–191. doi:10.120715327809jls1502_1

Menary, R. (2007). Writing as thinking. *Language Sciences*, *29*(5), 621–632. doi:10.1016/j.langsci.2007.01.005

Nückles, M., Hübner, S., Dümer, S., & Renkl, A. (2010). Expertise reversal effects in writing-to-learn. *Instructional Science*, *38*(3), 237–258. doi:10.100711251-009-9106-9

Papadopoulos, P., Demetriadis, S., Stamelos, I., & Tsoukalas, I. (2011). The value of writing-to-learn when using question prompts to support web-based learning in ill-structured domains. *Educational Technology Research and Development*, *59*(1), 71–90. doi:10.100711423-010-9167-0

Papert, S. (1993). *Mindstorms: children, computers, and powerful ideas* (2nd ed.). Basic Books.

Ponticorvo, M., Rubinacci, F., Marocco, D., Truglio, F., & Miglino, O. (2020). Educational Robotics to Foster and Assess Social Relations in Students' Groups. *Frontiers in Robotics and AI*, *7*, 78. doi:10.3389/frobt.2020.00078 PMID:33501245

Poulakis, E., & Politis, P. (2021). *Computational Thinking Assessment: Literature Review. In Research on E-Learning and ICT in Education: Technological, Pedagogical and Instructional Perspectives*. Springer International Publishing., doi:10.1007/978-3-030-64363-8_7

Rodriguez del Rey, Y. A., Cawanga Cambinda, I. N., Deco, C., Bender, C., Avello-Martínez, R., & Villalba-Condori, K. O. (2020). Developing computational thinking with a module of solved problems. *Computer Applications in Engineering Education*, 1–11. doi:10.1002/cae.22214

Rogers, C., & Portsmore, M. (2004). Bringing engineering to elementary school. *Journal of STEM Education: Innovations and Research*, *5*(3&4), 17–28.

Rosenshine, B., Meister, C., & Chapman, S. (1996). Teaching students to generate questions: A review of the intervention studies. *Review of Educational Research*, *66*(2), 181–221. doi:10.3102/00346543066002181

Saidin, D., Khalid, F., Martin, R., Kuppusamy, Y., & Nalini, A. P. (2021). Benefits and Challenges of Applying Computational Thinking in Education. *International Journal of Information and Education Technology*. *11*(5), 248–254. .11.5.1519 doi:10.18178/ijiet.2021

Sapounidis, T., & Alimisis, D. (2021). Educational Robotics Curricula: Current Trends and Shortcomings. In M. Malvezzi, D. Alimisis, & M. Moro (Eds.), *Education in & with Robotics to Foster 21st-Century Skills. Studies in Computational Intelligence* (Vol. 982). Springer. doi:10.1007/978-3-030-77022-8_12

Schmidt, H., Loyens, S., Van Gog, T., & Paas, F. (2007). Problem-based learning is compatible with human cognitive architecture: Commentary on Kirschner, Sweller, and Clark (2006). *Educational Psychologist*, *42*(2), 91–97. doi:10.1080/00461520701263350

Settle, A., & Perkovic, L. (2010). *Computational Thinking across the Curriculum: A Conceptual Framework*. Technical Reports.

Shanmugam, L., Yassin, F., & Khalid, F. (2019). Enhancing students' motivation to learn computational thinking through mobile development module (M-CT). *International Journal of Engineering and Advanced Technology*, *8*(5), 1293–1303.

Stewart, T., Myers, A., & Culley, M. (2010). Enhanced learning and retention through "writing to learn" in the psychology classroom. *Teaching of Psychology*, *37*(1), 46–49. doi:10.1080/00986280903425813

Stewart, W., Baek, Y., Kwid, G., & Taylor, K. (2021). Exploring Factors That Influence Computational Thinking Skills in Elementary Students' Collaborative Robotics. *Journal of Educational Computing Research*, *59*(6), 1208–1239. Advance online publication. doi:10.1177/0735633121992479

Sweller, J., Kirschner, P., & Clark, R. E. (2007). Why minimally guided teaching techniques do not work: A reply to commentaries. *Educational Psychologist*, *42*(2), 115–121. doi:10.1080/00461520701263426

Wecker, C., & Fischer, F. (2011). From guided to self-regulated performance of domain general skills: The role of peer monitoring during the fading of instructional scripts. *Learning and Instruction*, *21*(6), 746–756. doi:10.1016/j.learninstruc.2011.05.001

Wilhelm, J. (2007). *Engaging readers and writers with inquiry: Promoting deep understanding in language arts and the content areas with guiding questions*. Scholastic Teaching Resources.

Wing, J. M. (2006). Computational thinking. *Communications of the ACM*, *49*(3), 33–35. doi:10.1145/1118178.1118215

Wing, J. M. (2008). Computational thinking and thinking about computing. *Philosophical Transactions - Royal Society. Mathematical, Physical, and Engineering Sciences*, *366*(1881), 3717–3725. doi:10.1098/rsta.2008.0118 PMID:18672462

Wing, J.M. (2011). Research notebook: Computational thinking-What and why. *The Link Magazine, 6*, 20-23.

Wing, J. M. (2017). Computational thinking's influence on research and education for all. *Italian Journal of Educational Technology*, *25*(2), 7–14. doi:10.17471/2499-4324/922

Yadav, A., Zhou, N., Mayfield, C., Hambrusch, S., & Korb, J. T. (2011). Introducing computational thinking in education courses. *Proceedings of the 42nd ACM technical symposium on Computer science education-SIGCSE0000'11*, 465–470. 10.1145/1953163.1953297

KEY TERMS AND DEFINITIONS

Abstraction: The process of creating something simple from something complicated by leaving out the irrelevant details, finding the relevant patterns, and separating ideas from tangible details.

Algorithm: Algorithm is a step-by-step sequence of finite and unambiguous instructions for carrying out a process.

Collaboration Script: They organize the collaboration process of two or more people, entities, or organizations working together to complete a task or achieve a goal. Specifically, the collaboration scripts organize the task into a sequence of phases, assigning student roles and tasks, and explicating the specificities of role-playing and task development. They aim to trigger peer interaction and to engage team members in fruitful and productive learning.

Computational Thinking (CT): CT is the thought processes involved in formulating a problem and expressing its solution(s) in such a way that a computer -human or machine- can effectively carry out. It is a fundamental skill for all individuals and its key components are analysis, algorithm, decomposition (breaking down into parts), patterns, abstraction, generalization, evaluation, and debugging.

Debugging: Debugging is the process of detecting and resolving existing and potential errors that prevent the correct operation of a system

Educational Robotics (ER): ER is a powerful, flexible, teaching and learning tool, encouraging students to construct and control robots using specific programming languages. They help students to turn from passive into active learners as they construct new knowledge and develop fundamental skills by collaborating with their peers and acting as researchers.

Fading (Fade-Out): The fading strategy refers to decreasing the level of assistance or guidance needed to complete a task or activity. It is usually paired with prompts.

Generalization: Generalization is transferring a problem-solving process to a wide variety of problems.

Modularity: The development of autonomous processes which encapsulate a set of often used commands that perform a specific function and might be used in the same or different problems.

Chapter 6
Developing Storytelling Activities Supporting Computational Thinking Using Educational Robots

Karin Tengler
University of Teacher Education Lower Austria, Austria

ABSTRACT

Computational thinking and educational robotics are two terms that are increasingly encountered in the literature when discussing the introduction of digital education in primary school. Computer science education is becoming increasingly important, even for young learners. Since its planned implementation in the Austrian primary school curriculum, it became necessary to find appropriate educational required concepts and methods for its implementation. Therefore, a research project was developed to investigate the use of programmable robots combined with the method of digital storytelling. To combine theoretical knowledge with pedagogical practice, the approach of educational design research was chosen. The study was conducted in five classes of primary schools. The aim was to investigate how to design a robotics-based learning environment supporting computational thinking skills. The purpose of this chapter is to give an overview of the findings of this research project.

INTRODUCTION

Computer science education, digital education, and computational thinking have become important terms when discussing education in the 21st century. Digital education has been defined as a key competence for lifelong learning (Kids, 2019) that learners should possess today. Over the years, several competence models, such as DigcompEDU, TPACK (Brandhofer et al., 2018), or digikompP (BMBWF, 2021), have been established for both teachers and students. What is already common practice in many countries, familiarizing the youngest in the education system with computer science education (Rich et al., 2019), will become an issue with its anchoring as cross-curricular competence development in the following

DOI: 10.4018/978-1-7998-7443-0.ch006

Austrian primary school curriculum (Kern, 2020). For this reason, the implementation and didactics of digital education are becoming increasingly important. Still, the transfer of innovations and didactic approaches into the classroom is one of the biggest challenges. To introduce computer science education in schools, the playful use of educational robotics offers a possible approach by providing activities that combine analog and digital concepts (Tengler et al., 2021a; Tengler et al., 2021b).

Through research, experimentation, and creative design, students can simulate and visualize phenomena in playful learning environments. Educational robotics (ER) provides access and age-appropriate opportunities to explore students' lives and work digitally. Studies show that playful programming methods with educational robots can promote problem-solving thinking and student motivation (Atmatzidou & Demetriadis, 2016). In addition to traditional approaches to robotics, students are particularly motivated when robotics activities are offered interdisciplinary (Angeli & Valanides, 2020). Although there are already successful individual initiatives (Himpsl-Gutermann et al., 2017) in Austria, where robotics is used in the classroom to promote computational thinking, programmable robots are not currently widely used in schools. However, a close link between theory and practice seems important in computer science education. Initial concepts can be further developed to lead to sustainable innovation and implementation in the classroom (Esteve-Mon et al., 2019). For this reason, a research project with programmable robots is being conducted that focuses on the learning and teaching method of digital storytelling to support the implementation of computer science education using programmable robots and to integrate computational thinking into the classroom across subjects. Since the introduction of computer science into primary education requires practical and theoretical knowledge, educational design research was chosen for investigation, allowing close interaction between researchers and practitioners.

This chapter gives an overview of the theoretical framework. It describes the interdisciplinary approach of the teaching and learning method of digital storytelling. The next step presents the findings on the aspects of enhancing computational thinking skills identified in a previous study of this research project (Tengler et al., 2021b) and confirmed in the following sub-cycle. At last, it answers the research question by proposing how to design a robotics-based learning environment using storytelling supporting computational thinking skills.

THEORETICAL BACKGROUND

Digital Storytelling

The potential of stories to convey information, explain problems, and evoke emotions has found its way into various scientific disciplines under the term storytelling. Digital storytelling is a teaching and learning method that combines traditional storytelling with digital technologies (Robin, 2006). The aim is to create digital stories based on a combination of different artifacts. Although digital stories result from this process, digital storytelling's pedagogical value lies in the process rather than the final product (Otto, 2020). Robin (2006) classifies digital stories into three categories, "personal narratives, stories that examine historical events and stories, that are primarily used to inform or instruct" (p.2). Digital storytelling can be integrated into teaching practice in a variety of ways (Robin, 2006). Kordaki et al. (2017) analyzed the stages of digital storytelling. They proposed an initial framework that highlights the relationship between the stages of digital storytelling and the development of computational thinking skills. By creating interactive stories, students are enabled to improve their digital skills, which can help improve their problem-solving skills. In addition, this method supports the promotion of creativity (Robin, 2016) and other 21st century skills such as collaboration, communication, and critical thinking. There is evidence in educational research that digital storytelling positively affects students' motivation and performance in learning. Moreover, digital storytelling activities that promote problem-solving thinking create a connection between the real world and the classroom and make learning more fun (Parsazadeh et al., 2020). Furthermore, digital storytelling affects creative thinking, especially when the method is applied across disciplines (Yang et al., 2020). Stork (2020) showed in his study that the use of robots and digital storytelling promotes design thinking and contributes to the building of 21st century skills. Leoste et al. (2020) described the great potential of educational robotics in their study because it combines the advantages of gamification and storytelling.

Computational Thinking

Computational thinking (CT), first mentioned by Papert (1980), inventor of the programming language LOGO and founder of constructionism, is considered an essential aspect of developing problem-solving thinking skills. Since the publication of Jeannette Wing's article on "Computational Thinking" (Wing, 2006) many educators and researchers have agreed that computational thinking should be integrated into the teaching of students at all levels of education as a fundamental 21st century skill (Barr & Stephenson, 2011; Lee et al., 2014). In their literature review, Shute et al. (2017) emphasize that CT skills are not to be equated with programming skills. Computational thinking should provide suitable approaches ready to solve everyday problems, whether theoretical or practical. Meanwhile, several definitions and concepts of computational thinking have been proposed (Bitesize, 2017). However, computational thinking is about solving problems using concepts and strategies, whereby the formulation of the problem is seen as an important part of the problem-solving process (Grover & Pea, 2018). Wing (2006) mentions computational thinking "involves solving problems, designing systems, and understanding human behavior by drawing on the fundamental concepts of computer science" (p.33). Brennan and Resnick's (2012) framework categorizes computational thinking into three dimensions: "concepts, practices, and perspectives". Another popular definition of computational thinking developed by the International Society for Technology in Education and the Computer Science Teachers Association (ISTE & CSTA, 2011) states that "Computational thinking (CT) is a problem-solving process that includes (but is not limited to) the following characteristics" (ISTE & CSTA, 2011):

1. Formulating problems in a way that enables us to use a computer and other tools to help solve them
2. Logically organizing and analyzing data
3. Representing data through abstractions, such as models and simulations
4. Automating solutions through algorithmic thinking
5. Identifying, analyzing, and implementing possible solutions to achieve the most efficient and effective combination of steps and resources
6. Generalizing and transferring this problem-solving process to a wide variety of problems

Along the same lines, BBC Bitesize (2017, p.1) formulates "the four cornerstones of computational thinking": First, problems are broken down into smaller ones ("decomposition"), then it is considered whether a solution for a similar problem is available ("pattern recognition"). After that, only the basic information remains ("abstraction"). Next, a solution strategy can be designed ("algorithm").

Grover and Pea (2018) mentioned the following components that CT practices include:

- Problem decomposition
- Creating computational artifacts
- Testing and debugging
- Iterative refinement (incremental development)
- Collaboration & creativity (part of broader twenty-first century skills)

Computational thinking is increasingly becoming a key competence for future scientific and technological progress, and it is now more necessary than ever to familiarize learners with computational thinking (Kids, 2019). In the last decade, the definitions and components of CT have been expanded to include aspects of collaboration and creativity (Grover & Pea, 2018). For Humbert et al. (2019) in addition to an adequate implementation of computer science as a creative design area for problem-solving, corresponding didactic concepts are also required.

Researchers emphasize the need to implement algorithmic thinking and foster problem-solving skills in young learners to achieve successful computational thinking and develop an early interest in technical careers (Kong et al., 2019). Developing a definition or approach to computational thinking appropriate for K-12 is particularly challenging given that there is not yet a widely agreed definition of computational thinking. However, researchers agree that K-12 students can learn problem-solving skills in various disciplines (Barr & Stephenson, 2011). Although computational thinking is mostly situated in computer science, particularly in the context of programming, the approach to this competence is increasingly being pursued in several other fields. Lee et al. (2014) described that many everyday activities could be made more efficient when performing them can apply computational thinking.

To foster computational thinking even at the primary school level, problem-based learning (PBL) is one of the most common learning approach (Conde et al., 2021), primarily when technological tools are used to achieve a successful learning process. A similar positive outcome is obtained by integrating PBL into the teaching programming skills (Lye & Koh, 2014). Problem-based learning is an instructional method that uses real-life scenarios and topics familiar to students as a context to develop critical thinking and problem-solving skills (Eteokleous, 2018). Students are encouraged to collaborate and engage in problem-solving activities to gain problem-solving thinking skills and expertise in computer science education. Educational robotics allows teachers to create engaging lessons that foster interest in technology, student motivation, and computational thinking (Benitti, 2012). Problem-based learning activities enable students to design concepts and hypotheses and solve with the robots during the intervention. Students actively influence the learning process as they develop their problem-solving skills. In summary, active approaches benefit computer science education because of their flexibility (Conde et al., 2021).

Educational Robotics

Learning and innovation skills (Batelle for Kids, 2019) are increasingly recognized as those that prepare students for the more complex living and working environments of the 21st century. These skills provide students with the mental processes needed to adapt to and excel in a modern work environment. A focus on creativity, critical thinking, communication, and collaboration is essential to prepare students for the future (Fadel et al., 2015). Educational robotics is a good resource for developing problem-solving skills and students' digital literacy (Angeli & Valanides, 2020). The robot is described as a tangible object that interacts with the environment through programmed instructions, serves as a tool for developing cognitive skills, through play, creativity, or solving challenges (Esteve-Mon et al. (2019). To promote creativity and computational thinking, it is crucial to construct new content from existing. There should be the opportunity to allow own ideas, collaboration, and multiple solutions (Brandhofer, 2017). The use of robots contributes to motivation in the classroom. It enables designing activities that promote computational thinking and contribute to achieving competencies such as social skills, collaboration, and teamwork (Benitti, 2012).

Playful programming with programmable robots is seen as a promising method to promote problem-solving already in young students (Atmatzidou & Demetriadis, 2016) and is seen as a possible prospect for an interest in technical professions to develop later on (Himpsl-Gutermann et al., 2017). Besides, it is stated that children from the third grade onwards have no problems understanding and using the fundamental elements of a programming language for controlling activity flows and naming entities (Futschek, 2016; Weigend, 2009). The haptic use of programmable floor robots makes them particularly suitable for younger students (Himpsl-Gutermann et al., 2017). The use of programmable robots in education has become more and more important in recent years. It has been shown that robotics can increase students' motivation because it enables them to work in an integrative way while developing several additional skills as Ching et al. (2019) mentioned "Robotics activities in education offer opportunities for students to explore, create, and apply knowledge to solve real-world problems" (p.2).

Several research studies (Stork, 2020; Tzagkaraki et al., 2021) constitute that educational robotics is an attractive approach for developing young children's computational thinking because children can directly interact with a robot and observe the immediate effects of their interactions on the robot's behavior (Stoeckelmayr et al., 2011). The fact that the robot moves and carries out commands from the students strengthens an empathetic relationship towards it (Fanny et al., 2020). The use of robotics in education increases interest and curiosity and promotes additional skills such as initiative, responsibility, autonomy, creativity, and teamwork (Arís & Orcos, 2019). Students can create codified digital languages and work in groups (Kandlhofer & Steinbauer, 2016).

Angeli and Valanides (2020) presented a study that investigated the effects of learning with Bee-bots, which are programmable floor robots, on young learner's computational thinking in the context of scaffolding techniques. The term scaffolding is used in constructivist learning theories to describe the support of a learning process in an educational-psychological context providing instructions, thinking prompts, and other assistance (Stangl, W., 2021). The study reported statistically significant learning gains and "that both boys and girls benefited from the scaffolding techniques". However, it was also found "that boys benefited more from the individualistic, kinesthetic, and spatial activity with the cards [...], while girls benefited more from the collaborative writing activity" (Angeli and Valanides, p. 10). Nevertheless, there is very little research on scaffolding and interventions with educational robotics in early childhood education (Botički et al., 2018).

Dittert et al. (2019) observed that the structure given by this scenario supported creativity, inventing a story and creating characters, and collaboration, so it is planned to use this activity sequence as a first framework to support creativity and collaboration using technology in a kindergarten. Educational robotics (ER) becomes more and more popular at the research and educational level nowadays, with an increasing number of teachers and schools using learning activities with robots to teach Science, Technology, Engineering, and Mathematics (STEM).

Another interesting example is documented by Estivill-Castro (2019). The author has explored three lessons using learning robots based on inquiry and problem-based learning principles to highlight the role of mathematics in STEM. In addition, they are designed to pique students' curiosity and develop mathematical thinking and problem-solving skills.

An example of the programmable robots that are particularly suitable for primary education is the Ozobot, which will be described in more detail in the learning environment paragraph. Working with the Ozobot makes it possible to achieve competencies in media and computer science, and technology playfully and promote 21st century skills such as creativity and critical thinking. However, due to its "relatively simple application" (Picka et al., 2020), young learners can also work with the little robot from primary school onwards and gain their first insights into informatics concepts. Since the Ozobots can be used at different levels, they are suitable for simple programming to more complex tasks. Algorithmic thinking is particularly encouraged by using the Ozoblockly programming environment (Fojtik, 2017). Furthermore, there are no limits to the imagination when creatively dressing up the Ozobot. For example, he can slip into different roles of characters (Fanny et al., 2020). These creative activities provide numerous occasions for storytelling activities (Tengler, 2020).

However, despite the growing research interest and efforts in educational robotics, it seems that not enough attention is paid to important aspects such as didactic application and appropriate teaching methodology. For this reason, this research project aimed at designing a robotics-based learning environment that promotes computational thinking skills in primary schools, as far as developing didactic approaches and design principles for pre-service teachers and in-service teachers training. The following research question emerged:

How to design an interdisciplinary learning environment focusing on robotics and storytelling to support primary school students (aged 8-10) to promote their computational thinking skills?

THE METHODOLOGY

Educational Design Research

Educational design research is the systematic study of the design, development, and evaluation of educational interventions that contributes to the solution of complex problems in the educational landscape and aims to expand knowledge about the characteristics of these interventions and the processes of their design and development (Van den Akker et al., 2006). The commitment to developing theoretical insights and practical solutions simultaneously in real-world contexts distinguishes educational design research from other forms of research. In this way, different solutions to educational products, processes, and strategies can be developed and explored (McKenney & Reeves, 2018). "By its nature, design research is relevant for educational practice (and therefore also for educational policy) as it aims to develop research-based solutions for complex problems in educational practice or to develop or validate theories about processes of learning and teaching" (Plomp & Nieveen, 2013, p.17). The starting point for design research is educational problems for which no or only a few validated principles are available to structure and support design and development activities (Bakker, 2018). Educational design research aims to solve important problems facing educational practitioners and discover new knowledge that can be applied to the work of others facing similar issues. Working systematically and simultaneously toward these two goals is perhaps the most defining characteristic of educational design research (McKenney & Reeves, 2018). As reflected in the generic model, the entire design research process comprises all sub-cycles together. The learning environment is designed, developed, evaluated, and constantly improved (McKenney &

Figure 1. Design of the research study
(Source: Tengler)

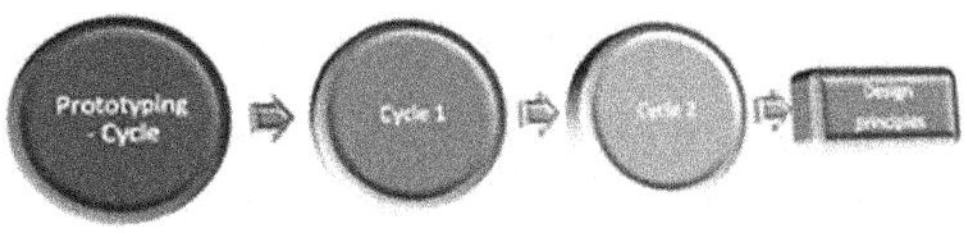

Reeves, 2018). This is exactly what the author wants to aim at, to learn about developing a robotics-based learning environment supporting computational thinking skills in primary school, developing an intervention, and passing on the knowledge gained to teachers in the context of teacher training.

If the design research is seen as a developmental study, the following phases are needed to be done (Plomp & Nieveen, 2013):

- preliminary research: needs and context analysis, review of literature, development of the theoretical framework for the study
- development or prototyping phase: designing and evaluating the prototype of the intervention
- assessment phase: evaluation to conclude whether the intervention meets the pre-determined specifications

Should an innovative intervention development is set as the research goal, the research design requires finding design principles to define the characteristics of an intervention against which the purpose or the outcome is to be determined (Plomp & Nieveen, 2007). That means that in the first phase of design conception, the design principles are generated representing theory-based principles of action concreted in the teaching arrangement (Hußmann et al., 2013). The design is further developed in testing, evaluation, and reflection until the final intervention is available. Preliminary design principles are necessary to create and develop a prototype of the intervention as a basis for the design-based research approach to assess whether and how these design principles would need to be modified based on the findings of the formative evaluation activities (Van den Akker et al., 2006). Although other terms have been introduced, the term 'design principles' is probably the most widely used to characterize the kind of prescriptive theoretical understanding developed through educational design research (McKenney & Reeves, 2018).

The Research Design of the Study

Educational Design Research (Bakker, 2018) was chosen because, in this study, the implementation of an innovation is to be investigated. This is preferably examined in the interplay of theory and practice. As mentioned above, design research is relevant to educational practice because it aims to develop research-based solutions to complex problems of educational practice (Van den Akker et al., 2006; McKenney & Reeves, 2018). The research design based on a prototype cycle and two further cycles of field tests in which the learning environment is tested in selected primary schools (Figure 1).

This research project is intended to contribute to the body of knowledge on an interdisciplinary introduction of computer science at the primary level. The robotics-based storytelling activities provide a viable approach, so the results are of practical importance to teachers and researchers. Possibly, this concept can also serve as a basis in studies that aim to measure computational thinking.

Prototyping Phase

The prototyping cycle included literature research, development of preliminary design principles (Tengler et al., 2021c) prototype development, and field testing. The evaluation of this cycle was conducted in the alpha testing phase (McKenney & Reeves, 2018). Primary school students (n=19) in a 4th grade class participated in the field test, which took place in a rural primary school. The purpose of the prototyping cycle was to evaluate whether the intervention met the requirements, was feasible, and could be implemented effectively. The results of the prototyping cycle showed that this learning environment, combined with the storytelling method, is a good way to implement computer science education in elementary schools. After the analysis, the intervention was further developed by researchers and practitioners, and small shortcomings were improved. Minor changes were made in the teaching materials of the learning environment. In addition, the research team decided to incorporate digital storytelling to promote computational thinking further. In addition, a case study conducted by the author during the prototyping cycle showed that aspects of creativity development could be observed through playful programming with the Ozobots (Tengler et al., 2019).

The First Cycle – the Beta-Testing

After completing the prototyping cycle, a tryout in selected classes was conducted. Tryouts investigate how interventions work, what participants think or feel about them, and what results they deliver. The evaluation of this cycle is called beta-testing (McKenney & Reeves, 2018), which aims to identify opportunities for further development and optimize the learning environment. Beta-testing is considered an important phase of the study regarding the further development of the learning environment. The first cycle of the design research consisted of two sub-studies. One study took place in three primary school classes. It explored how relevant and beneficial the intervention was perceived and experienced by practitioners and what intended and unintended learning activities could be observed. A total of 43 students, 24 male and 19 female students, took part in the study, aged 8-10 years. In addition, 12 teachers were involved as participant observers, and seven teachers were involved as non-participant observers. The schools were located in both rural and urban areas. Evaluation and reflection of this cycle show that teachers appreciate the use of learning robots combined with storytelling. The results of this sub-study can be found in Tengler et al. (2021a). Another sub-study of this research cycle, conducted as a case study in a primary school with 16 students, found that this interdisciplinary approach enhances the development of computational thinking skills (Tengler et al., 2021b).

The second cycle is currently being analyzed. Participants are primary school teachers. This quantitative study is about their perceptions and intention to use this interdisciplinary approach. Nevertheless, its first insights also showed that the method is well positively received. Evaluations and analyses will follow and be published at a later date.

The Participants and Data Collection

To evaluate the first two cycles of the research project, several iterative cycles of the design research process were carried out in three Austrian primary schools, consisting of students of three 4th grade classes (n=45) and two 3rd grade classes (n=33). In summary, a total of 78 students took part, 43 boys and 35 girls.

Classroom observations, expert interviews, student reflections, group discussion, and student-created artifacts provided the data for the research project. The observations were partially conducted as video recordings, but this was not permitted in all schools. The interviews were conducted as semi-structured interviews and evaluated using the content analysis (Mayring, 2015). Group discussion was conducted with teachers who were non-participating observers. The collected data and the artifacts were analyzed and interpreted. For the analysis, the framework of concepts and practices of computational thinking (ISTE & CSTA, 2011) was used.

THE LEARNING ENVIRONMENT

The robot-based learning environment developed and investigated in this research project is part of the "Fairy Tale Computer Science" project. This project aims to create and investigate didactic approaches and learning materials to be made available to teachers and pre-service teachers in the context of teacher training. The design of the intervention, which is the key element of the design-based research, aims to use programmable robots in an interdisciplinary approach in primary education. Furthermore, the problem-based approach (Conde et al., 2021) is used by giving the students an inquiry-based task. The development and evaluation of the first version of the learning environment, including two units, is described in Tengler et al. (2021a).

Since the focus was on promoting problem-solving skills and creative learning the familiar topic "fairy tales" was chosen. The project "Fairy Tale Computer Science" sees itself as a work in progress and is intended to provide teachers a collection of ideas to establish a connection between digital media and the promotion of literacy in the context of fairy tales. The "Four P's of Creative Learning" (project, peers, passion, play) (Resnick, 2014) was used as a framework. A playful approach, collaboration, and creative learning were important aspects for the design. The robot chosen for this learning environment was the Ozobot. An Ozobot (Figure 2) is a programmable robot about 2.5 cm small and moves on two wheels. Ozobots use sensors to follow lines and read color codes. There are two types of Ozobots, the Ozobot Bit and the Ozobot Evo. The Ozobot Evo differs by additional features, such as the distance sensor. You can draw or stick the codes with markers or stickers so that the robot can recognize and follow the commands. There are two different ways for coding: Programming with codes and programming with the visual programming language Ozoblockly. The Ozobot's simple programming makes it particularly suitable for primary school children. Its unique programming style especially encourages the students' creativity. Moreover, they can be used at different levels, from simple programming to more complex tasks and programming solutions (Brandhofer, 2017).

Programming With Color Codes

It can be programmed with the colors red, blue, green, and black. Lines and codes can easily be drawn with pens. With the help of the color codes, which can be found on a code table (Figure 3), the robot can change speed and direction and perform cool moves, like zigzag, and dance moves. The exact drawing of the lines and color codes is necessary for the successful execution of the commands. Instead of drawing the codes, it is also possible to stick blue, green, and red glue dots on the lines. Since a free Ozobot app for Android or iOS is available, drawing lines and codes on a tablet is also possible.

Figure 2. The robot Ozobot
(Source: Tengler)

Programming With Ozoblockly

Another possibility to program the Ozobot is the block-based coding language Ozoblockly similar to Scratch. The programming language works on a computer, tablet, or smartphone. It impresses with its relatively simple operation and the simple transfer from the screen to the Ozobot, where the commands are loaded onto the Ozobot using color signals.

The Lesson Plans

The intervention consists of three units (see Table 1), an introduction to Ozobot programming and a lesson with the fairy tale Little Red Riding Hood. In the third unit, the pupils can invent a fairy tale themselves.

Figure 3. Examples of Ozobot codes
(Source: Tengler)

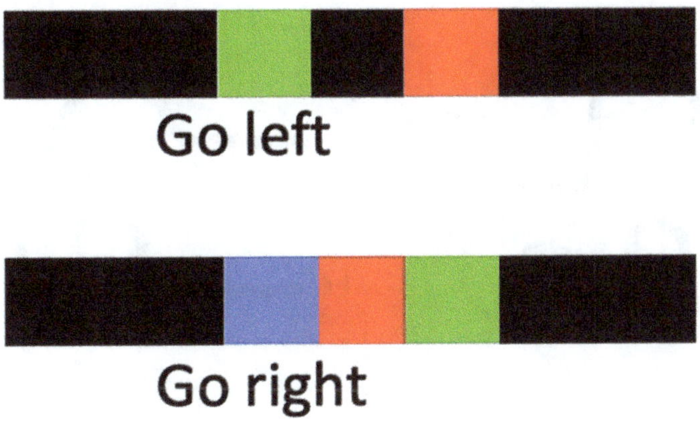

Table 1. Lesson plans

Units	Unit 1	Unit 2	Unit 3
Title	*Introduction – Ozobots*	*Fairy tale - Little Red Ring Hood*	*Creating a fairy tale*
Learning goals	• Getting to know the Ozobot robot • Getting to know how it works • Getting to know the programming	• Telling the fairy tale Little Red Riding Hood • Knowing the plot of the fairy tale • Graphically representing the plot • Programming the Ozobot according to the plot • Retelling the fairy tale using the programmed lines and codes	• Creating a fairy tale based on the given characters • Graphically representing the plot of the fairy tale • Including details of the plot in the plan (e.g., bridge, forest) • Programming the Ozobot according to the plot • Telling the fairy tale • Filming the story • Giving feedback to the other pupils
Activities	• Experimenting • Drawing lines • Using the programming codes	• Telling the fairy tale or solving a riddle • Drawing the path of Little Red Riding Hood • Using appropriate codes, such as turn right, slow down, or speed up • Retelling the fairy tale	• Reviewing the functions of the Ozobot • Making up a story • Drawing the story • Retelling the fairy tale • Filming the story • Giving feedback

(Sources: Tengler)

The first unit starts with a description of the Ozobot and how it works. This is elaborated with the students in a demonstration. After that, the first programming experiences are made with paper and pens. First, the lines are drawn, and the students observe how the Ozobot reacts. It is important to pay attention to precise drawing and a certain width (about 4-5 mm) of the lines. Among the most popular shapes drawn by the students are the circle, the spiral, and their own name.

After that, they begin to apply the codes. Learners are introduced to the first programming codes using the Ozobot basic card file (https://eis.ph-noe.ac.at) (see Figure 4). Mainly stickers are used for coding because the Ozobot responds better to them. To promote cooperation and communication, the students are divided into groups.

Figure 4. Ozobot-Basics Card file
(Source: Tengler)

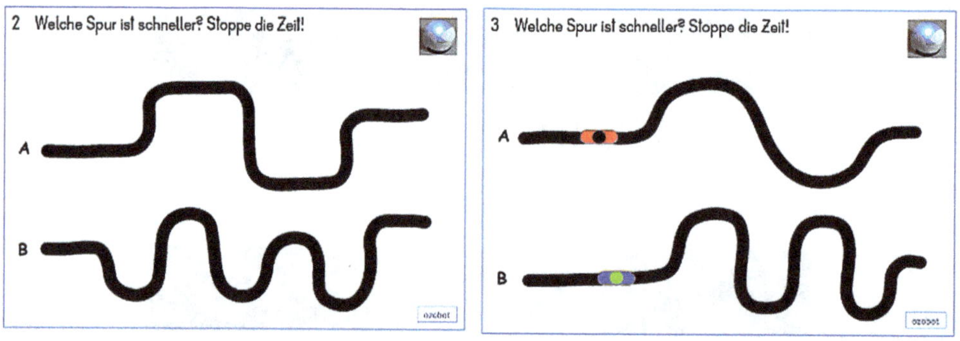

Figure 5. First experiences with the robot Ozobot
(Source: Tengler)

Afterward, the students could try out the lines and codes on shapes of their own choice and have their first hands-on experiences with the Ozobot (Figure 5).

In the second unit, the fairy tale plot is repeated to make sure that the students are familiar with the topic. Then the students get their problem-based task in which they have to depict the fairy tale plot graphically. Students are asked to draw lines for the Ozobot and the details of the story. Then the students' task is to program the Ozobot to follow Little Red Riding Hood's path (Figure 6) through the forest, past the wolf, to her grandmother. Several programming codes, such as turn right, speed up, etc., should be integrated. If there is time available, the students can also create appropriate costumes for the characters. Finally, the students decode the story because they retell the fairy tale using the lines and the codes. After that, each group can give feedback to the other groups and how they liked the lesson.

The third unit takes place according to the same principle. The aim is to make up a story and then depict it graphically and program it for the Ozobot. The students' assignment is to create the fairy tale based on the given characters (princess, dragon) and some facts (e.g., cross over the river, ...). How they design the environment around the Ozobot is up to the students.

Figure 6. Coding the Little Red Riding Hood
(Source: Tengler)

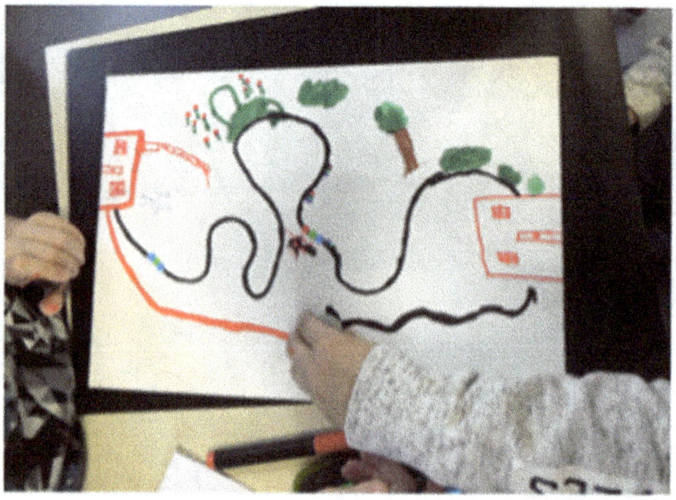

FINDINGS

Computational Thinking Practices

The data analysis shows that the use of learning robots combined with storytelling is well appreciated by both students and teachers and is a good possibility to implement computer science teaching in primary school in an interdisciplinary approach while supporting the development of computational thinking skills. Table 2 shows the CT practices (Grover & Pea, 2018), which can be observed as activities in this robotics-based learning environment.

When comparing the CT practices summarized by Grover and Pea (2018) with the identifiable activities during the study, it can be seen that these practices are recognizable in the activities, suggesting that computational thinking skills can be promoted by using this robotics-based learning environment. The subsequent paragraph shows aspects identified during the robot-based storytelling activities that promote computational thinking.

Table 2. CT practices - observed activities

CT Practices (Grover & Pea, 2018)	Observed Activities During the Storytelling and Programming Activities
Problem decomposition	Identifying the problem, analyzing which details of the plot are relevant to create the story
Creating computational artifacts	Depicting the story graphically using lines and codes for the Ozobot
Testing and debugging	Code and decode the story, trying out codes that best fit the plot
Iterative refinement (incremental development)	Intervening at any time in the graphical representation and designing the story, and changing details, carrying out the process of evaluation constantly
Collaboration & creativity (part of broader twenty-first century skills)	Work in groups, reaching a decision together, creating artifacts, putting ideas into practice

Table 3. Aspects enhancing CT skills

Aspect	Comments
Motivation and Enthusiasm	Through the playful programming of the Ozobot robot, the students worked with great motivation on their problem-based tasks. It was also clear from the children's feedback that they were much more enthusiastic about retelling or inventing the story by working with the robot. They said that telling stories without the robot was much more boring.
Orientation	What is particularly noticeable in the use of programmable floor robots is the promotion of orientation and spatial skills development. They have to imagine the design plan and decide how to plan the details of the plot on the sheet. Moreover, they determine where which storyline activity is placed.
Structuring	The task is broken down into individual work steps and executed one after another, corresponding to programming sequencing. An exact sequence of action in broad strokes had to be planned first.
Abstraction	The graphic representation abstracts the plot of the story. Since the students tell the story with the Ozobot and the codes, the story is decoded again. Programming and transferring the plot into a graphic representation stimulates algorithmic and logical thinking enormously.
Interactivity	An essential aspect is also that of interactivity. The pupils can intervene at any time in the presentation and design of the story and change details. As a result, the process of evaluation is constantly carried out.
Communication and Collaboration	Communication and collaboration are important aspects of achieving a common goal by developing a problem-solving strategy, discussing details of the story and its realization, and having an internal agreement within the group where responsibilities are defined and accepted.

Aspects Enhancing Computational Thinking Skills

Based on literature research and the findings of a previous study (Tengler et al., 2021b), the following aspects (Table 3) contributing to promoting computational thinking skills could be observed and identified in the first cycle.

Furthermore, two other competencies could be identified. Firstly, these were social competence, the ability to work in groups and divide, assign, and fulfill different roles in the process. The second was technological competence, using educational robots and digital media to record the story. Besides, it was possible to see that the graphic representation of the story particularly encouraged creativity, since the groups had ideas for the realization of the story.

Design Principles

To answer the research question on how an interdisciplinary learning environment focusing on robotics and storytelling to support primary school students to promote their computational thinking skills could be designed, the following design principles were identified. They should serve as guidelines to create a viable learning environment.

Choose Topics That Students Are Familiar With in a Cross-Curricular Approach

If the students work with familiar topics, they can concentrate better on acquiring new knowledge. This aspect also increases their motivation to solve their tasks and fosters their problem-solving thinking skills (Angeli & Valanides, 2020). The theme of fairy tales was well received and had a very inspiring effect.

Use Age-Appropriate Robots

An important decision is which robots are chosen for the learning environment (Leoste et al., 2020). It is important to decide that the chosen robot is easy to handle according to the age of the pupils. Since haptic perception is especially relevant for primary school students, floor robots are ideal for this purpose. The Ozobot robot used in this research project is very suitable for the primary school sector and meets all the requirements.

Provide Tasks That Encourage Problem-Solving Thinking

With problem-based learning and inquiry-based learning, two methods should be considered to plan such learning environments. Pupils should learn in a problem-oriented and inquiry-based way (Atmatzidou & Demetriadis, 2016), not by being given all the instructions, but by being able to discover solutions themselves, whereby it should be emphasized here to allow for a gradual increase in difficulty.

Consider Using a Playful Approach

When introducing computer science education, the playful approach has a special significance. It turned out that the programmable robots are particularly suitable for this purpose because they can teach complex topics playfully. The playful aspect enriches the combination of programming and storytelling (Leoste et al., 2020) and promotes collaboration and communication skills (Stork, 2020).

Encourage Learners to Create Their Own Ideas

It became apparent that working with the Ozobot and creating their stories inspired the students to develop their ideas. Creativity is perhaps the most important skill that students need to learn (Fadel et al., 2015), and it is the beginning of many innovations (Stork, 2020).

DISCUSSION

In designing robotics-based storytelling activities, the research team aimed to provide a viable approach for teachers and researchers to introduce computer science education and foster computational thinking. To combine practical with theoretical knowledge and derive appropriate insights for teaching, educational design research was chosen because it encourages continuous improvement with its iterative cycles (McKenney & Reeves, 2018). The intervention was evaluated during the prototyping cycle to determine whether it met the requirements, was feasible, and could be implemented effectively. In a further cycle using this approach, CT practices (Grover & Pea, 2018) were identified in the students' activities when they draw and code the stories. Previous studies of the iterative cycles of the research process showed that this learning environment is positively well-received by practitioners and researchers (Tengler, 2021), and it can activate creativity (Tengler et al., 2019).

The focus of the research was to develop a learning environment that supports children to gain CT skills to formulate problems, recognize patterns, find solutions (Bitesize, 2017) by playfully telling stories with programmable robots. As in the study of Parsazadeh et al. (2020), this learning environment also shows the increasing motivation when using the storytelling activities integrating a CT approach. Through the problem-based tasks, the graphical implementation of a fairy tale, a successful learning process (Conde et al., 2021) is also initiated, and critical thinking and problem-solving strategies developed (Eteokleous, 2018). The fact that robotics was chosen in connection with a topic familiar to the students contributes favorably. Due to the haptic perception (Himpsl-Gutermann et al., 2017) and that the robot obeys the commands of the students and moves (Fanny et al., 2020) through their graphic representation, this approach is particularly suitable for the primary level.

During the iterative cycles of the design-based research, aspects promoting CT could be identified and confirmed. These aspects include motivation, orientation, structuring, abstraction, interactivity and communication, and collaboration, which can contribute decisively to the development of problem-solving strategies. Similarly, what has been mentioned by other researchers (e.g. Esteve-Mon et al., 2019).

Finally, the design principles that have emerged from the research to date were noted. They provide guidelines to develop further robotics-based learning environments, possibly in a different context.

CONCLUSION AND FUTURE WORK

This chapter is intended to provide a constructive input to the design, construction, and programming of robots for learning. The described design-based research project was conducted to reach the following outcomes:

- to develop a robotics-based learning environment to provide an appropriate approach for educators to implement computer science education
- to show a connection of robotics-based storytelling activities with CT practices
- to identify aspects that can promote the development of CT skills using this method

Finally, the design principles that emerged from the study should serve as guidelines for the development of further didactic approaches in other subjects as well.

The research project results show that digital storytelling in combination with educational robots is indeed a promising didactic method to introduce computer science education in primary schools in an interdisciplinary approach. Furthermore, the design of this learning environment promotes the development of computational thinking and competences like creativity, communication, and collaboration.

It is planned to extend the implementation of the learning environment to several classes at primary school and conduct further research. In a future approach, it would be interesting to investigate to what extent digital storytelling is also possible with other programmable robots, what insights are gained from this, and whether these are comparable to the use of the Ozobot, especially concerning computational thinking.

REFERENCES

Angeli, C., & Valanides, N. (2020). Developing young children's computational thinking with educational robotics: An interaction effect between gender and scaffolding strategy. *Computers in Human Behavior, 105*, 105954. doi:10.1016/j.chb.2019.03.018

Arís, N., & Orcos, L. (2019). Educational robotics in the stage of secondary education: Empirical study on motivation and STEM skills. *Education Sciences*, *9*(2), 73. doi:10.3390/educsci9020073

Atmatzidou, S., & Demetriadis, S. (2016). Advancing students' computational thinking skills through educational robotics: A study on age and gender relevant differences. *Robotics and Autonomous Systems*, *75*, 661–670. doi:10.1016/j.robot.2015.10.008

Bakker, A. (2018). *Design research in education: A practical guide for early career researchers*. Routledge. doi:10.4324/9780203701010

Barr, V., & Stephenson, C. (2011). Bringing computational thinking to K-12: What is Involved and what is the role of the computer science education community? *ACM Inroads*, *2*(1), 48–54. doi:10.1145/1929887.1929905

Benitti, F. B. V. (2012). Exploring the educational potential of robotics in schools: A systematic review. *Computers & Education*, *58*(3), 978–988. doi:10.1016/j.compedu.2011.10.006

Bitesize, B. B. C. (2017). *Introduction to computational thinking*. Online verfügbar unter: http://www. bbc. co. uk/education/guides/zp92mp3/revision

BMBWF. (2021). *Digitale Kompetenzen-Informatische Bildung*. digikomp4. https://digikomp.at/index. php?id=530

Botički, I., Pivalica, D., & Seow, P. (2018). The Use of Computational Thinking Concepts in Early Primary School. *International Conference on Computational Thinking Education 2018*.

Brandhofer, G. (2017). Coding und Robotik im Unterricht. *Erziehung und Unterricht*, *167*(7–8), 630–637.

Brandhofer, G., Baumgartner, P., Ebner, M., Köberer, N., Trültzsch-Wijnen, C., & Wiesner, C. (2018). Bildung im Zeitalter der Digitalisierung. *Nationaler Bildungsbericht Österreich*, *2*, 307–362.

Conde, M. Á., Rodríguez-Sedano, F. J., Fernández-Llamas, C., Gonçalves, J., Lima, J., & García-Peñalvo, F. J. (2021). Fostering STEAM through challenge-based learning, robotics, and physical devices: A systematic mapping literature review. *Computer Applications in Engineering Education*, *29*(1), 46–65. doi:10.1002/cae.22354

Dittert, N., Robinson, S., Thestrup, K., & Knudsen, J. (2019). Supporting creativity and collaboration: Considerations for the development of a technologically enhanced toolkit for kindergartens. *Proceedings of the FabLearn Europe 2019 Conference*, 1–3. 10.1145/3335055.3335064

Esteve-Mon, F. M., Adell-Segura, J., Llopis Nebot, M. Á., Valdeolivas Novella, M. G., & Pacheco Aparicio, J. (2019). *The Development of Computational Thinking in Student Teachers through an Intervention with Educational Robotics*. Academic Press.

Estivill-Castro, V. (2019). Inviting teachers to use educational robotics to foster mathematical problem-solving. *International Conference on Robotics and Education RiE 2017*, 248–261.

Eteokleous, N. (2018). Employing educational robotics for the development of problem-based learning skills. In *Encyclopedia of Information Science and Technology* (4th ed., pp. 2492–2502). IGI Global.

Fadel, C., Bialik, M., & Trilling, B. (2015). *Four-dimensional education*. Academic Press.

Fanny, B., Julie, H., & Anne-Sophie, C. (2020). Developing a critical robot literacy for young people from conceptual metaphors analysis. *2020 IEEE Frontiers in Education Conference (FIE)*, 1–7.

Fojtik, R. (2017). The Ozobot and education of programming. *New Trends and Issues Proceedings on Humanities and Social Sciences, 4*(5).

Futschek, G. (2016). Computational Thinking im Unterricht. *Schule Aktiv*, 4–5.

Grover, S., & Pea, R. (2018). Computational Thinking: A competency whose time has come. *Computer science education: Perspectives on teaching and learning in school*, 19.

Himpsl-Gutermann, K., Brandhofer, G., Frick, K., Fikisz, W., Steiner, M., Bachinger, A., Gawin, A., Gawin, P., Szepannek, P., & Lechner, I. (2017). *Denken lernen–Probleme lösen (DLPL)*. Primarstufe.

Humbert, L., Herper, H., Best, A., Borowski, C., Freudenberg, R., Fricke, M., Haselmeier, K., Hinz, V., Müller, D., & Schwill, A. (2019). *Empfehlungen der GI–Kompetenzen für informatische Bildung im Primarbereich*. Informatik für alle.

Hußmann, S., Thiele, J., Hinz, R., Prediger, S., & Ralle, B. (2013). Gegenstandsorientierte Unterrichtsdesigns entwickeln und erforschen. In Der lange Weg zum Unterrichtsdesign. Zur Begründung und Umsetzung fachdidaktischer Forschungs-und Entwicklungsprogramme. Münster: Waxmann.

ISTE & CSTA. (2011). *Operational Definition of Computational Thinking*. https://cdn.iste.org/www-root/ct-documents/computational-thinking-operational-definition-flyer.pdf?sfvrsn=2

Kandlhofer, M., & Steinbauer, G. (2016). Evaluating the impact of educational robotics on pupils' technical-and social-skills and science related attitudes. *Robotics and Autonomous Systems*, *75*, 679–685. doi:10.1016/j.robot.2015.09.007

Kids, B. (2019). *Framework for 21st century learning definitions*. Battelle for Kids.

Kong, S.-C., Abelson, H., & Lai, M. (2019). Introduction to computational thinking education. In *Computational thinking education* (pp. 1–10). Springer. doi:10.1007/978-981-13-6528-7_1

Kordaki, M., & Kakavas, P. (2017). Digital storytelling as an effective framework for the development of computational thinking skills. *EDULEARN2017*, 3–5.

Lee, T. Y., Mauriello, M. L., Ahn, J., & Bederson, B. B. (2014). CTArcade: Computational thinking with games in school age children. *International Journal of Child-Computer Interaction*, 2(1), 26–33. doi:10.1016/j.ijcci.2014.06.003

Leoste, J., Pastor, L., López, J. S. M., Garre, C., Seitlinger, P., Martino, P., & Peribáñez, E. (2020). Using Robots for Digital Storytelling. A Game Design Framework for Teaching Human Rights to Primary School Students. *International Conference on Robotics in Education (RiE)*, 26–37.

Lye, S. Y., & Koh, J. H. L. (2014). Review on teaching and learning of computational thinking through programming: What is next for K-12? *Computers in Human Behavior*, *41*, 51–61. doi:10.1016/j.chb.2014.09.012

Mayring, P. (2015). Qualitative Inhaltsanalyse. Grundlagen und Techniken, 4, 58.

McKenney, S., & Reeves, T. C. (2018). *Conducting educational design research.* Routledge. doi:10.4324/9781315105642

Otto, D. (2020). Hochschullehre und Digitalisierung: Digital Storytelling als Lehr-Lernmethode für Kompetenzen in der digitalen Welt. *Bildung und Digitalisierung*, 135–152.

Papert, S. (1980). *"Mindstorms" Children. Computers and powerful ideas.* Academic Press.

Parsazadeh, N., Cheng, P.-Y., Wu, T.-T., & Huang, Y.-M. (2020). Integrating Computational Thinking Concept Into Digital Storytelling to Improve Learners' Motivation and Performance. *Journal of Educational Computing Research.* doi:10.1177/0735633120967315

Picka, K., Dosedla, M., & Stuchlikova, L. (2020). Robotic didactic aid Ozobot in Czech schools. *2020 18th International Conference on Emerging eLearning Technologies and Applications (ICETA)*, 525–533.

Plomp, T., & Nieveen, N. (2007). An introduction to educational design research. *Proceedings of the seminar conducted at the East China Normal University*, 23.

Plomp, T., & Nieveen, N. (2013). *Educational design research part A: An introduction.* SLO, Enschede.

Resnick, M. (2014). Give P's a chance: Projects, peers, passion, play. In *Constructionism and creativity: Proceedings of the third international constructionism conference.* Austrian Computer Society.

Rich, P. J., Browning, S. F., Perkins, M., Shoop, T., Yoshikawa, E., & Belikov, O. M. (2019). Coding in K-8: International trends in teaching elementary/primary computing. *TechTrends*, *63*(3), 311–329. doi:10.100711528-018-0295-4

Robin, B. (2006). The educational uses of digital storytelling. *Society for Information Technology & Teacher Education International Conference*, 709–716.

Robin, B. R. (2016). The power of digital storytelling to support teaching and learning. *Digital Education Review*, *30*, 17–29.

Shute, V. J., Sun, C., & Asbell-Clarke, J. (2017). Demystifying computational thinking. *Educational Research Review*, *22*, 142–158. doi:10.1016/j.edurev.2017.09.003

Stangl, W. (2021). *Stichwort: "Scaffolding".* Online Lexikon für Psychologie und Pädagogik: https://lexikon.stangl.eu/13399/scaffolding/

Stoeckelmayr, K., Tesar, M., & Hofmann, A. (2011). Kindergarten children programming robots: A first attempt. *Proceedings of 2nd International Conference on Robotics in Education (RIE).*

Stork, M. G. (2020). Supporting Twenty-First Century Competencies Using Robots and Digital Storytelling. *Journal of Formative Design in Learning*, *4*(1), 1–8. doi:10.100741686-019-00039-w

Tengler, K. (2020). Klein, kreativ, Ozobot: Förderung von Kreativität und informatischem Denken durch spielerisches Programmieren. *Resource (Saint Joseph, Mich.).*

Tengler, K. (2021). *Prototyping a robotics-based learning environment supporting computational thinking skills.* Submitted.

Tengler, K., Kastner-Hauler, O., & Sabitzer, B. (2021a). A Robotics-based Learning Environment Supporting Computational Thinking Skills – Design and Development. In *2021 IEEE Frontiers in Education Conference (FIE)*. IEEE.

Tengler, K., Kastner-Hauler, O., & Sabitzer, B. (2021b). Enhancing Computational Thinking Skills using Robots and Digital Storytelling. *CSEDU*, (1), 157–164.

Tengler, K., Kastner-Hauler, O., & Sabitzer, B. (2021c). Identifying preliminary design principles. *ICCSE 2021. The 4th International Conference on Computer Science & Education*, 733–738.

Tengler, K., Sabitzer, B., & Rottenhofer, M. (2019). "Fairy tale computer science"—Creative approaches for early computer science in primary schools. *ICERI Proceedings*, 8968-8974.

Van den Akker, J., Gravemeijer, K., McKenney, S., & Nieveen, N. (2006). Introducing educational design research. *Educational Design Research, 1*(3).

Wing, J. M. (2006). Computational thinking. *Communications of the ACM, 49*(3), 33–35. doi:10.1145/1118178.1118215

Yang, Y.-T. C., Chen, Y.-C., & Hung, H.-T. (2020). Digital storytelling as an interdisciplinary project to improve students' English speaking and creative thinking. *Computer Assisted Language Learning*, 1–23. doi:10.1080/09588221.2020.1750431

KEY TERMS AND DEFINITIONS

Computational Thinking: A competence that can be useful in solving complex problems of everyday life.

Design Principles: Guidelines for designing a successful learning environment.

Digital Storytelling: A teaching and learning method to promote narrative skills using digital media.

Educational Design Research: Research method that takes place in iterative cycles and aims to conduct research in school environment through the interaction of researchers and practitioners.

Educational Robotics: The use of robots to support students' learning processes.

Learning Environment: Learning package consisting of teaching methods, material, and media.

Ozobot: A programmable robot which follow lines and can be coded with colors.

Chapter 7
Towards the Development of a Game for Computational Thinking:
Identifying Students' Needs and Interests

Panagiotis Kosmas
Center for the Advancement of Research and Development in Educational Technology, Cyprus & School of Education, University of Nicosia, Cyprus

Andrea Philippou
Center for the Advancement of Research and Development in Educational Technology, Cyprus

Panagiotis Psomos
University of the Aegean, Greece

ABSTRACT

Computational thinking (CT) is now considered an essential approach for developing critical thinking and 21st-century skills. CT as a teaching methodological approach is more connected to STEM education as it provides clearer conceptual and practical considerations to understand science, computer, and mathematical concepts. Based on the recent literature, educational robotics, applications, and serious games are the means of applying CT in teaching practice. This study examines students' needs, interests, and motivations for using a game in the context of CT. Quantitative analysis from an online questionnaire to 394 students from secondary education in different five countries (Greece, Cyprus, Italy, Poland, United Kingdom) demonstrate the students' game interests and needs that guide us to develop a game for CT's implementation in the classroom. Essential insights, considerations, and implications are providing for the design, development, and use of games for the CT in an educational environment.

DOI: 10.4018/978-1-7998-7443-0.ch007

INTRODUCTION

Nowadays, in the digital society of the 21st century, the exponential onset of computers is forcing a transition in which digital literacy is now a necessary ability to cultivate (Shute, Sun, & Asbell-Clarke, 2017; Angeli & Giannakos, 2020). Most of us use computers regularly and we need to learn how to work with them to get the most out of their computing power (Shute et al., 2017).

It seems that CT is the new literacy. Wing (2006) acknowledged CT as a vital skill cultivated by all literate people attending compulsory education to supplement the other three key competencies: reading, writing, and mathematical skills. Since then, several research studies have been published and many scientific discussions among scholars have been started on how CT can be integrated into the school practice. CT is considered as a thinking process (or otherwise a human thinking ability) that uses analytical and algorithmic methods to formulate, evaluate and solve problems (Bocconi et al, 2016). CT, also, has been advocated by most educational policy makers as a capability that is equally important for all as numeracy and literacy (Bocconi et al, 2016). Not only it is the core for the STEM disciplines and courses (Science, Technology, Engineering and Mathematics), but it is also useful in daily life. The human brain itself is wired to think computationally; therefore, our development and future prospects need to learn how to use its full potential (Henderson, Cortina & Wing, 2007).

In an academic setting, the use of various game tools and educational robotics can be a fun and motivating technique and is recommended to support teaching and learning in the context of CT (Ioannou & Makridou, 2018). Specifically, serious games applications can support teachers' practices providing further understanding and meaningful experiences to students (Anastasiadis, Lampropoulos & Siakas, 2018). As educators continue to unlock their skills, serious and other mobile games, tools or applications are becoming increasingly widespread (Kazimoglu, Kiernan, Bacon, & Mackinnon, 2012). In parallel, students are getting used to gaming in their everyday lives, and technology is even more present around us. Minimizing or eradicating the "digital gap" is vital by promoting more significant involvement in the growing digital environment. Along the same lines, educational robotics is closely related to the CT approach, as it offers to students opportunities to think, develop, construct, communicate, collaborate, and critically reflect on their creations and solutions (Alimisis 2013; Bers, Flannery, Kazakoff, & Sullivan, 2014; Eguchi, 2010).

There are several grey areas in the literature and lots of definitions and explanations, including the definition of CT and mainly how CT can be incorporated into the school curriculum. The use of tools such as robotics posit CT as a very promising area to support learning outcomes at schools (Angeli & Giannakos, 2020). Despite that, different kinds of serious games have been proposed as another way to cultivate students' CT. Based on the existing empirical literature, this chapter aims to provide an important set of considerations regarding the development and the use of robotics and serious games to promote the development of CT in educational contexts. The innovation of this study is the provision of significant conclusions based on secondary education students' perspectives regarding the design and development of a game to be used in the classroom in the context of CT. Educators or researchers could use our findings and conclusions to develop an interesting, meaningful and attractive learning experience applying CT's concepts and approaches.

What follows is first the provision of some theoretical underpinnings, principles, and definitions in CT. After that, the chapter offers an overview of recent empirical studies on the application of games to enhance CT and discusses what the literature demonstrates on the implementation of CT in education. The following section describes the methodology used in this research and presents the most important findings based on the quantitative analysis. Finally, based on the students' responses, the chapter concludes with some important implications for the design, development and use of games for the CT in an educational environment.

COMPUTATIONAL THINKING: DEFINITIONS AND THEORETICAL UNDERPINNINGS

Having as the primary goal to 'foster the 21st century skills necessary to fully participate in the digital world' Bocconi et al., (2016) mentioned that CT is a concept that has been gaining attention recently. Closely linked to coding, programming, algorithmic thinking, CT has been promoted by educational stakeholders along with other skills that are regarded as fundamental for all, such as numeracy and literacy, as well as a means for developing new skills for integration into the employment market (Bocconi et al., 2016), for learning STEM (Weintrop et al., 2014) and for coping with the challenges of the twenty first century (Angeli & Giannakos, 2020). Since it can help youngsters develop skills linked to problem-solving (Henderson, Cortina, & Wing, 2007) and decision-making (Kules, 2016; Wing, 2008), CT has been a subject of attention in studies from various fields in recent years. As a result, CT is a necessary ability for everyone (Hsu, Chang, & Hung, 2018).

In the literature, there is not one current unanimous concept of CT that is being used. Wing (2006) defined the 'computational thinking' as it "*...involves solving problems, designing systems, and understanding human behavior, by drawing on the concepts fundamental to computer science. Computational thinking includes a range of mental tools that reflect the breadth of the field of computer science*" (p.33). Wing, (2006) also, asserted that CT "*represents a universally applicable attitude and skill set everyone, not just computer scientists, would be eager to learn and use*" (p. 33). She also claimed that, "*to reading, writing, and arithmetic, we should add computational thinking to every child's analytical ability*" (Wing, 2006, p. 33). Similarly, Wing (2011), introduced another definition of CT and defined CT as "*the thought processes involved in formulating problems and their solutions so that the solutions are represented in a form that can be effectively carried out by an information-processing agent*" (p. 1).

Although there is currently no distinct unanimous definition of CT, based on the research literature CT is a thought process that applies some core features (Angeli et al., 2016; Wing, 2006; Barr et al., 2011; Bers et al., 2014; National Research Council, 2010; Selby and Woolard, 2013). The features mutual among researchers are four, namely abstraction, decomposition, algorithmic thinking, and debugging. Table 1 defines the core features of computational thinking.

COMPUTATIONAL THINKING AND STEM EDUCATION

For many educators, CT is a relatively new concept. The goal of CT in education is to teach students how to think like computer scientists so that they can solve problems in a way that a computer might (Shuchi & Roy, 2013). The core features of CT as proposed by many researchers are critical and have a strong link with STEM education and learning because of their connection to the STEM disciplinary processes of modelling, reasoning, and problem solving (Foster, 2006; Henderson, Cortina, Hazzan & Wing et al, 2007; Sengupta et al., 2013).

Table 1. Core features of computational thinking

Feature	Definition
Abstraction	The ability to determine what data to preserve for an entity/object and what to neglect (Wing, 2011).
Decomposition	The ability to break down a complicated problem into easier parts in order to comprehend and solve it (National Research Council, 2010; Wing, 2011).
Algorithmic thinking **a. Sequencing** **b. Flow of control**	The capability to prepare a step by step series of actions to solve a problem (Selby, 2014). The ability to place actions in the right order (Selby, 2014). The sequence in which to perform instructions/actions (Selby, 2014).
Debugging	The ability to recognize, delete, and correct mistakes (Selby, 2014).

In addition, the STEM subjects provide a natural setting for CT instruction (Grover & Pea, 2018) and CT was acknowledged as a core scientific practice by the Next Generation Science Standards (NGSS Lead States, 2013). Through CT-embedded scientific inquiry, the integration of CT for science, technology, engineering, and mathematics (STEM) curriculum has the potential to improve science learning and boost student engagement in STEM learning (Yang, Swanson, Chittoori, & Baek, 2018). By promoting innovation and problem solving, incorporating CT into the classroom helps students prepare for the future (Fessakis, Gouli, & Mavroudi, 2013; Kosmas & Zaphiris, 2019). Computation is an indispensable component of STEM disciplines as they are practiced in the professional world (Jona et al., 2014). As a result, to sustain continuous discovery, thinking skills set among STEM educators and students must be developed (Swaid, 2015). Thus, research needs to focus on CT in STEM education (Tang, Yin, Lin, Hadad, & Zhai, 2020). The question that research in STEM education needs to answer is not why we must integrate CT, but how. As a result, it's vital to look for effective ways to include CT in our STEM teaching.

COMPUTATIONAL THINKING AND EDUCATIONAL ROBOTICS

The research in this field has led to many dialogues and thoughts on the best way to teach CT as learners face many academic difficulties (Bonar & Soloway, 1983; Coull & Duncan, 2011). There is a lot of interest in educational robotics currently, as seen by the constant introduction of new educational robots and articles in the media (Komis, Romero, Depover, & Karsenti, 2019). Naturally, academics have begun to investigate the contribution of educational robotics in promoting CT development (Bers, 2010; Grover & Pea, 2013; Kazakoff, Sullivan, & Bers, 2013; Lee et al., 2011). According to research, children who program robots acquire and apply key CT principles including abstraction, automation, analysis, decomposition, modularization, and iterative design (Bers, 2010, Kazakoff et al., 2013; Lee et al., 2011).

In addition to that, research involving younger children found that children as young as four years old may develop modest robotics projects while learning about important principles in engineering, technology, and computer programming (Bers et al., 2010; Bers et al., 2014). For example, Bers et al., (2014) used Lego WeDo robots and the CHERP (Creative Hybrid Environment for Robotics Programming) language in a study with 53 kindergarten children and found that the children were engaged in the process and understood basic programming and CT concepts relevant to sequencing and selecting the correct instructions. Also, in young children's education, the programming of toy robots (e.g., Bee-Bot) is widely applied (Atmatzidou & Demetriadis, 2016). For this activity, in particular, the learner needs to organize the actions (that wants the robot is to carry out) in a sequence of movements, paying attention to spot similar actions in different situations that can be repeated without re-programming them. Hence, the learner carries out useful practices of abstraction and decomposition. This resonates with the different affordances of physical and virtual environments supporting multiple pathways to CT. Angeli and Valanides (2019) looked at the effects of learning with Bee-Bot, a floor-programmable robot, on the computational thinking of young boys and girls. Because Bee-Bot does not provide a visual representation of the command's children use to program it, scaffolding was expected to play a crucial role in developing children's computational thinking skills when learning with it. The findings displayed CT gains among children.

Studies on the development of CT skills in older children have also yielded promising findings. Recent research by Angeli and Makridou (2018) demonstrated that educational robotics (using the kit LEGO WeDo) was an efficient technique for teaching elementary school students CT skills, even in a short intervention period. What is more, two studies were conducted by Constantinou and Ioannou (2018) at a primary and a secondary school in the Eastern Mediterranean to discuss CT gains of students related to their involvement in educational robotics activities. It was found that students participating in the ER interventions showed substantial development in their CT abilities.

Moreover, a research by Ioannou and Angeli (2016) described the efforts towards designing technology-enhanced instruction (using educational robotics and a 3D interactive programming environment) for teaching Computational and Algorithmic Thinking in 8th graders coming from different secondary education schools in Cyprus. Based on the results, the Technological Pedagogical Content Knowledge framework and the Technology Mapping approach, which guided the design of the instructional intervention, were effective in promoting the development and understanding of computational and algorithmic thinking skills and concepts by students, respectively. Ioannou and Makridou (2018) reviewed published literature explicitly focusing on the use of educational robotics to advance the CT skills of students in K-12. The articles reviewed illustrate empirical evidence indicating that educational robotics can promote students' cognitive and social skills.

However, while robotics appears to be an efficient tool for teaching and learning and a fascinating topic for students of all ages, robotics pedagogy is still in its infancy. Despite the fact that the use of Educational Robotics different kinds of other serious games activities has been proposed as a different way to cultivate students' CT. For example, Bellegarde, Boyaval, and Alvarez (as mentioned in Komis, Romero, Depover, & Karsenti, 2019) give a comparative analysis of cognitive mediations at work in the context of an educational device aimed at introducing robotics and computer science to kindergarten students (5-6 years old). The experimental device is a robot named «Blue Bot,» which features three modalities in the realm of serious games: the body, the robot, and the digital tablet (Komis, Romero, Depover, & Karsenti, 2019).

COMPUTATIONAL THINKING AND SERIOUS GAMES

Screening through limited literature, there was found that games existed long before CT was popularized and labelled as an essential skill. Moreover, games that were once disconnected from schools are now being adopted by teachers as a key teaching tool, while such serious games are incorporated into traditional lesson plans so that students learn concepts through playing (Kazimoglu et al., 2012). This comes as no surprise because computer games contain interactive, engaging and immersive elements that have educational affordances (Frazer, Recio, Gilbert, & Wills, 2014). Most of these games involve a scenario designed to cover a basic programming task and learn algorithmic thinking and help students communicate and collaborate with their classmates, whereas some games address more advanced learning objectives (Malliarakis, Satratzemi, & Xinogalos, 2014). The main idea is to close the gap between theory and practice, merge abstract concepts with practical experiences, and inspire students to learn (Vahldick, Mendes, & Marcelino, 2014).

The concept of 'serious games' has been first introduced by Abt (1970); in his book Serious Games Abt suggested that simulations and games can improve education in the classroom as well as in an informal environment. The definition of the' serious game' has been updated many times since 1970. Sawyer (2002) linked serious games with the connection between a serious purpose and the knowledge and technology now present in the video game industry. Nowadays, there is not one unanimous definition of "serious games". The term is currently established and is becoming increasingly popular. Zyda's (2005, p.26) definition of serious games, stated that entertainment is undoubtedly a component: "*Serious game: a mental contest, played with a computer in accordance with specific rules, that uses entertainment to further government or corporate training, education, health, public policy, and strategic communication objectives.*" Zyda (2005) states that serious games supplement pedagogy including activities that teach, thereby transferring information or skills and as a result makes games serious. Additionally, he emphasizes, however, that pedagogy must be secondary to the story and that the entertainment aspect comes first (Zyda, 2005).

The emphasis on the entertainment element contrasts with the definition of serious games proposed by Michael and Chen (2006). Michael and Chen (2006, p.21) defined Serious Games as "*games that do not have entertainment, enjoyment, or fun as their primary purpose*". This definition is consistent with that proposed by, e.g., PIXELearning (PIXELearning.com, 2006); "*The use of computer game and simulation approaches and/or technologies for primarily non-entertainment purposes*". However, Michael and Chen (2006) noted that this is not to suggest that serious games are not amusing, entertaining, or enjoyable, only that there is an added objective over and above the entertainment aspect.

According to more recent literature, Gouin-Vallerand, Ferreira, & Hotte, (2018) created a serious game to provide students with education as a supplement to everyday school learning. A mobile serious game was created and tested with children to introduce ideas in mathematics and English. The findings demonstrated the serious game's usefulness in persuading youngsters to participate in the educational activity and suggested that their understanding had improved. It also highlighted aspects of the game's design. Also, Hart Margheri, Paci, & Sassone, (2020) supported that serious games provide an engaging, enjoyable teaching environment in which participants learn cyber security theory and ideas and put them into practice while playing the game. During the last decade, the importance of using computer video games in secondary and tertiary education has grown significantly, with the dual focus of sharing theoretical and applied knowledge while delivering lessons, as well as offering a means to attract students and maintain their interest in core subjects at the same time (Kazimoglu et al., 2012). Meanwhile, few studies have explored the usefulness of serious games as technological means to enhance computational thinking.

Curricula that used serious games to specialize in learning programming have found positive effects on students as well as on learning outcomes (Ater-Kranov et al., 2010). In their work, "*A serious game for developing computational thinking and learning introductory computer programming*", Kazimoglu et al. (2012), created a serious game, through the use of video game technologies, which promotes the development of CT skills in order to make teaching and learning basic computer programming easier.

The main pedagogical advantages presented to emphasize the fact that 'games are engaging and motivational, thus students will be encouraged to learn programming constructs in an entertaining and potentially familiar environment. Then, students will be able to transfer their learning outcomes from that environment into learning introductory computer programming with a programming language'.

Additionally, a study by Wu and Richards (2011) examined the use of a digital game-based curriculum on the emergence of CT skills in middle school students in Taiwan. Specifically, researchers mentioned that the students would be able to perform the essential abilities of CT (decomposition, pattern recognition, pattern generalization and abstractions, algorithm design, and data visualization) and would demonstrate their ability to apply computational thinking skills to problems that were not within the scope of the game. Also, Cano and colleagues (2021) created a serious game (Perdi-Dogs) for children aged 7 to 11 with hearing impairment. The findings illustrated that children were highly motivated to play.

The most frequently occurring outcomes and impacts seem to be knowledge acquisition/content understanding and affective and motivational outcomes (Basawapatna, Koh, & Repenning, 2010; Combefis, Beresnevičius, & Dagienė, 2016). Moreover, playing serious games to learn programming is linked to a range of perceptual, cognitive, behavioral, affective and motivational impacts and outcomes (Theodoropoulos, Antoniou, & Lepouras, 2017). However, the CT education domain is still in its infancy and requires research for developing theories of the learning mechanisms occurring in computer games (Kazimoglu et al., 2012).

Although for the field of CT education, it is clear that serious games lead to a variety of positive outcomes and impacts; it is also acknowledged that the literature on such games is fragmented and lacks coherence. Hence, the impact of serious games on CT development has only been evaluated to a small extent. It should also be taken into account that CT education is particularly challenging for students underrepresented in computing and engineering, such as girls and other learners from nondominant groups (Eordanidis, Gee, & Carmichael, 2017). For these students, programming learning methods and digital games have been used together in such a way that one benefits from another. Last, although robotics and serious games seem to be different tools or techniques to achieve CT skills, both intend to encourage users to deal with tasks computationally through playing.

DESIGNING A SERIOUS GAME

When creating a serious game for children, it's crucial to think about the aspects of the game that will fulfil their demands. The idea that well-designed digital games can help students learn is supported by research (Prensky, 2001, 2006; Gee, 2007, 2008). According to previous studies, a fundamental issue in serious game design is to strike a balance between learning and fun (Arnab et al., 2015). Klopfer, Osterweil, and Salen (2009) advise that design teams should focus on both learning and game play simultaneously from the start of the design process. The balance between instruction and enjoyment can be achieved by "*connecting the learning goals and skills with the mechanics in a non-superficial way*", and by "*recognizing the minimum requirements to maintain user engagement with a given outcome/objective*" (Dimitriadou et al., 2021, p. 142). In order for the game to be entertaining and deliver "individualized feedback for learners," the games must also be interactive. Educators have voiced a desire to present learners with a simulation of reality within a serious game where they may apply what they've learned. Considerations about the best approach to display content in the game, adapting the numerous game components for the target audience, handling all age groups, having back-up activity options for different students, retaining engagement, and playtesting were among the other design issues (Dimitriadou et al., 2021).

Previous research has identified features of children's cognitive development that should be considered when building tablet applications for them, such as holding only one item in memory at a time (Chang, Tilahun, & Breazeal, 2014) or manifesting concentration, or focusing their attention on only one trait at a time. Other studies have discovered a number of concerns with children with tablets that can be linked to their ongoing development of motor skills, such as difficulty with multi-touch (McKnight & Fitton, 2010), drag and drop interaction (Humphries & McDonald, 2011), not recognized gestures (Anthony, Brown, Nias, & Tate, 2013), and misinterpreting touch movements (Anthony, Brown, Nias, & Tate, 2013; McKnight & Fitton, 2010).

Immediate feedback, a sandbox, customizability, and customizable complexity encourage players to work within their own zone of competency when dealing with the game's issue area (Gee, 2003). It's important to strike a balance between addressing these needs and the game's mechanics. When this balance is not achieved, one of two things can happen: (1) if the challenges are too complex for the individual's skills, the game becomes too difficult, causing anxiety; or (2) if the game is too simple, the player's skills are too far above the game's challenges, the child becomes bored and loses motivation. Furthermore, interaction with tangible objects piques children's interest and can motivate them; yet, digital apps have not yet to be widely adopted in children's learning. Children's social interactions provide a wealth of information for their learning, not just in terms of language but also in terms of behavioral, cognitive, and social dynamics (Cano et al., 2016). CT as a serious game can be a useful tool in the teaching–learning process for children. Designing a serious game necessitates a technique that entails the collaboration of several experts in the subject in order to identify goals that can be applied to the context of use.

METHODOLOGY

The study follows a quantitative approach (Roni, Merga, & Morris, 2020) with the aim to identify the needs and interests of the students in secondary education in order to start working on the designing of a serious game.

An online survey was distributed to secondary students in five countries (Cyprus, Greece, Poland, Italy, UK). Participants were 394 students (N=394) who attended five countries' lower and upper secondary education in the private and public sectors. Namely, 47 students from Cyprus, 39 students from Greece, 190 students from Poland, 98 students from Italy, and 20 students from the United Kingdom. The survey aimed to explore the gaming interests of the students in lower and higher secondary education to use this knowledge to develop an attractive game concept and learning environment in the context of CT.

The educators, through a link, administered the questionnaire to the participants for approximately 10-15 minutes. The questionnaire was created in English and it was translated to Greek, Italian and Polish. The survey included both close-ended and open-ended questions that allowed participants to express their satisfaction on specific aspects of game interests.

First of all, at the beginning of the questionnaire, participants were informed about the aim of the study, issues regarding GDPR and they were asked about their informed consent. After that, they were asked to answer some background information questions, for example, about age, sex and school grade. Then, participants were asked 11 closed ended questions regarding games: 1) if they have played an education game before, 2) if they think that solving problems related to mathematics in a game is important, 3) how much time they spend playing games, 4) why they are playing games, 5) their preferred device, 6) their preferred type of game and 7) setting as well as questions in terms of 8) interaction and 9) comparison with other players, 10) features of games that find most appealing and 11) if they would like to create their own character. To manipulate the game aspects that keep players immersed through the educational content of the game, participants were instructed to rate ten statements about games. For example, "I learn better if I can relate the experiences of an educational game to experiences in real life", on a scale from 1 ('Strongly Agree') to 5 ('Strongly Disagree'). Finally, one open-ended question was used to allow the participants to express their comments, suggestions, and opinions on any other game elements that they would consider important to be included in the game to make it more engaging and attractive.

The survey questionnaire was developed online using the Google Forms software to be able to be administered online. This option works well in terms of saving time as the online software automatically collects the questionnaires and allows visual representation of the data. Also, it offers convenience and easy access to participants as they can fill the questionnaire from everywhere and when they choose according to their own time schedules.

As for the analysis of the results, descriptive analysis was performed in some significant findings regarding the students' responses in the survey (Petscher, Schatschneider, & Compton, 2013).

RESULTS

In this section, we present the main results that emerged from the analysis of the online questionnaire, focusing on data regarding some demographics data, game characteristics and features and students' preferences and interests.

Starting with demographic data, it seems that the majority of students that answered the survey were female. This applies to all countries except the UK, where the majority of students were male. The gender of students that participated in the survey in all countries is shown in figure 1.

Next, the students were asked to report how much time per week they spend online or playing games or using technology devices. A lot of time is spent by children playing, which is as was actually expected. It seems that the majority of the students play games on average for 4 to 6 hours weekly. Specifically, students in the UK spend more time playing games (average time 6 hours) than the other countries that participated in the research. Students from Greece spend less time playing games (average 4 hours), while in Cyprus, Italy, and Poland the average time is 2-4 hours (see figure 2). Of course, that is a positive indicator that we should use learning through playing. Even better we believe that this could be done through hiding learning in the playing. Mobile or serious games is a way to achieve that goal.

In the question "Would you like the fact that you could compare your performance with that of other players?", the majority of students answered yes. From the responses, it is evident that students want their performance to be compared with other players during the game. As we can see in all countries, competition is an element that students prefer during playing (see figure 3). Thus, adding the competition factor in gaming enhances the motivation of students which is key to learning.

In the following question, students shared their opinion regarding the most appealing features in a game. Namely, students had to select among the following features: challenges, clear goal, feedback, graphics, reality, sound, and storyline. The storyline is the most important feature based on students' responses, followed by challenges and graphics. Notably, in Greece and Italy the most appealing feature is the storyline, while in UK and Cyprus the proffered feature is the possibility of playing the game

Figure 1. Participants' gender

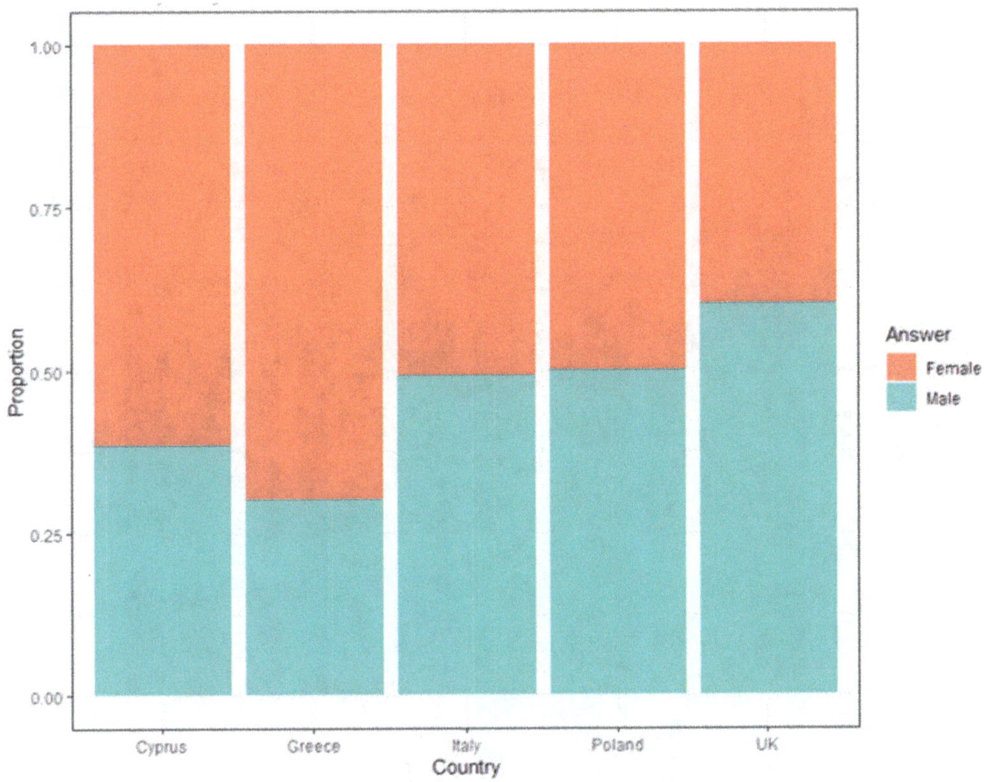

through challenges. One also important feature, according to the students, is the graphics that games have. A less important feature for students is the sound and the clear goal of the game (see figure 4). That's really important for the design and development of the game since the storyline seems to be an excellent feature for all students. Therefore, one consideration for the development is to add a good story in a game since our minds are hardwired to store information in a storytelling format.

In the question "Have you played an educational game before in the classroom?" the majority of students from all countries gave a positive answer. It seems that almost all participating students have previous experience with educational games. Only in Greece and Italy a high percentage of students said no compared to the students from Cyprus, the UK, and Poland (see figure 5). The real challenge when developing a game for learning purposes is to create a really engaging game and create a flow to students and not just presenting educational information in a media format.

As shown in figure 6, the smartphone is the most common answer regarding the preferred device for playing games, followed by PC. Specifically, students from Cyprus, Greece, and Italy preferred to play games using a smartphone, while students from Poland preferred to use a PC. Students from the UK claimed that the preferred device for them is Xbox (see figure 6).

One crucial characteristic/element of such educational games is the possibility to create your own character. As shown in figure 7, it is profound that the vast majority of participating students from all countries, would like to have the option to create their own character while playing a game. Only a small

Figure 2. Time spent per week on playing games

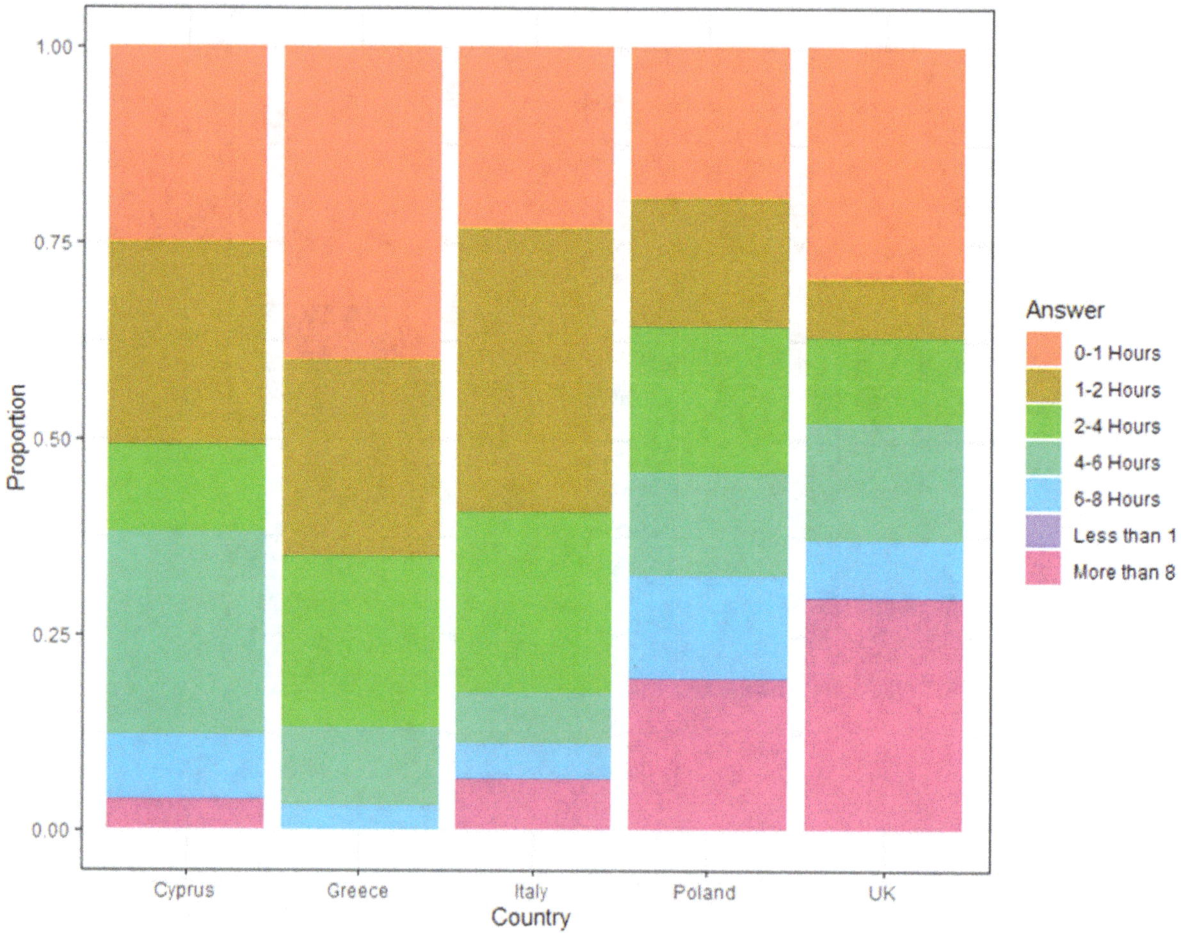

number of students said no. It is undoubtedly a feature that should be added in every online game since the game's personalization is enhanced in that way.

Furthermore, interactivity is one of the most important aspects that we need to consider when designing or developing a game to be used by students in the classroom. Students were asked to answer a question related to whether they prefer a single user or multiplier game. Students' options were three: playing the game alone, two players, more than two players. We report only the responses from students in Cyprus, Italy and UK. According to the responses (see figure 8), the majority prefer a game with multiple players (i.e. two or more players). However, many students still like a single user game (i.e. playing the game alone). It seems that the best option is to provide both single user and multi-user interaction to students to make the game appealing to all.

Last, in the question "what type of setting do you prefer?" students hat to choose whether they prefer an imagination setting or a Real World or both settings. As shown in figure 9, imagination is the most popular answer for all students except for Italian students who prefer real world settings. Almost none of the participating students chose the option for both settings (imagination and real world). Therefore, it seems that imagination with some real-world elements could be the best option when developing a game for students.

Figure 3. Competition during playing

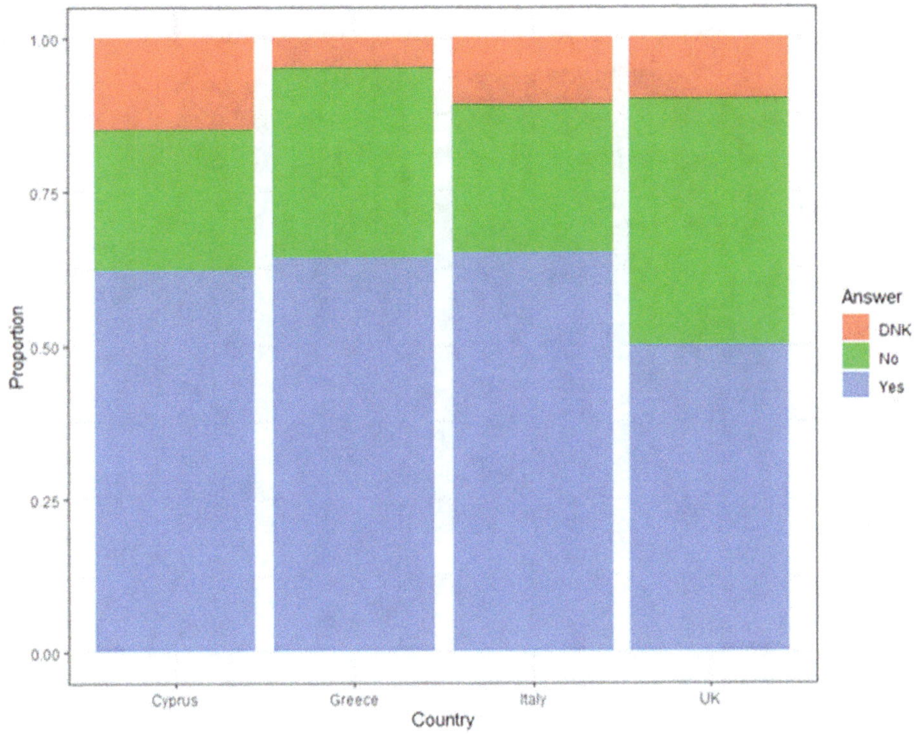

Figure 4. Appealing features in a game according to students

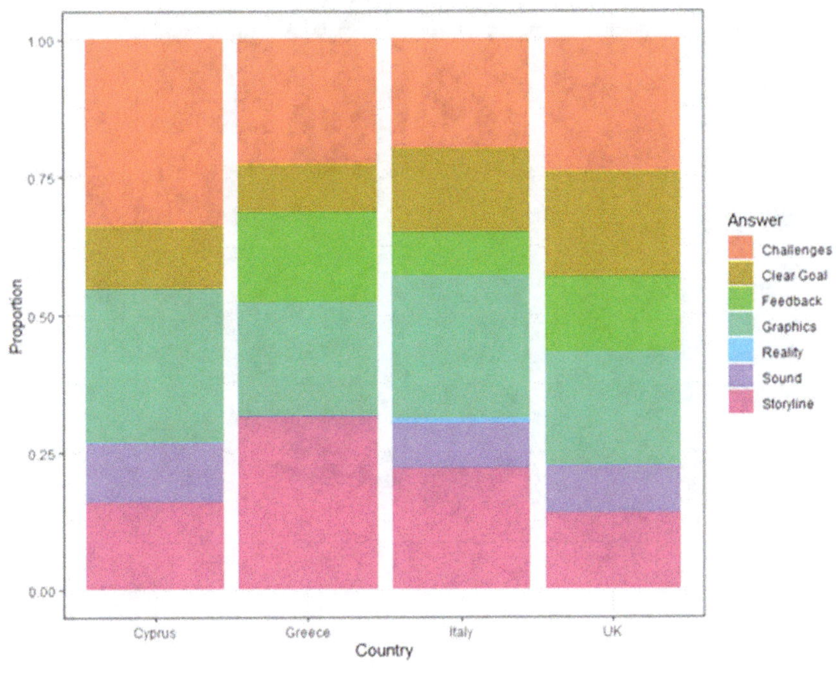

Figure 5. Previous experience with educational games

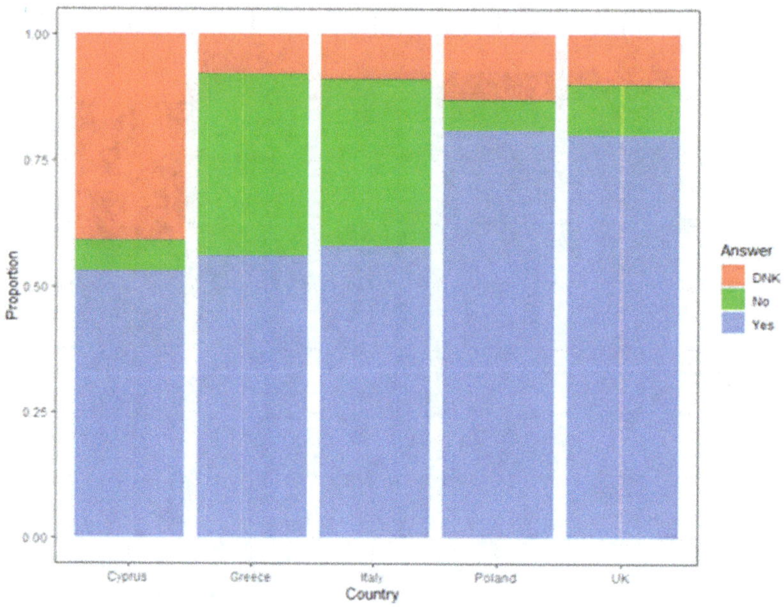

Figure 6. Students preferred devices for playing games

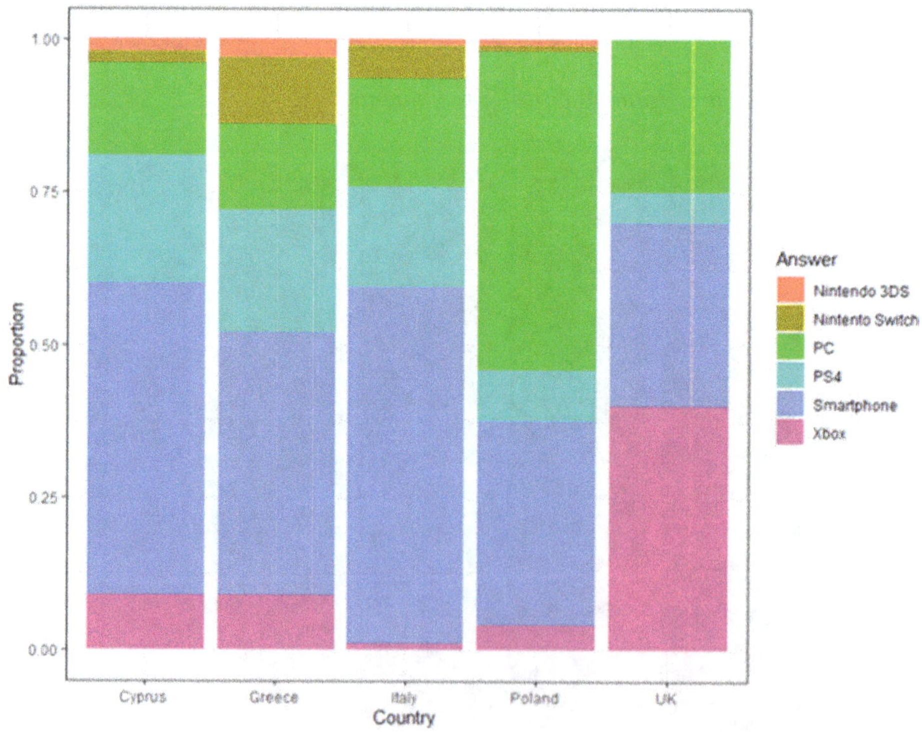

Figure 7. Would you like to create your own character?

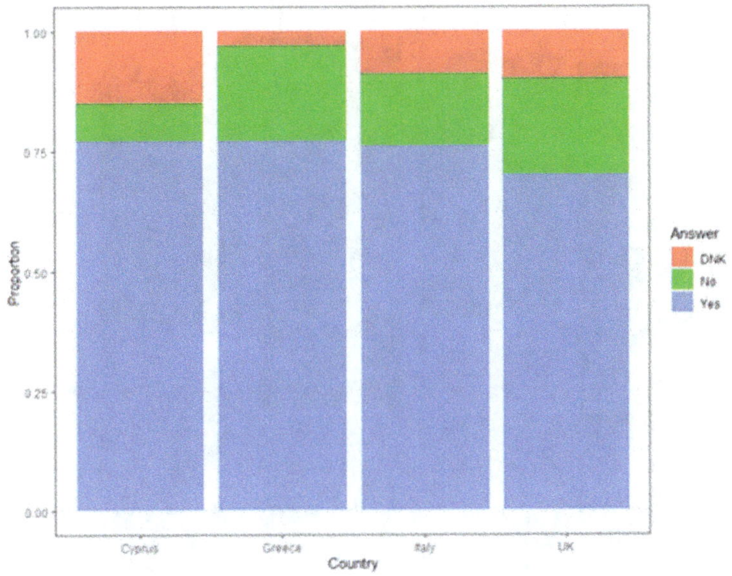

Figure 8. Most preferable interactivity option for students

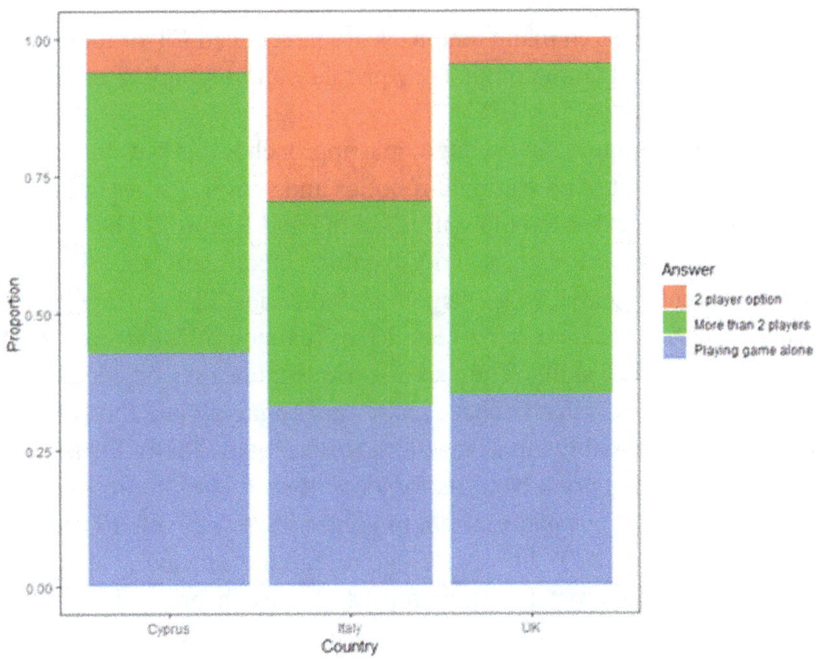

DISCUSSION

CT in education is emerging worldwide. It seems to be a promise for the education of the new genera-

Figure 9. Type of setting in a game

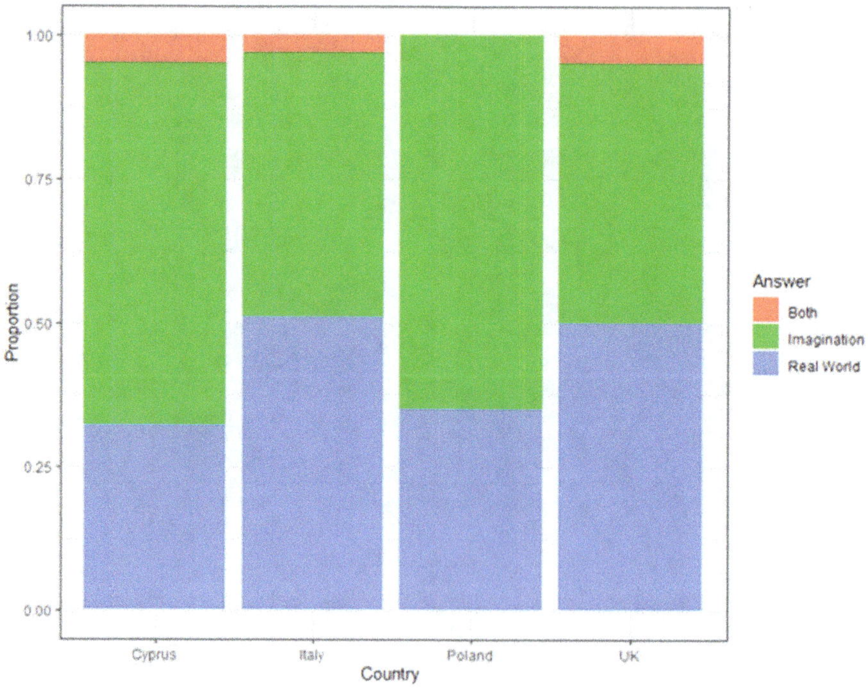

tion of children. Based on the research many countries are introduced CT or they currently planning to introduce it and/or related concepts into compulsory primary and secondary education and proceed to the curricula reforms.

Moreover, teaching CT may require new pedagogical approaches that put students at the centre of the learning process. The employment of educational robotics and serious games in an educational gaming setting are some techniques advocated to help with teaching and learning. The research was conducted investigating the use of Educational Robotics and CT gains. Specifically, the use of educational robotics is demonstrated by recent research (e.g. Angeli & Valanides, 2019; Angeli & Makridou, 2018; Constantinou & Ioannou, 2018; Ioannou & Angeli, 2016; Ioannou & Makridou, 2018) as an efficient technique for teaching students CT skills. The use of serious games is considered a helpful technique as well. The reasoning behind this is that because games are engaging and motivating, children will be motivated to learn in a fun and possibly familiar setting (Frazer et al, 2014). Furthermore, a curriculum that uses serious games has proven good benefits for students and learning results (Ater-Kranov et al., 2010; Kazimoglu et al.,2012). Serious games seem to contribute effectively to this experience (Gouin-Vallerand et al., 2018; Hart et al., 2020), however more research is needed given that the CT education domain is still in its infancy and requires research for developing theories of the learning mechanisms occurring in computer games.

This chapter conducted a literature review on the role of CT, the role of educational robotics and serious games and the relationship between CT and the state-of-the-art of educational robotics and serious games as a didactical approach for students to learn CT ways and principles. Also, this chapter conducted a field research on students' game interests. Specifically, based on students' responses, a game's specific features and characteristics will make the learning more interesting and engaging for students. This will aim to create a serious game concept that will improve students' CT skills.

Considering everything, it is evident that most of the students have played a game in the classroom before. The majority of students spend playing games on average 4 to 6 hours per week, but a significant percentage of students play more than 8 hours. The majority of the students preferred to use a smartphone to play a game and they prefer to play action, strategy games, sports games, and FSP games in a real-world or imagination setting. In terms of interaction, they prefer to play a game with two or more players, and they would like to compare their performance with that of other players. Most of them they would like to create their own character. The students, also, mentioned that, the most appealing features of the games are challenges, graphics, storyline and clear goal. The majority of them play for pleasure, excitement, competition, challenge, relaxation and leisure.

Additionally, students supported that they learn better if they can relate the experiences of an educational game to experiences in real life and when each new piece of knowledge builds on pre-existing knowledge. Furthermore, they enjoy games that seem too hard and they find feedback on their actions in-game help them to progress. They prefer playing games that have clear goals to achieve. They feel that they learn more when they are engaged in a role they play in a game and they can understand a subject being taught to them if they can experiment with the ideas that are taught. Finally, they are more engaged in games when using knowledge about the game's story and the world to solve problems and if the rewards/bonuses are adjusted to the difficulty of the performance.

Limitations

Although the research has reached its aims, there are some limitations to be addressed in future research. First, our findings are mainly focused on the gaming characteristics that students choose for a serious game, which is a major limitation of this study. Given that the serious game is intended for instructional reasons, additional factors such as students' learning styles may have an impact on their gaming preferences. Future research could gather qualitative data on students' gaming experiences, habits, and interests in order to conduct more extensive analyses and get new insights. On the other hand, obtaining this information is vital and will be the next step for us to fit the game to its educational objective successfully. Also, another limitation of the study is that the questionnaire didn't include topics or questions in relation to STEM or CT. We wanted to create a simple questionnaire for students as a preliminary part of our research. The aim was to examine the students' interests and motivation in playing an educational game in the classroom. Our future intention is to involve teachers in our analysis to see their perspective on the implementation of CT through a game.

FUTURE RESEARCH DIRECTIONS AND IMPLICATIONS

This research provides some important quantitative data regarding the game interests of secondary education students. Future directions involve designing or developing a game by applying all of the above preferences raised by the research participants. Some of the essential considerations when designing a game for CT that need to be taken into account include:

- *Interactivity*. It seems that this is one of the most important features that a serious or mobile game provides. The level of interaction is crucial for the game's success. According to the students, the game must offer opportunities for interaction with other players, but not as a mandatory choice. Players could have the possibility to choose if they want to interact with other players or not.
- *Storyline*. According to students, it is one attractive game feature. Narrative and storytelling attract students' attention and curiosity. Interesting storytelling also motivates students to continue and complete the game to investigate what happened and why. Students can write a story about a topic and act it out with various robots.
- *Challenge-based approach*. A game should have some real-world challenges. During the game students must be faced with a real problem that needs to be solved. This would allow students to critically think, identify the possible alternatives, investigate, and finally propose solutions. The challenges will enable students to gain deep knowledge and develop their skills. The challenge-based approach can be used for designing educational robotics as well.
- *Creativity*. The game should offer possibilities to the students to create or develop something, for example, a character of the game. This is also something that could attract students' interest to play an educational game. Educational Robotics foster students' creativity as it gives the opportunity them to create and develop something.
- *Competition*. Students claimed that it is interesting for them to compete with other players during the game. When a competitive environment is created in a serious game, students' motivation and academic performance improve significantly (Cagiltay, Ozcelik, & Ozcelik, 2015). Robotics Competitions are organized as well in order to increase students' motivation.
- *Educational robotics*. Some numerous games and applications engage students in robotics, teaching them how to code with robotics. Students through the use of robots involved in problem-solving situations and systems thinking. It is now time to consider how a serious game could adopt the logic of robots and provide valuable learning experiences to students. Many characteristics of educational robotics could serve as principles in the development of a game for CT. For example, characteristics such as teamwork, collaboration, problem-solving approaches etc. Could be some basic principles when designing a game in the context of CT.
- *Programming*. The process of producing a set of instructions that inform a computer how to complete a task.

In addition, future endeavors need to be implemented on how educational robotics and serious games could be connected. Although educational robotics and serious games are different techniques to achieve CT gains, both could be considered as methods to achieve 21st century skills through playing. Future studies could explore methods, strategies and approaches on how serious games and robotics can be integrated in the classroom.

Finally, to give a systematic and structured evaluation of the game concept, a set of rigorous experiments needs to be constructed. These will provide analytic data to see if the game successfully supports the development of CT abilities and, as a result, if the game aids students in learning and applying the essential ideas in their lessons. The results need to be analyzed to be able to determine the impact of this game approach and any advantages that can be derived from it.

CONCLUSION

CT has become a buzzword that seems to promise the education of a new generation of children with a much deeper understanding of our digital world. Through the above field research, we looked at the role of CT in education, the relationship between CT and the serious games as a didactical approach for students to learn CT ways and principles. The information collected stresses a clear tendency; the educational landscape is changing fast, and we are now at a tipping point. Various initiatives centered on CT in education are emerging worldwide. Serious games seem to contribute effectively to this experience; however, more research is needed given that the CT education domain is still in its infancy and requires research for developing theories of the learning mechanisms occurring in computer games. We believe that this research offers a clearer understanding of CT in education and provides useful implications for researchers and educators who want to apply CT concepts using games in their practices.

ACKNOWLEDGMENT

This paper was created within Erasmus+ Project CTApp: Teaching Students Computational Thinking Through a Mobile Application [Project Number: 2020-1-PL01-KA201-081924]

REFERENCES

Abt, C. C. (1970). *Serious Games*. Viking Press.

Alimisis, D. (2013). Educational robotics: Open questions and new challenges. *Themes Science and Technology Education*, *6*(1), 63–71.

Anastasiadis, T., Lampropoulos, G., & Siakas, K. (2018). Digital game-based learning serious games in education. *International Journal of Advances in Scientific Research and Engineering*, *4*(12), 139–144. doi:10.31695/IJASRE.2018.33016

Angeli, C., & Giannakos, M. (2020). Computational thinking education: Issues and challenges. *Computers in Human Behavior*, *105*, 106185. doi:10.1016/j.chb.2019.106185

Angeli, C., & Makridou, E. (2018). Developing Third-Grade Students' Computational Thinking Skills with Educational Robotics. In E. Langran & J. Borup (Eds.), *Proceedings of Society for Information Technology & Teacher Education International Conference* (pp. 1-8). Association for the Advancement of Computing in Education (AACE). Retrieved January 20, 2021 from https://www.learntechlib.org/primary/p/182493/

Angeli, C., & Valanides, N. (2020). Developing young children's computational thinking with educational robotics: An interaction effect between gender and scaffolding strategy. *Computers in Human Behavior*, *105*, 105954. doi:10.1016/j.chb.2019.03.018

Angeli, C., Voogt, J., Fluck, A., Webb, M., Cox, M., Malyn-Smith, J., & Zagami, J. (2016). A K-6 Computational Thinking Curriculum Framework- Implications for Teacher Knowledge. *Journal of Educational Technology & Society*, *19*(3), 47–57.

Anthony, L., Brown, Q., Nias, J., & Tate, B. (2013). Examining the need for visual feedback during gesture interaction on mobile touchscreen devices for kids. *Proceedings of the 12th International Conference on Interaction Design and Children -IDC '13*, 157–164. 10.1145/2485760.2485775

Arnab, S., Lim, T., Carvalho, M. B., Bellotti, F., De Freitas, S., Louchart, S., Suttie, N., Berta, R., & De Gloria, A. (2015). Mapping learning and game mechanics for serious games analysis. *British Journal of Educational Technology*, *46*(2), 391–411. doi:10.1111/bjet.12113

Ater-Kranov, A., Bryant, R., Orr, G., Wallace, S., & Zhang, M. (2010). Developing a community definition and teaching modules for computational thinking: accomplishments and challenges. *Proceedings of the 2010 ACM conference on Information technology education.* 10.1145/1867651.1867689

Atmatzidou, S., & Demetriadis, S. (2016). Advancing students' computational thinking skills through educational robotics: A study on age and gender relevant differences. *Robotics and Autonomous Systems*, *75*(Part B), 661–670. doi:10.1016/j.robot.2015.10.008

Basawapatna, A. R., Koh, K. H., & Repenning, A. (2010). Using scalable game design to teach computer science from middle school to graduate school. *Proceedings of the Fifteenth Annual Conference on Innovation and Technology in Computer Science Education*, 224–228. 10.1145/1822090.1822154

Bers, M. U. (2010). The TangibleK robotics program: Applied computational thinking for young children. *Early Childhood Research & Practice*, *12*(2), 2.

Bers, M. U., Flannery, L., Kazakoff, E. R., & Sullivan, A. (2014). Computational thinking and tinkering: Exploration of an early childhood robotics curriculum. *Computers & Education*, *72*, 145–157. doi:10.1016/j.compedu.2013.10.020

Bocconi, S., Chioccariello, A., Dettori, G., Ferrari, A., & Engelhardt, K. (2016). *Developing computational thinking in compulsory education-Implications for policy and practice* (No. JRC104188). Joint Research Centre (Seville site).

Bonar, J., & Soloway, E. (1983, January). Uncovering principles of novice programming. In *Proceedings of the 10th ACM SIGACT-SIGPLAN symposium on Principles of programming languages* (pp. 10-13). ACM.

Cagiltay, N. E., Ozcelik, E., & Ozcelik, N. S. (2015). The effect of competition on learning in games. *Computers & Education*, *87*, 35–41. doi:10.1016/j.compedu.2015.04.001

Cano, S., Arteaga, J. M., Collazos, C. A., Gonzalez, C. S., & Zapata, S. (2016). Toward methodology for serious games design for children with auditory impairments. *IEEE Latin America Transactions*, *14*(5), 2511–2521. doi:10.1109/TLA.2016.7530453

Chang, A., Tilahun, L., & Breazeal, C. (2014). Visualisations of Data from the Literacy Tablet Reading Project in Rural Ethiopia. In *Proceedings of the Electronic Visualisation and the Arts (EVA)*, 142–149. 10.14236/ewic/EVA2014.35

Combefis, S., Beresnevičius, G., & Dagienė, V. (2016). Learning programming through games and contests: Overview, characterisation and discussion. *Olympiads in Informatics*, *10*(1), 39–60. doi:10.15388/ioi.2016.03

Constantinou, V., & Ioannou, A. (2018). Development of Computational Thinking Skills through Educational Robotics. In *EC-TEL*. Practitioner Proceedings.

Coull, N. J., & Duncan, I. M. (2011). Emergent requirements for supporting introductory programming. *Innovation in Teaching and Learning in Information and Computer Sciences, 10*(1), 78–85. doi:10.11120/ital.2011.10010078

Dimitriadou, A., Djafarova, N., Turetken, O., Verkuyl, M., & Ferworn, A. (2021). Challenges in serious game design and development: Educators' experiences. *Simulation & Gaming, 52*(2), 132–152. doi:10.1177/1046878120944197

Eguchi, A. (2010). What is Educational Robotics? Theories behind it and practical implementation. In Society for information technology & teacher education international conference (pp. 4006-4014). Association for the Advancement of Computing in Education (AACE).

Eordanidis, S., Gee, E., & Carmichael, G. (2017). The effectiveness of pairing analog and digital games to teach computer science principles to female youth. *Journal of Computing Sciences in Colleges, 32*(3), 12–19.

Fessakis, G., Gouli, E., & Mavroudi, E. (2013). Problem solving by 5–6 years old kindergarten children in a computer programming environment: A case study. *Computers & Education, 63*, 87–97. doi:10.1016/j.compedu.2012.11.016

Foster, I. (2006). 2020 Computing: A two-way street to science's future. *Nature, 440*(7083), 419–419. doi:10.1038/440419a PMID:16554785

Frazer, A., Recio, A., Gilbert, L., & Wills, G. (2014). Profiling the educational value of computer games. *Interaction Design & Architecture (s) Journal-IxD&A, 19*, 1–19.

Gee, J. P. (2003). *What Video Games Have to Teach Us about Learning and Literacy*. Palgrave Macmillan. doi:10.1145/950566.950595

Gee, J. P. (2007). *Good video games + good learning: Collected essays on video games, learning and literacy*. Peter Lang. doi:10.3726/978-1-4539-1162-4

Gee, J. P. (2008). *What video games have to teach us about learning and literacy, revised and updated.* Palgrave Macmillan.

Gouin-Vallerand, C., Ferreira, S. M., & Hotte, R. (2018, November). Towards a mobile serious game environment for children self-learning. In *Proceedings of the 4th EAI International Conference on Smart Objects and Technologies for Social Good* (pp. 100-105). 10.1145/3284869.3284901

Grover, S., & Pea, R. (2013). Computational thinking in K–12: A review of the state of the field. *Educational Researcher, 42*(1), 38–43. doi:10.3102/0013189X12463051

Grover, S., & Pea, R. (2018). Computational thinking: A competency whose time has come. In S. Sentance, E. Barendsen, & C. Schulte (Eds.), *Computer science education: Perspectives on teaching and learning* (pp. 19–38). Bloomsbury Academic. doi:10.5040/9781350057142.ch-003

Hart, S., Margheri, A., Paci, F., & Sassone, V. (2020). Riskio: A serious game for cyber security awareness and education. *Computers & Security*, *95*, 101–827. doi:10.1016/j.cose.2020.101827

Henderson, P. B., Cortina, T. J., & Wing, J. M. (2007). Computational thinking. In *Proceedings of the 38th SIGCSE Technical Symposium on Computer Science Education-SIGCSE* '07, 195–19. 10.1145/1227310.1227378

Hsu, T. C., Chang, S. C., & Hung, Y. T. (2018). How to learn and how to teach computational thinking: Suggestions based on a review of the literature. *Computers & Education*, *126*, 296–310. doi:10.1016/j.compedu.2018.07.004

Humphries, L., & McDonald, S. (2011). Emotion faces: the design and evaluation of a game for preschool children. *Proceedings of the 2011 annual conference extended abstracts on Human factors in computing systems (CHIEA '11)*, 1453. 10.1145/1979742.1979790

Ioannou, A., & Makridou, E. (2018). Exploring the potentials of educational robotics in the development of computational thinking: A summary of current research and practical proposal for future work. *Education and Information Technologies*, *23*(6), 2531–2544. doi:10.100710639-018-9729-z

Ioannou, I., & Angeli, C. (2016). A Framework and an Instructional Design Model for the Development of Students' Computational and Algorithmic Thinking. In MCIS (p. 19). Academic Press.

Jona, K., Wilensky, U., Trouille, L., Horn, M. S., Orton, K., Weintrop, D., & Beheshti, E. (2014, January). Embedding computational thinking in science, technology, engineering, and math (CT-STEM). In Future directions in computer science education summit meeting, Orlando, FL.

Kazakoff, E. R., Sullivan, A., & Bers, M. U. (2013). The effect of a classroom-based intensive robotics and programming workshop on sequencing ability in early childhood. *Early Childhood Education Journal*, *41*(4), 245–255. doi:10.100710643-012-0554-5

Kazimoglu, C., Kiernan, M., Bacon, L., & Mackinnon, L. (2012). A serious game for developing computational thinking and learning introductory computer *programming. Procedia: Social and Behavioral Sciences*, *47*, 1991–1999. doi:10.1016/j.sbspro.2012.06.938

Klopfer, E., Osterweil, S., & Salen, K. (2009). *Moving learning games forward.* The Education Arcade.

Komis, Romero, Depover, & Karsenti, (2019). Robotics in Primary Education–Robotique en Éducation Primaire: Introduction. Review of Science. *Mathematics and ICT Education, 13*(1), 3-6.

Kosmas, P., & Zaphiris, P. (2019, September). Embodied interaction in language learning: Enhancing students' collaboration and emotional engagement. In *IFIP Conference on Human-Computer Interaction* (pp. 179-196). Springer. 10.1007/978-3-030-29384-0_11

Kules, B. (2016). Computational thinking is critical thinking: Connecting to university discourse, goals, and learning outcomes. *Proceedings of the Association for Information Science and Technology*, *53*(1), 1–6. doi:10.1002/pra2.2016.14505301092

Lee, I., Martin, F., Denner, J., Coulter, B., Allan, W., Erickson, J., Malyn-Smith, J., & Werner, L. (2011). Computational thinking for youth in practice. *Acfm Inroads*, *2*(1), 32–37. doi:10.1145/1929887.1929902

Malliarakis, C., Satratzemi, M., & Xinogalos, S. (2014). Educational games for teaching computer programming. In *Research on e-Learning and ICT in Education* (pp. 87–98). Springer.

McKnight, L., & Fitton, D. (2010). Touch-screen Technology for Children: Giving the Right Instructions and Getting the Right Responses. *Proceedings of the 9th International Conference on Interaction Design and Children -IDC '10*, 238. 10.1145/1810543.1810580

Michael, D., & Chen, S. (2006). *Serious games: Games that educate, train, and inform.* Thomson Course Technology.

National Research Council. (2010). *Committee for the workshops on computational thinking: Report of a workshop on the scope and nature of computational thinking.* National Academy Press. doi:10.17226/12840

NGSS Lead States. (2013). *Next Generation Science Standards: For States, by States.* The National Academies Press.

Petscher, Y. M., Schatschneider, C., & Compton, D. L. (Eds.). (2013). *Applied quantitative analysis in education and the social sciences.* Routledge. doi:10.4324/9780203108550

Prensky, M. (2001). *Digital game-based learning.* McGraw-Hill.

Prensky, M. (2006). *Don't Bother Me, Mom, I'm Learning! How Computer and Video Games Are Preparing Your Kids for 21st Century Success and How You Can Help!* Paragon House.

Roni, S. M., Merga, M. K., & Morris, J. E. (2020). *Conducting quantitative research in education.* Springer. doi:10.1007/978-981-13-9132-3

Sawyer, B., & Rejeski, D. (2002). *Serious Game Initiative.* Academic Press.

Selby, C. C. (2014). *How can the teaching of programming be used to enhance computational thinking skills?* (Unpublished doctoral dissertation). University of Southampton, Southampton, UK.

Sengupta, P., Kinnebrew, J. S., Basu, S., Biswas, G., & Clark, D. (2013). Integrating computational thinking with K-12 science education using agent-based computation: A theoretical framework. *Education and Information Technologies, 18*(2), 351–380. doi:10.100710639-012-9240-x

Shuchi, G., & Roy, P. (2013). Computational Thinking in K–12: A Review of the State of the Field. *Educational Researcher, 42*(1), 38–43. doi:10.3102/0013189X12463051

Shute, V. J., Sun, C., & Asbell-Clarke, J. (2017). Demystifying computational thinking. *Educational Research Review, 22*, 142–158. doi:10.1016/j.edurev.2017.09.003

Swaid, S. I. (2015). Bringing computational thinking to STEM education. *Procedia Manufacturing, 3*, 3657–3662. doi:10.1016/j.promfg.2015.07.761

Tang, X., Yin, Y., Lin, Q., Hadad, R., & Zhai, X. (2020). Assessing computational thinking: A systematic review of empirical studies. *Computers & Education, 148*, 103798. doi:10.1016/j.compedu.2019.103798

Theodoropoulos, A., Antoniou, A., & Lepouras, G. (2017). How do different cognitive styles affect learning programming? Insights from a game-based approach in Greek schools. *ACM Transactions on Computing Education, 17*(1), 3. doi:10.1145/2940330

Vahldick, A., Mendes, A. J., & Marcelino, M. J. (2014). A review of games designed to improve introductory computer programming competencies. *2014 IEEE Frontiers in Education Conference (FIE) Proceedings*, 1–7.

Weintrop, D., Beheshti, E., Horn, M. S., Orton, K., Trouille, L., & Jona, K. (2014). *Interactive assessment tools for computational thinking in High School STEM classrooms.* Paper presented at the international conference on intelligent Technologies for interactive entertainment. 10.1007/978-3-319-08189-2_3

Wing, J. (2008). Computational thinking and thinking about computing. *Proceedings of the IPDPS 2008: 22nd IEEE International Parallel and Distributed Processing Symposium*, 14–18.

Wing, J. M. (2006). Computational thinking. *Communications of the ACM*, *49*(3), 33–35. doi:10.1145/1118178.1118215

Wing, J. M. (2011, March). *Computational thinking.* Retrieved from https://csta.acm.org/Curriculum/sub/CurrFiles/WingCTPrez.pdf

Yang, D., Swanson, S., Chittoori, B., & Baek, Y. (2018). Work-in-Progress: Integrating computational thinking in STEM education through a project-based learning approach. *Proceedings of 2018 American Society for Engineering Education (ASEE) Annual Conference and Exposition.* 10.18260/1-2--30091

Zyda, M. (2005). From visual simulation to virtual reality to games. *Computer*, *38*(9), 25–32. doi:10.1109/MC.2005.297

KEY TERMS AND DEFINITIONS

Algorithmic Thinking: It is a method of solving by clearly defining the procedures required. Instead of coming up with a single solution to a problem, students create algorithms. When followed precisely (by a person or a computer), they are instructions or rules that lead to solutions to the original and related problems.

Coding: It is the process of using programming languages to create instructions for computers.

Computational Thinking: Computational thinking is a collection of problem-solving techniques that describes problems and solutions in a way that a computer can understand.

Digital Literacy: It refers to a person's capacity to find, assess, and clearly transmit information on a variety of digital platforms using typing and other media. It is based on a person's grammar, composition, typing skills, and ability to use technology to create text, images, audio, and designs.

Educational Robotics: Educational robotics, also known as pedagogical robotics, is a discipline that aims to teach children robotics and programming through hands-on activities starting at a young age.

Mobile Games: A mobile game is a video game that is played on a mobile device.

Serious Games: Serious games are games in which their primary goal is to study or practice a skill rather than to have fun.

Section 3

Educational Robotics for Inclusive Education

The third section includes two chapters that discuss the use of robotics in promoting inclusive education. The first one focuses on using robotics to enhance social skills in children and adolescents with autism. The second chapter discusses assistive robotics integration to empower people with disabilities.

Chapter 8

The Use of Robotics in Enhancing Social Skills in School and Therapeutic Settings in Children and Adolescents With Autism Spectrum Disorder

Maria Georgiadi
https://orcid.org/0000-0002-6846-0156
University of Crete, Greece

Stefanos Plexousakis
University of Crete, Greece

Potheini Vaiouli
University of Cyprus, Cyprus

Maria Lithoxopoulou
Frederick University, Cyprus

ABSTRACT

The current review focuses on the approaches that have been implemented in the development of the appropriate communication skills to children and adolescents with ASD (autism spectrum disorder) through their interaction with therapeutic robots. This study aims to emphasize the importance of the fact that robots can play a major role in enhancing social skills in students with ASD. The authors present the latest evidence on research and the most popular robots which have been used in developing social and communicational skills to children and adolescents with ASD. In particular, they focus on the use of robots in imitation, attention, eye contact, emotional regulation, initiative, and language development. It is crucial to realize that the use of robots cannot be the only educational or therapeutic procedure for ASD, but they can be used in combination with therapy and education. Robots can be a valuable tool in order to assist educators and therapists to help students with ASD developing their communication and social skills.

DOI: 10.4018/978-1-7998-7443-0.ch008

INTRODUCTION

Autism Spectrum Disorder (ASD) involves a variety of difficulties including social and communicational deficits that can provoke several obstacles in their everyday life (Plexousakis et al., 2020). Current literature review aims to describe the role of technology and in particular the role of the robotics in the development of social skills for students who have been diagnosed with Autism Spectrum Disorder (ASD). Specifically, children in the autism spectrum seem to have a unique developmental path, presenting various degrees of complexity in stereotypical behaviors, lack of social interests, and challenges in the emotional and social domain (American Psychiatric Association [APA], 2013).

Research has shown that children with ASD have difficulty acquiring conventional and symbolic meaning through actions of joint attention (Kasari, et al., 2006; Leekam & Ramsden, 2006) and symbolic play (Kasari et al., 2006; Keen & Halle, 2016). Additionally, research on emotional processing provides evidence for individuals' with ASD difficulties in emotional reactivity and emotional regulation (Samson et al., 2012) and challenges in describing the emotional content of verbal, auditory, and visual stimuli (Allen, et al, 2013). Challenges in reciprocal communication usually impact language acquisition and further deprive young children with autism of learning new skills through interacting with people in their environment (Mundy & Neal, 2001). This is because reciprocal interactions, either at the preverbal or verbal stage of communication of the child, may lead to enhanced episodes of social initiations, jointly engaged time on a shared activity, and overall advancements on children's play skills, language, and social skills (Shire et al., 2016; Venker et al., 2012). Gradually, young children acquire shared meaning and social experiences through such spontaneous actions and interactions with others in their everyday lives. As such, offering contexts for learning that are engaging and promote the interplay between children and their social environment are of critical importance for young children with ASD and constitute an important target for intervention.

The scope of current review is to highlight the significant role of technology and robotics in the enhancements of the social skills and other lacking areas of students with ASD. More specifically current chapter emphasizes the beneficial aspects of robots which can deliver the appropriate cognitive and behavioral interventions to teach communication and social skills and bring a positive change to their lives. Chapter illustrates the significance of robots in offering rich stimuli like different kind of sounds and pictures which facilitate a kind of simulation with real life stimuli and act as vehicle to improve ASD children's functionality (expression of feelings, verbal expressions, social skills) and benefit their lives.

BACKGROUND

Children with ASD could be benefited from cognitive and behavioral interventions which include teaching of the appropriate communication and social skills so as to improve several aspects of their emotional and social deficits and consequently to be able to live a more functional life. Several studies (Scassellati et al., 2012), emphasize the significance of early initiation of therapeutic intervention programs so as to achieve better results. Toys and other activities which resemble games could bea useful tool so as to stimulate and facilitate the interaction of children with other people. Although studies (Scassellati et al., 2012) emphasize the general goals like the improvement of social and communication skills it is important that children receive different and more rich stimuli like different kind of sounds and pictures which operate as a vehicle to their shift of emotions and skills (Pennisi et al., 2016). Recent studies attempt to explore different roots to create new situations and stimuli to these children, in order to produce better results both in cognitive and social level and development (Roberts-Yates & Silvera-Tawil, 2019).

Several games and robots could be used to stimulate and facilitate the development of social skills in different levels like verbal, face expression, general imitation as well as tactile sensitivity (Cabibihan et al., 2013; Kim et al., 2013). These research applications have investigated the use of the above kinds of stimuli and how these stimuli could be applied effectively in psychotherapeutic settings and interventions (Duquette et al., 2008; Michaud et al., 2003). Robots are able to function effectively and stimulate social and cognitive skills and abilities in children with ASD. They can be categorized into anthropomorphic, non-anthropomorphic, and non-biometic (Cabibihan et al., 2013; Scassellati et al., 2012).

First category resembles some human features in its appearance, such (anthropomorphic) robots are KASPAR, ROBOTA, NAO (Cabibihan et al., 2013; Dautenhahn, 2003; Robins and Dautenhahn, 2014).

Second category does not resemble human features, but biological features such as animals and examples of those (non-anthropomorphic) robots are PLEO, Keepon and Paro (Kim et al., 2013; Kozima et al., 2009; Scassellati et al., 2012).

In the third category are included (non-biomimetic) robots which are not resembling to any biological creature like Roball (Michaud et al., 2003).

The use of the above robots are essential in the therapeutic interventions for ASD children, facilitating them in the psychoeducation of emotions as well as teaching them appropriate social skills and interaction with other children and outside world. In the area of robotics, these devices are known as social-assistive robots, which are major tool to help children to express their feelings and simulate a human like interaction (Scassellati et al., 2012).

MAIN FOCUS OF THE CHAPTER

The Use of Robotics in ASD

Children with autism can benefit from their interaction with tasks that are using technology (Weiner et al., 2014). Robotics it has been developed as a promising branch in the educational and therapeutic procedure in the work with students with Autism Spectrum Disorder (ASD). Using robotics can be very appealing to students with ASD as it can engage them in the educational and therapeutic goals. Another advantage is that robotics are working in a strictly structure context which minimizes the stress of students with ASD, especially because one of the characteristics of the disorder is that those students cannot handle sudden changes. (Scassellati, et al. 2012).

Part of successful application of robots in ASD children are due to robots predictability, in contrast with the unpredictable humans actions where ASD children are struggling to adjust themselves (Cabibihan et al., 2013; Duquette et al., 2008). Part of beneficial aspect of robot interaction to these children are their pedagogical aspect and facilitation of both emotional development and social skills (Duquette et al., 2008; Kim et al., 2013; Kozima et al., 2009).

ASD research in Canada has used a toy-like robotic platform which is called ROBUS (ROBot of Université de Sherbrooke) (Michaud et al., 2003). Due to the beneficial use of robots in therapy there is whole emerging area where many researches (Cabibihan et al., 2013; Kim et al., 2013; Kozima et al., 2009; Scassellati et al., 2012) have demonstrated examples of robotic assisted therapeutic interventions.

Research related to robots in ASD therapy have received great attention last few years, emphasizing the variety of different roles and purposes as well as the resulting benefits to ASD children.

Literature has highlighted different roles of the use of robots such as:

1. A diagnostic agent for early detection of autism
2. Friendly playmate that facilitates child's engagement in play activities.
3. A behavior-eliciting agent to facilitate children's behaviors including imitation, eye contact, turn-taking and self-initiation.
4. A social mediator between child and the therapist promoting the appropriate social skills aiming in generalizing learned behaviors to children's peers.
5. A social actor aiming at assisting the individual to educate in the appropriate behaviors for specific social situations.
6. A personal therapist providing personalized therapy for the child, in accordance with their preferences and needs.
7. A social skills educator teaching social interactions eliciting certain behaviors to improve children's social skills and communication
8. A facilitator of Triadic Interaction, prompting and reinforcing the collaboration of participants to build for example a LEGO construction (Cabibiham et al., 2013; Chung, 2019; Damianidou et al., 2020; Ferrari et al., 2009; Pennisi et al., 2016).

The Use of Robots in Enhancing Social Skills to Students With ASD- Socially Assistive Robots (SAR)- Research Findings

Robotics are part of the everyday life and it has been used in education as well. In the previous years there has been an interest in the use of "social robotics" (Scassellati et al., 2012). This is an application where robots are executing a simulation of social interaction with humans. They can imitate feelings expressions, they can use no verbal signs and reproduce dialogues. Assistive robotics are also another use of robotics that have been used for people with Disabilities. The combination of Social Robotics and Assistive Robotics is The Social Assistive Robotics (S.A.R), which gives emphasis to social skills but also aims to create motivation and produce change to dysfunctional behaviors (Tapus et al. 2013). An advantage of this approach compared to virtual reality is that children can "touch" the robot and they can use repetition that can help imitation more easily (Huijnen et al., 2004). Educators and therapist can use robots individually as well as in group sessions. The sessions are based in three stages: designing, interaction and evaluation.

Pennisi et al. (2016) in their systematic review on social robots and their contribution in autism therapy found out that the use of social robots can be beneficial in enhancing social skills and spontaneous language. Also they mentioned that stereotyped behaviours were reduced.

The Appearance of the Robots- Presentation of Some of the Most Popular Robots That They Have Been Used in ASD

In their study Cabibiham et al. (2013) found out that the appearance of the robots is playing a crucial role to their efficacy in therapy. We have to mention that not all the children that have been diagnosed with autism have the same reactions towards the appearance of robots. The size should adjust to the child's height in order to help interactive communication and eye contact. The child must know that the robot is a mechanical creation and not a human, although its appearance looks like a human.

According to the types of robots that they have been used in therapeutic settings with children with ASD, they can be categorized in a) anthropomorphic when the appearance imitates human characteristics, b) non anthropomorphic when the robots are animals for example or c) non biomimetic when the robots are not representing biological species such as cars for example. Please read again the following sentence: Also except the appearance it is very important and the motion of a robot to the impression that gives to children with autism (Kumazaki et al., 2020).

Some of the most popular robots that they have been applied for students with ASD are:

Anthropomorphic Robots

AuRoRa (Autonomous Robotic platform as a Remedial tool for children with Autism). This robot aims to imitation and understanding of cause and effect, recognition of feelings and expressions and in learning socially acceptable behaviors in the interaction among child and teacher and among students (Wainer et al., 2010).

CHARLIE (CHild-centered Adaptive Robot for Learning in an Interactive Environment), is a social robot that has been created in order to help children with autism participate in interaction with human for effectively. It is a low cost robot that also can inform the researcher or the therapist automatically for the interactions (type and time) of the child and give an objective measuring of the responses of the child (Boccanfuso & O'Kane, 2011).

FACE (Facial Automaton for Conveying Emotions), a female android, has been used for enhancing the social, emotive skills and emotion recognition for children with autism (Pioggia et al., 2007).

Infanoid, is an upper torso humanoid with the appearance of a child 3 to 4 years old that can express a variety of gestures and enchases social introduction through gaze and facial emotional expressions (Kozima et al., 2004; Kumazaki et al., 2020; Pioggia et al., 2007). According to a survey that has been contacted by Kozima et al. (2004) children with autism through their interaction with the Infanoid, enhanced their social ability because of the predictable behavior of the robot.

IROMEC (Interactive Robotic Social Mediators as Companions). IROMEC robot is using play scenarios that they are aiming to help children with autism in sensory, cognitive and motor development, communication and social interaction (Ferrari et al., 2009).

KASPAR. This is a social robot simulating real human size. It has been created from Adaptive Systems Research Group, University of Hertfordshire, in order to research the interaction between child and robot. KASPAR according to Weiner et al. (2014) can play a crucial role in the participation of children with autism in playing collaborative games with their peers. In their study they conducted 23 controlled play sessions and the results indicated an improvement in social behaviors and collaborative skills in children with autism (Huijnen et al., 2019; Wainer et al., 2014).

Kismet is an expressive robot that expresses approval, attention, prohibition, and soothing and helps the communication between the child and the therapist/care giver (Breazeal & Aryananda, 2002).

MARIA. This is an anthropomorphic robot (acronym for "Mobile Automatic Robot For Interaction with Autistics"). The idea behind the construction of MARIA was to customize a mobile robot which was able to offer a playful appearance which could facilitate the interaction and trigger their social skills like eye gaze, touch as well as imitation areas which lacking in children with ASD. An essential side of this robot is that has a child-size height making it easier for children to interact and it is also mobile so catching children's attention and it is balancing a style between human and robot (Cabibihan et al., 2013).

NAO, is a S.A.R that it has been used for several purposes in health and in education. (Shamsuddin et al., 2012). Tabus et al. (2012) in a series of single case experiments found out that the interaction with NAO in some cases can be beneficial for students with autism in certain domains such as eye gaze, initiations of movements and expression of happiness. The NAO humanoid robot can perform a variety of behaviors, such as movements, walking and can also recognize speech, sounds, and objects (Chung, 2019).

ROBOTA, is a doll that consists of LEGO parts with human characteristics and aims to the cultivation of empathy and imitation and social skills, such as imitation, turn-taking, and role-switching (Billard, 2007; Kumazaki et al., 2020).

TITO. Tito is 28 inches tall and haw colored clothes (red, yellow and blue), tt can move (walk and rotate the head, move mouth and eyes). Tito can helps in social interactions but it seems to have better results in students with high-functioned autism (Duquette et al., 2008)

TREVOR (Triadic Relationship EVOking Robot). Trevor is a humanoid LEGO robot and before the therapy session you can programme the necessary actions that their required for the particular session (Giullian et al., 2010)

Troy. Troy has the size of a 4 years old child and has over than 30 actions that can be used in the therapeutic setting. These actions include "raising both arms straight up, using an arm to push an object, and bouncing its forearm to tap a tambourine or xylophone" (Giullian et al., 2010, p.2560)

Qtrobot, tries to captivate children's attention in order to teach them social skills and improve their emotional recognitions and reactions (Costa et al., 2017)

Milo, is a robot which is able to give orders to students but also at the same time in the screen that it has in the face projects visual material. It helps children to maintain the levels of attention, educate them in social skills and in the expression and recognition of feelings. Also is able to narrate stories to children but also listening and participating in children's stories (Yousif et al., 2019).

Non Anthropomorphic Robots

Bobus is a non-anthropomorphic robot that is being used in increasing the social interaction skills to students with autism (e.g. it plays music when the child is approaches the robot (Cabibihan et al., 2013).

Diskcat, This robot has the appereance of a cat and is equipped with dancing and visual effects with LEDs as eyes and resistive bend sensors which are used as whiskers. She has a routine when she enters the recreation room starting right away to follow the walls and repeating that over and over again. (Michaud et. al., 2003)

Jumbo. This is a robot with elephant shape and is equipped with a moving head and trunk, one pyroelectric sensor and an infrared range sensor. Robot is programmed to approach the child and to stop at a distance of 20 cm and then asks the child to touch one of the three buttons associated with pictograms located on its back. If the child is successful robot raises its trunk and rewarding the child playing some music (Michaud et al., 2003)

Keepon, has been created from Hideki Kozima, in Japan. This robot is equipped with two cameras in the eyes and a microphone in the nose. This allows robot to detect visual and audio stimuli and its adjusting his behavior according to these stimuli. (Kozima, et al., 2009). Keepon is used to create a triadic interaction between the therapist, the child with autism and the robot (Tapus et al., 2012).

NeCoRo, has the appearance of a cat. It has seven sensors that helps it to adjust his behavior to the environment (Shamsuddin et al., 2012),

Ozobot™ .This is a hand held robot that is using sensors in order to help students with ASD to acquire basic academic skills and enhance the collaboration of student and teacher as well as student with peers (Knight et al., 2019).

Pleo, this robot looks like a dinosaur pet toy and is equipped with a camera based vision system for light detection and navigation, ground foot sensors, touch sensors and an orientation tilt sensor for body. Pleo is able to express emotions with the use of motions and sounds responding to children's touch or imitating other human interactions. (Kim et al., 2013).

Probo, is a robot which is designed to replicate and offer a natural interaction with humans and is controlled by a user friendly Robotic User Interface (Vanderborght et. Al., 2012).

Non Biomimetic

-Roball, is a robot which can interact using vocal messages and movements patterns like spinning, shaking or pushing. Roboll is a spherical robot which is capable of navigating in all kinds of environments without getting stuck somewhere or falling on the side (Michaud et al., 2003).

INTERVENTIONS IN SCHOOL AND IN THERAPY –THE ROLE OF ROBOTS

Robots are used to communicate, express and perceive emotions, maintain social relationships, interpret natural cues, and develop social competencies. (Peng et al. 2014; Pérez-Vázquez et al., 2020; Pop et al, 2013). Educators and therapist can use robots individually as well as in group sessions, targeting a variety of goals that include but are not limited to: imitation, eye contact, joint attention, turn taking, emotion recognition and expression, self-initiated interactions, and triadic interactions (Knight et al., 2019). In essence, robots in therapeutic and teaching sessions may be designed to take up different roles, even during the same session, they can be designed to function autonomously, generate interesting and interactive interplays, and sustain attention through various ways (Fachantidis et al., 2020; Michaud et al, 2007). As such, they may be used to facilitate games and interactive communicative episodes, provide encouragement and immediate positive reinforcement to the child's actions. Fachantidis et al. (2020) implemented a programme with a 3D LEGO robot in a typical class of 22 students (one students was diagnosed with ASD). After the implementation of the programme the social, emotional and communicational skills of the student with ASD were improved and the attitudes of the typical students to their classmate with ASD were more positive.

So et al. (2020) in their study with preschool children diagnosed with ASD implemented a robot-based play-drama intervention to promote joint attention and functional play behaviors. Results showed that there was improvement in these skills something that was also mentioned from their parents.

In their review, Cabibihan et al. (2013) identify the following categories on the roles of robots:

Robots as diagnostic agents. Robotics technology can provide a method for early autism detection, specifically focusing on non-verbal social behaviors. Based on the robot's ability to reproduce the exact same actions from one interaction to another, they can be an efficient addition to the clinicians' process of repeating actions during their interactions with children. A series of robots have been designed for this purpose.

Robots as friendly playmates. In one-child-one-robot scenario, social robots can be developed for offering personalized, play interactions to children with ASD, tailored to their interests and needs. Hence, the use of robots may help in the creation of an engaging, safe, and enjoyable environment, one that enhances the children's social and cognitive development.

Robots as behaviour eliciting agents. Important target behaviors that are considered pre-requisite for children's social development, include imitation, eye contact, turn-taking and self-initiation. Robots can be successfully used to promote such behaviours in engaging and fun activities that may include teaching a child to initiate greetings, turn taking (i.e. when throwing a ball), changing gazes between the robot and an object of common interest, and imitating the robot's movements in dance activities. In a recent research (David et al., 2020), which used at the same time robots and human intervention in turn taking activities, the authors mention that in both interventions contexts children presented progress but they tended to look more at their robotic partner.

Robots in the role of social mediators and social actors. Robots help enact context-appropriate social experiences that help children with ASD practice important social skills as well as generalize the learnt social skills in their environment. In essence the predictable but progressively changing actions of robots provide a structured and supportive environment, supportive of children's growth according to their level of abilities.

Robots in the role of personal therapists. Robots can be designed to provide personalized therapy for every child, in accordance to the child's preferences, disabilities, and needs by taking advantage of a robot's modular features and customizable appearance.

Overall, socially interactive robots hold important benefits for children with autism. For one, they are more complex than humans and they may provide structured, predictable, and easier to follow instructions, compared to humans. In the case of robots, we may avoid the complexities of verbal and nonverbal communication, facilitating the participation of children with ASD. At the same time, it is possible to make social interactions more engaging and interesting for children, due to the robots' physical characteristics (including gesture, speech, and touch). Finally, robots can be programmed to meet the specific, individualized needs of the child that is using them while they offer predictability and repetition—all-important characteristics for enhancing children's with autism social and emotional development. Marino et al. (2020) in their randomized control trial implemented a cognitive behavioural therapy intervention with a social robot. They found out that the socio-emotional understanding of children with ASD (aged 4 to 8 years old) was improved.

Research of Begum et al (2016) showed possible goals of robots application in ASD Interventions.

Social and emotional Development. Teaching appropriate life skills necessary for social interaction like joint attention, friendship skills, symbolic play, social engagement, social problem skills, participation in group activities and interpersonal skills, emotional regulation. Kumazaki et al. (2019) found out that young children (aged 3 years old) with ASD seemed to motivate more to social interactions with the use of robots than humans.

Communication Development: Eye contact, verbal expression, initiative questions, requesting, labeling, receptive and expressive language, conversation, greetings, speech, pragmatics, non-verbal language.

Behavioral skills Development: Reducing maladaptive behavior, eliminating aggression and tantrums, repetitive behaviors, depression, anxiety, non-functional patterns of behaviors, developing new interests or activities.

Academic development: Teaching academic learning skills related to schools readiness such as learning readiness, higher cognitive functions, sensor and motor skills required for a specific job (Begum et al., 2016).

To conclude, child-robot interaction sessions hold the potential to help children gain a better understanding of the world, despite the challenges they may face. Such interactions may include specific social skills, emotional awareness abilities, and their communication with the environment and people around them. Hence, social robots can be an important teaching and therapeutic tool for teaching new skills to children with autism, enhancing playful moments, and for eliciting positive social behaviours from them (Fachantidis et al., 2018).

THE USE OF ROBOTS FOR CHILDREN WITH ASD IN GREECE

Although treatments based on robotics are the subject of countless studies around the world, in Greece unfortunately, the literature on robotics in the therapeutic process of autism, has been limited. However, scientists in Greece are trying to create and use robotics in certain therapeutic settings in Greece.

The Department of Education and Social Policy of the University of Macedonia (PAMAK) proceeded in the creation of a SAR system with the collaboration of Sociobotix (a team which works with robotics). It is a stuffed toy - robot in the shape of a daisy (Social Flower), which is exploited by the innovative application - Arrow (Assist Relations Robotic Workfellow) method. The method incorporates "features of speech, expression and movement" in the toy - robot, in order to enhance the child's motivation for social interaction. Results of a study by Pliasa and Fachantidis (2019), showed that the literature on enhancing social skills through robots is confirmed. Indeed, the social Flower robot appeared to be the first social assistance robot, featuring a daisy presentation (rather than a humanoid) and participating in activities that combine both digital and board games (Knight et al., 2019; Kumazaki et al., 2019; Pliasa & Fachantidis, 2019). The robot in question, therefore, aims to increase social skills and the maximum possible inclusion of children with autism in groups or classes with other neurotypical children. This method can be used by teachers, therapists and parents.

The particular appearance of the robot piques the attention of children leading to participation in triune and binary activities that promote co-social abilities. The interest, enthusiasm and attention of children is promoted in this way. The analysis of the results showed the improvement in children's performance after the interventions with Daisy in three of the four observed skills (activity in a group, information, turn taking). The Daisy robot, with its stable and not impulsive behavior, seems to be able to create an environment that respects the deficits of children on the autism spectrum, where children are mobilized to become part of the group, to express themselves verbally and to participate in activities and develop social skills (Pliasa & Fachantidis, 2019).

Another recent research on the treatment involving robots for children on the autism spectrum, despite providing limited results due to the spread of the Covid 19 pandemic, highlights, while examining modern and asynchronous tele-education, the advantages of modern tele-education. The main finding is that the modern educational process with the exploitation of humanoids piques enhances the attention of children, achieving better academic results. At the same time, according to the research findings, psychoeducation under these conditions is more effective (Huijen et al.2019; Lytridis et al., 2020; So et al., 2020).

Following the asynchronous digital material, a specifically designed scenario for the live session was developed, specifically for the robot-assisted group. The live session was supervised and guided by the therapist. The therapist made the conference call, controlled the robot and initiated the various activities. The objective of the synchronous sessions was the investigation of the children's response to the social robot during a video conference call for a given therapeutic scenario. Moreover, results from the live sessions were used to compare the asynchronous and the synchronous methods.

The advantages of the synchronous approach STEM from the application of central aspects of the original intervention protocol, i.e., the enhancement of the children's engagement, to provide the motivation to actively participate in the various activities, to allow interaction using direct verbal or nonverbal communication, as well as implement imitation and joint attention-oriented activities. The sessions were held via Zoom. The activities selected were based on the originally designed scenario, which was adapted to the needs of the children (Lytridis et al., 2020):

Greeting: The robot greets the child using gestures and gives instructions on how to conduct the video call. Instructions include the way the child should stand in front of the camera and how the child should speak so that effective visual and auditory interaction can be achieved.

Communication skills: In the second part the robot initiates an imitation exercise designed to determine if the child remembers the imitation skills obtained during the in-person sessions. At the same time, the scenario is attempting to teach the child how to conduct under the special circumstances of the pandemic. For this reason, embedded in the actions to be imitated are explanations why these actions are needed. For example, the child is shown how to say hello from a distance using gestures, or explanations why wearing a mask is needed.

Dance: The robot asks the child whether it wants it to perform a dance and requests that the child replicates the movements as an exercise. The dance is chosen from the list of dances that are actually part of the original intervention scenario, so that the child is already familiar with the movements and music.

Relaxation: The relaxation scenario involves breathing exercises guided by the robot, conducted in the same manner as the in-person meetings. At the end of the relaxation activity, the robot says goodbye and the session is concluded by the therapist.

The study has shown that there are benefits in providing personalized support material to children with ASD during the pandemic where there are no in-person sessions, both in the form of asynchronous as well as in the form of synchronous delivery of the material. While it has been shown that even digital material with no interaction properties can potentially mitigate the negative effects of abrupt cessation of an intervention, the addition of a certain amount of interaction in the live sessions with the robot-assisted group resulted in higher connectedness, and has shown to be the parents' preferred method of psychoeducation material delivery compared to the asynchronous approach, which involved videos and presentations (Marino et al., 2020; Lytridis et al., 2020; So et al., 2020).

Karakosta, Dautenham et al. (2019) studied the humanoid Kaspar in a Greek special school that offered educational provision only to children diagnosed with ASD and Cognitive Disabilities. The main purpose was to support the social and communication skills of children on the autism spectrum. Research-based robots create a familiar and predictable environment in which children with autism feel comfortable to play and move (Karakosta et al., 2019; So et al., 2020).

The results from this study indicate that the play sessions with Kaspar appear to have positively influenced the behaviors of some of the children in specific domains such as communication and interaction, prompted speech, unprompted imitation and focus/attention. Furthermore, the children's teachers expressed positive views regarding the impact of the play sessions on the children and offered interesting suggestions about the ways in which the robot could potentially be used in everyday teaching tasks and were eager to obtain a Kaspar for their classroom activities (Karakosta et al., 2019).

The play sessions with the robot took place in a quiet room in the school and were conducted by the researcher.The room was designed so as not to distract the child's attention from external factors during the sessions. Kaspar was placed on a table where the child would sit facing the robot, whilst the researcher was sitting next to the table, near the child. Two video cameras were placed in the play sessions room: one camera was placed close to the robot facing the child in order to capture the facial expressions of the child and one was placed lateral to the scene in order to record a wider view of the interactions between the child, Kaspar and the researcher (Karakosta et al., 2019).

Although Kaspar is an educational tool weighted in English culture and although it was not weighted in Greek data, the results of the children's interactions with Kaspar were encouraging. The majority of participants highlighted the unique and innovative role of the robot mainly in enhancing the social and communicative abilities of children on the autism spectrum (Huijnen et al., 2019; Karakosta et al., 2019;Wainer et al., 2014).

Throughout the course of the survey, questionnaires were completed by teachers, on the basis of which the development of children was evaluated and highlighted. In cases of low-functioning children the sessions were harder to achieve. In the majority of cases, however, the ability to express improved, the children began to mimic the movements of the robot as well as the sequence of movements and facial expressions improved. The current research highlights the lack of relevant Greek research. This work has demonstrated that the approach towards robot-assisted therapy for children with autism with the Kaspar robot, which has been studied extensively in schools and family homes in the UK, can also be applied to educational settings in other countries (Karakosta et al., 2019).

In a survey by Dardani, Kaburlasos and Amanatidis (2020) on the application of interactive social robots in the educational process for children on the autism spectrum observed that the interest of children increases resulting in more committed to the task they undertake. The aim of the research was to design and test humanoid interactive robots in a series of structured activities targeting areas of neurocognitive function within three distinct conditions: a. robots as facilitators, b. robots as participants, and c. existence of a facilitator robot and a participant robot within the same session. The two-robot condition enabled children to observe expected behaviors by creating a playful context. The research exploited the known 56cm humanoid NAO that entails qualities which are extremely well-adapted to the context of special education. Its 25 degrees of freedom allow a great versatility for rich movements making the intervention with children more accessible. A range of play-based activities targeting areas of behavioural and neurocognitive functioning were selected. Two major categories were the following: 1. activities targeting at social and communicative skills, and 2. activities targeting at executive functioning. This was a preliminary study that led to the first research initiative in Greece aiming towards designing, expanding and testing the efficacy of interactive social robotics applications in educational interventions of children with ASD (Dardani, et al., 2020).

In a survey of Holeva et al. (2019) presenting Robot-Assisted psychosocial intervention for children on the autism spectrum, emphasis is placed on presenting the effectiveness of social robots by engaging a large sample of students and testing the effectiveness of robot (NAO)-based treatment of ASD children compared to conventional human (therapist)-based treatment. The protocol was developed by the collaboration of a clinical scientific team with a technical scientific team. The modular structure of the protocol allows for implementing parametrically a number of tools and/or theories such as the theory-of-mind account from psychology. This paper focuses on the methodological and design details of the proposed intervention protocol (Holeva et al. 2019).

It is clear, therefore, that the research activity in Greece on robotics for children with autism has started. However, future research is important in order to improve the educational conditions and to ensure the permanence of acquired skills. In recent years the use of social robots to assess and support people on the autism spectrum has been the subject of much research. However, research on the use of robots in the evaluation process is limited. The study of the effect of robots on therapeutic support has seen a greater impact. In any case, the research in Greece is still at a fairly early stage, therefore it is necessary to carry out further research that highlights the positive effects on social and communication skills of individuals.

DISCUSSION

The ever-growing use of technology in education and therapy of children with ASD has highlighted the different roles that robots can play in the educational and social growth of children with ASD. Current literature presented in this chapter highlights that robots may create predictive and controlled learning environments, adjusted to the needs of children with ASD due to the nature of autistic disorder (Ferrari et al., 2009; Lorenzo, Vasquez, Lledo, et Lledo., 2020; Michaud et al., 2003; Pennisi et al., 2016). Specifically, robots can produce beneficial results to children with ASD and improve many areas of their lives. For example, central place in the literature holds the use of robots for the development of social skills, with emphasis on creating motivation and modifying challenging behaviors (Scassellati et al. 2012), the cultivation of imitation and empathy (Kumazaki et al., 2020), and the teaching of academic skills related to school readiness (Begum et al., 2016).

Although the structure of the interventions vary among different studies, the activities are mainly categorized in two conditions: interaction child-robot-human and interaction human-child (i.e. Lorenzo et al., 2018) with emphasis on the repeating exposure to the stimuli as it is produced by robots. Each robot is equipped with different characteristics but they are all designed to stimulate the cognitive and social-emotional abilities of children with ASD. Increased interactions facilitated by robots hold the potential to enhance the emergence and improvement of crucial social behaviors like verbal initiation and interaction, eye gaze, touch, intimacy and general imitation of a variety of skills. Verbal initiation for example although at first place is becoming a mechanical imitation for the child, the educational and repeating nature of the application could function as a 'vehicle' to copy this behavior with their peers. When the child is able to establish the verbal initiation the robotic friend can be substituted by a human friend so child is now in better position to generalize the skill he/she has trained with the robot and finally acquired.

Robot NAO for example is one of the most famous robots and many researches (Tapus eat al., 2012) have shown the gradual improvement of eye gaze in children with ASD. Other researches (Costa et al, 2017) have mentioned that robots can act like a social mediator between ASD child and other individual, producing positive results, having being trained to a gradual familiarization of the child eye gaze from robot to the human. Robot Probo which has the figure of an animal is able to express its self verbally using also facial expressions prompting children to imitate and establish a better eye contact adding to previous researches which highlighted the benefits of robots in developing eye gaze.

Robot Pleo (Kim et al., 2013) has shown that children with ASD exhibited significant increase in their verbal abilities and communication in the interaction with it rather than the human interaction. Some researches (Kumazaki, 2020) have claimed that the younger the child the better is trained and producing positive results, although that cannot verified from current researches and more future data will needed to examine the role of age in the language development of children with ASD.

Similarly, other robots have proved to be positive for ASD children in their attempt to familiarize with tactile stimuli allowing itself to be aware of his/her presence and the presence of others. While ASD children are often overwhelmed by tactile interactions, many robots, like Kaspar, are designed to prompt these interconnections while at the same time they offer measuring on the quality of the interaction in terms of frequency and intensity (Wainer et al., 2010).

Symbolic play, an important area of intervention for children with ASD can, also, be supported through the use of robots. Current research (see Pop et al., 2014) have shown that the use of robots like Probo hold the potential to facilitate the improvement of symbolic play and consequent instances of social interactions, promoting children's with ASD social inclusion and psychic retreat. Of interest to the use of robots in the design of interventions with robots, participation of children in a collaborative and functional play was even more productive when the robot acted like a colleague (Pop et al., 2014). Also, research has also shown that stereotypical behaviors, one of main characteristics of ASD children, were reduced with the use of robots compared to children's interaction with human colleague (Ferrari et al., 2009).

Overall, growing evidence indicates the potential for robotics to function and work complementary as a valuable colleague in evaluation and implementation of innovative interventions and significantly improve many areas of ASD children. However, it is important to highlight that despite the benefits of the use of robots (i.e. Coecklbergh et al, 2016; Penissi et al, 2016) they cannot replace the vital role of the psychologist/special educator. Rather, it is the presence of humans in the implementation of interventions along with the use of robots that makes it for the successful results—otherwise, children cannot imitate and generalize their new skills. Similarly, emphasis is given on ethical issues when robots are called to replace the human presence of therapist or special educator with concerns centering on possible emotional and physical damage, due to the lack of human interaction. Therefore, we suggest a careful examination, design and implementation of the use of robots in educational and therapeutic settings as well as on the use of autonomous robots or affect-aware interactive robots so as to explore the potential risks from the implications of their use. Within this lens, robotics may be included carefully and harmoniously in the therapeutic and pedagogical process in two ways: 1) By creating understanding of where the strengths or advantages of robots might be in an intervention for children, and 2) By informing the delivery of mediated interventions, ones that use research insights as well as the gained awareness and knowledge of the professionals in the field as well as the people with ASD (Mostajo et al., 2021).

CONCLUSION

Our review has come to show the critical role of technology and in particular the role of the robotics in the development of social skills for students who have been diagnosed with Autism Spectrum Disorder (ASD). Robots are able to transfer cognitive and behavioral interventions, which are beneficial for children with ASD, while they include teaching of the appropriate communication skills with consequent improvement of several aspects of their emotional and social deficits and finally having an impact so as to live a more functional and satisfactory and life. Taking into account the social and emotional difficulties of the ASD children it is of great importance that children receive different and richer stimuli like different kind of sounds and pictures which function as a vehicle to bring a major change and improvement of emotions and skills. Our review showed that games and robots are able to stimulate and facilitate the development of social skills in different levels like verbal, face expression, general imitation as well as tactile sensitivity aspects which are crucial for ASD children functionality. A significant part of robots effective quality is coming from their anthropomorphic appearance which resembles some human features but also non anthropomorphic which can deliver and stimulate social and cognitive skills and abilities in children with ASD. Robots, especially the so called social-assistive, have been prooved essential in the facilitation of expression of ASD children emotions as well as in teaching the appropriate skills, simulating human like social interactions, and finally improving their communication and quality of interaction with their peers and the outside world.

REFERENCES

Allen, R., Davis, R., & Hill, E. (2013). The effects of autism and alexithymia on physiological and verbal responsiveness to music. *Journal of Autism and Developmental Disorders*, *43*(2), 432–444. doi:10.100710803-012-1587-8 PMID:22752845

American Psychiatric Association. (2013). *Diagnostic and statistical manual of mental disorders* (5th ed.). Author.

Begum, M., Serna, R. W., & Yanco, H. A. (2016). Are Robots Ready to Deliver Autism Interventions? A Comprehensive Review. *International Journal of Social Robotics*, *8*(2), 157–181. doi:10.100712369-016-0346-y

Billard, A., Robins, B., Nadel, J., Dautenhahn, K. (2007). Building Robota, a Mini Humanoid Robot for the Rehabilitation of Children With Autism. *Assistive Technology, 19*, 37-49. . doi:10.1080/10400435.2007.10131864

Boccanfuso, L., & O'Kane, J. M. (2011). CHARLIE: An adaptive robot design with hand and face tracking for use in autism therapy. *International Journal of Social Robotics*, *3*(4), 337–347. doi:10.100712369-011-0110-2

Breazeal, C., & Aryananda, L. (2002). Recognition of affective communicative intent in robot-directed speech. *Autonomous Robots*, *12*(1), 83–104. doi:10.1023/A:1013215010749

Cabibihan, J. J., Javed, H., Ang, M., & Aljunied, S. M. (2013). Why Robots? A Survey on the Roles and Benefits of Social Robots in the Therapy of Children with Autism. In International Journal of Social Robotics (Vol. 5, Issue 4, pp. 593–618). doi:10.100712369-013-0202-2

Chung, E. Y. (2019). Robotic Intervention Program for Enhancement of Social Engagement among Children with Autism Spectrum Disorder. *Journal of Developmental and Physical Disabilities*, *31*(4), 419–434. doi:10.100710882-018-9651-8

Coeckelbergh, M., Pop, C., Simut, R., Peca, A., Pintea, S., David, D., & Vanderborght, B. (2016). A survey of expectations about the role of robots in robot-assisted therapy for children with ASD: Ethical acceptability, trust, sociability, appearance, and attachment. *Science and Engineering Ethics*, *22*(1), 47–65. doi:10.100711948-015-9649-x PMID:25894654

Costa, A., Steffgen, G., & Ziafati, P. (2017). *Socially assistive robots for teaching emotional abilities to children with autism spectrum disorder.* CIFIL: Characterization of Railway Infrastructure using Lidar View project Quality of work View project. www.luxai.eu

Damianidou, D., Eidels, A., & Arthur-Kelly, M. (2020). The Use of Robots in Social Communications and Interactions for Individuals with ASD: A Systematic Review. *Advances in Neurodevelopmental Disorders*, *4*(4), 357–388. doi:10.100741252-020-00184-5

Dardani, C., Kaburlasos, V., & Amanatidis, A. (2020). *Enhancing the Application of Interactive Social Robots in Educational Interventions for Autism Spectrum Disorders: A Research Initiative in Greece.* E-Poster Presented in INSAR 2020. https://insar.confex.com/insar/2020/techapp/eposterview.cgi?eposterid=278

Dautenhahn, K., Nehaniv, C. L., Walters, M. L., Robins, B., Kose-Bagci, H., Mirza, N. A., & Blow, M. (2009). KASPAR – A Minimally Expressive Humanoid Robot for Human–Robot Interaction Research. *Applied Bionics and Biomechanics, 6*(3–4), 369–397. . doi:10.1155/2009/708594

David, D. O., Costescu, C. A., Matu, S., Szentagotai, A., & Dobrean, A. (2020). Effects of a Robot-Enhanced Intervention for Children With ASD on Teaching Turn-Taking Skills. *Journal of Educational Computing Research*, *58*(1), 29–62. doi:10.1177/0735633119830344

Duquette, A., Michaud, F., & Mercier, H. (2008). Exploring the use of a mobile robot as an imitation agent with children with low-functioning autism. *Autonomous Robots*, *24*(2), 147–157. doi:10.100710514-007-9056-5

Fachantidis, N., Syriopoulou-Delli, C. K., Vezyrtzis, I., & Zygopoulou, M. (2020). Beneficial effects of robot-mediated class activities on a child with ASD and his typical classmates. *International Journal of Developmental Disabilities*, *66*(3), 245–253. doi:10.1080/20473869.2019.1565725 PMID:34141387

Fachantidis, N., Syriopoulou-Delli, C. K., & Zygopoulou, M. (2018). The effectiveness of socially assistive robotics in children with ASD. *International Journal of Developmental Disabilities*. Advance online publication. doi:10.1080/20473869.2018.1495391 PMID:34141373

Ferrari, E., Robins, B., & Dautenhahn, K. (2009). Therapeutic and educational objectives in robot assisted play for children with autism. *Proceedings - IEEE International Workshop on Robot and Human Interactive Communication*, 108–114. 10.1109/ROMAN.2009.5326251

Foster, J. M. Jr, I. R. C., & Romesberg, F. E. (2014). The Clinical use of robots for individuals with Autism Spectrum disorder. *Research in Autism Spectrum Disorders*, *509*(7500), 385–388. doi:10.1016/j.rasd.2011.05.006.The

Fujimoto, I., Matsumoto, T., de Silva, P. R. S., Kobayashi, M., & Higashi, M. (2011). Mimicking and evaluating human motion to improve the imitation skill of children with autism through a robot. *International Journal of Social Robotics*, *3*(4), 349–357. doi:10.100712369-011-0116-9

Giullian, N., Ricks, D., Atherton, A., Colton, M., Goodrich, M., & Brinton, B. (2010). Detailed requirements for robots in autism therapy. *Conference Proceedings - IEEE International Conference on Systems, Man and Cybernetics*, 2595–2602. 10.1109/ICSMC.2010.5641908

Holeva, V., Nikopoulou, V. A., Papadopoulou, M., Vrochidou, E., Papakostas, G. A., & Kaburlasos, V. G. (2019). Toward Robot-Assisted Psychosocial Intervention for Children with Autism Spectrum Disorder (ASD). In Lecture Notes in Computer Science: Vol. 11876. *Social Robotics. ICSR 2019*. Springer. doi:10.1007/978-3-030-35888-4_45

Huijnen, C., Lexis, M., & Witte, L. (2017). Robots as New Tools in Therapy and Education for Children with Autism. *Robots as New Tools in Therapy and Education for Children with Autism.*, *4*(4), 1–4. doi:10.4172/2376-0281.1000278

Huijnen, C. A. G. J., Lexis, M. A. S., Jansens, R., & de Witte, L. P. (2019). Roles, Strengths and Challenges of Using Robots in Interventions for Children with Autism Spectrum Disorder (ASD). *Journal of Autism and Developmental Disorders*, *49*(1), 11–21. doi:10.100710803-018-3683-x PMID:30019273

Karakosta, E., Dautenhahn, K., Syrdal, D. S., Wood, L. J., & Robins, B. (2019). *Using the humanoid robot Kaspar in a Greek school environment to support children with Autism Spectrum Condition*. De Gruyter.

Kasari, C., Freeman, S., & Paparella, T. (2006). Joint attention and symbolic play in young children with autism: A randomized controlled intervention study. *Journal of Child Psychology and Psychiatry, and Allied Disciplines*, *47*(6), 611–620. doi:10.1111/j.1469-7610.2005.01567.x PMID:16712638

Keen, D., & Halle, J. W. (2016). *Prelinguistic and Minimally Verbal Communicators on the Autism Spectrum* (H. Meadan & N. C. Brady, Eds.). Springer Verlag. doi:10.1007/978-981-10-0713-2

Kim, E.S., Berkovits, L.D., Bernier, E.P., Leyzberg, D., Shic, F., Paul, R., & Scassellati, B. (2013). Social robots as embedded reinforcers of social behavior in children with autism. *Journal of Autism Development Disorder, 43*, 1038–1049. . doi:10.1007/s10803-012-1645-2

Knight, V. F., Wright, J., & DeFreese, A. (2019). Teaching Robotics Coding to a Student with ASD and Severe Problem Behavior. *Journal of Autism and Developmental Disorders*, *49*(6), 2632–2636. doi:10.100710803-019-03888-3 PMID:30734176

Knight, V. F., Wright, J., Wilson, K., & Hooper, A. (2019). Teaching Digital, Block-Based Coding of Robots to High School Students with Autism Spectrum Disorder and Challenging Behavior. *Journal of Autism and Developmental Disorders*, *49*(8), 3113–3126. doi:10.100710803-019-04033-w PMID:31055684

Kozima, H., Michalowski, M., & Nakagawa, C. (2009). Keepon:A Playful Robot for Research, Therapy, and Entertainment. *International Journal of Social Robotics*, *3*(18), 3–18. Advance online publication. doi:10.100712369-008-0009-8

Kozima, H., Nakagawa, C., Kawai, N., Kosugi, D., & Yano, Y. (2004). A humanoid in company with children. *2004 4th IEEE-RAS International Conference on Humanoid Robots, 1*, 470–477. 10.1109/ICHR.2004.1442138

Kroiss, A., Sonego, D., & Rollins, P. (2014). *Using a Humanoid Robot as a Co-therapist with Children with ASD Mean Percent Engaged Child instructs Milo.* . doi:10.1109/TNSRE.2012.2230188.Diehl

Kumazaki, H., Muramatsu, T., Yoshikawa, Y., Matsumoto, Y., Ishiguro, H., Kikuchi, M., Sumiyoshi, T., & Mimura, M. (2020). Optimal robot for intervention for individuals with autism spectrum disorders. *Psychiatry and Clinical Neurosciences*, *74*(11), 581–586. doi:10.1111/pcn.13132 PMID:32827328

Kumazaki, H., Warren, Z., Swanson, A., Yoshikawa, Y., Matsumoto, Y., Yoshimura, Y., Shimaya, J., Ishiguro, H., Sarkar, N., Wade, J., Mimura, M., Minabe, Y., & Kikuchi, M. (2019). Brief Report: Evaluating the Utility of Varied Technological Agents to Elicit Social Attention from Children with Autism Spectrum Disorders. *Journal of Autism and Developmental Disorders*, *49*(4), 1700–1708. doi:10.100710803-018-3841-1 PMID:30511126

Leekam, S. R., & Ramsden, C. A. (2006). Dyadic orienting and joint attention in preschool children with autism. *Journal of Autism and Developmental Disorders*, *36*(2), 185–197. doi:10.100710803-005-0054-1 PMID:16502142

Lorenzo, G., Lledó, A., Arráez-Vera, G., & Lorenzo-Lledó, A. (2018). The application of immersive virtual reality for students with ASD: A review between 1990–2017. *Education and Information Technologies*, *24*(1), 1–25. doi:10.100710639-018-9766-7

Lytridis, C., Bazinas, C., Sidiropoulos, G., Papakostas, G. A., Kaburlasos, V. G., Nikopoulou, V. A., Holeva, V., & Evangeliou, A. (2020). Distance special education delivery by social robots. *Electronics (Basel)*, *9*(6), 1–13. doi:10.3390/electronics9061034

Marino, F., Chilà, P., Sfrazzetto, S. T., Carrozza, C., Crimi, I., Failla, C., Busà, M., Bernava, G., Tartarisco, G., Vagni, D., Ruta, L., & Pioggia, G. (2020). Outcomes of a Robot-Assisted Social-Emotional Understanding Intervention for Young Children with Autism Spectrum Disorders. *Journal of Autism and Developmental Disorders*, *50*(6), 1973–1987. doi:10.100710803-019-03953-x PMID:30852783

Michaud, F., Duquette, A., & Nadeau, I. (2003). Characteristics of mobile robotics toys for children with pervasive developmental disorders. *Proceedings of the IEEE International Conference on Systems, Man and Cybernetics*, *3*(May), 2938–2943. 10.1109/ICSMC.2003.1244338

Michaud, F., Salter, T., Duquette, A., Mercier, H., Lauria, M., Larouche, H., & Larosse, F. (2007). Mobile robots engaging children in learning. *CMBES Proceedings*, 30.

Mostajo, S. T., Legaspi, O. M., Camarse, M. G., & Salva, R. A. (2021). Exploring the potentials of robotics in supporting children with autism spectrum disorder. *IAFOR Journal of Education*, *9*(1), 77–93. doi:10.22492/ije.9.1.05

Mundy, P., & Neal, R. (2001). Neural plasticity, joint attention and autistic developmental pathology. *International Review of Research in Mental Retardation*, *23*, 139–168. doi:10.1016/S0074-7750(00)80009-9

Peng, Y. H., Lin, C. W., Mayer, N. M., & Wang, M. L. (2014, November). Using a humanoid robot for music therapy with autistic children. In *2014 CACS International Automatic Control Conference (CACS 2014)* (pp. 156-160). IEEE. 10.1109/CACS.2014.7097180

Pennisi, P., Tonacci, A., Tartarisco, G., Billeci, L., Ruta, L., Gangemi, S., & Pioggia, G. (2016). Autism and social robotics: A systematic review. *Autism Research*, *9*(2), 165–183. doi:10.1002/aur.1527 PMID:26483270

Pérez-Vázquez, E., Lorenzo, G., Lledó, A., & Lorenzo-Lledó, A. (2020). Evolution and Identification from a Bibliometric Perspective of the Use of Robots in the Intervention of Children with ASD. Technology. *Knowledge and Learning*, *25*(1), 83–114. doi:10.100710758-019-09415-8

Pioggia, G., Sica, M. L., Ferro, M., Igliozzi, R., Muratori, F., Ahluwalia, A., & De Rossi, D. (2007). Human-robot interaction in autism: FACE, an android-based social therapy. *Proceedings - IEEE International Workshop on Robot and Human Interactive Communication*, 605–612. 10.1109/ROMAN.2007.4415156

Plexousakis, S., Georgiadi, M., Kourkoutas, E., Gkintoni, E., Halkiopoulos, K., & Roumeliotou, V. (2020). Enhancing sexual awareness in children with Autism Spectrum Disorder: A case study report. In *Cases on Teaching Sexuality Education to Individuals with Autism* (pp. 79–98). IGI Global. doi:10.4018/978-1-7998-2987-4.ch005

Pliasa, S., & Fachantidis, N. (2019). Using Daisy Robot as a Motive for Children with ASD to Participate in Triadic Activities. *Themes in ELearning*, *12*(12), 35–50.

Pop, C. A., Simut, R., Pintea, S., Saldien, J., Rusu, A., David, D., Vanderfaeillie, J., Lefeber, D., & Vanderborght, B. (2013). Can the social robot Probo help children with autism to identify situation-based emotions? A series of single case experiments. *International Journal of HR; Humanoid Robotics*, *10*(03), 1350025. doi:10.1142/S0219843613500254

Roberts-Yates, C., & Silvera-Tawil, D. (2019). Better education opportunities for students with autism and intellectual disabilities through digital technology. *International Journal of Special Education*, *34*(1), 197–210.

Samson, A. C., Huber, O., & Gross, J. J. (2012). Emotion regulation in Asperger's syndrome and high-functioning autism. *Emotion (Washington, D.C.)*, *12*(4), 659–665. doi:10.1037/a0027975 PMID:22642342

Scassellati, B., Admoni, H., & Matari, M. (2012). Robots for Use in Autism Research. *Annual Review of Biomedical Engineering*, *14*(1), 275–294. doi:10.1146/annurev-bioeng-071811-150036 PMID:22577778

Shamsuddin, S., Yussof, H., Ismail, L. I., Mohamed, S., Hanapiah, F. A., & Zahari, N. I. (2012). Humanoid robot NAO interacting with autistic children of moderately impaired intelligence to augment communication skills. *Procedia Engineering, 41*(Iris), 1533–1538. . doi:10.1016/j.proeng.2012.07.346

Shire, S. Y., Gulsrud, A., & Kasari, C. (2016). Increasing responsive parent–child interactions and joint engagement: Comparing the influence of parent-mediated intervention and parent psychoeducation. *Journal of Autism and Developmental Disorders*, *46*(5), 1737–1747. doi:10.100710803-016-2702-z PMID:26797940

So, W. C., Cheng, C. H., Lam, W. Y., Huang, Y., Ng, K. C., Tung, H. C., & Wong, W. (2020). A Robot-Based Play-Drama Intervention May Improve the Joint Attention and Functional Play Behaviors of Chinese-Speaking Preschoolers with Autism Spectrum Disorder: A Pilot Study. *Journal of Autism and Developmental Disorders*, *50*(2), 467–481. doi:10.100710803-019-04270-z PMID:31655965

Tapus, A., Peca, A., Aly, A., Pop, C., Jisa, L., Pintea, S., Rusu, A., & David, D. (2012). Children with Autism Social Engagement in Interaction with Nao, an Imitative Robot-A Series of Single Case Experiments. *Interaction Studies: Social Behaviour and Communication in Biological and Artificial Systems*, *13*(3). Advance online publication. doi:10.1075/is.13.3.01tap

Vanderborght, B., Simut, R., Saldien, J., Pop, C., Rusu, A. S., Pintea, S., Lefeber, D., & David, D. O. (2012). Using the social robot Probo as a social story telling agent for children with ASD. *Interaction Studies: Social Behaviour and Communication in Biological and Artificial Systems*, *13*(3), 348–372. doi:10.1075/is.13.3.02van

Venker, C. E., McDuffie, A., Ellis Weismer, S., & Abbeduto, L. (2012). Increasing verbal responsiveness in parents of children with autism: A pilot study. *Autism*, *16*(6), 568–585. doi:10.1177/1362361311413396 PMID:21846665

Wainer, J., Dautenhahn, K., Robins, B., & Amirabdollahian, F. (2010). Collaborating with Kaspar: Using an autonomous humanoid robot to foster cooperative dyadic play among children with autism. *2010 10th IEEE-RAS International Conference on Humanoid Robots*.10.1109/ICHR.2010.5686346

Yousif, J. H., Kazem, H. A., & Chaichan, M. T. (2019). Evaluation Implementation of Humanoid Robot for Autistic Children. International. *Journal of Computation and Applied Sciences*, *6*(1), 412–420.

KEY TERMS AND DEFINITIONS

Assimilation With Real Life: The use of technology in order to help children with communication difficulties face real life circumstances in control environments.

Autism Spectrum Disorder: Developmental Disorder with deficits in communication level and in social skills.

Communication Deficits: Lack of vital elements which facilitate the interaction with others in everyday life.

Development of Communication: Helping children gain certain communicational skills in order to function in school, family as well as in outside world.

Robots in ASD: Using Robotics in ASD can minimize children's stress and support the implementation of therapeutic and social goals.

School Interventions: Tasks and goals that they have to be achieved to develop children's emotional/ social level and their general functionality in the school environment.

Social Skills: The capabilities which children need so as to be functional in their everyday life.

Chapter 9
Integrating Assistive Robotics in STEM Education to Empower People With Disabilities

Muhammet Demirbilek
Faculty of Education, Suleyman Demirel University, Turkey

Tarık Talan
https://orcid.org/0000-0002-5371-4520
Gaziantep Islam Science and Technology University, Turkey

ABSTRACT

Science, technology, engineering, and math (STEM) education plays a vital role in order to prepare students for the twenty-first century economy. Unfortunately, people with disabilities in countries throughout the world are vastly underrepresented in scientific and technological fields. Closing the gap in the STEM is vital for improving fairness and opportunities for the people with disabilities. The STEM industry is critical in a world where there are various environmental, health, and development issues. It is essential not only for increasing productivity and competitiveness, but also for ensuring individual and societal well-being. The STEM workforce is expected to rise exponentially in the future year in today's burgeoning knowledge-intensive economy. This chapter explores a current state-of-the-art literature review on the use of assistive robots in STEM education people with disabilities.

INTRODUCTION

Today's individuals are the workforce of tomorrow. The jobs that will be available to our students when they graduate will require a higher understanding of technology, a greater confidence in problem-solving, and more collaboration with professionals around the world. STEM in education plays a critical role in preparing students for careers as adults. Almost 80% of future careers will require some STEM skills. Therefore, connecting STEM with real data in real problems might require educational institutions to be equipped with labs and highly qualified and effective teachers, updated curricula, and connections to

DOI: 10.4018/978-1-7998-7443-0.ch009

higher education institutions and enterprises. Similarly, students with special needs should be prepared for the workforce of the future. Therefore, integrating STEM into the special education curriculum is essential to integrating them into life and developing the basic skills for workforce readiness in the 21st century (Moeller, 2012). The process of integrating STEM into authentic special education contexts is not an easy task. Special education teachers struggle to make connections across the STEM disciplines (Kutlu, Schreglmann & Cinisli, 2018). Global challenges such as migration, climate change, overpopulation, health, agriculture, and declining water resource and energy requires to rethink current approaches to STEM education in both general education and special education. Therefore, the STEM industry is critical in a world where there are various environmental, health, and development issues. Not only it is critical for enhancing productivity and competitiveness, but it also helps to ensure the continued well-being of individuals and societies. In today's burgeoning knowledge-intensive economy, the STEM workforce is projected to grow exponentially in the coming years.

Unfortunately, however, individuals with special needs are vastly underrepresented in scientific and technological fields in countries throughout the world. When it comes to pursuing successful and long-lasting STEM careers, they often face multiple barriers including a lack of access to STEM tools, STEM curriculum, and methodologies to teach STEM to special needs (Kutlu, Schreglmann & Cinisli, 2018). Closing the gap in STEM is critical to improving fairness and opportunity for people with disabilities. Education should aim at the full development of human potential and personality, talents and creativity to enable people with disabilities to participate effectively in society. Robotics can be a tool for individuals with special needs to integrate them in society. Therefore, educational robotics can be used in the classroom as a tool for cooperative learning to enhance the relational skills of all students, including those with physical and cognitive disabilities, and to create an emotionally proximal learning environment. A robot designed to help to improve certain ability of a person with special needs can be called assistive robot. Assistive robots have the potential to serve as surrogates and helpers for the people with special needs (Chen et al., 2013). For instance, children with autism are more interested in communication therapy when assistive robots involve the therapy (Papakostas et al., 2021; Robins et al., 2006).

The United Nations (UN) Convention on the Rights of Persons with Disabilities in Article 24 (UN, 2021) enshrines their right to education without discrimination and on an equal opportunity basis. Education must aim at the full development of human potential and personality, talents and creativity, enabling people with disabilities to participate effectively in society. In modern society, teachers need to update and increase their digital skills with increasing frequency. The tools that can be available to an educator are numerous and he must therefore develop the skills necessary to select these products and create effective educational activities.

Disability can be explained as physical impairment or a health condition that prevents an individual from taking full advantage of life's opportunities such as activities of daily living, recreation, vocation, and education. An individual with disabilities can experience problems arising from health condition (WHO Report, 2011).

A robot is called assistive when the device is able to acquire information from the context, process it and carry out actions that can bring benefits to people with disabilities and in old age during daily activities (Stranger & Cawley, 1996; Uluer et al., 2015). An assistive robot is therefore a tool able to compensate for the difficulties of the disabled user, commonly understood as physical ones. Educational robotics can be used in the classroom as a tool for cooperative work aimed at learning and at the same time to enhance the individual abilities of all students, including those with difficulties. Assist therefore wants to spread this use of educational robotics and new technologies in a social sense and as a tool for strengthening the cognitive and basic functions of disabled students (Di Battista et al., 2020).

Assistive robotics technology has made significant progress in recent decades. With the advent of improved computing power and hardware capabilities, new tool and platform capabilities assistive robots are now much more complicated and better (Campeau-Lecours et al., 2016). Assistive robotics is attracting the interest of special education institutions. Considering the STEM education, robotics coding skills are important competencies for individuals with special needs. Robotic applications are in high demand in many industrial sectors and fields like healthcare, military, engineering, service, industrial, entertainment and space exploration. The positive results achieved in these fields have also attracted the attention of educators. Robots can be seen in all aspects of our daily lives. They have a promising future to find their place in the field of education. Robotic applications have been used in many levels of education from K-12 to university level. The robotics has become one of the most popular areas of educational research. The use of robotic technologies as educational materials has spread to in many disciplines such as information technologies, mathematics and science, foreign languages, music, sign language teaching, and special education. Recently, educational robotics applications have been promoted by many institutions and organizations around the world. Robotics is a part of summer school STEM curriculum in many educational institutions.

The definition of robots is varied. It can be said that there is no consensus has been attained to define the robot. In general, autonomous or pre-programmed electromechanical vehicles are called robots. Usually, they can be controlled by a computer. It is also possible to control the robots by remote control. According to the definition of robots made by the American Robot Institute in 1979, robots are programmable, multifunctional machines that can move objects, parts, tools or special devices from one place to another or to perform specific tasks (Jablonski & Posey, 1985).

Czech author Karel Capek Rossum first used the word robot in the 1921 play "Universal Robots of Rosssum" (Hemal & Menon, 2018). The word "robota" in the Czech language means "forced labor, slave".

Isaac Asimov used and popularized the term robotics in science fiction novels and short stories. Asimov was a visionary in the 1930s who dreamed of a positronic brain to control robots. Asimov stated that the purpose of robots in the Law of Three Robotics is to serve people and that a robot must never prefer its own goals to the goals of humans. The provisions of this law are as follows:

1. A robot cannot hurt or injure a person.
2. A robot must obey the orders given to it, as long as it does not conflict with the first law.
3. A robot must survive as long as it does not conflict with the first and second provisions.

Robots are generally preferred for the following reasons. However, cost is always considered in this choice.

- Jobs with high labor costs
- Jobs when extreme attention is required
- Risky and dangerous jobs in terms of health and work safety when done by human
- Jobs requiring very high temperatures
- Non-dangerous but repetitive and boring jobs

Nowadays, the fields of application of robots are increasing. This situation also triggers fears among people that this will lead to job losses. Robots can be classified according to their fields of application.

Robots in Various Fields

There are different types of robots designed to perform different types of tasks. For example, in medicine and healthcare, robots are commonly used in surgical procedures and in physical therapy and rehabilitation. On the other hand, industrial robots are mainly used in dangerous, tiring, repetitive and demanding industrial production tasks. Robots in entertainment are the robots designed to entertain people and children. Robots in transportation were usually developed to provide logistical support. Today, self-driving vehicles are an example of this category. Robots in agriculture and livestock; these types of robots were developed to save time and money in agricultural fields and gardens. They can perform many tasks such as picking fruit, watering, spraying, harvesting and fertilizing. Operational robots are designed to perform tasks that are too dangerous to employ by humans. Operational robots are generally controlled by remote devices. For example, mine cleaning robots. Educational robots are generally preferred to teach basic skills to students. Robotics are commonly used in language teaching, mathematics teaching, and coding in STEM education (Crnokić et al., 2017).

Robots in Education

Robots are gradually becoming more integrated into people's daily life, both at home and at school (Smyrnova-Trybulska et al., 2020). Various ways of employing robots in education have been discussed in numerous articles and books. The use of robots in the field of education began in the 1960s with Seymour Papert's Logo (programming language) Turtle (robot). It is a computer-based environment in which Lego / Logo design activities can be carried out in the classroom (Hixon, 2007). Logo is a material where students design their own machines with Lego parts consisting of motors and sensors and write a computer program to control the machine. Resnick, Ocko and Papert have noted that for Logo, there are not only lego pieces and software, but also a specific approach to learning and thinking (Curto & Moreno, 2016).

The use of robotics in education has revealed the concept of educational robotics. It is well known that especially in recent years, robotic applications are being used all over the world. These applications occupy a wide place in the field of educational sciences. Educational robotics can be defined as a system that consists of computers, electrical materials and programs that essentially aim to acquire information (Junior et al, 2013). Educational robotics applications involve a process that integrates engineering, technology, science and mathematical applications. Robotic applications have come to the forefront in recent years, especially in terms of teaching 21st century thinking skills to students (Blikstein, 2013). The purpose of projects that involve robotic applications is to provide teachers with curricula in which science and technology are together, to make learning more permanent and meaningful by combining robotic applications with education (Wood, 2003). Employing robotics in the classroom can motivate students to choose future careers in fields related to STEM subjects (Curto & Moreno, 2016).

The literature highlights that educational robotics attracts the attention of educators and researchers to support interdisciplinary learning activities in the development of cognitive and social skills and learning STEM fields (Alimisis, 2013; Mataric, 2004). Although the field of educational robotics fundamentally takes engineering education as a model, it also encompasses science, technology and mathematics-based instructional activities. It can be stated that robots that are easily accessible to individuals, easier to use, and have lower costs are an integral part of the engineering and science education process.

The interest in using robots in education is increasing day by day. Therefore, sales of educational robots are on the rise (Lindh & Holgersson, 2007; Weinberg & Yu; 2008). Many countries around the world are conducting extensive research on educational robots. Activities such as robotics competitions and robotics camps are being organized to provide cognitive and social benefits to students around the world. Literature state that robots have grown significantly in the field of education. It is expected that market share will grove in recent years and will grow even faster in the coming years (Barreto & Benitti, 2012).

When the literature is examined, it can be said that educational robotics has four main goals (Barak & Assal, 2018; Chaudhary et al., 2016; Ching et al., 2019; Karaahmetoğlu, 2019; Talan, 2020; Ucgul & Cagiltay, 2014; Yolcu & Demirer, 2017). The first goal is to support the teaching of subjects such as engineering applications, design, robot production and programming, and artificial intelligence. The second is to acquire and develop STEM knowledge and skills at an early age. The third is to develop broad learning skills such as exploration, engineering design, product-oriented thinking, creative thinking, analytical thinking, more willingness to research and explore, teamwork. The fourth main goal of educational robotics is to increase the willingness of individuals to engage in mathematics, science and technology, and to break down psychological or cultural barriers to engaging in these areas.

There are basically two different areas for the use robots in education (Chandra, 2014; Malec, 2001; Yolcu & Demirer, 2017). If desired, both electronic and mechanical product development and programming processes can be used for both purposes. The first of these areas is the use of robots in postgraduate education at university. Educational robotics at this level; engineering applications are used in teaching subjects such as artificial intelligence, design and robotic production. The other area is the application of pedagogical achievements within the K-12 education level, and their use for representing the learning experience through doing and experiencing. In this area, prefabricated learning robot kits are utilized. The use of educational robot kits at the K-12 level is intended to develop high-level thinking skills such as problem solving, creativity and cooperation.

Advantages and Disadvantages of Employing Robotics in Education

Robotic applications are an integral part of the engineering and science education process and can contribute to education in many ways (Aksu, 2019; Kilinc, 2014; Talan, 2020; Yavuz-Konokman & Cukurbasi, 2019). The development of educational robotics has accelerated adopting robotics in educational context. Students Abstract concepts can be transformed into a concrete understanding by students while employing robotic in daily life situations (Sullivan et al., 2013). In this way, students assimilate the building blocks of technological developments at an early age, learn advanced technological applications at an early age, and realize more effective development of technology with the participation of more people in technology (Chandra, 2014). In this context, educational robotic applications can positively contribute to students' attitudes towards STEM carriers, and contribute to the development of many skills that students did not have before. In the literature, it has been revealed that robots are motivating and engaging for students. They are also an effective tool to increase student motivation and learning, and facilitate understanding of scientific concepts (Chang et al., 2010; Chen & Wang, 2011; Mitnik et al., 2009). Robotic activities in the classroom provide students with many skills such as problem solving, critical thinking, discovering their own abilities, learning by doing, and increasing their level of proficiency with technology (Costa & Fernandes, 2004; Gulbahar, 2018; Kim et al., 2019). Therefore, educational robotic applications have a great part in the acquisition of basic skills that the 21st century brings (Barreto & Benitti, 2012; Eguchi, 2010). Integrating robotics in the classroom contributes to the development of students' fine motor skills and eye-hand coordination (Bers et al., 2014). Robotic activities increase students' creativity due to its interdisciplinary nature that brings together the fields of educational robotics, computers, engineering and science (Rockland et al., 2010). Educational robotic applications also help learners relate to daily life, enable permanent learning, provide interdisciplinary understanding, ensure learning by doing and experiencing, and active participation (Kaygisiz et al., 2020). Robotic applications also stand out as a realistic tool to improve students' skills such as project management, teamwork and communication (Ersoy et al., 2011).

Although the use of robots offers many benefits in terms of learning process and outcomes, there are studies in the literature that point out limitations (Talan, 2020). The cost of equipment: robots and its parts are expensive. School may not afford to buy robots. Robots need electricity and battery to operate. Robots and its parts are susceptible to be broken. Teachers and student should be careful while handling the robots (Aksu, 2019; Erten, 2019). Robot sets are very expensive tools and it is difficult for any school to buy and use the necessary educational and technical equipment (Akca, 2020; Eroglu & Bektas, 2016; Hacker, 2003). Furthermore, the interruptions in the Bluetooth connection, the slowing down of the motor speed due to the weakening of the batteries and the failure of the sensors in the recognition process may cause some problems during the activity (Akman-Selcuk, 2019; Kilinc, 2014). -In addition, difficulties in the robot design process, lack of textbooks, and focus on extracurricular competitions can be mentioned as disadvantages of robotic activities (Gundogdu, 2020). Teachers' inability to provide adequate classroom supervision during implementation and lack of interest and support from school administrators can be referred to as another important problem related to the use of robotic applications in the classroom (Alimisis, 2013; Rahman et al., 2017; Yang et al., 2008). Such problems have been found in the literature to cause a decrease in motivation and an increase in aggressive behavior among students (Lykke et al., 2014). However, considering the contributions to the widespread use of STEM education and the experience gained in the field by improving students' programming skills, it is again evident how useful educational robotics applications are (Talan, 2020; Yang et al., 2008).

Educational Robots and STEM Education

STEM education can be used to give students a powerful educational program if it is based on learner-centered instruction. However, children must find ways to thrive in dynamic, hands-on learning situations in order to be successful. This is especially true for students with learning disabilities. Although the STEM classroom naturally provides support for students with special needs, it is beneficial for them to have support outside of the classroom from close family members and friends. This is what inspires children not to give up and pushes them to keep trying. One of the areas where robots can be used in the classroom is in national and international competitions. The main purpose of robotics competitions; apart from encouraging students to pursue careers in science and technology, it also allows them to do robotics projects and showcase their projects. One of the most important features of robotics applications in education is that they lend themselves to an interdisciplinary approach to teaching (Catlin, 2016). Especially with the introduction of the STEM approach, interdisciplinary studies have increased and educational robots are now commonly used in classrooms. Some robotics competitions are even designed to provide a fun, hands-on way for students to learn and apply concepts and principles from STEM. Robotics activities may improve students 'competencies in math and science addressing math and science concepts through trial and error testing and problem solving. Therefore, one of the uses of educational robotics kits is STEM (Chandra, 2014). Educational robotics activities are a pedagogical tool that has a positive impact on the development of STEM skills and arouses curiosity (Mead et al., 2012). Some studies show that robotics is an effective tool for the acquisition and development of basic knowledge and skills related to STEM (Barreto & Benitti, 2012; Eguchi, 2010; Jim, 2010; Mataric et al., 2007). Most studies that investigated robotics activities found that students create better learning environments for mathematics and science lessons and learning is more motivating (Robinson, 2005; Rogers & Portsmore, 2004). In addition, research has demonstrated that learning robotics sets have positive effects on student achievement in STEM (Barreto & Benitti, 2012; Eguchi 2014; Sullivan & Heffernan, 2016). Costa and Fernandes (2004) found that supporting science and technology instruction with robotics activities increased students' skills such as awareness of their abilities, problem solving, critical thinking, willingness to use technology, and learning by doing and experimental learning. In this case, it can be said that educational robotics activities occupy an important place in the context of teaching STEM and 21st century skills. Moreover, the increasing use of robots in the classroom has increased students' interest in science and technology in the world (Mubin et al., 2013).

Robotic applications provide a rich environment for students to learn many aspects of mathematics, physics, engineering, and computer science (Stripling & Simmons, 2016). The use of such tools in education is an ideal tool to teach STEM concepts to children. In robotics applications, students conduct studies on how to interact with materials such as gears, motors, and sensors to explore abstract concepts in a concrete way. When using such materials, students learn STEM concepts in a fun and creative way while engaging in the projects they design (Bers et al., 2006). In other words, robotics lessons can be used for different purposes depending on the creativity of the teacher in the classroom. In this way, students become individuals who absorb technology at an early age, develop original ideas, and direct technology in the future (Karthik & Chandra, 2014).

Educational Robot Sets

Robotics is a very difficult and expensive process that requires mechanical, electronic, and programming skills. However, with the rapid development of technology, many companies have produced various learning robot sets that student can make with a plug-and-join approach without having to deal with steps such as cables and soldering (Eguchi, 2014). Learning robot sets with a variety of functions are easy to install and disassemble and play an active role in teaching students the basic concepts of technology and engineering. Robot sets made in different set groups for all age groups and needs increase their usefulness in teaching. These sets consist of plastic and electronic parts in different colors and sizes that can be plugged together, include a variety of sensors and motors, and allow students to design programmable robots. Educational robotics kits, which allow real-world objects to be controlled, give students the opportunity to learn freely and create original designs. These technologies, when combined with the right teaching environments, are believed to contribute significantly to students' problem-solving skills (Kucuk & Sisman, 2017).

Educational robot sets, often used in the application of various disciplines, are widely used in education today. Examples of educational robotics sets used to teach students robotics and coding, as well as the environments used to program these sets, are listed in the following table (Akman-Selcuk, 2019). A typical robotics training set contains a programmable brick or controller that operates different types of sensors, such as color sensor, touch sensor, gyro sensor, ultrasonic distance sensor, rotation sensor, wheels, and motors. In particular, it can be said that some robot sets do not require much electronics or hardware knowledge and their use is simple (Ozer, 2019; Tatlisu, 2020). However, although the use of these prefabricated technological materials is effective and interesting for STEM applications, it can be expensive and difficult to obtain (Koc, 2019).

One of the most popular robotics sets for the classroom are those developed by the LEGO company Education Mindstorms. The LEGO Mindstorms set, which has more extensive technology and equipment than other LEGO series, is a robot set used primarily in robotics competitions. Developed in the laboratories of Massachusetts Institute of Technology (MIT), LEGO robots consist of programmable building blocks based on constructivism. Moreover, in recent years, Arduino-compatible sets, which are considered more useful in terms of benefits and costs compared to other prefabricated robot sets in studies on robot coding, have been incorporated into the classroom through their increasing use (Gezici et al., 2017). Examples of Arduino-compatible sets include Aduino Uno mini-sumo robots, Arduino obstacle avoidance robots with MZ80 series, Arduino line follower robots, and Arduino light follower robots. Moreover, Bee-Bot, KiwiRobotic, RoboRobo, CeBot, Cubelets, KIBO Robotics, Lego MoretoMath, O- BOT, Dfrobot can be mentioned as examples of robot sets. Some of the learning robot sets mentioned here can be programmed using a block-based programming environment, while others can be programmed using common text-based programming languages such as Java and Python.

Assistive Technologies

Robots and assistive technologies can be used to promote the personal development of people with special needs, achieve a better quality of life, improve their functional abilities by involving them in educational processes, and train individuals who will contribute to life in society (Bers et al., 2014; Citil, 2016; Lancioni et al., 2013; Reed & Bowser, 2005; Valadao et al., 2015). Generally, any device or method used in the field of special education is referred to as assistive technology (Tekinarslan & Yikmis, 2005). In other words, assistive technologies are all types of products, tools, and devices that are specifically designed or adapted to maintain, enhance, or improve the functional abilities of individuals with special needs (Borg et al., 2009; Pettersson & Fahlstrom, 2010). Therefore, these technologies not only help individuals learn more successfully and productively, but also increase their confidence and independence (Sani-Bozkurt, 2017). In this case, assistive technologies are any tools used to both overcome and compensate for individuals' barriers, inadequacies, and difficulties to facilitate their participation in activities and to enhance and support their existing potential (Cay et al., 2020; Hersh & Johnson, 2008; Pettersson & Fahlstrom, 2010; Raskind & Higgins, 1999). As can be seen from the definition of assistive technologies, these technologies can be used to improve the skills of individuals with special needs, facilitate their participation in activities, and enhance and maintain their quality of life. They overcome the difficulties they may encounter (Hersh & Johnson, 2008; Lancioni et al., 2013). Assistive technologies help individuals with special needs increase their interaction with their environment and gain individual independence (Fok et al., 2011; Johnston et al., 2009). These technologies provide methods and tools for individuals with special needs to complete their tasks and solve problems independently (Raskind & Higgins, 1999). With the development of technology, tablets, computers, phones, virtual and augmented reality, cloud technology, smartboard, mobile tools, and game-based applications in particular have found their place among assistive technologies. Thanks to these technologies, individuals with special needs have the ability to perform many operations that they have difficulty performing without depending on someone and improve themselves, such as facilitating communication, manipulation, assistance with walking, standing up, guidance, and behavioral imitation.

Assistive Robots

Robots have entered human life in a variety of ways in recent decades. The importance of robotics is increasing in various fields, including its application as an assistive technology. The assistive robot is a robot that can perceive and process sensory information and perform actions that benefit disabled and elderly people. According to Van der Loos et al. (2016), an assistance robot is a device that can perceive and process sensory information and perform actions that benefit people with disabilities and the elderly. Assistance robots represent a new type of robot as they share a workspace and interact directly with humans. They are characterized by the fact that they can use sensors and sophisticated algorithms to recognize their environment and people, communicate multimodally with humans, navigate autonomously and make decisions on their own. Despite the fact that there have been few research studies in this area, the results of the studies conducted by researchers show that people, especially the elderly and people with disabilities, benefit from the use of robots in their daily lives. As the population of elderly and people with special needs grows in the coming decades, there will be a strong need for assistive robots to improve movement and promote independence.

Assistive robots were not developed to "take over" teaching, but to support, assist, and supplement it STEAM and other subjects. Therefore, they provide more opportunities for students with special needs to feel comfortable learning. In this case, the role of the teacher is to assess, observe and analyze the student's performance.

Benefits of Assistive Technology Use and Problems During Implementation

Robots are increasingly being developed to help teach and support areas such as social and emotional learning, social skills, and activities of daily living. Teachers can help their students explore the wonders of technology and take advantage of the opportunities it offers by creating a dynamic, hands-on atmosphere. In a traditional approach to learning, individuals with autism face barriers to learning that are overcome through robotic technology. They have access to a specially prepared curriculum for children with autism and other special needs. You can interact with a friendly, engaging robot who encourages children to take academic chances, persevere, and learn from their mistakes. They are a social companion that autistic children can trust and feel comfortable with. They also provide constant positive reinforcement, which can be very beneficial for students.

In recent years, with the integration of information communication technologies in education, the number of assistive technologies used in special education environments is increasing day by day. When used effectively, these technologies enhance the learning of individuals with special needs, make their lives easier, support their communication and social adaptation skills, and ultimately help them live independently. Assistive technologies help individuals to systematically receive feedback as a result of their activities, to repeat more, and to increase their rate of learning. By using these technologies, the learning process becomes more enjoyable and students can participate more actively in class. In addition, assistive technologies facilitate learning and recall, contribute to communication skills, and provide an opportunity for a learning environment that allows for repetition (Bahceci, 2019). Assistive technologies increase students' academic success and motivation by facilitating rapid and sustained learning (Kim et al., 2003). Assistive technologies also have significant benefits for skills such as supporting learning strategies and skills of individuals with special needs, providing more effective work opportunities, and acquiring social skills (Campigotto et al., 2013; Citil, 2016; Dautenhahn & Billard, 2002; Green, 2011; Ozdamar, 2016; Sani-Bozkurt, 2017). For this reason, it can be said that such technologies that facilitate the daily life of people with special needs increase their self-confidence, their ability to succeed, and improve their quality of life.

Professionals and teachers working in the field of special education can also benefit from assistive robotics to support the daily life and STEM skills of students with special needs (Kim et al., 2003). However, there are several studies in the literature that show factors that negatively affect the use of robotics in the classroom (Alammary et al., 2017; Alkahtani, 2013; Cay et al., 2020; Copley & Ziviani, 2004; Kutlu et al., 2018; Zhou et al., 2011). As a result of the findings from these studies, it was found that teachers have problems in using assistive technologies, they do not have sufficient knowledge and equipment about these technologies, and they have difficulties in accessing technical support. These problems lead teachers to have negative attitudes towards the use of assistive technology. In addition, lack of expert support in assistive technologies, technological inadequacies, limited technological capabilities, costly tools and equipment, and complexity of tools and software are other problems. Such problems can have a negative impact on the quality of education provided. Considering that assistive technologies occupy an important place in the education of persons with special needs, it can be said that such problems should be solved. In this regard, it is important to provide teachers with the necessary technical support so that they can use these technologies correctly and effectively. It is considered that it would be beneficial to organize in-service trainings for the use of these technological tools when needed. In addition, considering the skills of teachers in the use of these technological tools, it is necessary to work on the development of these skills and their pedagogical use. In addition, the training of teachers and professionals who provide the necessary financial support for the acquisition of assistive technologies is crucial to improve the effectiveness and use of these technologies.

Assistive Technology Types

Children, particularly those with impairments, have had numerous educational avenues opened to them thanks to technological advancements in assistive technologies. Alternative technology solutions can help people with physical, sensory, or cognitive disabilities in a variety of ways. For children with mobility, hearing, or vision impairments, assistive technology can help level the playing field. Much of the assistive technology we use on a daily basis was created to help people with disabilities. For example, Curb cuts and curb slopes, which were created to aid people with orthopedic disabilities, are now utilized more frequently by families with strollers or individuals with grocery carts than by people in wheelchairs or walkers. The optical character reader, which was created to help those who couldn't read written text, has been used in the workplace to convert printed papers into editable computer files, saving a lot of time and effort. Students with special needs often feel better about themselves as a result of using technology. For instance, a student who physically unable to speak can communicate with spoken language can use a portable voice synthesizer, the student can ask and respond to questions in the "regular" classroom, overcoming a physical obstacle. Improvements in sensor controls allow for tiny motor motions to control mobility equipment such as electric wheelchairs, allowing students and community members to move about independently. Text and graphics enhancement software can magnify portions of a monitor so that people with vision impairments can see them. For a blind individual, text can be read electronically using a digital voice synthesizer. Amplification devices can filter out background noise or pick up an FM signal from a microphone on a teacher's lapel for people with hearing difficulties. Teachers working with children with impairments now have more powerful and efficient tools because to advancements in assistive technologies. Teachers can use these tools to provide new and more effective ways of learning while also tailoring lessons to a wide range of student learning demands. Educators are employing computers as instruments to deliver and promote learning beyond drill and practice, to create learning environments, and to ensure that all students have access to improved and fair learning opportunities.

Depending on the complexity (usability, development and practicality) of the technology used, assistive technologies can be examined in three basic categories: low-tech, mid-tech and high-tech (Beard et al., 2011; Blackhurst, 1997; Cakmak, 2016; Elicin, 2016; Sani-Bozkurt, 2017; Smith, 2008):

- Low-tech auxiliary technological tools are traditional and simple tools that do not have electronic functions and are generally available in many classes. Examples of these are tools such as pencil holders, visual cards / pictograms, reading magnifiers, visual charts, highlighter markers and pens, adapted pens and worksheets, adapted scissors and a page-turner device. These tools are easy to obtain and cost effective.
- Mid-tech assistive technology devices are vehicles that operate with a low energy (usually battery). Timers, talking calculators, reading pens, tools such as voice recorders, and translation software can be given as examples.
- High-tech assistive technological tools are electronic ones. These tools include computers, tablets, smart boards, phones, smartwatches, virtual and augmented reality applications, digital books, mobile applications. These technological tools are considered to be advanced technological tools. Because the use of these tools is more complex than tools at other levels, training may be required to use them.

In addition, the classification of tools used in assistive technologies into seven categories has been discussed in the literature (Bryant & Bryant, 2003; Bryant et al., 2008; Sola-Ozguc, 2017). These are as follows;

1. **Positioning and Seating:** It means the most appropriate posture that individuals with special needs should take in order to function. It helps individuals to take the appropriate position to fulfill functions such as eating, moving from one place to another, sleeping and studying.
2. **Mobility:** These are tools that allow individuals with special needs to move between environments. Wheelchair, navigation devices, white canes and walkers allow people to move around in various environments.
3. **Augmentative and alternative communication:** These are tools that help individuals with special needs communicate with each other. These tools; it is used to develop difficult-to-understand speech and to provide an alternative communication for individuals who have difficulties in communication. Picture Exchange Communication System, Voice Output Communication Aid, communication boards, iPad applications are sample tools for augmentative and alternative communication systems.
4. **Computer Access:** These are systems that provide computer access to individuals with special needs. An example of this is that individuals with visual impairment can surf the internet with speech programs.

5. **Adaptive toys and games:** These are technologies that enable children with special needs to play with some games and toys in early childhood. These technologies contribute to the cognitive skill development of children. For example, a child pressing a button that makes a sound will be able to establish the cause-and-effect relationship between pressing the button and the sound.
6. **Aids to daily living:** These are tools that support individuals with special needs to fulfill their daily life skills such as transportation, going to school, and working in a job. Systems that help individuals with physical disabilities turn the lights on and off without leaving their seats can be given as examples of these tools.
7. **Instructional aids:** These are tools and equipment that help individuals who are academically affected by disability in the school or work environment. These tools can compensate the inadequacies of individuals. Screen reader programs that allow access to information, software that transcribe the sound can be given as examples of these tools.

CONCLUSION

Since the birth of STEM, robotics has taken a central role in the practice of STEM education. Robotics is an intersection of modern science, technology, engineering, and mathematics. Robotics has significant potential as a teaching tool and can have a positive impact on student engagement and learning in STEM subjects if schools master the effective and meaningful use of robotics. Therefore, the promotion of assistive robotics technologies in STEM and special education aims to improve students' skills and recognize their potential to meet the challenges of the future. Therefore, in order to improve experience and confidence, it is necessary to incorporate assistive robotics into educational settings. This could be achieved with an extended professional development workshop to train in-service educators. In addition, more robotics-related curricula should be included in pre-vocational education.

Assistive robotics may help students with disabilities to improve their spatial and tactile learning skills are an essential tool for students to obtain professional employment in STEM fields. It is vital that instructors continue to employ and test robotic technologies to make STEM education accessible for students with special needs so that students can learn STEM educational content through hands on experiences.

Robotic technologies have transformed learning environments to build computational thinking, computational thinking practice, understanding of engineering as a craft, and the engineering design process utilizing coding practices (Alimisis & Boulougaris, 2014). Robotic practices may expand on the students' digital confidence rather than requiring them to work with unfamiliar tools. In addition, facilitating robotics in STEM teaching practices to teach specific knowledge in physics, mathematics, and programming may improve students reasoning skills in STEM.

Robotics can be an effective tool to enhance 21st century skills such as creativity, teamwork, collaboration, and communication skills (Khanlari, 2013). Teaching with robotics can improve students' tacit knowledge and creativity in STEM related subjects (Gaudiello & Zibetti, 2016). However, simply using robotics in the STEM classroom when teaching students with special needs does not guarantee the expected student learning outcomes.

The content and teaching method plays a very important role in teaching robotics in STEM class for students with special needs (Chan Lye et al., 2013). Therefore, the pedagogical approach in teaching robotics in STEM is a key factor. Pedagogical robotics in STEM education for students with special needs can be used as a motivational tool. This is because students are motivated when they use new tools and technologies (Kraetzschmar, 2009). In addition, robotics as a socialization tool can provide opportunities for students to acquire a technology-oriented habitus and stimulate their interest in technology-related careers. Building robots requires the ability to work together as a team. Therefore, robotics can be promoted as a social playground for students with special needs. Engaging students in programming robots to complete specific tasks can also enhance their critical thinking and problem-solving skills.

Robotics technology has changed the way of STEM teaching to improve teachers' decision making and logical thinking so that they are able to turn ideas into reality. Robotics has the potential to change the way students learn and ultimately create a more knowledgeable and balanced student body and an engaging introduction to STEM educational topics. By using robotics in the classroom, students with special needs can achieve a new level of learning. In addition, problem solving and computational thinking skills are among the benefits of robotics. The most substantial benefit of robotics is that it serves as a stepping stone into the subjects of coding and STEM, where students learn to apply their science and math skills. The use of robotics in special education may improve communication and social skills for students with special needs. In addition, the biggest advantage of robotics is that it can be programmed to meet the needs of each individual student, providing special education in a much easier, accessible format. To realize the full potential of robotics in learning environments, educational institutions must first gain access to robotics technology and then take the time to learn how to deploy, operate, maintain, and utilize this unique technological opportunity.

The development of robotic technologies enabled educators to implement robots in special education processes and STEM related applications. Providing students and teachers with an understanding of the application and potential of this new type of technology will be important in preparing them for a world in which similar technologies will be increasingly commonplace, especially in STEM contexts and in special education. In addition, robotics gives students the opportunity to interact with robots in a variety of ways, giving them the opportunity to fail and know that they can try again. This helps to increase their confidence and imagination. To meet the needs of a modern technological society and to develop the desire and ability of the population, especially young people with special needs, to engage with science and technology, STEM education should be developed in schools, based on a formal and informal basis.

Robotics in the STEM classroom is an ongoing opportunity for students to use problem solving skills and higher-level thinking in a different way. It is critical to introduce students with special needs to STEM education because the majority of future jobs will involve one of these subjects. Students with special needs are not only able to participate in STEM activities, but they also enjoy them. If a student enjoys doing something, he/she will likely want to do it more often.

REFERENCES

Akca, O. F. (2020). *The effect of inquiry based robotic activities on students' conceptual success, logical thinking and science process skills in science centers* [Unpublished master's dissertation]. Gazi University, Ankara, Turkey.

Akman-Selcuk, N. (2019). *The investigation of educational robotics activities on middle school students' course motivation, robotic attitudes and success* [Unpublished doctoral dissertation]. Istanbul University, Istanbul, Turkey.

Aksu, F. N. (2019). *Robotic coding and robotic competitions from the perspective of the information and communication technology teachers* [Unpublished master's dissertation]. Balıkesir University, Balıkesir, Turkey.

Alammary, J., Al-haiki, F., & Al-muqahwi, K. (2017). The impact of assistive technology on down syndrome students in Kingdom of Bahrain. *The Turkish Online Journal of Educational Technology, 16*, 103–119.

Alimisis, D. (2013). Educational robotics: Open questions and new challenges. *Themes in Science and Technology Education, 6*(1), 63–71.

Alimisis, D., & Boulougaris, G. (2014). Robotics in physics education: Fostering graphing abilities in kinematics. In D. Alimisis, G. Granosik, & M. Moro (Eds.), *Proceedings from the 4th international workshop teaching robotics: Teaching with robotics & 5th international conference robotics in education* (pp. 2–10). Padova, Italy: Academic Press.

Alkahtani, K. D. F. (2013). Teachers' knowledge and use of assistive technology for students with special educational needs. *Journal of Studies in Education, 3*(2), 65–86. doi:10.5296/jse.v3i2.3424

Bahceci, B. (2019). *To examine attitudes of teachers working in the field of special education towards assistive technologies and individual innovation levels* [Unpublished master's dissertation]. Konya Necmettin Erbakan University, Konya, Turkey.

Barak, M., & Assal, M. (2018). Robotics and STEM learning: Students' achievements in assignments according to the P3 Task Taxonomy—practice, problem solving, and projects. *International Journal of Technology and Design Education, 28*(1), 121–144. doi:10.100710798-016-9385-9

Barreto, F., & Benitti, V. (2012). Exploring the educational potential of robotics in schools: A systematic review. *Computers & Education, 58*(3), 978–988. doi:10.1016/j.compedu.2011.10.006

Beard, L., Carpenter, L., & Johnston, L. (2011). *Assistive technology: Access for all students* (2nd ed.). Merrill Prentice-Hall.

Bers, M. U., Flannery, L., Kazakoff, E. R., & Sullivan, A. (2014). Computational thinking and tinkering: Exploration of an early childhood robotics curriculum. *Computers & Education, 72*, 145–157. doi:10.1016/j.compedu.2013.10.020

Bers, M. U., Rogers, C., Beals, L., Portsmore, M., Staszowski, K., Cejka, E., . . . Barnett, M. (2006, June). Innovative session: Early childhood robotics for learning. In *Proceedings of the 7th International Conference on Learning Sciences, International Society of the Learning Sciences*, (pp. 1036-1042). Academic Press.

Blackhurst, A. (1997). Perspectives on technology in special education. *Teaching Exceptional Children, 29*(5), 41–48. doi:10.1177/004005999702900509

Blikstein, P. (2013). Digital fabrication and "making" in education: The democratization of invention. In J. Walter-Herrmann & C. Büching (Eds.), FabLabs: Of machines, makers and inventors. Bielefeld: Transcript Publishers.

Borg, J., Lindstrom, A., & Larsson, S. (2009). Assistive technology in developing countries: National and international responsibilities to implement the convention on the rights of persons with disabilities. *Lancet*, *374*(28), 1863–1865. doi:10.1016/S0140-6736(09)61872-9 PMID:19944867

Bryant, D. P., & Bryant, B. R. (2003). *Assistive technology for people with disabilities*. Allyn and Bacon.

Bryant, D. P., Smith, D. D., & Bryant, B. R. (2008). *Teaching students with special needs in inclusive settings*. Allyn & Bacon.

Cakmak, S. (Ed.). (2016). *Özel eğitim ve yardımcı teknolojiler* [Lexical characteristics of Turkish language]. Vize Publishing.

Campeau-Lecours, A., Maheu, V., Lepage, S., Lamontagne, H., Latour, S., Paquet, L., & Hardie, N. (2016, July). JACO assistive robotic device: Empowering people with disabilities through innovative algorithms. *Rehabilitation Engineering and Assistive Technology Society of North America (RESNA) Annual Conference*.

Campigotto, R., McEwan, R., & Demmans Epp, C. (2013). Especially social: Exploring the use of an iOS application in special needs classrooms. *Computers & Education*, *60*(1), 74–86. doi:10.1016/j.compedu.2012.08.002

Catlin, D. (2016). 29 Effective ways you can use robots in the classroom. In *International Conference EduRobotics* (pp. 135-148). Springer.

Cay, E., Yikmis, A., & Sola-Ozguc, C. (2020). Experiences and opinions of special education teachers regarding the use of technology. *Journal of Qualitative Research in Education*, *8*(2), 629–648. doi:10.14689/issn.2148-624.1.8c.2s.9m

Chan Lye, N., Wonga, K. W., & Chiou, A. (2013). Framework for educational robotics: A multiphase approach to enhance user learning in a competitive arena. *Interactive Learning Environments*, *21*(2), 142–155. doi:10.1080/10494820.2012.705853

Chandra, V. (2014). Developing students' technological literacy through robotics activities. *Literacy Learning: The Middle Years*, *22*(3), 24–29.

Chang, C. W., Lee, J. H., Chao, P. Y., Wang, C. Y., & Chen, G. D. (2010). Exploring the possibility of using humanoid robots as instructional tools for teaching a second language in primary school. *Journal of Educational Technology & Society*, *13*(2), 13–24.

Chaudhary, V., Agrawal, V., Sureka, P., & Sureka, A. (2016, December). An experience report on teaching programming and computational thinking to elementary level children using lego robotics education kit. In *IEEE Eighth International Conference on Technology for Education (T4E)* (pp. 38-41). IEEE. 10.1109/T4E.2016.016

Chen, G. D., & Wang, C. Y. (2011). A survey on storytelling with robots. In *International Conference on Technologies for E-Learning and Digital Entertainment* (pp. 450-456). Springer.

Chen, T. L., Ciocarlie, M., Cousins, S., Grice, P. M., Hawkins, K., Hsiao, K., ... Takayama, L. (2013). Robots for humanity: Using assistive robotics to empower people with disabilities. *IEEE Robotics & Automation Magazine*, *20*(1), 30–39. doi:10.1109/MRA.2012.2229950

Ching, Y. H., Yang, D., Wang, S., Baek, Y., Swanson, S., & Chittoori, B. (2019). Elementary school student development of STEM attitudes and perceived learning in a STEM integrated robotics curriculum. *TechTrends*, *63*(5), 1–12. doi:10.100711528-019-00388-0

Citil, M. (2016). Bilişsel yetersizliği olan çocuklar için yardımcı teknolojiler. In Özel eğitimde yardımcı teknolojiler (pp. 152-180). Ankara: Vize Publishing.

Copley, J., & Ziviani, J. (2004). Barriers to the use of assistive technology for children with multiple disabilities. *Occupational Therapy International*, *11*(4), 229–243. doi:10.1002/oti.213 PMID:15771212

Costa, M. F., & Fernandes, J. (2004). Growing up with robots. *Proceedings of Hsci2004*. http://www.hsci.info/hsci2004/PROCEEDINGS/FinalPapers/E00461377837.pdf

Crnokić, B., Grubišić, M., & Volarić, T. (2017). Different applications of mobile robots in education. *International Journal on Integrating Technology in Education*, *6*(3), 15–28. doi:10.5121/ijite.2017.6302

Curto, B., & Moreno, V. (2016). Robotics in education. *Journal of Intelligent & Robotic Systems*, *81*(1), 3–4. doi:10.100710846-015-0314-z

Dautenhahn, K., & Billard, A. (2002). Games children with autism can play with Robota, a humanoid robotic doll. In *Universal access and assistive technology* (pp. 179–190). Springer London. doi:10.1007/978-1-4471-3719-1_18

Di Battista, S., Pivetti, M., Moro, M., & Menegatti, E. (2020). Teachers' opinions towards educational robotics for special needs students: An exploratory Italian study. *Robotics*, *9*(3), 72. doi:10.3390/robotics9030072

Eguchi, A. (2010). What is educational robotics? Theories behind it and practical implementation. In *Proceedings of Society for Information Technology & Teacher Education International Conference* (pp. 4006-4014). Academic Press.

Eguchi, A. (2014, July). Robotics as a learning tool for educational transformation. *4th international workshop teaching robotics, teaching with robotics & 5th international conference robotics in education*, 27–34.

Elicin, Ö. (2016). *Özel eğitimde teknoloji destekli öğretim*. Vize Publishing.

Eroglu, S., & Bektas, O. (2016). Ideas of science teachers took STEM Education about STEM based activities. *Journal of Qualitative Research in Education - JOQRE*, *4*(3), 43-67.

Ersoy, H., Madran, R. O., & Gulbahar, Y. (2011, February). *Programlama dilleri öğretimine bir model önerisi: Robot programlama*. XIII. Akademik Bilişim Konferansı, Malatya, Turkey.

Erten, E. (2019). *A case study on coding and robotic teaching* [Unpublished master's dissertation]. Balıkesir University, Balıkesir, Turkey.

Fok, D., Polgar, J. M., Shaw, L., & Jutai, J. W. (2011). Low vision assistive technology device usage and importance in daily occupations. *Work (Reading, Mass.)*, *39*(1), 37–48. doi:10.3233/WOR-2011-1149 PMID:21815311

Gaudiello, I., & Zibetti, E. (2016). *Learning robotics, with robotics, by robotics: Educational robotics*. Wiley-ISTE. doi:10.1002/9781119335740

Gezici, H., Kocaoglu, S., Cosgun, E., Yilmazlar, E., & Tuna, M. (2017). Planning the Arduino embedded programming course with robot project application for mechatronics programs. *Electronic Journal of Vocational Colleges, 7*(1), 1–7.

Green, J. L. (2011). *The ultimate guide to assistive technology in special education: Resources for education, intervention, and rehabilitation*. Prufrock Press.

Gulbahar, Y. (2018). *Bilgi işlemsel düşünmeden programlamaya*. Pegem Akademi. doi:10.14527/9786052411117

Gundogdu, B. (2020). *The effect of algorithm teaching on vocational high school students' computational thinking, cognitive load and achievement* [Unpublished master's dissertation]. Atatürk University, Erzurum, Turkey.

Hacker, L. (2003). *Robotics in education: ROBOLAB and robotic technology as tools for learning science and engineering* [Unpublished Senior Thesis]. Department of Child Development, Tufts University.

Hemal, A. K., & Menon, M. (2018). *Robotics in genitourinary surgery*. Springer. doi:10.1007/978-3-319-20645-5

Hersh, M. A., & Johnson, M. A. (2008). On modelling assistive technology systems. Part I: Modelling framework. *Technology and Disability, 20*(3), 193–215. doi:10.3233/TAD-2008-20303

Hixon, R. (2007). Teaching software engineering principles using Robolab and Lego Mindstorms. *International Journal of Engineering Education, 23*(5), 868–873.

Jablonski, J., & Posey, J. (1985). Robotics terminology. In *Handbook of industrial robotics*. Wiley.

Jim, C. K. W. (2010). *Teaching with LEGO mindstorms robots: Effects on learning environment and attitudes toward science* [Unpublished master's dissertation]. The University of Texas at Dallas, United States. Available from ProQuest Dissertations and Theses database. (UMI No. 1489918)

Johnston, C., Thurlow, M., Altman, J., Timmons, J., & Kato, K. (2009). Assistive technology approaches for large-scale assessment: Perceptions of teachers of students with visual. *Impairments Exceptionality, 17*(2), 66–75. doi:10.1080/09362830902805756

Junior, L. A., Neto, O. T., Hernandez, M. F., Martins, P. S., Roger, L. L., & Guerra, F. A. (2013). A low-cost and simple arduino-based educational robotics kit. *Cyber Journals: Multidisciplinary Journals in Science and Technology Journal of Selected Areas in Robotics and Control, 3*(12), 1–7.

Karaahmetoğlu, K. (2019). *The effect of project-based Arduino educational robot applications on students' computational thinking skills and their perception of basic STEM skill levels* [Unpublished master's dissertation]. Amasya University, Amasya, Turkey.

Karthik, K. P., & Chandra, R. P. (2014). An overview of agricultural robots. *International Journal & Magazine of Engineering, Technology, Management and Research*. http://www.yuvaengineers.com/an-overview-of-agricultural-robots-p-koteswara-karthik-p-ravi-chandra

Kaygisiz, G. M., Uzumcu, O., & Ucar, F. M. (2020). The case of prospective teachers' integration of coding-robotics practices into science teaching with STEM approach. *Elementary Education Online, 19*(3), 1200–1213. doi:10.17051/ilkonline.2020.728020

Khanlari, A. (2013). Effects of robotics on 21st century skills. *European Scientific Journal, 9*(27), 26–36.

Kilinc, A. (2014). *The using of robotic technology in teaching of light unit 7th grade* [Unpublished master's dissertation]. Erciyes University, Turkey.

Kim, C., Yuan, J., Kim, D., Doshi, P., Thai, C., Hill, R., & Melias, E. (2019). Studying the usability of an intervention to promote teachers' use of robotics in STEM education. *Journal of Educational Computing Research, 56*(8), 1179–1212. doi:10.1177/0735633117738537

Kim, Y. W., Kim, Y. G., Kim, N. J., & Woo, Y. G. (2003). A study on the Korea teacher's perception in using assistive technology. *Journal of Asia-Pacific Special Education, 3*(1), 35–48.

Koc, A. (2019). *The comparison of stem implementations with robotic-assisted and simple materials in preschool and basic science education* [Unpublished doctoral dissertation]. Erciyes University, Kayseri, Turkey.

Kraetzschmar, G. K. (2009). Educational robotics: On the role of robotics in learning and education. In C. Zacharias, K. W. ter Horst, K.-U. Witt, V. Sommer, M. Ant, U. Essmann, & L. Mülheims (Eds.), *Forschungsspitzen und Spitzenforschung* (pp. 83–90). Springer. doi:10.1007/978-3-7908-2127-7_9

Kucuk, S., & Sisman, B. (2017). Instructor experiences in one-to-one robotics instruction. *Elementary Education Online, 16*(1), 312–325. doi:10.17051/io.2017.12092

Kutlu, M., Schreglmann, S., & Cinisli, N. A. (2018). The opinions of special education teachers on the use of assistive technologies in special education. *YYU Journal of Education Faculty, 15*(1), 1540–1569. doi:10.23891/efdyyu.2018.115

Lancioni, G. E., Sigafoos, J., O' Reilly, M. F., & Singh, N. N. (2013). Instructional technology for promoting writing, work, and leisure skills. In *Assistive Technology* (pp. 73–105). Springer. doi:10.1007/978-1-4614-4229-5_4

Lindh, J., & Holgersson, T. (2007). Does Lego training stimulate pupils' ability to solve logical problems? *Computers & Education, 49*(4), 1097–1111. doi:10.1016/j.compedu.2005.12.008

Lykke, M., Coto, M., Mora, S., Vandel, N., & Jantzen, C. (2014, April). Motivating programming students by problem based learning and LEGO robots. In *2014 IEEE Global Engineering Education Conference (EDUCON)* (pp. 544-555). 10.1109/EDUCON.2014.6826146

Malec, J. (2001). Some thoughts on robotics for education. *Proceedings of the AAAI Spring Symposium on Robotics and Education.*

Mataric, M. J. (2004, March). Robotics education for all ages. In *Proceeding of the American Association for Artificial Intelligence (AAAI) Spring Symposium on Accessible, Hands-on AI and Robotics Education* (pp. 22-24). Retrieved from https://www.aaai.org/Papers/Symposia/Spring/2004/SS-04-01/SS04-01-004.pdf

Mataric, M. J., Koenig, N. P., & Feil-Seifer, D. (2007, March). Materials for enabling hands-on robotics and STEM education. *AAAI Spring Symposium: Semantic Scientific Knowledge Integration* (pp. 99-102). AAAI.

Mead, R. A., Thomas, S. L., & Weinberg, J. B. (2012). From grade school to grad school: An integrated STEM pipeline model through robotics. In *Robots in K-12 Education: A New Technology for Learning* (pp. 302–325). IGI Global. doi:10.4018/978-1-4666-0182-6.ch015

Mitnik, R., Nussbaum, M., & Recabarren, M. (2009). Developing cognition with collaborative robotic activities. *Journal of Educational Technology & Society, 12*(4), 317–330.

Moeller, P. (2012). *Where the jobs will be in 2020.* U.S. News and World Report. http://money.usnews.com/money/careers/articles/2012/09/10/where–the–jobs–will–be–in–2020

Mubin, O., Stevens, C. J., Shahid, S., Al Mahmud, A., & Dong, J. J. (2013). A review of the applicability of robots in education. *Technology for Education and Learning., 1*(1), 1–7. doi:10.2316/Journal.209.2013.1.209-0015

Ozdamar, O. (2016). *The views of special education teachers about the use of assistive technology in lessons* [Unpublished master's dissertation]. Anadolu University, Eskisehir, Turkey.

Ozer, F. (2019). *Effect of using robotics in teaching coding on achievement, motivation and problem solving skills of middle school students* [Unpublished master's dissertation]. Hacettepe University, Ankara, Turkey.

Papakostas, G. A., Sidiropoulos, G. K., Papadopoulou, C. I., Vrochidou, E., Kaburlasos, V. G., Papadopoulou, M. T., Holeva, V., Nikopoulou, V.-A., & Dalivigkas, N. (2021). Social robots in special education: A systematic review. *Electronics (Basel), 10*(12), 1398. doi:10.3390/electronics10121398

Pettersson, I., & Fahlström, G. (2010). Roles of assistive devices for home care staff in Sweden: A qualitative study. *Disability and Rehabilitation. Assistive Technology, 5*(5), 295–304. doi:10.3109/17483100903100285 PMID:19968568

Rahman, S. M. M., Krishnan, V. J., & Kapila, V. (2017, June), Exploring the dynamic nature of TPACK framework in teaching STEM using robotics in middle school classrooms. *2017 ASEE Annual Conference & Exposition.* 10.18260/1-2--28336

Raskind, M. H., & Higgins, E. L. (1999). Speaking to read: The effects of speech recognition technology on the reading and spelling performance of children with learning disabilities. *Annals of Dyslexia, 49*(1), 251–281. doi:10.100711881-999-0026-9

Reed, P., & Bowser, G. (2005). Assistive technology and the IEP. In Handbook of special education technology research and practice (pp. 61–77). Knowledge by Design.

Robins, B., Dautenhahn, K., & Dubowski, J. (2006). Does appearance matter in the interaction of children with autism with a humanoid robot? *Interaction Studies: Social Behaviour and Communication in Biological and Artificial Systems, 7*(3), 479–512. doi:10.1075/is.7.3.16rob

Robinson, M. (2005). Robotics-driven activities: Can they improve middle school science learning? *Bulletin of Science, Technology & Society, 25*(1), 73–84. doi:10.1177/0270467604271244

Rockland, R., Bloom, D. S., Carpinelli, J., Burr-Alexander, L., Hirsch, L. S., & Kimmel, H. (2010). Advancing the. *The Journal of Technology Studies*, *36*(1), 53–64.

Rogers, C., & Portsmore, M. (2004). Bringing engineering to elementary school. *Journal of STEM Education: Innovations and Research*, *5*(3), 17–28.

Sani-Bozkurt, S. (2017). Özel eğitimde dijital destek: Yardımcı teknolojiler. *Açıköğretim Uygulamaları ve Araştırmaları Dergisi*, *3*(2), 37–60.

Smith, D. W. (2008). *Assistive technology competencies for teachers of students with visual impairments: A delphi study* [Unpublished doctoral dissertation]. Texas Tech University.

Smyrnova-Trybulska, E., Staniek, D., & Zegzuła, D. (2020). Robotics in education a survey report: Case study. *International Journal of Research in E-learning*, *6*(1), 1–18. doi:10.31261/IJREL.2020.6.2.01

Sola-Ozguc, S. (2017). Özel eğitimde yardımcı teknolojiler. In Özel Eğitim içinde (pp. 391-404). Vize Akademik Publishing.

Stanger, C. A., & Cawley, M. F. (1996). Demographics of rehabilitation robotics users. *Technology and Disability*, *5*(2), 125–137. doi:10.3233/TAD-1996-5202

Stripling, T., & Simmons, B. (2016). Get students revved up! Robotics brings excitement to STEM. *Tech Directions*, *75*(7), 13–17.

Sullivan, A., Kazakoff, E. R., & Bers, M. U. (2013). The wheels on the bot go round and round: Robotics curriculum in pre-kindergarten. *Journal of Information Technology Education*, *12*, 203–219.

Sullivan, F. R., & Heffernan, J. (2016). Robotic construction kits as computational manipulatives for learning in the STEM disciplines. *Journal of Research on Technology in Education*, *48*(2), 105–128. do i:10.1080/15391523.2016.1146563

Talan, T. (2020). Eğitsel robotik uygulamaları üzerine yapılan çalışmaların incelenmesi. *Yaşadıkça Eğitim Dergisi*, *34*(2), 503–522. doi:10.33308/26674874.2020342177

Tatlisu, M. (2020). *The effect on problem solving skills of elementary schools students of problem based learning in educational robotic* [Unpublished master's dissertation]. Bursa Uludağ University.

Tekinarslan, E., & Yikmis, A. (2005). Özel eğitim kurumlarında çalışan öğretmenlerin teknoloji kullanımına yönelik görüşleri ve beklentileri. *Journal of Abant Social Sciences*, *2*, 211–220.

Ucgul, M., & Cagiltay, K. (2014). Design and development issues for educational robotics training camps. *International Journal of Technology and Design Education*, *24*(2), 203–222. doi:10.100710798-013-9253-9

Uluer, P., Akalın, N., & Köse, H. (2015). A new robotic platform for sign language tutoring. *International Journal of Social Robotics*, *7*(5), 571–585. doi:10.100712369-015-0307-x

UN. (2021). *Article 24 – Education.* https://www.un.org/development/desa/disabilities/convention-on-the-rights-of-persons-with-disabilities/article-24-education.html

Valadao, C. T., Loterio, F., Cardoso, V., Bastos, T., Frizera-Neto, A., & Carelli, R. (2015). Robotics as a tool for physiotherapy and rehabilitation sessions. *IFAC-PapersOnLine*, *48*(19), 148–153. doi:10.1016/j.ifacol.2015.12.025

Van der Loos, H. M., Reinkensmeyer, D. J., & Guglielmelli, E. (2016). Rehabilitation and health care robotics. In B. Siciliano & O. Khatib (Eds.), *Springer Handbook of Robotics*. doi:10.1007/978-3-319-32552-1_64

Weinberg, J. B., & Yu, X. (2008). Low-cost platforms for teaching integrated systems. *IEEE Robotics & Automation Magazine*, 4–6.

Wood, S. (2003). Robotics in the classroom: a teaching tool for K-12 educators. *Symposium of Growing up with Science and Technology in the 21st Century*.

World Health Organization & World Bank. (2011). *World report on disability 2011*. World Health Organization. https://apps.who.int/iris/handle/10665/44575

Yang, X., Zhao, Y., Wu, W., & Wang, H. (2008, September). Virtual reality based robotics learning system. In *IEEE International Conference on Automation and Logistics* (pp. 859-864). IEEE. 10.1109/ICAL.2008.4636270

Yavuz-Konokman, G., & Cukurbasi, B. (2019). Effects of designing LEGO robotics instructional practices on the prospective science teachers' resistive behaviors towards technology supported instruction. *Malaysian Online Journal of Educational Technology*, *7*(3), 57–71. doi:10.17220/mojet.2019.03.005

Yolcu, V., & Demirer, V. (2017). A review on the studies about the use of robotic technologies in education. *SDU International Journal of Educational Studies*, *4*(2), 127–139.

Zhou, L., Parker, A. T., Smith, D. W., & Griffin-Shirley, N. (2011). Assistive technology for students with visual impairments: Challenges and needs in teachers' preparation programs and practice. *Journal of Visual Impairment & Blindness*, *105*(4), 197–210. doi:10.1177/0145482X1110500402

KEY TERMS AND DEFINITIONS

Arduino: Arduino refers to an open-source prototyping platform for users to create interactive electronic tools.

Assistive Robot: An assistive robot is aa robot that can perform a certain task or action that benefit an individual with disability.

Assistive Technology: Refers to products, equipment, and systems that enhance learning, working, and daily living for persons with disabilities.

Coding: Refers to the language used by computers to understand our commands and, therefore, process our requests.

LEGO: Refers to a line of plastic construction toys that are manufactured by The Lego Group.

mBot: A STEM education robot for beginners, that makes teaching and learning robot programming simple and fun.

Robotics: An interdisciplinary field that integrates computer science and engineering which involves design, construction, operation, and use of robots.

STEM: Science, Technology, Engineering, and Mathematics.

Section 4

Different and Emergent Approaches to Educational Robotics

The final section consists of four chapters focused on different and emergent approaches to educational robotics such as online learning, making, and the use of low-cost materials to build robots. The first chapter discusses the delivery of educational robotics in online learning environments. The next one relates educational robotics, teacher training and making. The last two chapters focus on educational robotics competitions and the use of robots as tour guides.

Chapter 10
Designing Tools and Activities for Educational Robotics in Online Learning

Lucio Negrini
SUPSI, Switzerland

Christian Giang
https://orcid.org/0000-0003-2034-9253
SUPSI, Switzerland

Evgeniia Bonnet
EPFL, Switzerland

ABSTRACT

The COVID-19 pandemic and the subsequent school closures created several challenges for teachers and students. From one day to the next, teachers had to rethink their educational activities and move to remote learning. Especially with regard to educational robotics activities, which makes large use of physical artefacts, this abrupt shift towards online learning represented a major change in how activities had to be designed and implemented. In this chapter, some experiences of online educational robotics activities carried out in compulsory schooling and teacher training are presented. The experiences are then discussed using a model for the development of educational robotics activities in order to reflect on how to design such activities that can be carried out online. The examples presented in this chapter showed there is great potential for educational robotics in online learning.

INTRODUCTION

The Covid-19 pandemic has affected many areas of our lives on an unprecedented scale, the educational system included. Almost overnight, educational institutions had to move from the traditional classroom in person setting to emergency remote learning (Hodges, Moore, Lockee, Trust, & Bond, 2020). Teachers

DOI: 10.4018/978-1-7998-7443-0.ch010

as well as students faced the challenge of adapting to this new reality in order to enable a continuation of courses in online learning formats. On the one hand, both teachers and students had to cope with technical challenges caused by the immediate and increased use of sometimes unfamiliar technological media such as new video conferencing tools, online discussion platforms or lecture recording software. On the other hand, and even more importantly, teachers had to think how these tools could effectively be used for teaching and learning activities, and redefine their practices. This was especially in regards to educational robotics activities, with the abrupt shift towards online learning representing a major change in how activities had to be designed and implemented.

Educational robotics (ER) builds heavily on the idea of having tangible objects that learners can use and manipulate following the learning theories of constructionism and social constructivism (Papert, 1991; Resnick et al., 1998). Another important aspect of educational robotics classroom activities is the social interactions within work groups, allowing peers to learn with and from each other, following an educational approach to projects where collaboration between group members is required to achieve the final result (Ardito, Mosley, & Scollins, 2014). Replicating these elements in online learning, where students and teachers are working from home poses significant challenges. Nevertheless, different efforts have been made before and during the pandemic to enable the implementation of educational robotics activities in online learning scenarios. To design meaningful ER online activities, it can be helpful to build on existing models and conceptualizations such as the Educational Robotics Learning System (ERLS) model (Giang, 2020). It was devised to support developers and educators in the design of ER tools and activities is based on the idea of instructional alignment (Cohen, 1987; Biggs, 2003).

The objective of this chapter is to provide an overview of different approaches to ER online activities explored in recent years. The advantages and challenges of these approaches are discussed on the basis of six examples, with a particular emphasis on instructional alignment as described by the ERLS model. Specifically, we illustrated how tools and activities have been designed to facilitate meaningful online learning experiences. The next section provides a brief introduction to the concept of instructional alignment and the ERLS model. Thereafter, we present three examples describing ER online activities designed for compulsory school students. Finally, another three examples illustrate how online learning can be leveraged to train teachers in the field of ER.

BACKGROUND

Instructional alignment is a well-established principle that has guided curriculum planning for a significant amount of time. It considers the interplay of three components: intended outcomes, instructional and assessment processes (Cohen, 1987). Previous work has shown that consciously aligning these components can result in significant improvements of student achievement (Mitchell, 1999; Wonder-McDowell et al., 2011). By linking the ideas of instructional alignment with the learning theory of constructivism, Biggs (1996) introduced the concept of *constructive alignment*. He particularly emphasized the importance of considering teaching and learning as a system, in which ideally "all aspects of teaching and assessment are tuned to support high-level learning" (Biggs, 2003, p.1).

Based on the ideas of constructive alignment, the Educational Robotics Learning System (ERLS) model was devised as a conceptualization of instructional alignment in the context of ER activities in classroom education (Giang, 2020). The model consists of four main components (Figure 1): intended learning outcomes, instructional activities, assessment activities and ER artifacts (i.e., robots, interaction/

Figure 1. Alignment model for Educational Robotics Learning Systems (ERLS)

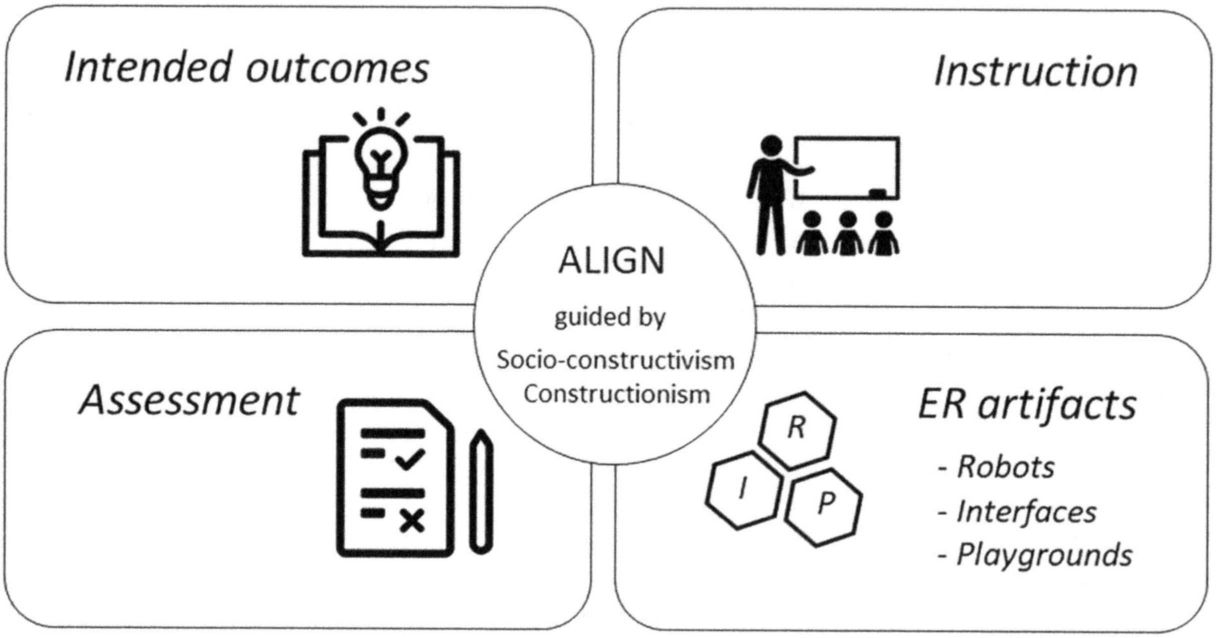

programming interfaces and playgrounds). It stipulates for pedagogically meaningful ER activities, it is important the designs of the four components are well aligned with each other.

In traditional instructive alignment, the design of the instructional activities should allow the learners to attain the intended learning outcomes, in order to then be evaluated in the assessment activities. The assessment activities on the other hand, should be designed in a way, such that they can capture how well they reached the intended learning outcomes. As opposed to traditional instructive alignment, the ERLS model also integrates a fourth component, namely the ER artifacts, which need to be appropriately designed and selected in order to effectively support the entire learning system. Specifically, they should facilitate the attainment of the intended learning outcomes through an appropriate integration as learning artifacts in the instructional activities. This can be achieved most efficiently, when the affordances of the tools meet the needs of the teachers and students (Antonenko, Dawson, & Sahay, 2017). In many cases, the ER artifacts can also be part of the assessment activities, for instance, when the ER activity is based on a goal-oriented learning approach. In addition to the alignment of its four components, the ERLS model explicitly emphasizes the importance of two learning theories underlying ER activities, namely constructionism and social constructivism. Papert's constructionism describes the idea that children's learning can be enhanced when they are involved in activities allowing them to construct artifacts that they can share with their environment (Papert, 1991). Social constructivism on the other hand, considers the social aspect of learning, for instance, that children construct knowledge through their interactions with teachers and peers (Vygotsky, 1978).

While the ERLS model was originally devised to support developers and educators in the context of ER classroom activities (Giang, 2020), the core ideas of the model can also be adopted in the context of ER in online learning. Equivalent to the situation where ER activities are performed in the classroom, it is essential to design tools and activities that facilitate the alignment of the different components of the model when moving to the online space. Ideally, such solutions should also provide possibilities to preserve the constructionist and social constructivist learning approaches underlying ER activities (Papert, 1991; Resnick et al., 1998). However, finding such solutions is not trivial, as the nature of the ERLS components may alter in online learning scenarios.

One element that is heavily affected by the transition to the online space, is the instructional activities component. In online learning, instruction is usually delivered through pre-recorded video lectures (asynchronous) or live video conference tools (synchronous). These are often complemented by different learning resources, such as readings or tutorials, requiring learners to take more responsibility and self-regulate their learning (Pelikan, et al., 2021) Moreover, the means of instruction used in online learning changes the way students interact with their teachers and peers. This makes it necessary to rethink how, for instance, teacher support can be provided or group work can be organized.

Another major challenge to overcome in the context of ER in online learning settings concerns the ER artifacts. In most classroom-based learning scenarios, students and teachers benefit from their schools' infrastructure, providing them with access to educational robots, computers as well as materials for crafting the development of the artifacts. While online learning is generally considered an interesting approach to democratize education, providing more individuals with access to ER artifacts and has always been of great interest (Eguchi, 2017). An important consideration is that ER builds on the idea of providing learners with means to learn with physical objects, hence finding affordable and accessible solutions for ER artifacts in online learning is essential.

ER in online learning also poses challenges with regards to assessment. Since ER often involves transversal, interdisciplinary and hands-on activities, finding appropriate assessment activities is already a challenge in the normal classroom setting. However, performing assessment of learning (both formative and summative) in ER online activities can be considered even more challenging and requires new methods that allow to define and capture the intended learning outcomes. In fact, the latter may also change when ER activities are performed online. Due to the altered circumstances under which learning takes place, it may not be possible to design for the same intended learning outcomes in online learning as in the classroom setting (Castro & Tumibay, 2021).

MAIN FOCUS OF THE CHAPTER

This chapter focuses on the design of tools and activities in online learning settings for ER activities. The objective is to provide developers, educators and researchers with an overview of different approaches explored in recent years while highlighting advantages and challenges. Six examples are presented, each highlighting a different aspect with respect to ER activities in online learning.

The first example is a remote educational robotics activity for compulsory school students where learners and robots are spatially separated. This Remote Rescue with Thymio II (R2T2) mission allows a student to remotely control a robot in a fictional scenario of a space mission. Together with participants connected from all over the world, the student collaboratively solves the mission by correctly programming the robot while at the same time communicates and coordinates his/her actions with the other students. Hence, it represents an ER online activity that can be performed even when no local robot is available. The second example presents an activity where the physical robots (BeeBots) are substituted with paper-based learning artifacts to organize an ER online activity. The objective of this approach was to conserve the tangible nature of ER activities even when robots are not available. Students in this case programmed their robots using paper and pencil and then sent their algorithms to the teacher via social media for testing with the real robot. The third example describes CreroBot, a do-it-yourself educational robot based on low-cost electronic components and accessible construction materials such as cardboard and paper. The objective of this project was to provide teachers and students with maker-based activities that they could easily perform from home, while preserving the constructionist idea of ER as best as possible. The robot was then used in a continuing professional development program for in-service teachers as well as in a robotics summer school for high school students, both organized in online learning settings.

In the final three examples the Covid-19 pandemic did not only affect compulsory school students but also their teachers and teacher training activities also needed to be continued online. Specifically, this chapter will present an example of a massive open online course (MOOC) that was created to introduce learners to the educational robot Thymio II. Another example describes a professional development course in ER for in-service teachers that started in face-to-face mode but had to be continued online due to the Covid-19 pandemic. Then, to illustrate another potential of ER in online learning an online robotic teacher community will be presented. Such online communities provide a platform for teachers, developers and researchers to exchange ideas and activities and can therefore facilitate the integration of ER into formal education, both in presence and online formats.

RECENT ADVANCES

ER Online Activities

Most previous work that explored ER online activities were implemented in synchronous learning settings. Often taking the form of remote robotics laboratories, such environments leverage the increasing capabilities of the internet, allowing learners to control robotic equipment from afar. Video streams provide feedback to the learners and at the same time enable a direct interaction between teachers and students.

Already in the late nineties, Després and George (1999) developed a multi-agent architecture to implement an online learning environment for ER. Secondary school students elaborated the theory using electronic course books and then remotely programmed micro-robots while receiving support from a teacher at distance. In the following years, several works have adopted similar approaches to implement remote robotics laboratories for topics such as control and path planning (Farias et al., 2019; Kulich et al., 2012; Mester, 2006) or to introduce students to programming (Almeida, Netto, & Rios, 2017; Dos Santos Lopes, Gomes, Trindade, Da Silva, & Lima, 2016). While these works represent interesting approaches to implement ER activities online and at distance, it could be argued that the physical separation of learners and the main learning artifact (i.e., the robot) may create a hindrance in constructionist learning. Since most remote robotics laboratories do not involve the manipulation of physical robots, they may not be able to take advantage of one of the most powerful concepts underlying ER.

The following three examples are presented with the objective to expand the notion of ER online activities to include new ideas that go beyond the conventional idea of remote robotics laboratories. For each project, it will be discussed how specific design choices impacted the different elements of the ERLS model in order to highlight the peculiarities of each project.

Figure 2. Example of R2T2 Mars Mission's set-up installed at the EPFL. The map is divided into four sectors with 4 Thymio robots per sector. Each team controls one robot. The teams see the result of their programs executed by the robots via video streams hosted on YouTube. The Aseba software package allows participants to connect to the robot via the internet.

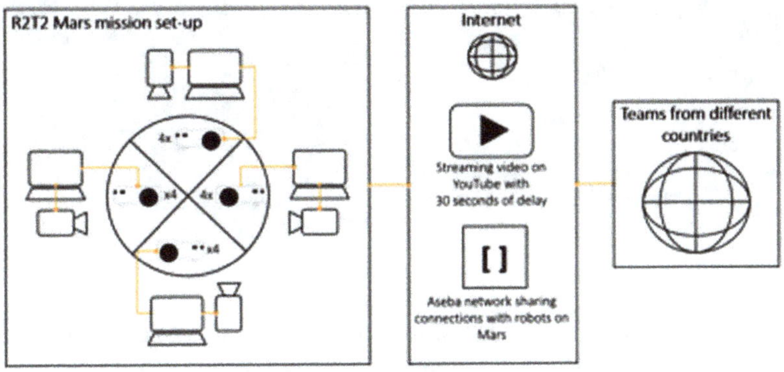

The Remote Rescue With Thymio II Project (R2T2)

The R2T2 (Remote Rescue with Thymio II) project (Mondada, Bonnet, Davrajh, Johal, & Stopforth, 2016) offers the possibility for children worldwide to collaborate and solve a fictional space mission by remotely controlling real Thymio II robots (Mondada et al., 2017) via the internet. The event is completely hosted online. It allows students as young as 8 years old to develop and apply skills related to computational thinking, creative problem solving as well as developing other 21st century skills associated with ER (Negrini & Giang, 2019).

This project was initiated by the EPFL (Ecole Polytechnique Fédérale de Lausanne) in 2015, and since then over 1500 students from 11 different countries have had a chance to take on the roles of space engineers who have to program Thymio robots in different fictional narratives playing on Mars, the Moon or a space station. The missions' set-ups with 4-16 robots and 2-5 web-cameras are installed at the university campus of the EPFL. The R2T2 Mars Mission for instance (Figure 2), can include 16 participating teams that are grouped into 4 sectors with 4 teams each. The grouping was introduced to facilitate the collaboration and direct communication between teams.

While solving the R2T2 missions, the teams see the result of their programming via video streams (Figure 3).

To make the mission more challenging and authentic, the video streams have a time delay. In the narrative, this is explained by the time that the signal needs to travel from Earth to space and vice versa. When a team sends a program to their remote robot, the robot executes it without any delay. However, the team will see what happens on the remote set-up only after around 30 seconds. Thus, if the program was not written correctly, the team will not have any chance to correct for it. For instance, during these 30 seconds, their robot could have gone outside of the cameras' view, bumped into another robot or a wall. A small mistake in these conditions can create a huge problem and change the situation completely for all teams. Therefore, it is very important the students thoroughly reflect on their solutions before sending them to the robots for execution, and this fosters the development of their computational thinking skills. This point represents the main intended learning outcomes of the activity, together with the development of communication and collaboration skills. This is further emphasized by the fact that all teams have a

Figure 3. Example of the R2T2 Mars Mission with 16 Thymio robots involved. There are four sectors with four teams controlling each one robot. On the left: Sector view with 4 Thymio robots that participants see in their schools. On the right: Top view with all 16 teams (16 robots).

common goal: successfully completing the mission requires all teams to correctly program their robots. This condition helps to prevent competitive behaviour on the one hand and fosters communication and collaboration on the other hand. To facilitate the interactions, an online chat is available to coordinate the actions between the different teams, helping them to find solutions that would suit all.

Until 2020, there were two R2T2 set-ups, the R2T2 Mars and R2T2 Moon missions. As a response to the Covid-19 pandemic, a third scenario was developed, the R2T2 Corona mission. This mission was specifically created for teachers in order to support them during the lockdown, allowing them to learn and train their programming skills over the internet and with the live support of a coach. The activity was accompanied by pedagogical material, specifically designed for online use, explaining the main programming tasks and the robot's characteristics.

The R2T2 project illustrates that the idea of remote robotics laboratories can be successfully adopted to the context of compulsory school education. It affords students an opportunity to explore and develop their computational thinking and programming skills, even when no physical robots are available. Moreover, the design of this ER online activity allows students to develop their communication and collaboration skills, without the need for them to be in the same physical space. The latest advances in the development of this project moreover showed that the approach could also be useful to remotely train teachers in ER.

BeeBot Online Activity

BeeBot is a programmable floor robot that is mainly used in primary schools. Thanks to four direction arrows (forward, backward, left and right), a pause button, and a start button, the children can create sequences of movements that the robot executes. BeeBot is normally used for activities done face-to-face where children have to program the robot to move it on a defined path. During the Covid-19 pandemic it was not possible to propose face-to-face activities and teachers had to find alternatives in order to continue to work with BeeBot. In this context, one effort made during this period was the BeeBot online activity (Giang & Negrini 2021). In a small village in Switzerland a teacher proposed, on a voluntary basis, to nine primary school students (9-12 years old) the following activity: the students were asked to program sequences of instructions using paper and pencil in order to solve a given programming task and then send their algorithms to the teacher that tested them with the BeeBots. During the lockdown, at the beginning of each week the children at home received new assignments and learning materials that were then discussed in a video conference. After these first instructions, children could work on the

Figure 4. Paper grid and triangles to locally simulate the behavior of the BeeBots (a), set of instructions for the robot (b), and an example of a solution proposal (c).

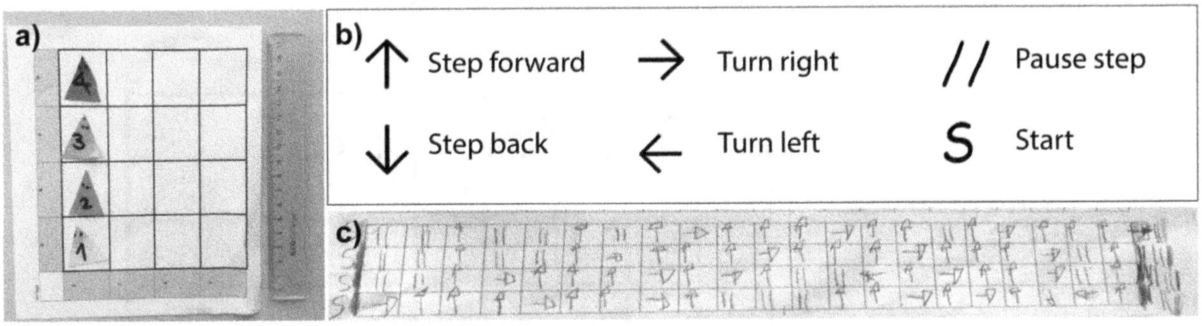

assignments received and submit their results to get feedback from the teacher. Children did not have access to the BeeBots but used coloured triangles as robots, paper strips as programming interfaces and a paper grid as the playground (Figure 4).

The triangles were used to simulate the behaviours of the robots. Each triangle represents a robot that is placed on a grid (a). Using the set of instructions (b) students had to draw their sequences of commands on a paper strip (c) and then simulate the behaviour moving the triangles. When satisfied with their solutions, students submitted a picture sequence of the commands to the teacher. The teacher, having access to the robots, programmed the robots according to the students' instructions and tested them. The behaviors of the robots were filmed and sent back to the students as feedback together with a written or oral comment (Figure 5a). Students could check if their instructions were correct and in case they were not, start with the debugging phase and send the new sequence to the teacher. Each week a new task was sent to the students. The tasks consisted in creating a choreography with four BeeBots complying with some constraints like the starting points of the robots, the steps that they need to do, the synchronism between them or the fact that they had to start all together and so on. The 12 tasks were of different levels of difficulty. For example, in the activity illustrated in the grid of Figure 5b, the four robots had to be placed on the four coloured starting positions and then be programmed in order to start all together and move around the outer square of the grid (marked with "X"). In order to successfully implement this activity for instance, students had to realize that in order to avoid robots crashing in another they had to make use of the pause button. In the activity illustrated in Figure 5c, the robots were placed in the four corners and the students had to program all robots with the same instructions in order to perform a choreography without crashing into one another.

Considering the ERLS model, the described activity has been modified on different aspects with respect of the same activity carried out face-to-face.

If in presence, this activity could be used to work on the concept of computational thinking and more in detail on the dimensions of the decomposition of the problem, the definition of an algorithm and the debugging of the created algorithm. All dimensions can be developed also with the activity at distance. In some cases, the absence of the robot can also represent a didactic advantage: students have to define an algorithm without the possibility to check it immediately with the robots. As previous work has shown, the direct feedback provided by robots often entices students to solve programming tasks by blind trial and error approaches in which they define the algorithms a posteriori, preventing them from desired reflection processes (Chevalier, Giang, Piatti & Mondada, 2020). However, what is missing in

Figure 5. Example of a message exchanged between student and teacher (a), assignment in which four robots have to be programmed to drive around the outer circle of the grid (b) and assignment in which all four robots have to perform the same choreography without crashing into each other (c).

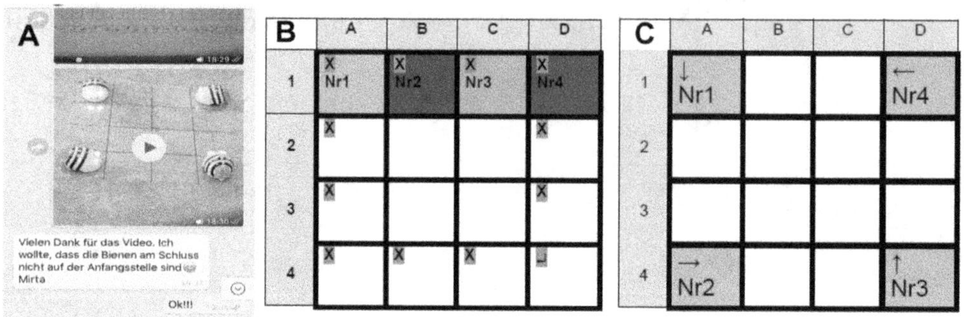

the home activity is the transversal competences like the communication or the collaboration that are present when students are working in groups face-to-face and this is more difficult to reproduce online.

In the presented activity the teacher used a video conference tool to give instructions to the students in a synchronous way and a messaging application for giving feedback in an asynchronous way. The mix of synchronous and asynchronous approaches is seen as having a positive impact to online learning (Rapanta, Botturi, Goodyear, Guardia, & Koole, 2020). When such distance learning approaches are well designed, they allow to shift the responsibility for learning to the student, while the teacher takes the role of a facilitator and tutor. In this case, students at home have to find their solutions without any help, teachers can give feedback and accompany the students in developing their competences by choosing the correct tasks to give them.

The assessment of the tasks has been done via social media. The teacher recorded the behaviour of the robots and sent it together with an audio or textual feedback to the students. Also in this case, working via distance and online could represent an advantage for students since teachers are forced to give individual detailed feedback, which is seen as an important element for learning (Hattie, 2012). As a matter of fact, providing individual feedback to a class with 20 students is not always feasible in the traditional classroom setting. On the other hand, the home activities are also more time consuming for teachers, since they have to collect all the algorithms, test them and send individual feedback to the children.

In order to propose this activity, alternative artifacts have been chosen. The use of paper and pencil materials not only had the positive aspect of allowing students to continue working with ER from home, but at the same time also reduced their active screen time during the lockdown. Indeed, one concern during the pandemic was to reduce active screen time, in order to address the limited online attention span of young children (Dong et al., 2020).

CreroBot

The CreroBot project (Mehrotra et al., 2021) was an approach to implement ER online activities that was devised during the Covid-19 pandemic. To allow learners to conduct ER activities from afar, new tools and activities were developed based on a maker approach. The use of accessible construction materials and affordable electronic components enabled learners to locally build their own robots and use them as artifacts for the learning activities. The devised online learning approach was tested with high school

Figure 6. Main building components (a) and an example of a CreroBot in the basic version (b).

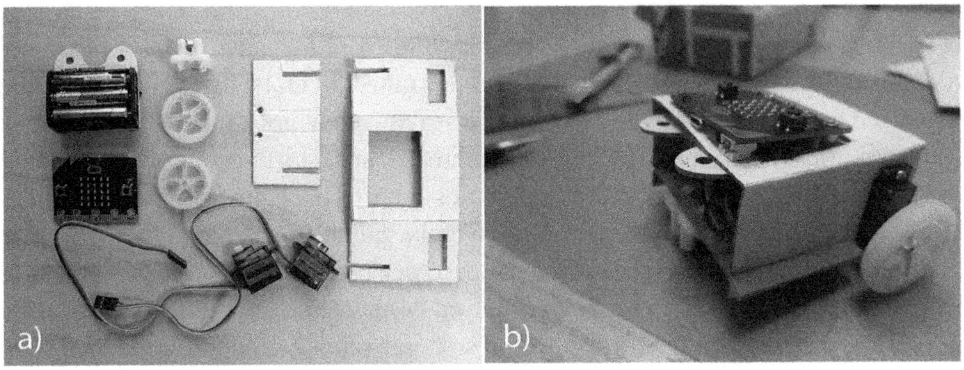

students participating in a mobile robotics summer school (16-18 years old) as well as in the context of a professional development course for in-service teachers.

CreroBot is a do-it-yourself educational robot that was devised during the Covid-19 pandemic as a tool to facilitate ER activities in distance learning settings. The objective of the ER activities was to introduce teachers and students to the basics of mobile robotics as well as to foster the development of a maker mindset. For the system to be reproducible from home, a particular emphasis was put on the use of low-cost and easily accessible construction components. Due to its growing popularity in computing education (Sentence, Waite, Hodges, MacLeod, & Yeomans, 2017) as well as its affordable price tag, the BBC micro:bit was chosen as the core for the CreroBot. The micro:bit has a compact design and incorporates many inbuilt sensors (e.g. magnetometer, temperature sensor, light sensor, accelerometer) allowing for a variety of introductory maker activities. Together with other low-cost components (e.g., micro-servo motors) it was used to build the basic version of CreroBot (see Figure 6). To enable the construction of CreroBot without using sophisticated tools such as 3D printers or laser cutters, cardboard was chosen as the principal material for its body. The body of the basic version consists of two cardboard pieces that could be cut out and assembled using a slit system. The basic version was proposed to teachers and students as an initial design to get started with the CreroBot activities. However, to promote the idea of making, teachers and students were encouraged to explore their own CreroBot designs as the activities progressed. Such designs would not only have a different look, but they could also incorporate additional sensors and actuators compatible with the micro:bit.

To program CreroBot, Microsoft Makecode (block-based) and MicroPython (textual), were used. Furthermore, to align with the intended outcome of introducing the maker mindset through the activities, a third, cardboard-based programming platform was developed: CreroBot PaPL is a tangible programming platform that was devised based on development frameworks presented in previous research (Mussati, Giang, Piatti & Mondada,2019; Mehrotra et al. 2020). It allows users to craft their own programming blocks from cardboard and arrange them in a puzzle-like fashion to create sequences of instructions. By taking a photo of the instructions using the PaPL Android application the commands are identified through computer vision algorithms and then sent to the micro:bit via Bluetooth Low Energy (BLE) for execution.

To allow teachers and students to perform the CreroBot learning activities during the Covid-19 pandemic, all the necessary material was provided sent as a parcel to their homes. Each parcel contained all the electronic components and cardboard material to construct the CreroBot and the PaPL programming tiles.

As a first use case, CreroBot was leveraged as a learning tool for a mobile robotics summer school offered to students aged between 16-18 years old. Due to the Covid-19 pandemic, the summer school (normally organized on the campus of a university), was realized using a mix of synchronous and asynchronous online activities. Each day of the four-day summer school started with a live online video lecture given by one of the workshop leaders, followed by hands-on activities during which the students worked independently with the materials provided to them. Web-based tutorials guided them through the different steps of building and programming their CreroBot to perform tasks of increasing difficulty. During the hands-on activities, the workshop organizers stayed connected, in order to provide live support to students that had questions. On the final day of the summer school, students presented their results in online slide presentations and live video demonstrations of their final robot designs.

As a second use case, the CreroBot activity was proposed to in-service teachers as an alternative to complete the module of a professional development course during the Covid-19 pandemic. Similar to the student summer school, a mix of synchronous and asynchronous approaches was performed to organize the learning activities. The lecturers of the course prepared a series of video lectures and tutorials that were shared with the teachers, who were given two weeks to independently build and program their CreroBot. . In case of issues, teachers could organize a video call with one of the lecturers to discuss their problems synchronously. After the two weeks, individual video interviews were scheduled with each teacher. During it they could present their final results and share their experiences.

The final assessments performed with both the students from the online summer school and the teachers following the professional development course, showed that the CreroBot activities appeared to be successful and helped them attain the intended learning outcomes, i.e. to introduce them to the basics of mobile robotics and the core ideas of the maker movement. Many students and teachers enjoyed building their own robots using cardboard and showed different facets of the maker mindset, modifying the initial design of the CreroBot to adapt it according to their preferences. While the participants acknowledged that cardboard lacks the robustness for the realization of more sophisticated designs, both students and teachers appreciated that the use of accessible materials facilitated their participation during the Covid-19 pandemic. Though most of the participants would have preferred to perform the activities on site, they also stated that the prepared materials were sufficient to clearly convey the instructions of the learning activities. The CreroBot project hence represents an interesting approach towards ER online activities, that in addition to robotics education also incorporates the idea of making.

ER Teacher Training Online

As previous work has highlighted, providing teachers with appropriate, well-adapted training is a crucial factor to successfully integrate ER in formal education (El-Hamamsy et al. 2020). While designing and implementing ER teacher training is a challenge in itself, moving the training to the online space can be considered even more challenging. Nevertheless, previously approaches have been made to explore this idea.

Similar to the ER activities online that were implemented for students, remote robotics laboratories have also served as a means for teacher training. Alimisis and Plessas (2011) explored in-service teacher training with Lego Mindstorms robots through synchronous audio conferencing. In a series of different activities, the teachers participated in discussions using online whiteboards and received instruction

Figure 7. Screenshot of a video about Thymio's characteristics from the Chapter 2 of the MOOC "The Thymio robot as a tool for discovering digital sciences".

through slide presentations by the trainer. Subsequently, they used simulation tools to virtually design a robot, before being provided with the possibility to send their programs to a remote robot and observe the actions through a video feed. In another work, Conchinha and de Freitas (2015) performed ER workshops in distance learning for teachers working in special needs education. The authors highlighted the high dropout rate with only 11 teachers out of 24 completing the workshops, indicating that such approaches need to be carefully designed and well thought-out to be successful.

The following three examples are presented with the aim of illustrating how the online space can be further leveraged to support teacher training in ER. The first example describes the design and implementation of a massive open online course (MOOC) dedicated to ER. The second example reports on the experiences made when moving an in-service teacher training to the online space during the Covid-19 pandemic. Finally, the last example introduces a platform that allows teachers to share and exchange activities, experiences and other kinds of useful resources encompassing the topic of ER.

The Thymio MOOC

Thymio is an open-source robot that was created for education (Mondada et al., 2017) as a tool for students to discover robotics and do first steps in programming. It has many different sensors and actuators (Figure 7) and it can be programmed using five different programming interfaces: Thymio VPL (Visual Programming Language), Thymio VPL3, Scratch, Blockly and Aseba Studio.

While the robot can be used as a useful learning tool in many classroom activities, it is often not easy for teachers to discover this potential. Teacher training programs in ER are still not widespread, making it hard for many interested teachers to take the first steps in ER. To support these teachers, a massive open online course (MOOC) was created for this purpose. Launched in 2018, the MOOC "*The Thymio robot as a tool for discovering digital sciences*" is a result of a collaboration between EPFL and Inria, who have involved experts in ER and computer science for primary and secondary school levels from France and Switzerland. The MOOC is organized in six chapters:

- In Chapter 1 "*Introduction to computer science and robotics*", interested teachers can discover the history of these two fields and the related concepts.

- Chapter 2 "*Thymio*" allows participants to discover Thymio's pre-programmed behaviors and helps them to learn more about sensors, actuators and other technical aspect of robotics
- Chapter 3. "*First steps in programming with Thymio*" guides the learners in creating their first programs using the Thymio VPL (Visual Programming Language) interface.
- Chapter 4. "*Program Thymio with Scratch*" introduces how Thymio can be programmed with the graphical programming platform Scratch.
- In Chapter 5, "*Program Thymio with Aseba Studio*", learners will then be introduced to the textual Aseba programming language.
- In the final Chapter 6, "*Activities with Thymio in class*", several examples are presented, illustrating how Thymio can be integrated into classroom activities with students of different age groups and covering different school subjects.

It is a self-paced MOOC that includes 70 short videos (of max. 6 minutes length), illustrations, readings as well as quizzes for self-assessment at the end of each chapter. Online forums, regularly monitored by experts in ER, were used to provide the learners with support in case they had questions. The MOOC was published on three different platforms: *fun-mooc.fr*, *courseware.ch*, *edx.org*. Until today, more than 10,000 learners have subscribed to the course on one of the three platforms. For instance, the *edx.org* platform counts 1300 subscribers from 97 different countries and different age groups (18,5% < younger than 25 years old, 40.4% between 26 and 40 years old, 41% older than 41 years old). These numbers highlight the accessibility of those resources and interest to the subject given in the course, regardless of origin and age of the learners. The MOOC therefore represents a promising approach to make training in ER accessible to a wide mass of interested learners. This could be especially important for many teachers, who otherwise would not have the possibilities to participate in ER training programs. Interestingly, more than half of the participants indicated that they followed the course without having a physical robot, but using the Thymio simulator, which is freely available online.

CAS in Educational Robotics

The Certificate of Advanced Studies (CAS) in educational robotics is a two year part-time professional development course for in-service teachers. It is carried out at the University of Applied Sciences and Arts of Southern Switzerland (SUPSI). The CAS has three modules:

- A theoretical module that includes an introduction to ER (e.g., the history of ER, the different learning approaches in ER, the competences that are developed with ER and aspects of robotics and ethics), the concept of computational thinking (Wing, 2006), learning theories and teaching methods like Papert's constructionism (Harel & Papert, 1991), the project method (Kilpatrick, 1918), and technical knowledge about the proposed ER (see also Negrini, 2019).
- A programming module where participants can learn different programming languages for the educational robots Thymio and LEGO Mindstorms EV3.
- A module where participants introduce the ER in their classroom teaching and develop didactic activities.

When the lockdown in March 2020 started, participants of the CAS were following the second module about the robot Thymio II. Thymio II is a robot with a wide range of sensors (nine IR proximity sensors, a three-axis accelerometer, a microphone, a temperature sensor, a remote-control receiver, an SD-card slot, and five capacitive buttons) as well as two motors, a loud-speaker, and 39 LEDs spread all over its body (Chevalier et al., 2016). By default, Thymio II has six pre-programmed behaviours allowing young students to explore basic robotics concepts even without the use of a computer. Older children on the other hand, can engage in programming activities with Thymio II using languages of varying complexity: beginners can get started with the visual programming language (VPL), intermediate students can code using Scratch and advanced students can program using the textual language ASEBA.

The first two lessons on the VPL basic programming language could be completed in face-to-face lessons prior to the lockdown. During these two meetings, each participant also received an individual Thymio robot. In order to face the challenges of distance learning and continue with the CAS during the lockdown, the teachers made some changes in the course. First, they decided to use more intensively the Moodle platform and a video conference system in order to share materials, give instructions, communicate with participants and so on. On the Moodle platform they uploaded short video lessons for the participants with some introduction and tutorials on the different new programming languages: VPL advanced, Scratch for Thymio and ASEBA (a text-based language). Participants had to watch the videos and then try to solve some programming exercises with the Thymio II robot. They were then asked to share their code with the teacher or to send a video showing the desired behaviour executed by the robot. The main change related to a face-to-face situation was during the feedback phase. In class, teachers could follow more closely the participants as they solved the programming challenges and intervene more often. Via distance they received only the final version or saw the final video and could hence not follow the process on how participants came to the solution. Short video conference sessions were organized to share feedback to the participants and monitor their needs and issues. Individual feedback was however given via email. Having also synchronous moments allowed maintaining a closer relationship with the participants. The intended outcomes of the CAS were not changed. However, it was noticed that participants could not reach a comparable skill level as participants of the previous editions carried out face-to-face. This is because working via distance requested more time.

Roteco

Roteco is a Swiss "Robotic Teacher Community" based on the online platform *Roteco.ch* (Negrini et al., 2021). In this community, teachers can exchange their experiences and activities, get in touch with other teachers, be informed about the most recent developments, events and courses in the field of educational robotics and computational thinking. The main aim of this community is to support teachers to carry out ER activities in their classes. Previous experiences have namely pointed out that in order to integrate technologies or ER in schools it is not enough to train teachers but there is a need for further long-term support (Negrini et al., 2021). A community can give this support to teachers since it can help foster the collaboration among teachers. In a community, for example teachers can share their didactic materials and use the one of colleagues; they can get advices from other peers on how to design new activities; or they can learn new aspects related to ER and change their attitudes and practices (Dionne et al., 2010; Hedges, 2010; Vescio et al., 2008).

By March 2021, *Roteco.ch* hosted more than 630 learning activities and had more than 800 active teachers registered on the platform. Since their start in February 2019 *Roteco* has had 115,000-page views.

A teacher community is not only important to support teachers for face-to-face activities but also for online learning. During the COVID-pandemic *Roteco* organised different webinars for teachers on the

Figure 8. Page views of the roteco.ch platform before, during and after the lockdown.

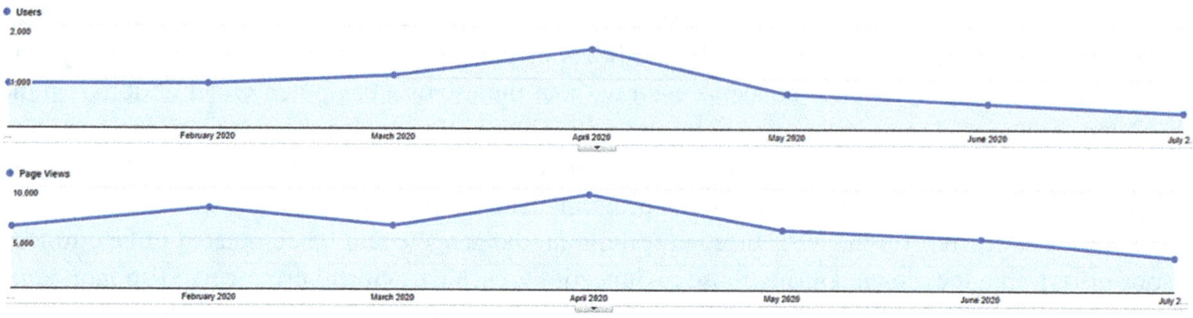

use of ER in classes. These webinars were an opportunity to remain informed and to get some training in ER even during the closure of the schools and of the Universities. Despite this formative/informative webinars, *Roteco* also created and shared different "learn at home activities". These activities were conceived to be done at home and did not need any special material or robots. This support in creating distance learning activities was much appreciated by teachers. This can also be noticed by comparing the number of users and page views before, during and after the first lockdown (Figure 8).

In April 2020, in the middle of the lockdown in Switzerland, *Roteco* observed the number of page views and users was higher than during the other months. A teacher community can therefore be important to support teachers in preparing and designing ER activities. They can be particularly pertinent in situations such as during the pandemic when teachers were forced to stay at home and had to face the new challenge to design ER online activities without having the possibility to exchange face-to-face with other colleagues.

SOLUTIONS AND RECOMMENDATIONS

While the potential of online learning to make education more accessible in terms of both convenience and cost has been acknowledged before (Starr-Glass, 2020), it still seems to be under explored in the context of ER. The examples presented in this chapter show there is great potential for ER in online learning. As highlighted, leveraging the online capabilities even allows for students worldwide to work together and collaboratively solve a space mission for example. It in turn this also supports interested teachers with free online courses to learn about educational robots and how to use them in the classroom and provides ways to create online communities where teachers, researchers and developers can exchange activities, ideas and news around the topic of ER. The examples presented in this chapter illustrate that there are many possibilities to perform ER activities in online learning. However, equivalent to ER classroom activities, it is essential that developers and educators make the necessary efforts to design tools and learning activities adapted to the context of online learning. For these activities to be pedagogically meaningful, it is not sufficient to simply transfer existing ER classroom activities to the online space. Instead, it must be reconsidered how intended learning outcomes, instructional activities, assessment activities and ER learning artifacts should be reshaped, in order to meaningfully support the online learning experience.

With regards to instructional activities, ER online activities designs should consider an appropriate balancing of synchronous and asynchronous learning approaches. Synchronous moments are crucial for students to keep in touch with their teachers and peers, allowing them to directly ask questions, receive feedback as well as share their experiences. The latest advances in the development of online tools offer students and teachers many new engaging opportunities to interact with each other. Although such tools cannot fully replace personal interactions, they can be considered valuable alternatives, contingent upon being appropriately used. Yet educators should carefully consider how much weight to give to synchronous approaches, since an excess can quickly lead to fatigue associated with active screen time. This is especially important when working with young children, who generally have a shorter online attention span (Dong et al., 2020). Asynchronous approaches on the other hand, provide both students and teachers with more flexibility, but also require teachers to invest more time in the preparation and students to take more responsibility to self-regulate their learning. However, in the context of ER, the latter is not necessarily a drawback. If the learning activities are appropriately prepared and implemented, asynchronous approaches could actually support constructionist learning, since it provides the learners with enough time to deal with the subject matter. The examples presented in this chapter illustrated a mix of synchronous and asynchronous approaches was helpful to implement ER online activities for in-service teachers, high school students and even primary school students. A crucial element is to find the right balance between those approaches, which may depend on different factors, such as the age range of the learners, their prior knowledge as well as the time that educators are able to invest for the design and preparation (Rapanta et al., 2020).

Another major challenge when organizing ER online activities, concerns the question of the learning artifacts. ER is built on the idea of learners working with physical manipulatives and finding alternative solutions, given this it is essential for ER online activities to be meaningful. The presented examples show that there are multiple ways to deal with this issue. Remote robotics laboratories have a considerably long history, especially in university education. In recent years, such approaches have been adopted to also implement remote labs for younger students and novices. Efforts such as the R2T2 project have made remote labs more accessible for beginners, by integrating less complex robots and programming platforms, as well as engaging narratives and collaborative challenges. They allow learners to develop important competences underlying ER activities, such as programming and computational thinking skills, as well as communication and collaboration skills, even if they do not involve the direct manipulation of physical objects. Alternatively, ER online activities can aim at finding suitable ways to substitute the ER artifacts in online learning. As shown in the BeeBot online activity, primary school students used paper and pen to represent the program commands they wanted to send to their BeeBots. Moreover, they used paper figures to simulate and predict the behaviours of their programmed robots. As well in this case, it was possible to attain the intended learning outcomes, i.e., to introduce students to sequential programming, even without the use of "real" robots and programming interfaces. Finally, the CreroBot project also illustrated that there are possibilities to provide learners with proper ER artifacts even in online learning. Based on accessible electronic components and cardboard, learners can easily reproduce their own CreroBot at home and at low cost. Not only does this approach provide individuals with a local robot that they can manipulate, it also introduces the idea of the making, which in recent years has garnered more and more attention (Martinez & Stager, 2019).

With respect to assessment activities, online quizzes are one of the most prevalent approaches in online learning (Kearns, 2012). Therefore, it is not surprising that they have also been used to assess learners in the context of ER online activities. However, the examples have also illustrated that different approaches can be taken, providing different insights into students' learning. With the increased capabilities of the internet, many approaches which have shown to be useful in the context of classroom education can also be adopted in online learning. For instance, to perform summative assessment, learners can be asked to prepare pictures, videos and/or presentation slides to document and share their ER projects online. Moreover, live video calls provide means for learners to directly interact with their teachers and peers as well as to share their experiences and receive feedback. Online forums and social media represent other means for students to ask questions and help each other. Furthermore, such platforms can support teachers in performing formative assessment, giving them valuable insight into the problems their students encounter and helping them to evaluate their current learning state.

As for the intended learning outcomes, the presented examples showed that overall, moving to the online space did not alter them significantly with respect to the classroom setting. In most ER online activities, the focus remains on the development of competences such as programming and computational thinking skills as well as general digital literacy skills. Projects such as R2T2, moreover provide means to promote the development of communication and collaboration skills, competences that are also often associated with ER. However, a major difference with respect to the classroom setting is that ER online activities require learners to take more responsibility and self-regulate their learning. As seen from the examples, this does not necessarily represent a drawback. Instead, if the ER activities are well designed, it can actually support constructionist learning, making the online learning experience more personal and hence more relevant to the learners. Although, when not carefully designed, this can also quicker lead to learning frustration and higher dropout rates. Fostering self-regulated learning represents a learning outcome implicit to the online learning context and should therefore be carefully considered when designing ER online activities.

CONCLUSION

In this chapter, we presented several approaches to ER in online learning settings with the goal to support educators, developers and researchers in designing future tools and activities as well as implementing them. The presented examples illustrate that there are a broad variety of ways to implement ER activities in online learning: from remote robotics laboratories, to pen-and-paper approaches to maker-based activities allowing learners to build their own robots from home. Nevertheless, moving ER activities to the online space changes the nature of how students interact with the learning artifacts, their peers and their teachers. For such approaches to be effective, it is therefore crucial that educators, developers and researchers reflect on how the tools and activities should be specifically designed for online learning settings. Equivalent to ER activities in classrooms, it is essential for ER online activities are well aligned regarding: the learning objectives, the ER learning artifacts, as well as the instructional and assessment activities. The recent technological advances have yielded a myriad of tools and software facilitating the preparation and implementation of synchronous and asynchronous ER activities in online learning. This makes communication and collaboration between students and teachers time and location independent. The challenge now is to make proper use of these tools to design ER online activities that are pedagogically meaningful. By effectively harnessing the benefits of both ER and online learning, such activities could ultimately represent a valuable complement to traditional ER classroom activities with areas of application going far beyond the context of pandemic situations.

ACKNOWLEDGMENT

This research received no specific grant from any funding agency in the public, commercial, or not-for-profit sectors.

REFERENCES

Alimisis, D., & Plessas, A. (2011). Teacher distance training in educational robotics through synchronous audiographic conferencing: towards a learner-centered approach. In *Proceedings of EdMedia+ Innovate Learning* (pp. 901–906). Association for the Advancement of Computing in Education AACE.

Almeida, T. O., Netto, J. F. d M., & Rios, M.L. (2017). Remote robotics laboratory as support to teaching programming. In *Proceedings of IEEE Frontiers in Education Conference* (pp. 1–6). Indianapolis, IN: IEEE. 10.1109/FIE.2017.8190472

Antonenko, P. D., Dawson, K., & Sahay, S. (2017). A framework for aligning needs, abilities and affordances to inform design and practice of educational technologies. *British Journal of Educational Technology, 48*(4), 916–927. doi:10.1111/bjet.12466

Ardito, G., Mosley, P., & Scollins, L. (2014). We, robot: Using robotics to promote collaborative and mathematics learning in a middle school classroom. *Middle Grades Research Journal, 93*(3), 73–88.

Biggs, J. (1996). Enhancing teaching through constructive alignment. *Higher Education, 32*(3), 347–364. doi:10.1007/BF00138871

Biggs, J. (2003). Aligning teaching and assessing to course objectives. Teaching and Learning in Higher Education. *New Trends and Innovations, 2*(April), 13–17.

Castro, M. D. B., & Tumibay, G. M. (2021). A literature review: Efficacy of online learning courses for higher education institution using meta-analysis. *Education and Information Technologies, 26*(2), 1367–1385. doi:10.100710639-019-10027-z

Chevalier, M., Giang, C., Piatti, A., & Mondada, F. (2020). Fostering computational thinking through educational robotics: A model for creative computational problem solving. *International Journal of STEM Education, 7*(1), 1–18. doi:10.118640594-020-00238-z

Chevalier, M., Riedo, F., & Mondada, F. (2016). Uses of Thymio II: How Do Teachers Perceive Educational Robots in Formal Education? *IEEE Robotics & Automation Magazine, 23*(2), 16–23. doi:10.1109/MRA.2016.2535080

Cohen, S. A. (1987). Instructional Alignment: Searching for a Magic Bullet. *Educational Researcher, 16*(8), 16–20. doi:10.3102/0013189X016008016

Conchinha, C., Osório, P., & de Freitas, J. C. (2015). Playful learning: Educational robotics applied to students with learning disabilities. In *Proceedings of International symposium on computers in education (SIIE)* (pp. 167-171). Setubal, Portugal: IEEE. 10.1109/SIIE.2015.7451669

Després, C., & George, S. (1999). *Computer-supported distance learning-an example in educational robotics*. Université du Maine.

Dionne, L., Lemyre, F., & Savoie-Zajc, L. (2010). Vers une définition englobante de la communauté d'apprentissage (CA) comme dispositif de développement professionnel. *Revue des Sciences de l'Education, 36*(1), 25–43. doi:10.7202/043985ar

Dong, C., Cao, S., & Li, H. (2020). Young children's online learning during covid-19 pandemic: Chinese parents' beliefs and attitudes. *Children and Youth Services Review, 118*, 105440. doi:10.1016/j.childyouth.2020.105440 PMID:32921857

Dos Santos Lopes, M. S., Gomes, I. P., Trindade, R. M., da Silva, A. F., & Lima, A. C. (2016). Web environment for programming and control of a mobile robot in a remote laboratory. *IEEE Transactions on Learning Technologies, 10*(4), 526–531. doi:10.1109/TLT.2016.2627565

Eguchi, A. (2017). Bringing robotics in classrooms. In M. Khine (Ed.), *Robotics in STEM education* (pp. 3–31). Springer. doi:10.1007/978-3-319-57786-9_1

El-Hamamsy, L., Chessel-Lazzarotto, F., Bruno, B., Roy, D., Cahlikova, T., Chevalier, M., & Mondada, F. (2020). A computer science and robotics integration model for primary school: Evaluation of a large-scale in-service K-4 teacher-training program. *Education and Information Technologies*, 1–31. PMID:33162777

Farias, G., Fabregas, E., Peralta, E., Vargas, H., Dormido-Canto, S., & Dormido, S. (2019). Development of an easy-to-use multi-agent platform for teaching mobile robotics. *IEEE Access: Practical Innovations, Open Solutions, 7*, 55885–55897. doi:10.1109/ACCESS.2019.2913916

Giang, C. (2020). *Towards the alignment of educational robotics learning systems with classroom activities* (Ph.D. dissertation). EPFL.

Giang, C., & Negrini, L. (2021). Educational Robotics in Online Distance Learning: An Experience from Primary School. In M. Merdan, W. Lepuschitz, G. Koppensteiner, R. Balogh, & D. Obdržálek (Eds.), *Robotics in Education. RiE 2021. Advances in Intelligent Systems and Computing* (Vol. 1359). Springer.

Harel, I., & Papert, S. (Eds.). (1991). *Constructionsim*. Ablex Publishing Corporation.

Hattie, J. (2012). *Visible learning for teachers: maximizing impact on learning*. Routledge. doi:10.4324/9780203181522

Hedges, H. (2010). Blurring the boundaries: Connecting research, practice and professional learning. *Cambridge Journal of Education, 40*(3), 299–314. doi:10.1080/0305764X.2010.502884

Hodges, C., Moore, S., Lockee, B., Trust, T., & Bond, A. (2020). *The difference between emergency remote teaching and online learning*. Retrieved August 2021 from https://er.educause.edu/articles/2020/3/the-difference-between-emergency-remote-teaching-and-online-learning

Kearns, L. R. (2012). Student Assessment in Online Learning: Challenges and Effective Practices. *Journal of Online Learning and Teaching, 8*(3), 198–208.

Kilpatrick, W. H. (1918). *The Project Method: The Use of the Purposeful Act in the Educative Process*. Teachers College, Columbia University.

Kulich, M., Chudoba, J., Kosnar, K., Krajnik, T., Faigl, J., & Preucil, L. (2012). Syrotek—Distance teaching of mobile robotics. *IEEE Transactions on Education, 56*(1), 18–23. doi:10.1109/TE.2012.2224867

Martinez, S. L., & Stager, G. S. (2019). *Invent to learn: Making, tinkering, and engineering in the classroom*. Constructing Modern Knowledge Press.

Mehrotra, A., Giang, C., Duruz, N., Dedelley, J., Mussati, A., Skweres, M., & Mondada, F. (2020). Introducing a Paper-Based Programming Language for Computing Education in Classrooms. In *Proceedings of the 2020 ACM Conference on Innovation and Technology in Computer Science Education* (pp. 180-186). ACM. 10.1145/3341525.3387402

Mehrotra, A., Giang, C., El-Hamamsy, L., Guinchard, A., Dame, A., Zahnd, G., & Mondada, F. (2021). Accessible Maker-Based Approaches to Educational Robotics in Online Learning. *IEEE Access: Practical Innovations, Open Solutions, 9*, 96877–96889. doi:10.1109/ACCESS.2021.3094158

Mester, G. (2006). Distance learning in robotics. In *Proceedings of The Third International Conference on Informatics, Educational Technology and New Media in Education* (pp. 239–245). Academic Press.

Mitchell, F. M. (1999, April). *All Students Can Learn: Effects of Curriculum Alignment on the Mathematics Achievement of Third-Grade Students*. Paper presented at the Annual Meeting of the American Educational Research Association, Montreal, Quebec, Canada.

Mondada, F., Bonani, M., Riedo, F., Briod, M., Pereyre, L., Rétornaz, P., & Magnenat, S. (2017). Bringing robotics to formal education: The Thymio open-source hardware robot. *IEEE Robotics & Automation Magazine, 24*(1), 77–85. doi:10.1109/MRA.2016.2636372

Mondada, F., Bonnet, E., Davrajh, S., Johal, W., & Stopforth, R. (2016). R2T2: Robotics to integrate educational efforts in South Africa and Europe. *International Journal of Advanced Robotic Systems, 13*(5). doi:10.1177/1729881416658165

Mussati, A., Giang, C., Piatti, A., & Mondada, F. (2019). A tangible programming language for the educational robot Thymio. In *Proceedings of 10th International Conference on Information, Intelligence, Systems and Applications (IISA)* (pp. 1-4). IEEE. 10.1109/IISA.2019.8900743

Negrini, L. (2019). Teacher Training in Educational Robotics. An Experience in Southern Switzerland: The PReSO Project. In W. Lepuschitz, M. Merdan, G. Koppensteiner, R. Balogh, & D. Obdrzalek (Eds.), *Robotics in Education: Methods and Applications for Teaching and Learning* (pp. 92–97). Springer. doi:10.1007/978-3-319-97085-1_10

Negrini, L., & Giang, C. (2019). How do pupils perceive educational robotics as a tool to improve their 21st century skills? *Journal of e-Learning and Knowledge Society, 15*(2), 77-87.

Negrini, L., Reyes-Mury, S., & Moonnee, D. (2021). A robotic teacher community to foster the integration of educational robotics in school. In W. Lepuschitz, M. Merdan, G. Koppensteiner, R. Balogh, & D. Obdrzalek (Eds.), *Robotics in Education: Advances in Intelligent Systems and Computing* (pp. 354–359). Springer. doi:10.1007/978-3-030-67411-3_32

Papert, S. (1991). Situating constructionism. In S. Papert & I. Harel (Eds.), *Constructionism*. Ablex Publishing Corporation.

Pelikan, E. R., Lüftenegger, M., Holzer, J., Korlat, S., Spiel, C., & Schober, B. (2021). Learning during COVID-19: The role of self-regulated learning, motivation, and procrastination for perceived competence. *Zeitschrift für Erziehungswissenschaft*, *24*(2), 393–418. doi:10.100711618-021-01002-x PMID:33686344

Rapanta, C., Botturi, L., Goodyear, P., Guardia, L., & Koole, M. (2020). Online university teaching during and after the covid-19 crisis: Refocusing teacher presence and learning activity. *Postdigital Science and Education*, *2*(3), 923–945. doi:10.100742438-020-00155-y

Resnick, M., Martin, F., Berg, R., Borovoy, R., Colella, V., Kramer, K., & Silverman, B. (1998). Digital manipulatives: new toys to think with. In *Proceedings of the SIGCHI Conference on Human Factors in Computing Systems* (pp. 281-287). 10.1145/274644.274684

Sentence, S., Waite, J., Hodges, S., MacLeod, E., & Yeomans, L. (2017). "Creating cool stuff" pupils' experience of the bbc micro:bit. In *Proceedings of the 2017 ACM SIGCSE technical symposium on computer science education* (pp. 531–536). ACM. 10.1145/3017680.3017749

Starr-Glass, D. (2020). Encouraging engagement: Video-conference augmentation of online distance learning environments. *On the Horizon*, *28*(3), 125–132. doi:10.1108/OTH-06-2020-0020

Vescio, V., Ross, D., & Adams, A. (2008). A review of research on the impact of professional learning communities on teaching practice and student learning. *Teaching and Teacher Education*, *24*(1), 80–91. doi:10.1016/j.tate.2007.01.004

Vygotsky, L. S. (1978). Mind in Society: The Development of Higher Psychological Processes. Harvard University Press.

Wing, J. (2006). Computational Thinking. *Communications of the ACM*, *49*(3), 33–35. doi:10.1145/1118178.1118215

Wonder-McDowell, C., Reutzel, D. R., & Smith, J. A. (2011). Does instructional alignment matter? Effects on struggling second graders' reading achievement. *The Elementary School Journal*, *112*(2), 259–279. doi:10.1086/661524

KEY TERMS AND DEFINITIONS

BeeBot: A programmable floor robot that is mainly used in primary schools.

Constructionism: A learning theory that foresees that children's learning can be enhanced when they are involved in activities allowing them to construct artifacts.

CreroBot: A do-it-yourself educational robot.

Educational Robotics: The use of robots with educational intentions.

MOOC: A massive open online course.

Social Constructivism: A learning theory that foresees that children construct knowledge through their interactions with teachers and peers.

Thymio: An open-source robot that was created for education.

Chapter 11

Creating Educational Robots:
Basics From Teacher Training to Making

Martin Fislake
https://orcid.org/0000-0003-3189-8741
University of Koblenz, Germany

ABSTRACT

After more than 30 years of development, the designing, constructing, and programming of educational robots is still enjoying increasing popularity in formal, non-formal, and informal educational settings. Although building instructions and required technical components are easily available and accessible, the realization of own teaching projects is a special challenge and is subject to decisive influences. This includes the content-related training of teachers as well as their attitudes and ways of thinking and acting. Therefore, the first section of this chapter spans an arc from the didactic concept of the extracurricular project technikcamps related to robotics. The experiences gained from it and the consequences for teacher training to the philosophical roots of technical education follows. In connection with this, the main part deals with the technological basics of creating educational robots in general. It leads from manufacturing single parts through the creation of a support structure and automation to the application of the engineering design process.

INTRODUCTION

One of the most challenging tasks preparing learners and teachers for a future we can't foresee is to prepare them in a general way regarding the current's educational robotics and the announced developments offered by tech labs, manufacturer and others. In the case of teachers, the goal is to train and educate future teachers to make them enlightened decision makers, who are independent from tech labs promises and manufacturers advertises but also from their own preconceptions to avoid a training for a single specific tool.

One of these steps to go is to familiarize them with as many different robot systems as possible during their training so that they develop an open approach by learning about similarities and particularities. In combination with appropriate instructions for abstraction and transfer, this gives them a flexible,

DOI: 10.4018/978-1-7998-7443-0.ch011

pragmatic and generalist way of thinking and acting when dealing with technical aspects of and about educational robots.

Another step is to explain and present the technical basics of educational robots. This includes not only considering the different robotic systems as given, but also abstracting and explaining them in a generalized manner in such a way that a basic transferability and a deeper understanding is possible. This also includes the examination or understanding of the structure-giving physical elements in connection with the different possibilities of the necessary programming. These two steps together demystify, changes the view and understanding of robots as automated machines.

There are two sections in this paper dedicated to this topic. One presents and explains the concept and the experiences in the technikcamps project in connection to the training of student teachers at the University of Koblenz in this subject, while a second section is purely dedicated to technical components and the engineering design process (EDP) of educational robots including designing, building and programming of educational robots. This will be the main section.

While for teachers it is a matter of preparing for a conceivable teaching assignment, the teaching goals for pupils are more directed towards typical ways of thinking and acting literacy in the sense of an enlightened and emancipatory approach to technology, computer sciences and other contents after the time with educational robotics. To illustrate how this can be achieved with pupils and young people, the didactic concept of the technikcamps and its consecutive structure is described in a separate section.

Since professional teaching with educational robotics is not only based on the existence of technical artefacts, guiding to declarative knowledge and arbitrary activities, guiding to procedural knowledge, in the concrete encounter with pupils, it is finally important to consider the attitudes and views of the teachers in addition to the teaching methodological approaches in the operative engagement with pupils. It is the difference of having only the know-how and know-what or also understanding the know-why in a deeper way.

To make it clear: teaching methods themselves belong to the classic tools of the trade of teachers, which they should have received like a vocational training at least enumerative during their training. It is the attitude, view and conviction of a teacher that shape concrete action, but can hardly be conveyed instructively themselves.

This is the adjusting screw that decides whether a teaching scenario is more instructive, constructionist or constructivist and how educational robotics are used. In this respect, this chapter is dedicated to two further sections that are formative for the way teachers think and act. One section deals with the problematic relationship between science and technology education while another section discusses the relationship between educational robotics and learning through designing. Both passages can shake the foundations and teaching beliefs and are intended to stimulate reflection.

EDUCATIONAL ROBOTICS IN THE TECHNIKCAMPS PROJECT

The idea for the technikcamps project (Fislake, 2010) was conceived by the Department of Technology at the University of Koblenz in Germany and fulfils several functions at the same time. As an out of school project, it is an extracurricular place of learning. Its focus is on technology education which is complementary to the established canon of subjects in schools, especially in Rhineland-Palatinate, Germany.

It is also used as a research laboratory and considered as a teacher training centre within the framework of teacher education. This allows prospective teachers to gain both practical and valuable experience in a kind of laboratory school through the supervised organization, execution and reflection of courses, while at the same time the project is researched by students and colleagues from different disciplines within the framework of diverse scientific questions at any level.

The Consecutive Course Structure

This section is intended to describe how the technikcamps project gives children and young people the opportunity to devote themselves to a technical subject and technical ways of thinking and acting under expert guidance. In addition, there is a description of the underlying didactic concept, the structure and the resulting experiences, which are taken up again at the appropriate points in this contribution.

Nevertheless, all individual camps and courses are offered in a suitable environment, with sufficient time and in community with like-minded people. They differ from each other in terms of concept and content in order to be able to do justice to the respective subject matter and the respective didactic demands. In doing so, situation-specific attempts are made to promote the independence of the participants, independent learning and technical maturity. By having fun and actually encountering technology, participants should be able to test their own talents in order to develop and, if necessary, deepen a possible lasting interest, including in technical professions.

A well-known problem in many extracurricular learning places and also in research papers is that robotics courses are often only seen and considered as a single session. Long-term studies or courses that build on each other over the long term are rarely found in the literature or in the course-programs from organizer of non-formal out of school activities. This is regrettable in so far as acatech (2009, 2011) rates this as one of the missing success factors for more interest in technical professions and courses of study.

The technikcamps project therefore offers a course structure in which one proceeds from the simplest manually made artefacts to pure computer science courses with an ascending degree of difficulty and increasing abstraction. This begins with learning simple handicraft activities such as working with wood, soldering small electronic circuits and assembling the first small vehicles with comparator circuits for line tracking (Figure 1). The use of child-friendly microcontrollers then leads smoothly into digitally driven automation and avoids the increasing use of wood as a structural element.

At the beginning of the technikcamps project in 2003, a concept for children aged 8 to 16 years was already developed and offered. The consecutive course structure started 2 years later with four courses that build on each other using the first-generation Mindstorms (LEGO RCX). The numbering of the courses with roman I, II, and III, however, hardly expresses the increasing level of the participants in handling the robotics system.

At the end of these three one-week-courses, the participants were generally quite open about their desire to program in other ways than just with colorful pictures. Consequently, a course rounded off this sequence upwards, which still used the same robotics system, but switched to a first but easy-to-use line-based programming language with NQC.

Years later, with the robotics kits then available for younger ones, such as the WeDo Robotics System and those from other manufacturers, additional offerings for 6–8-year-olds could be added to the program. Since the courses take place in close proximity to the courses for the older ones and a meeting during lunch and the accompanying social program is unavoidable, there is a lively exchange between the age groups. As a result, the youngest ones always look forward to the courses with the robots from the group with older participants. Here, after the NXT, the EV3 has now also been introduced and supplemented by comparable systems such as the TXT from fischertechnik in alternative courses.

Figure 1. Line follower assembled on a wooden structure

As before, these courses are offered in consecutive formats and also introduce the use of line-based programming languages and microcontroller programming in a playful way, for example, through the use of Scratch. In the meantime, it is possible to build on the previous knowledge of the participants in some consecutive courses and in conjunction with the optional offer of electronics, programming, CAD-CAM and 3D-printing courses. Advanced participants can finally design, construct, build and program their own concept of educational robots. Finally, few of these senior participants now switched their role to course instructors, which proves the lasting impact and the creation of a community and thus the success of the project.

Importance of Supervisors in Robotics Courses

The research conducted internally and externally within the framework of the technikcamps project at the University of Koblenz underlines the aforementioned laboratory character of the project. Through the years many kinds of data were generated through different empirical methods and statistical analysis were performed. But most of the evaluations were mainly descriptive in nature. Examples of questionnaires and result presentations derived from these activities can be seen in Figure 2 and Figure 3.

The results repeatedly provided feedback on individual questions to the management team, whereby the identification of success factors for the project in particular fulfilled an important task. Beside it, the study of learning effects in the technikcamps project regarding teaching declarative knowledge or in direct encounter with children is rather unproblematic. It follows to a large extent of what is taught in youth work preparation courses and in teacher training as technical tools and didactic methods.

Research Conducted

While the technikcamps project was generally open and used for a lot of external researches from different departments at the University of Koblenz ranging from psychology to management and at different research levels including bachelor and master level only a small selection of internal researches is considered in the following report.

In the case of the success factors, a distinction was made once again between the learning success of the participants and those factors that ensure the survival of the course-program and the project. For this purpose, insights were to be gained in particular about the motivation to participate, about satisfaction with the organization and the contents, but also about the cooperation of the participants in the course and the satisfaction with the course instructors. Sometimes another survey was added to get feedback from the course instructors about their motivation to participate, how to improve the support from the team and other things.

Finally, the responses of instructors, participants and their parents conducted internally for the purpose of self-evaluation produced revealing findings, which were generally confirmed by external researchers if they took such questions into account. The results were also used to give valuable feedback to the instructors and to be reflected within the team for further improvement.

Regarding common ways of developing surveys, the original questionnaire was first generated in 2009 by the organizing team of the technikcamps project as a pilot for internal use. It was repeatedly adapted until 2016 and was then used when no other surveys or research projects were conducted as those shown below.

The team consisted of student teachers and student assistants and was under the academic direction of Dr. Martin Fislake. It included closed and opened items and considered former experiences made by student research. A pre-test was made to check the quality criteria of empirical research and the possible correlations. In practice the title and the instructional text of each survey was adapted to each course to meet necessary requirements. Even though there were occasionally longer intervals between surveys, the basic results were regularly confirmed.

In 2011 a totally new survey was developed and evaluated using SPSS in cooperation with the Institute of Social Sciences University of Stuttgart under the chair of Dr. Uwe Pfenning. He was also responsible for the acatech studies (acatech 2009; 2011) mentioned in this chapter. The survey included pre- and post-surveys for participants and parents but confirmed all findings made in former evaluations. It was conducted only in one season but covering hundreds of questionnaires.

Another survey was conducted in 2013 by a research group with a focus on questions how to make the technikcamps project a spin off. The questions of the questionnaires were similar in content, but supplemented by the focus on how to market the educational project as a product. The use of an online tool for the survey was completely new but generating similar results again to the old questions.

Summarized Results

With a focus on the importance of supervisors in robotics courses the generated results from the surveys showed clear indications of how important a competent and friendly course leader is for the participants in the robotics courses within the technikcamps project. To illustrate it, after a successful course, younger participants in particular often reacted exuberantly and positively to the question of whether the course instructor had been nice. They usually acknowledged this at the same time with a clear desire to participate again. The feedback on the materials, contents and the course progress clearly lagged behind.

The expectation that the central motivation to participate again was the desire to build certain things, such as robots or in particular LEGO robots, could therefore not be confirmed in the assumed way for

Figure 2. Examples of questionnaires that have been used

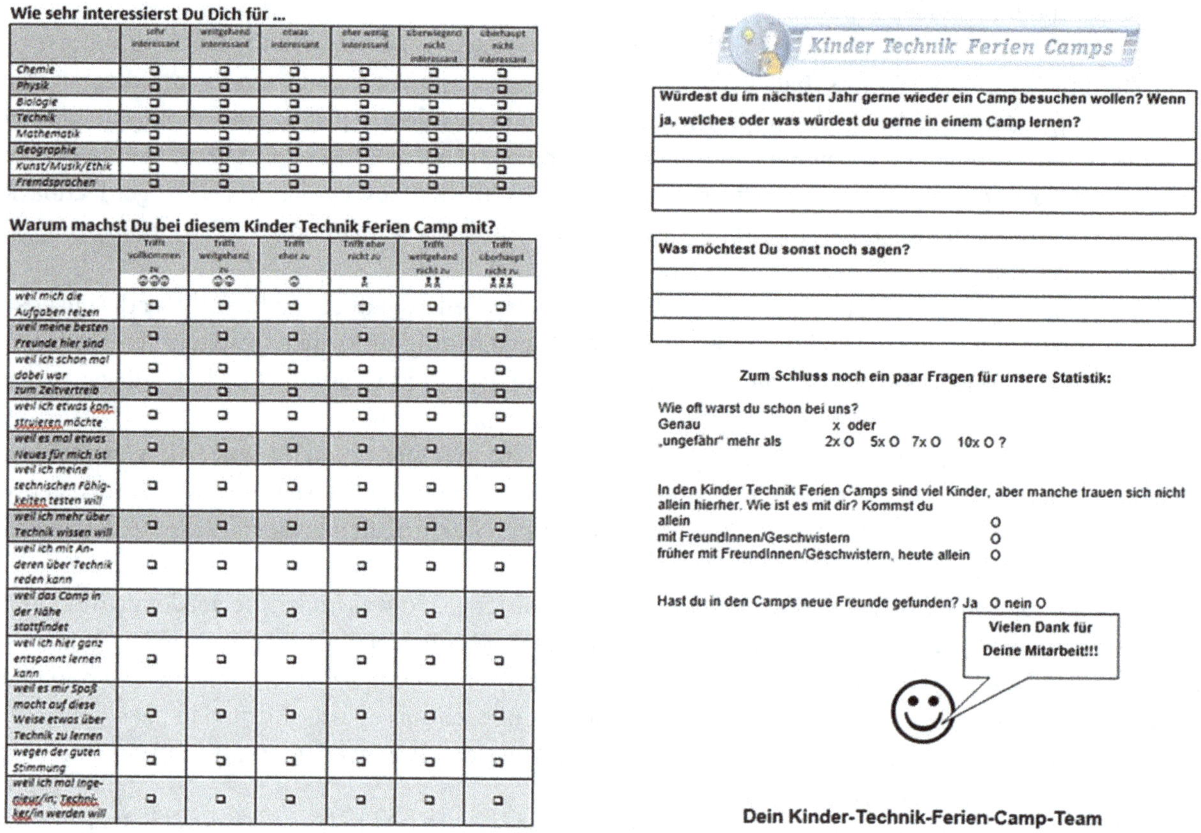

Wie sehr interessierst Du Dich für ...

	sehr interessant	weitgehend interessant	etwas interessant	eher wenig interessant	überwiegend nicht interessant	überhaupt nicht interessant
Chemie	☐	☐	☐	☐	☐	☐
Physik	☐	☐	☐	☐	☐	☐
Biologie	☐	☐	☐	☐	☐	☐
Technik	☐	☐	☐	☐	☐	☐
Mathematik	☐	☐	☐	☐	☐	☐
Geographie	☐	☐	☐	☐	☐	☐
Kunst/Musik/Ethik	☐	☐	☐	☐	☐	☐
Fremdsprachen	☐	☐	☐	☐	☐	☐

Warum machst Du bei diesem Kinder Technik Ferien Camp mit?

	Trifft vollkommen zu	Trifft weitgehend zu	Trifft eher zu	Trifft eher nicht zu	Trifft weitgehend nicht zu	Trifft überhaupt nicht zu
weil mich die Aufgaben reizen	☐	☐	☐	☐	☐	☐
weil meine besten Freunde hier sind	☐	☐	☐	☐	☐	☐
weil ich schon mal dabei war	☐	☐	☐	☐	☐	☐
zum Zeitvertreib	☐	☐	☐	☐	☐	☐
weil ich etwas konstruieren möchte	☐	☐	☐	☐	☐	☐
weil es mal etwas Neues für mich ist	☐	☐	☐	☐	☐	☐
weil ich meine technischen Fähigkeiten testen will	☐	☐	☐	☐	☐	☐
weil ich mehr über Technik wissen will	☐	☐	☐	☐	☐	☐
weil ich mit Anderen über Technik reden kann	☐	☐	☐	☐	☐	☐
weil das Camp in der Nähe stattfindet	☐	☐	☐	☐	☐	☐
weil ich hier ganz entspannt lernen kann	☐	☐	☐	☐	☐	☐
weil es mir Spaß macht auf diese Weise etwas über Technik zu lernen	☐	☐	☐	☐	☐	☐
wegen der guten Stimmung	☐	☐	☐	☐	☐	☐
weil ich mal Ingenieur/in; Techniker/in werden will	☐	☐	☐	☐	☐	☐

Kinder Technik Ferien Camps

Würdest du im nächsten Jahr gerne wieder ein Camp besuchen wollen? Wenn ja, welches oder was würdest du gerne in einem Camp lernen?

Was möchtest Du sonst noch sagen?

Zum Schluss noch ein paar Fragen für unsere Statistik:

Wie oft warst du schon bei uns?
Genau x oder
„ungefähr" mehr als 2x O 5x O 7x O 10x O ?

In den Kinder Technik Ferien Camps sind viel Kinder, aber manche trauen sich nicht allein hierher. Wie ist es mit dir? Kommst du
allein O
mit FreundInnen/Geschwistern O
früher mit FreundInnen/Geschwistern, heute allein O

Hast du in den Camps neue Freunde gefunden? Ja O nein O

Vielen Dank für Deine Mitarbeit!!!

Dein Kinder-Technik-Ferien-Camp-Team

younger participants. A more important factor was the course-instructor, which lost relevance with increasing age.

The results correspond well with those of the Hattie study (Hattie, 2009), emphasizing the importance of teacher personality and those of acatech (acatech, 2011) emphasizing that teaching of technology requires professionally and didactically trained teaching staff, their own technology didactics and appropriate equipment for learning success.

Therefore, the question arises as to how strong the motivating effect of learning robots for learning really is and how it should be evaluated compared to other factors.

TEACHER EDUCATION FOR EDUCATIONAL ROBOTICS

As a consequence of the experiences of the technikcamps project as a teaching lab, prospective technology teachers are also trained in a consecutive structure on different educational robots. In the first part of this section, it will be shown that at the beginning (bachelor-phase) of the teacher training, the very simplest (construction) materials are used, and then they move on to the master level via simple electrification and the use of commercially available robots and robot construction kits. School-specific microcontrollers

Figure 3. Examples of result presentations that have been generated

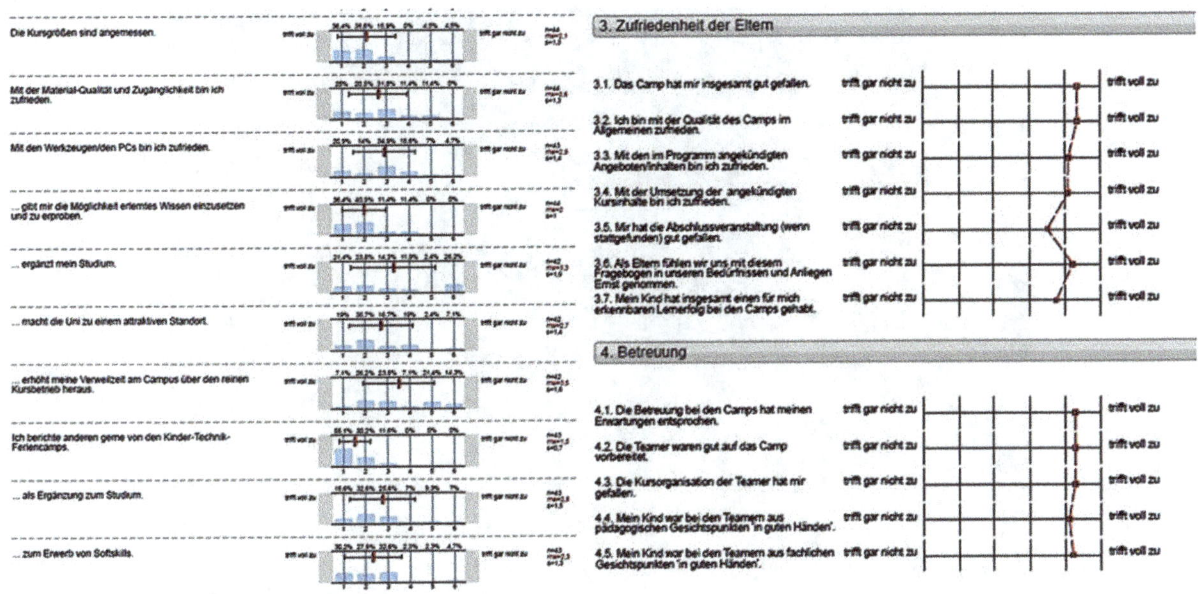

and simulations for grades 1-6 are also considered. The master's phase picks up with more complex robotics kits and continues the education with more powerful microcontrollers and programming tools.

The focus in the second section then switches to aspects from the background that influences teaching with educational robots. It's because the work with educational robots in school or in extracurricular projects is not only based on the available robotics materials and other factual framework conditions, but also on the phantasy, attitudes, beliefs and teaching repertoire of the teachers.

Teaching Robotics Consecutive

The teacher training program at the University of Koblenz comprises a total of four modules and starts with the very simplest constructions made of paper and cardboard, but seemingly unsolvable tasks. During the courses, the students experience how much their prior knowledge and their own ideas of what is technically possible influence their thinking and making. In contrast to younger children, their ideas are readily and often limited to reproducing old, familiar concepts instead of following the idea and looking for ways to realize it. Recognizable for example through they ask for prefabricated building instructions.

The idea behind this concept is that the students learn to put aside their acquired receptive ways of acting and turn to problem-solving ways of thinking and making, which later lead to concepts such as the engineering design concept, which will be taken up again later in this chapter. The aim is to get them away from just doing or copying what others tell them to do - even if demonstration and imitating are used in the meantime as phases of didactic methods for teaching individual learning objectives.

They should develop motivation and willingness to realize their own ideas. This may also include a change of perspectives. Because people who believe only things will work, they already know will not be able to generate ideas and innovations in the end as Lauxmann illustrated in a dialogue between his Art figures Sokrates and Technikos in one of his short stories (1996).

Figure 4. Switching roller of a traffic light control

As the story goes Technikos always wanted to have examples and precedents to follow, while Sokrates tried to explain him someone has to be the first to do it, without a manual but driven by his own willing. This applies to technical things as well as to teaching methods and relies on receptive rather than constructivist approaches to learning with educational robots in the sense of Papert (1980) and others.

For this reason, the tasks are formulated in such a way that the crafting requirement initially appear easy and the first handicraft experiences with materials can be made. As the tasks become more difficult, it becomes necessary to make systematic decisions and also to develop the willingness to use additional equipment such as tools, fixtures and the like. To support them in addition, the students also receive guiding step by step introductions to the methodical procedure, such as the EDP and the use of technical drawings from the hand sketch to the 3-D model.

These experiences are followed by the first mechanical constructions, which are first operated by hand and later electrified with motors and gears. The model of a traffic light control based on a rotating tin can as switching roller (Figure 4) is a typical example and provides an early introduction to digitalization and programming.

The concept is taken up again with the construction of simple vehicles made of wood and continued with the use of self-soldered electronic circuits (Figure 1). As with the technikcamps, the use of several different user-friendly microcontrollers such as the BBC micro:bit then leads seamlessly into digital-electronic automation and avoids the increasing use of wood as building elements.

Then the introduction to familiar robot construction kits does begin with the sequence being based on the age of the target group in schools and the typical distribution of the species in regional schools. Accordingly, the educational sequence begins with robot construction kits such as the well-known WeDo set from LEGO and then continues with an NXT/EV3 sequence, without neglecting the alternatives from fischertechnik and other manufacturers.

Figure 5. Robot-car with 3D-printed body and single-board-computer inside

The sequence concludes with an introduction to the development and construction of one's own semi-professional concepts for educational robots using single-board computers such as the Raspberry Pi, manufacturer-independent system construction kits, manually manufactured components made of metal or plastic, 3D printing solutions (Figure 5) or others.

Only any form of ready-assembled robots is not used, because their options for getting to know coding editors for programming and their interfaces are sufficiently covered by the use of various programming tools in the robotics construction kits and self-made robots controlled by microcontrollers.

Teaching Robotics Influenced by Philosophy

Philosophy of technology is a discipline that has much to offer for technology education de Vries and Tamir stated (1997). They believe that insights into the real nature of technology and its relationship with science and society can help technology educators to build a subject that helps pupils get a good concept of technology and to learn to understand and use concepts in technology.

In this respect the potential role of the philosophy of technology in technology education is important for the way educators gain their position and appreciation of technology education. Because of the history of sciences and universities the position of technology education is substantially different from that of natural science education. There the academic roots are much clearer and the philosophy of natural science has been an established discipline at all levels of education for almost a century worldwide while the paradigm of technology as applied sciences fails and is no longer seen as a valid description of the nature of technology (de Vries & Tamir, 1997).

As a result, natural sciences like biology or physics belong to the established subject canon of the school curriculum, while technology education does not. Consequently, pupils in schools today learn for example concepts from phenomenology natural sciences and by reconstructing pre-concepts that they picked up from daily-life experiences but minor concepts from technology.

Snow (1967) went a step further and compared humanities and literature on the one hand with natural sciences and technology on the other. Therefore, he illustrates by the way the common misunderstanding that reduces technology to applied natural sciences, while he did not fail to point out the limitations of his own classification, in this case the conflict between natural sciences and technology.

Thus, the first problem area already arises here, according to which natural sciences have already been simplistically summarized with technology for this purpose and are commonly juxtaposed with the humanities and literature. A juxtaposition that can later be formulated formally as a decision between the positions of Fischer (2000) and Schwanitz (1999) which kind of education is the most valuable.

In this respect, the question is about the equivalence of technology education respectively of technical education in general and its relationship to the natural sciences. The last-mentioned comparison then opens up the discussion about the different paradigms as well as about the science-theoretical findings and contexts. It is to see against the background of experiences made in everyday life through the encounter in occasional confrontations. These experiences show clearly the impact of the misunderstanding that reduces technology to applied natural sciences.

This includes the engineering community, but also those who want to offer support to projects like that of the technikcamps project through internet-platforms and community-networks. Trying to support those projects through promotional activities they all ask for experiments that are taught in the laboratories instead to ask for the challenges which are offered to the participants in the workshop. These challenges include tasks and projects that have to be solved through a problem-solving design process and technology artefacts to be created. They have to understand that scientific experiments transport par excellence the causal based paradigm of natural sciences along with them while the paradigm of technology education is in contrast final based.

The widely cultivated misunderstanding (Gaycken, 2005) can also be seen as the cause here, according to which the image of a technical science, the mere application of the existing natural sciences to technical problems, still prevailed until the 1970s - hence the designation as applied sciences that is still common today (Gaycken, 2005).

This can only be explained by looking for example at the history of German Technical Universities. Until 1899, the Engineering Schools went through a phase of realism, which, although scientific, was closely oriented to the practical application of technology in the emerging industry and to the practical needs of engineering. But in order to gain the right to award doctorates, they had to go through a development according to which the study of engineering was not only recognized as practically useful, but also as scientific from a different point of view. Almost inevitably, the criterion of scientific was no longer linked to engineering practice, but to the proportion of natural sciences and mathematics in the scientific study (Fislake 2001). The roots of which, in turn, lay in philosophy itself.

Instead of establishing mathematics and natural sciences as supporting auxiliary sciences of an independent technical science with, for example, a more holistic focus proposed by Riedler (Riedler, 1919) or an economic orientation presented by Schenk (Schenk, 1920), the paradigm dispute of the years 1890 to 1920 was primarily about using the shares of mathematical and natural science subjects to secure proof of the scientific nature of engineering science courses.

With that in mind, many activities which aim to attract the next generation to technical and engineering subjects seem all the more paradoxical. Especially when they take their cue in addition to more mathematics from subjects like physics. Because, based on its phenomenology approach and instructional teaching methods, which are as traditional as they are unsuccessful physics is the most unpopular school subject at all.

In addition, experiments in science subjects at schools are often trivial and associated with strong didactic reductions. The purpose is often not transparent for the experimenting students while references to their everyday life or to their lifeworld are often only difficult to recognize as Euler (2009) elaborated.

Unglaube (1997) speaks in this context of the strict view of the subject natural sciences and their didactics, according to which the subject lessons of secondary schools, even with their experiments, merely aim at existing explanations and prevent the child's engagement with the facts, which actually triggers learning. The key moments that are actually important and provide orientation are missing.

At the same time, the receptive approach conveys to the learners implicitly or sometimes explicitly a highly simplified image of science as Euler (2009) stated. The impression is given that observations and experiments lead in a straight line to the formulation of laws (Euler, 2009).

But those follow a causal-oriented paradigm neglecting the final-oriented one applied in technology although the relationship of these two paradigms is widely discussed and cleared in the philosophy of technology education by Lassahn (1979), Schmayl (1981), de Vries (2005) and ITEEA (2020). The result is an ongoing empirical misunderstanding of scientific methods in natural sciences and technology, according to Euler (2009).

Teaching Robotics influenced by Learning Through Designing

In contrast Martinez and Stager (2019) believe that in the classroom, iteration through multiple design cycles like the EDP is useful and relevant. They further claim out how important it is to understand that the iterative, cyclical nature of modern, real-world design models looks nothing like the linear, step-by-step design models or scientific methods found in most textbooks. Going through multiple design circles instead of linear methods they will develop a better understanding of the requirements, tools and materials as they improve their prototypes (Martinez & Stager, 2019).

This makes it all the more important for teacher education to enlighten the backgrounds of our days teaching to make enlightened decision makers as well as for out-of-school activities to mediate between deficiencies of the established educational ideal and a modern knowledge society with fairly distributed educational opportunities.

All in all, this is a very important adjusting element that decides whether teaching follows a more instructive or a more constructionism learning scenario and how educational robotics are involved.

BASICS OF CREATING EDUCATIONAL ROBOTS

This section will uncover the declaratively knowledge of technical basics of educational robots and tear down the curtain to help understanding what is inside a robot and how it is designed and build out of mechanic, electric and electronic parts in connection with programming tools. It is followed by a short introduction into the procedurally knowledge of the EDP to explain how an educational robot can be realized in learning settings.

How to Construct and Build an Educational Robot

Anyone who wants to use education robots in the classroom can of course opt for a ready-assembled and ready-to-use version. The range is large and covers a wide range of proven robot systems at different

Figure 6. From a physical device to behavior

levels of difficulty. There is a number of papers in the literature like (Berger & Müller, 2018; Fislake, 2018; 2021, Vandervelde et al., 2013) that address this issue of the correct selection of education robotics.

Nevertheless, such a decision should be thoroughly considered, because fully assembled education robots are little more than simulations with physical manifestations and essentially limit their use in teaching to content such as coding and computational thinking. This creates the danger that the physical manifestations are taken for preexisting and possible solutions to problems and their strategies are limited to digital approaches.

Robot simulations itself are an exception, because the disembodied variants of robots relativize an aspect that is often overrated and less pragmatically assessed when purchasing robots, although they do not in themselves generate any functional added value.

In contrast and corresponding to Figure 6 all educational robots can first be defined as physical devices that are based on a mechanical structure build out of mechanical components completed by electric and electronic parts. As illustrated in Figure 13, the automation of this structure can finally be done by means of mechanical elements or digital controls and ensures the robot's functionality in a way that Resnick called behaving machines (Resnick, 1993).

From Components to Structure

The actual construction of an educational robot in learning settings depends on the available possibilities and materials as well as the pedagogical decisions of the teacher. But basically, the realization of a mechanical structure can be divided into two sequences, the manufacturing of components and the assembly, independent of didactic decisions.

Both sequences are included in the task area that is called the design and manufacturing task in technology didactics (Henseler & Höpken, 1996). It includes the assembly of the mechanical structure, the design and the production of components, in which the available raw materials are processed so that they can be used for the intended purpose.

The diagram in Figure 7 show a selection of original source materials and production areas that are used to build educational robots. The spectrum ranges from filaments for 3D printing and free materials to classic materials such as wood, plastic, metal and cardboard, which are also used as semi-finished products.

The manufacturing area ranges from 3D-printing via tinkering up to handcrafting and includes different manufacturing processes. For example, 3D printing is an additive manufacturing process in which components are made from plastics by means of templates or CAD designs. In the design and handcrafted manufacture of components from classic materials, the manual manufacturing processes differ only marginally, whereby the use of semi-finished products simplify and anticipate necessary design and manufacturing steps.

Figure 7. From scratch to a mechanical structure for an educational robot

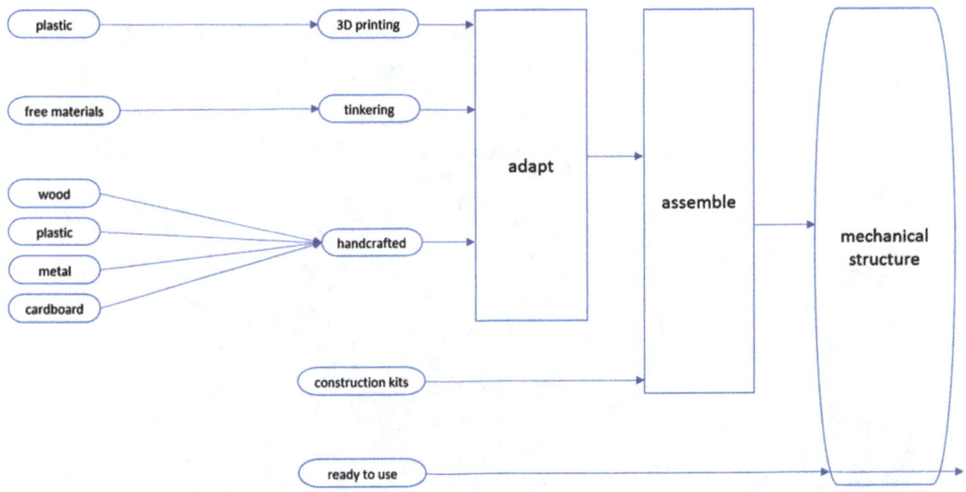

Finally, the use of free materials can be assigned to an area in which the widest range of applied manufacturing processes can be used. It ranges from simple handicrafts and tinkering including the upcycling of recycled materials with a focus in kindergartens and primary schools to professional manufacturing processes in makerspaces.

Jeretin-Kopf et al. (2015) investigated the effects of the selection of such different basic materials on motivation and creativity within the design process. In a two-stage study of 246 children between the ages of 8 and 10, LEGO and fischertechnik construction sets, the UMT semi-construction system and free materials from the DIY store were used.

The results showed that the design of the learning environment seems to be more important than the differences in learning conditions and that in addition to the experiences of the participants, the age of the children, gender and especially the type of material system played a role.

With regards to interest as well as challenges, the participants in the DIY store group seem to be the most interested and challenged, followed by the UMT, LEGO and fischertechnik group. Moreover, the interest in such challenges seems to be more pronounced in boys than in girls, while the fear of failure decreases with age.

Even in terms of the variety and properties of the objects built, the participants in the DIY group made the most complex designs, followed in alternating order by the LEGO or UMT group and the fischertechnik group in last place. The fischertechnik group, on the other hand, was able to present the best scores in functionality, while the UMT system managed to do so in the category of perseverance and determination.

By using construction systems, the scope of tasks is reduced to design and a so-called assembly task (Wilkening 1982), in which already prefabricated components are put together to form assemblies and finally the complete mechanical structure as can be seen in Figure 8. Both tasks, the manufacturing task as well as the assembling task may by supported by building instructions or manuals and is dependent on didactic decisions for use in the classroom. With ready-assembled robots, these tasks and activity areas are omitted.

Figure 8. Assembled mechanical structure

To design a mobile robot, suitable child friendly CAD systems are available. They can be used to assemble designs in advance and test them for missing components, functions and other design questions. As a rule, they are adapted to a specific construction system and sometimes even enable photorealistic animations or collision tests of moving parts. Figure 9 displays such a situation where a child in the age of 8 years regarding available bricks first used a CAD tool to design a mobile before assembling.

Classic paper and pencil drawings, on the other hand, are neither common nor contemporary in this area of application, although hand sketches still make sense as a quick tool for exchange between the people involved.

According to the experiences from the technikcamps project, children prefer the trial-and-error method when confronted with a construction, building and programming task that involves a robot that is supposed to show a certain behavior. They just love to build and tend to anticipate certain construction steps only with increasing experience.

The use of CAD systems, on the other hand, must be prescribed, introduced and supervised. The playful connection of real and virtual objects in this proximity trains and promotes spatial thinking and the ability of mental rotation. At the same time, it provides a visible link to professional ways of thinking and acting, as it makes both construction tools and procedures for a specific area of the EDP tangible and understandable.

After or already during the assembly of an educational robot, the mechanical elements are supplemented by electrical and electronic components. They themselves can be load-bearing elements and must be selected for the intended purpose. Figure 10 shows how only the interaction of actuators, sensors, energy source and a data processing unit completes a purely mechanical structure into a physical device.

Figure 9. Child friendly CAD-tool before assembling

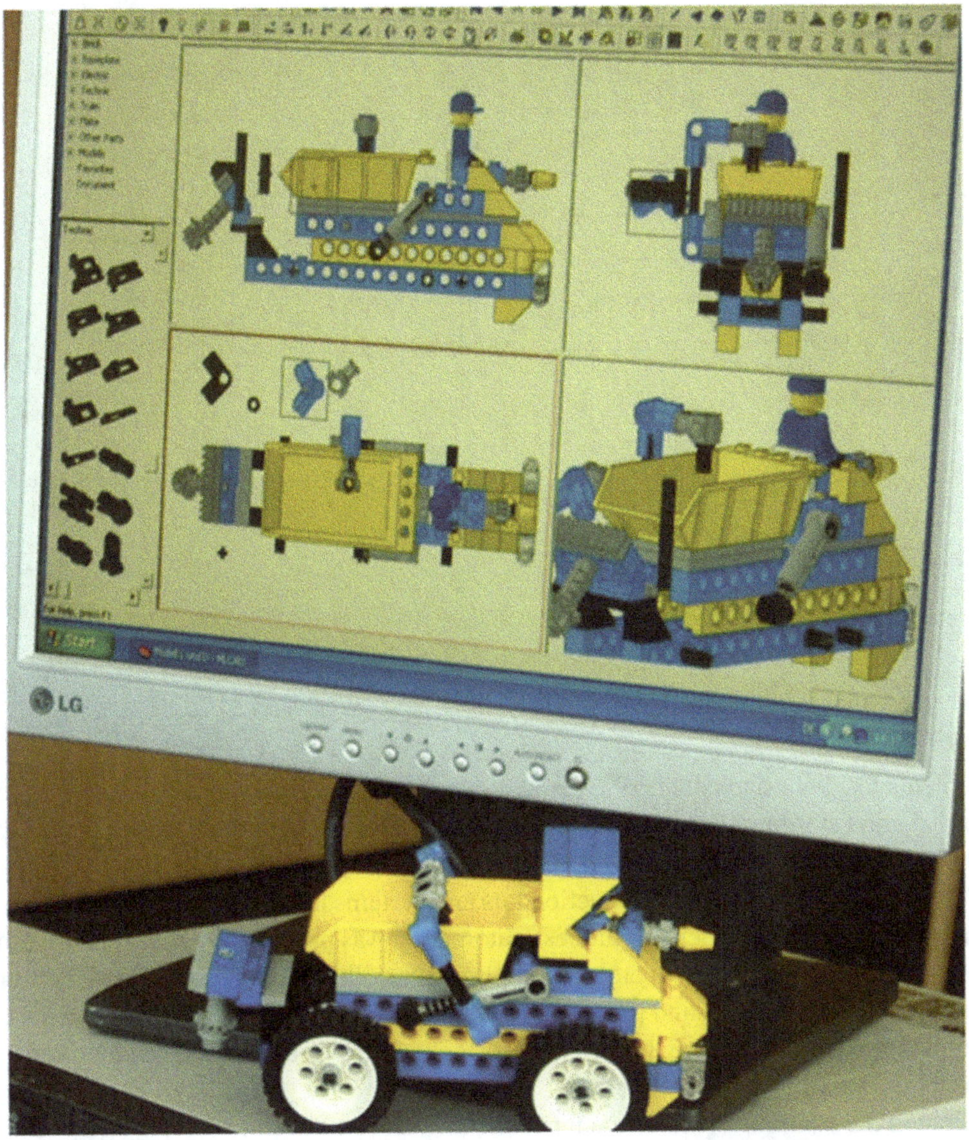

Figure 1 shows such an example of a completed line follower made of mechanic, electric and electronic parts assembled on a wooden structure. It is based on a common learning kit from the teaching materials trade, in which the students have to make the supporting structure and to solder the electronic circuit before they connect it to the mechanical elements.

One of the next steps to enhance this simple mobile robot is to replace the self made circuit by a BBC micro:bit, an Arduino or something quite similar as control unit. At this point, at the latest, it becomes clear that the use of wooden components has reached its limits and no longer allows the necessary precision for more demanding tasks.

The use of standard parts together with construction systems or semi-finished products is only a consequence of this. For that, construction systems bring most used standard parts like motors, trans-

Figure 10. From a mechanic structure to a physical device

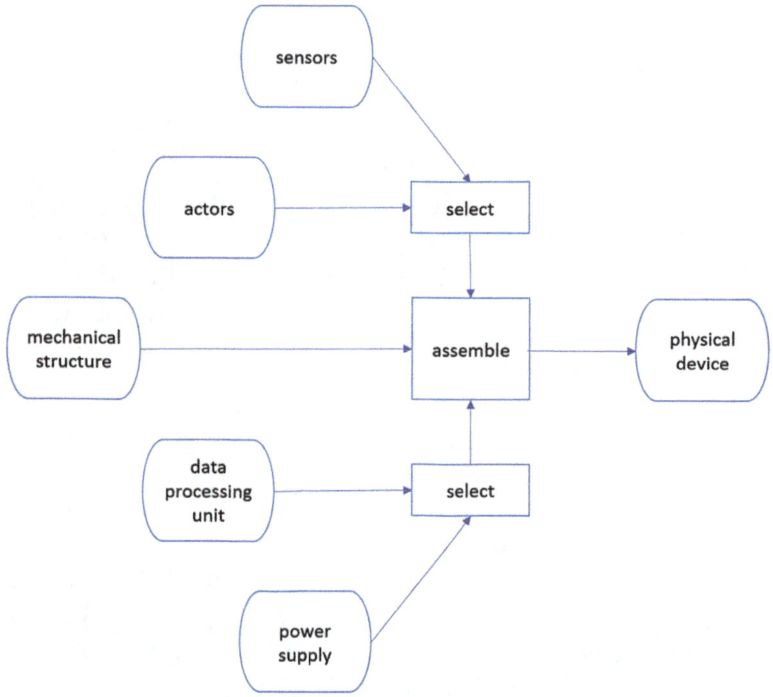

mission elements like belts and connecting elements such as screws, nuts and bolts along with it. Many of these construction systems even include the necessary tools, some of them even a central processing unit for robotics tasks too.

An early systematic overview of construction kits and systems in general and their use in conjunction with programmable logic controllers for process automation was provided by Fast (2006). Among other things, he describes how the first papers on computer-controlled models were published in the early 1980s. With the use of interfaces and programmable logic controllers, these eventually paved the way for first mobile robots such as the BBC Buggy (Bostock, 1983) as a teaching tool and later to current educational robots based on construction kits.

Depending on the equipment and character of the kits, such a structure can be easily or more elaborately supplemented and expanded with electronic components and microprocessors as shown in Figure 11. If necessary, mounted existing motors and other components can also be replaced by those that are more suitable for the intended purpose. A possible example is the replacement of simple DC motors with servo or stepper motors from third-party suppliers.

An alternative to usual construction system is the use of the UMT semi-finished construction system. In addition to the structure-giving semi-finished products made of plastic, it also contains a large number of standard parts and special tool sets for use in schools. Benjes et al (1994) characterize the system as a synthesis between the workbench method and the construction kit principle, which lowers the demand on pupils' manual skills and reduces the need for the school's equipment. Figure 12 shows a line follower assembled from such worked semi-finished products, purchased standard parts and a custom-made circuit board.

Figure 11. Construction kit enhanced with non-proprietary components
(Fislake 2018)

Figure 12. Line-follower based on adapted semi-finished plastic products

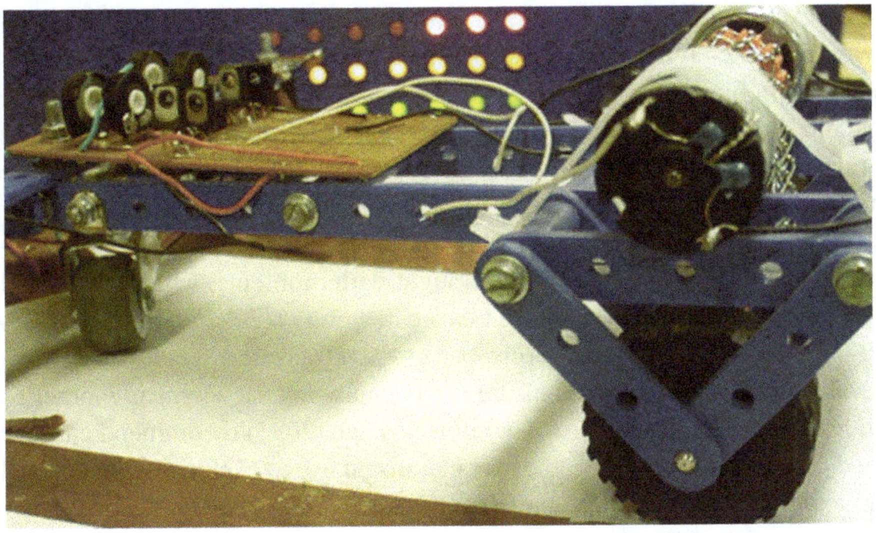

From Structure to Motion

This subsection explains absolutely pragmatic how the program code can get from the user interface (UI) into the controlling element. According to the IPO principle (input-processing-output) this central processing unit (CPU) processes received data and signals according to the program code into actions. The way the code can be programmed and transmitted into the CPU depends on the appearance of the UI and the connecting elements. Figure 13 gives a schematic overview of different appearances that are raw divided into a bodyless-digital section at the left side and a physically-analogue section (right).

The difference between these appearances is that the execution code in the physically section is fixed and hard implemented in the electrical circuit or the mechanic structure while a CPU uses bodiless digitally code that is saved on any memory. It doesn't matter if the digitally code was generated via keyboards, RFID-Tags or something else and digitally transmitted via connecting cables or other transmission elements which are used occasionally. Signals to be processed in this section may also have an analogue character, but must be digitized. All physically based versions need a physically conversion to change the processing code. Beside it signals to be processed in here may well also have an analogue or a digital character. A typical example of this are switches that only ever change between the states on and off.

Consequently, building a robot means taking into account the specifics of each transmission path and to anticipate and weigh their advantages and disadvantages when designing a robot concept.

It includes issues of

- reliability,
- usability,
- compatibility and
- connectivity.

Anyway, with most prefabricated robots and robot kits, this is limited to reliability and usability. These systems today often offer different transmission paths simultaneously, but these are fixed. In this case, the compatibility and connectivity between transmitter and receiver, i.e., the code-transmitting unit and the code-receiving unit, should have been sufficiently ensured by the manufacturer.

It always becomes problematic when there is no compatibility between the code-generating unit and the connecting element, as it is the case with the Edison robot. Here, the actual programmed code is transmitted by means of a special aux-cable (the EdComm programming cable) and tones. However, it requires a computer that can also generate these tones in the intended way, which needs mandatory an audio-out jack and is not possible with many Android tablets.

Connectivity, on the other hand, can still be fraught with problems, which was not always the case with the first versions of the education robots as Fislake mentioned (2021). The first generation of LEGO Mindstorms (RCX) was a good example of this, because the transmission of the code via infrared light (IR) was not always stable. Similar problems are occasionally reported by using larger numbers of current robots with Bluetooth (BT) interfaces. Therefore Fislake (2021) recommend cable-based solutions with standardized cables, such as USB cables for classroom use to exclude technical problems.

The freedom to decide how the programmed code is generated and transmitted is closely related to the technical openness of the systems.

While it is not given in the case of ready-assembled robots or is merely reduced to the selection of programming tools, construction kits can additionally offer the freedom to add or replace relevant haptic

Figure 13. From coding to action

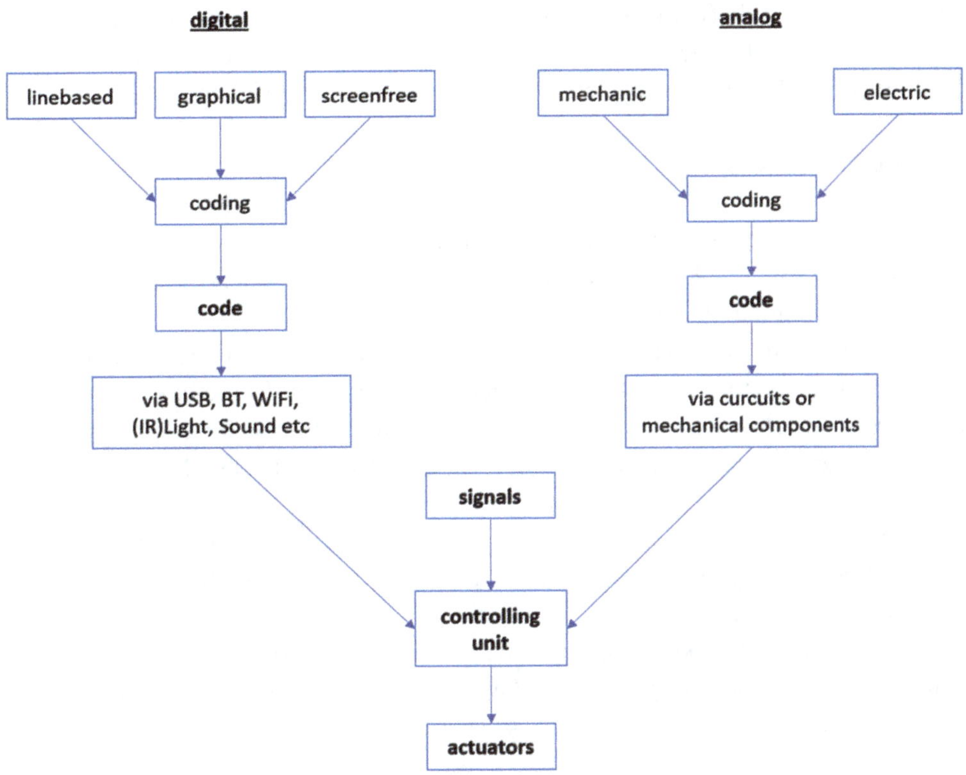

components and thus significantly change the type of programming, the processing of data and the way to execution.

If changes of components remain within the respective system of one manufacturer, the individual technical components may remain compatible with each other, which is essential to consider when selecting the necessary components for self-designed and built robots. For that a distinction must be made between:

- free materials for self-constructions,
- construction systems and
- robot construction kits.

While construction systems like the goBILDA (goBilda, 2021) - building system are designed openly in a kind that they have to be supplemented with the necessary electrical and electronic components, these are usually already available in the case of robot construction kits.

Depending on the manufacturer, the components contained therein are compatible, partially compatible or not compatible at all with each other over generations. The same applies to the proprietary software. Here, too, there are suppliers whose software remains compatible with every subsequent generation of building sets for years, as fischertechnik impressively demonstrates with its ROBO Pro programming software until 2021.

As a result, technical upgrades in their systems, for example to more powerful controllers, are associated with comparatively little financial and didactic effort because the introduced software and many parts can be reused. Nevertheless, from this point on, the assembled artefacts can move, but these physical devices (Figure 6) still lack the necessary automation to really be viewed as robots.

From Motion to Behavior

The definition of robots varies somewhat in detail depending on the nations and institutions. Nevertheless, a generally valid characteristic of robots is that their movement sequences are freely programmable while other approaches neglect this condition.

As a consequence, the artefacts assembled up to this point may already look like robots, but they do not yet count in every respect as such due to the lack of automation. In the case of robots for learning these hardware-based structures are often called and traded as platforms for educational robots like the PiBot (Vega & Cañas 2018), the Duckiebot (Paull et al, 2017) and Turtlebot 3 (Amsters &Slaets, 2020) to name a few only.

Various automated machines like cars and hoovers for example, on the other hand, may not look like supposed robots, although they clearly are. At the same time, the type of automation is not really fixed, so that it is not limited to computer-based versions alone.

For example, to automate a tinkered robot, a string with knots at different distances can be enough, because the distances represent a simple code that automates the artefact (Figure 14). Sometimes a strip of paper with holes or a switching roller (Figure 4) is enough to program and realize simple functional sequences.

As with a secret language, messages are encoded and decoded, trigger actions and can be changed by the programmer. In more complicated systems, additional conditions such as if-then queries and subroutines are added. Classic examples are barrel organs and looms. Clocks and the chime of bells are also illustrating examples.

No matter what the type of programming ends up being, it is automation that can turn an assembled structure, a platform or a simple construction kit into a behaving machine in the sense of Resnick (1993). He not limited programming to computation, but emphasized the advantages of programmable devices for controlling robotic creatures guiding directly to computational thinking.

The Engineering Design Process

The engineering design process (EDP) is ultimately the concept that guides the development, construction and realization of educational robots in a systematic way. It contains the procedural knowledge for problem solving and is defined as a decision-making process (often iterative) to meet desired needs. The roots of the concept may go back to Leonardo da Vinci as Pahl et al. (2007) stated, because he is known as one of the first who used a systematic variation of possible solutions.

The current used systematic structured design process is based on verifiable development of important contributions that started at the end of the nineteenth century (Feldhusen &Grote, 2013). Redtenbacher and his pupil Reuleaux developed these early ideas while first Riedler, Wögerbauer, Kesselring and later many others continued the process still today (Pahl et al., 2007, Feldhusen & Grote, 2013).

Therefore, all design models from the world of digital products like the iterative development approach, the spiral design model related to Boehm (1988) or the kindergarten approach to learning mentioned by

Figure 14. Parts of a robot-kit automated by a string with knots

Resnick (2007) are quite similar but clearly newer. They all are used to avoid the traditional waterfall design method that is a hard step-by-step process of stages.

Today the EDP is an essential tool for engineering and product development. It is taken into account and applied in school curricula for technology education and in educational standards for technology literacy (NGSS, 2013; ITEEA, 2020). Because it is not a rigidly prescribed approach it is a kind of generally principle that helps in the search for solutions and thus guides the development of strategic ways of thinking and acting as a part of general education.

Therefore, the EDP will be described here in principle before it is transferred to the application in the development and construction of robotics in schools.

Following the current VDI Guideline 2222/Sheet 1 the EDP starts with clarification and specification of the task including a variety of realistic constraints, such as economic factors, safety, usability, reliability, aesthetics, ethics and social impact and then leads through the seven steps solving process as follows often illustrated in a circle, known as the engineering design circle:

- Clarification and specification of the task;
- Determination of functions and their structures;
- Search for solution principles and their structures;
- Structuring into feasible modules;
- Design of the relevant modules;
- Design of the entire product;
- Elaboration of the design and use specifications.

Figure 15. Adapted engineering design process

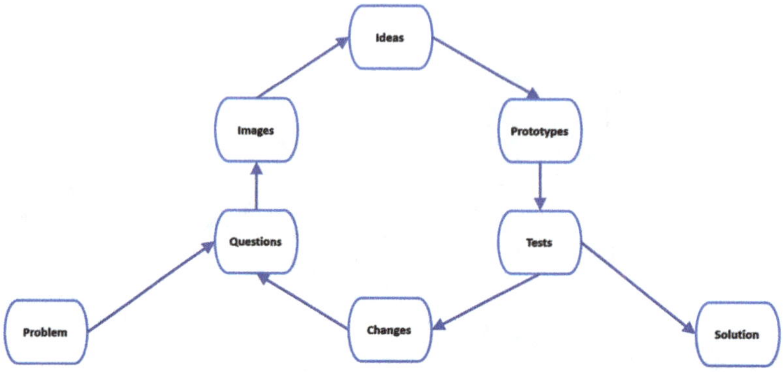

Commonly the results do not lead to a satisfactory outcome within one pass and have to go through the relevant steps or the circle has to start again (Feldhusen J. & Grote K.H., 2013). It is also common to make additional interim decisions that are not shown in the schedule. Even the termination of a development that proves to be no longer worthwhile is possible at different points in time (Wittel et al., 2015). In school contexts, the EDP can be adapted or simplified by omitting boundary conditions (Figure 15) in order to focus on the purely technical design issues and teaching the EDP.

Ziaeefard et al. (2016) used to simplify the EDP regarding the NGSS (2013) framework and adapted it for their middle-school robotics project to engage students from early ages in engineering practice to learn the engineering process. Their engineering design concept begins with identifying the problem and asking the right questions because it addresses the first component of the engineering core idea, which is defining and delimiting engineering problems.

Then their creativity and problem-solving phase starts with the second step where possible solution have to be imagined and different aspects of the problem explored by utilizing different resources ranging from informative websites to journal papers and textbooks.

After collecting relevant data and first vague solutions were identified a brainstorming follows as third step to generate new ideas. First prototypes are planned before they are built in the fourth step while tested in the fifth step to validate the performance against requirements. While different criteria associated with the problem and the solution have to be analyzed after the brainstorming the plan has to be projected in terms of cost, time and deliverable. If the prototype does not meet the requirements through testing, design changes and improvements follow together as sixth step before steps 1-5 have to be repeated until a desired solution was found.

With these remarks in mind, the duality of the EDP is expressed in a special way. On the one hand, it appears as an educational added value to learn it in general as a meaningful procedure and, on the other hand, to apply it at the same time as a constructive method in designing, constructing, and programming robots for learning. While the educational added value is characterized by a systematic problem-solving strategy, the procedure as a constructive method helps to achieve better results.

Using the example of educational robots, the previous sections described the basic design and construction of a structure, which was then equipped with the necessary electrical and electronic components, without dealing with the actual methodological procedure. In addition, Figure 16 shows the flow chart of a conceivable test routine in the sense of the EDP from the assembly of the available components to the use of the robot, including the redesign loops in the event that the intended functions were not achieved.

Figure 16. Creating and testing procedure

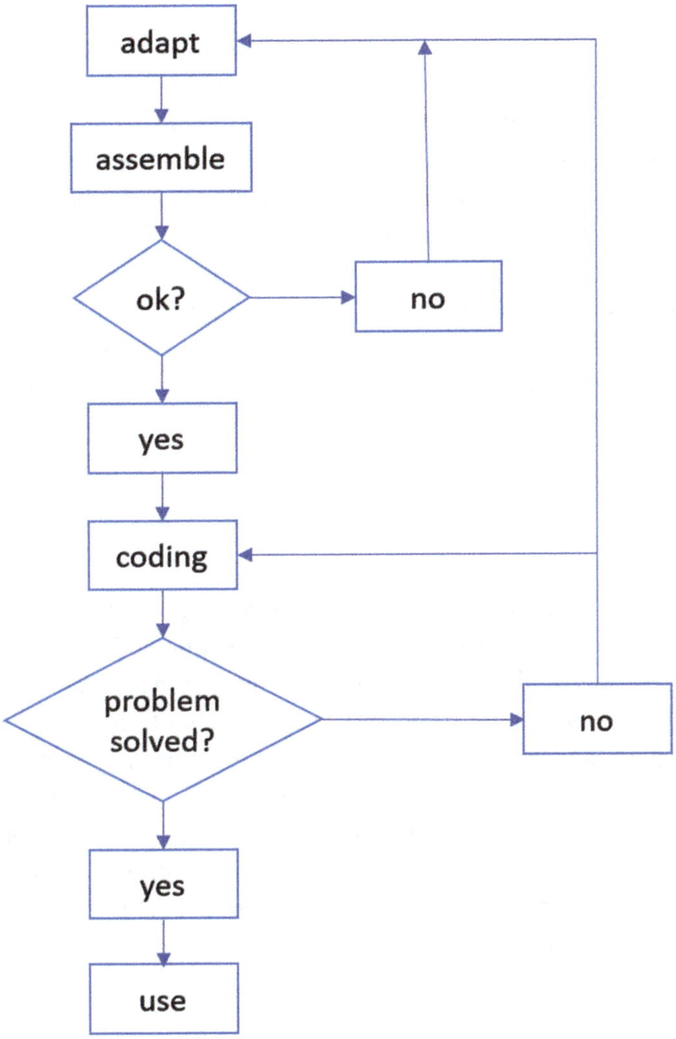

Accordingly, the first stage begins with the assembly of the selected and prepared components and a check of the mechanical functionality. If this is not the case, the loop leads back to the production of new components or the adaptation of existing components and a new assembly. Only when the assembly has led to a satisfying result does the programming of the educational robot take place.

Excluded from possible pre-tests and studies it is in the nature of things that programming only makes sense after the hardware has been assembled, because it must take into account the technical specifics of the robot at hand. Once the program code has been transferred to the robot, the programming is checked in conjunction with the mechanical functionality. If faults occur during this process, a further redesign loop is started. Here it is important to check whether the problem is due to the programming and can therefore be remedied comparatively easily, or whether the mechanical structure and its additional components need to be adapted to reach the desired behavior.

At this point at the latest, it becomes clear that concepts such as computational thinking have methodological similarities to the EDP, but can be presented as an integrated part of it in terms of content.

CONCLUSION

This chapter contributes to the book since it illuminates didactic and technical basics for learning through designing, constructing and programming robots for learning. The seemingly separate presentation of technical didactic considerations with the comparatively neutral presentation of the construction of education robotics should have made it clear afterwards how close and important the connection is between the didactic intentions and the teaching process in the construction of robots.

In the end, it is decisive for whether a lesson has a more instructional or constructivist orientation and how the actual process of building an education robot takes place. It also determines whether realizing educational robots will be reduced to a tool to facilitate computational thinking or used as a learning setting to teach engineering as a broader approach of thinking and making. In sum, the chapter shows that educational robotics is nothing more than an automated machine that only acquires its meaning in interaction with teaching decisions.

SUMMARY

Before this chapter deals with the basics of designing, constructing, and programming robots for learning a short and general introduction into the topic is given. This is followed by a theoretical outline from the didactic concept of the extracurricular project technikcamps dealing with robotics and the experiences gained from it. A subsequent look at the consequences for teacher training and the philosophical backgrounds of technical education is than presented. It reflects any reasons for misunderstandings between technology and natural sciences that triggers teaching in relevant subjects in any cases the wrong way. This section is followed by a general overview about educational robots as making projects and shows the genesis from single parts to a final object. It ranges from components to structure, from structure to motion and from motion to behavior and gives at the end a short introduction into the engineering design process that guides the realization.

REFERENCES

acatech. (2011). *Monitoring von Motivationskonzepten für den Techniknachwuchs (MoMoTech)*. Springer.

acatech und VDI. (2009). *Nachwuchsbarometer Technikwissenschaften*. Acatech und VDI.

Amsters, R., & Slaets, P. (2020). Turtlebot 3 as a Robotics Education Platform. In M. Merdan, W. Lepuschitz, G. Koppensteiner, R. Balogh & D. Obdržálek (Eds.), *Robotics in Education. RiE 2019. Advances in Intelligent Systems and Computing* (Vol. 1023). Springer. doi:10.1007/978-3-030-26945-6_16

Benjes, H., Zuschlag, F., & Welschehold, E. (1992). *Fertigen, Konstruieren und Steuern im Technikunterricht: Lehrerhandbuch für die Arbeit mit dem Halbzeugsystem UMT*. NLI.

Bergner, N., & Müller, K. (2018). Fachempfehlung Informatiksysteme. In N. Bergner, H. Köster, J. Magenheim, K. Müller, R. Romeike, U. Schroeder, & C. Schulte (Eds.), *Frühe informatische Bildung – Ziele und Gelingensbedingungen für den Elementar- und Primarbereich* (pp. 268–301). Barbara Budrich.

Boehm, B. W. (1988). A spiral model of software development and enhancement. *Computer, 21*(5), 61–72. https://doi.org/10.1109/2.59

Bostock, M. (1983). BBC Buggy. *Your Computer, 3*(4), 51–53.

De Vries, M. (2005). Teaching about Technology: An Introduction to the Philosophy of Technology for Non-philosophers. In *Science & Technology Education Library* (Vol. 27). Springer. doi:10.1007/1-4020-3410-5

De Vries, M., & Tamir, A. (1997). Shaping Concepts of Technology: What Concepts and How to Shape Them. *International Journal of Technology and Design Education, 7*, 3–10.

Euler, M. (2009). Schülerlabore. Lernen; forschen und kreative Potenziale entfalten. In Kursbuch 2010. Schülerlabore in Deutschland. Tectum.

Fast, L. (2006). Zur Genese von Baukästen im Technikunterricht. *Unterricht - Arbeit + Technik, 8*(29), 50-53.

Feldhusen, J., & Grote, K. H. (2013). Der Produktentstehungsprozess (PEP). In J. Feldhusen & K. H. Grote (Eds.), Pahl/Beitz Konstruktionslehre (pp. 11–14). Springer Vieweg. https://doi.org/10.1007/978-3-642-29569-0_2.

Fischer, E. (2000). *Die andere Bildung: Was man von den Naturwissenschaften wissen sollte.* Ullstein.

Fislake, M. (2001). *Determinanten und Interdependenzen Technischer Bildung aus didaktischer und bildungsökonomischer Sicht.* Peter Lang.

Fislake, M. (2010). Kinder Technik Ferien Camps in Rheinland-Pfalz - Bildungsarbeit zwischen Markt und Mission. Bildungsprozesse im MINT. In C. Quaiser-Pohl & M. Endepohls-Ulpe (Eds.), *Bildungsprozesse im MINT-Bereich. Interesse, Partizipation und Leistungen von Mädchen und Jungen* (pp. 127–141). Waxmann.

Fislake, M. (2018). Robotics in Technology Education. In M. de Vries (Ed.), Handbook of Technology Education (pp. 361–384). Springer. https://doi.org/10.1007/978-3-319-44687-5_28.

Fislake, M. (2021). Educational Robotics Between Coding and Engineering Education. In S. Papadakis & M. Kalogiannakis (Eds.), Handbook of Research on Using Educational Robotics to Facilitate Student Learning (pp. 107–139). IGI Global. https://doi.org/10.4018/978-1-7998-6717-3.ch004.

Gaycken, S. (2005). *Was ist technisches Wissen?: Philosophische Grundlagen technischer Wissenschaften.* https://events.ccc.de/congress/2005/fahrplan/attachments/575-paper_technischesWissen.pdf

goBilda. (2021). *What is goBILDA?* https://www.gobilda.com

Hattie, J. (2009). *Visible Learning - A synthesis of over 800 meta-analyses relating to achievement.* Routledge.

Henseler, K., & Höpken, G. (1996). *Methodik des Technikunterrichts.* Klinkhardt.

ITEEA, International Technology and Engineering Educators Association. (2020). *Standards for Technological and Engineering Literacy: Defining the Role of Technology and Engineering in STEM Education*. Pre-Publication Copy. Author.

Jeretin-Kopf, M., Kosack, M., & Wiesmüller, M. (2015). Technikdidaktische Medien – Einfluss verschiedener technikdidaktischer Medien auf die kindliche Motivation, problemlösendes Denken und technische Kreativität. In G. Graube, M. Jeretin-Kopf, W. Kosack, I. Mammes, O. Renn & C. Wiesmüller (Eds.), Wissenschaftliche Untersuchungen zur Arbeit der Stiftung "Haus der kleinen Forscher". Stiftung Haus der kleinen Forscher.

Lassahn, R. (1979). Technik und Bildung. In G. P. Bunk & R. Lassahn (Eds.), *Technik und Bildung*. Quelle & Meyer.

Lauxmann, F. (1999). *Mit Hegel auf der Datenautobahn. Über die Freude an geistigen Zusammenhängen*. DTV.

Martinez, S. L., & Stager, G. (2019). *Invent to learn: Making, tinkering, and engineering in the classroom*. Academic Press.

NGSS Lead States. (2013). *Next Generation Science Standards: For States, By States*. The National Academies Press.

Pahl, G., Beitz, W., Feldhusen, J., & Grote, K. H. (2007). Introduction. In G. Pahl, W. Beitz, J. Feldhusen & K.H. Grote (Eds.), *Engineering Design. A Systematic Approach* (pp. 1-25). Springer. doi:10.1007/978-1-84628-319-2_1

Papert, S. (1980). *Mindstorms: Children, computers, and powerful ideas*. Basic Books.

Paull, L., Tani, J., Ahn, H., Alonso-Mora, J., Carlone, L., Cap, M., Chen, Y. F., Choi, Ch., Dusek, J., Fang, Y., Hoehener, D., Liu, S.-Y., Novitzky, M., Okuyama, I. F., Pazis, J., Rosman, G., Varricchio, V., Wang, H.-Ch., Yershov, D., . . . Censi, A. (2017). Duckietown: An open, inexpensive and flexible platform for autonomy education and research. In 2017 IEEE International Conference on Robotics and Automation. ICRA. https://doi.org/10.1109/ICRA.2017.7989179.

Rahwan, I., Cebrian, M., Obradovich, N., Bongard, J., Bonnefon, J.-F., Breazeal, C., Crandall, J. W., Christakis, N. A., Couzin, I. D., Jackson, M. O., Jennings, N. R., Kamar, E., Kloumann, I. M., Larochelle, H., Lazer, D., McElreath, R., Mislove, A., Parkes, D. C., Pentland, A., Roberts, M. E., Tenenbaum, J. B., & Wellman, M. (2019). Machine behaviour. *Nature*, *568*(7753), 477-486. doi:10.1038/s41586-019-1138-y

Resnick, M. (1993). Behavior construction kits. *Communications of the ACM*, *36*(7), 64–71. https://doi.org/10.1145/159544.159593

Resnick, M. (2007). All I really need to know (about creative thinking) I learned (by studying how children learn) in kindergarten. In Proceedings of the 6th ACM SIGCHI conference on Creativity & cognition (C&C '07) (pp. 1–6). Association for Computing Machinery. https://doi.org/10.1145/1254960.1254961.

Riedler, A. (1919). *Wirklichkeitsblinde in Wissenschaft und Technik*. Springer.

Schenk, J. (1920). *Zur Reform des Unterrichtes des Maschinenbauwesens an den Technischen Hochschulen*. Dülfer.

Schmayl, W. (1981). *Das Experiment im Technikunterricht - methodologische und didaktische Studien zur Grundlegung einer Unterrichtsmethode*. Franzbecker.

Schwanitz, D. (1999). *Bildung. Alles was man wissen muß*. Eichborn.

Seng, J. (2021). Constructing Robots for Undergraduate Projects Using Commodity Aluminum Build Systems. In M. Merdan, W. Lepuschitz, G. Koppensteiner, R. Balogh, D. Obdržálek (Eds.), Robotics in Education. RiE 2021. Advances in Intelligent Systems and Computing. Academic Press.

Snow, C. P. (1967). *Die zwei Kulturen. Literarische und naturwissenschaftliche Intelligenz*. Klett.

Unglaube, H. (1995). Experimente im Sachunterricht. In R. Meier, H. Unglaube & G. Faust-Siehl (Eds.), Sachunterricht in der Grundschule (pp. 224-236). Frankfurt a. M.

Vandevelde, C., Ciocci, M.-C., Saldien, J., & Vanderborght, B. (2013). Overview of Technologies for Building Robots in the Classroom. *4th International Conference on Robots in Education*. Lodz, Poland.

Vega, J., & Cañas, J. (2018). PiBot: An Open Low-Cost Robotic Platform with Camera for STEM Education. *Electronics (Basel)*, *7*(12), 430. https://doi.org/10.3390/electronics7120430

Wilkening, F. (1982). *Unterrichtsverfahren. Lernbereich Arbeit und Technik*. Neckar.

Wittel, H., Muhs, D., Jannasch, D., & Voßiek, J. (2015). Allgemeine und konstruktive Grundlagen. In H. Wittel, D. Muhs, D. Jannasch, & J. Voßiek (Eds.), Roloff/Matek Maschinenelemente (pp. 1–26). Springer Vieweg. https://doi.org/10.1007/978-3-658-09082-1_1.

Ziaeefard, S., Mahmoudian, N., Miller, M., & Rastgaar, M. (2016). *Engaging Students in STEM Learning through Co-Robotic Hands-On Activities (Evaluation)*. Paper presented at 2016 ASEE Annual Conference & Exposition. doi:10.18260/p.26975

ADDITIONAL READING

ABET - Accreditation Board for Engineering and Technology. (2017). *Criteria for accrediting engineering programs*. Engineering Accrediting Commission.

Blikstein, P. (2013). Gears of our childhood: constructionist toolkits, robotics, and physical computing, past and future. *Proceedings of the 12th International Conference on Interaction Design and Children (IDC '13)*, 173–182. 10.1145/2485760.2485786

Blikstein, P. (2015). Computationally Enhanced Toolkits for Children: Historical Review and a Framework for Future Design. Foundations and Trends. *Human-Computer Interaction*, *9*(1), 1–68. doi:10.1561/1100000057

Eguchi, A. (2014). Educational Robotics for Promoting 21st Century Skills. *Journal of Automation. Mobile Robotics & Intelligent Systems*, *8*(1), 5–11. doi:10.14313/JAMRIS_1-2014/1

NAGB, National Assessment Governing Board. (2018). *Technology and Engineering Literacy Framework for the 2018 National Assessment of Educational Progress*. U.S. Department of Education.

Riedo, F. (2015). *Thymio: a holistic approach to designing accessible educational robots*. École Polytechnique Fédérale de Lausanne.

KEY TERMS AND DEFINITIONS

Assembly Kit: Also known as assembly set, model kit, or kit it follows a closed approach, which is characterized through an assorted content including counted components and a building instruction that serves the realization of only one model. Included components as a rule are not compatible with others.

Assembly Task: The assembly task as a teaching method originates from technology didactics and is considered a variant of the work task. In order for the problem-solving process to concentrate more on inventing and re-inventing, it preferably uses modular elements with precise fitting shapes and quick connection possibilities by plugging and screwing.

Behavior Machine: A term coined by Resnick 1993 while working with the first generations of programmable construction kits. As machine behavior (Rahwan et al., 2019) this field is concerned with the scientific study of intelligent machines, not as engineering artefacts, but as a class of actors with particular behavioral patterns and ecology. This field overlaps with, but is distinct from, computer science and robotics.

Building System/Construction System: In contrast to a construction kit with its assorted content, the building system is characterized by an open approach. The components are not counted and do not serve the realization of one or more predefined construction proposals, but primarily the free design.

Coding: Coding stands colloquially for programming. In this paper and commonly used to describe how beginner program without using control structures, models, or algorithms.

Construction Kit: Based on a building system a construction kit is characterized by a half open approach. It has an assorted content including counted components and a building instruction that serves the realization of one or more predefined construction models. The components of a construction kit may be used as a building system, also extended through compatible components.

Design and Manufacturing Task: In technology education, the didactic characteristics of the teaching methods design task and manufacturing task are combined together and defined into one coherent teaching method. In a lesson that follows this method pupils plan and design a technical object and manufacture it. At the end, their manufactured products are assessed and compared with the previously set requirements.

Design Process: A design process is a general concept that guides the development, construction, and realization of projects in a systematic manner. It contains the procedural knowledge for problem solving and is defined as a decision-making process (often iterative) to meet desired needs. Depending on the area of application, the basic shape is adapted to the requirements.

Educational Robotics: Educational Robotics are defined by technology artefacts ranging from programable robot-like toys to professional automated systems and the way they are used in educational settings. They include robot simulations, constructional robotic kits as well as prefabricated systems. Educational Robotics play also an active role to enhance learning experience through the creation, implementation and validation of pedagogical activities, tools and technologies.

Engineering Design Process: The engineering design process (EDP) is an iterative, not linear thinking strategy. It was developed since the end of the nineteenth century to solve technical problems systematically and uses creative as well as rationale methods. It requires numerous decisions and includes a variety of realistic constraints, such as economic factors, safety, usability, reliability, aesthetics, ethics and social impact to achieve satisfactory results. In educational applications it helps to develop a kind of engineering thinking for general use that includes the development of student's creativity and their problem-solving ability.

Chapter 12
Brushbots for Educational Robotics:
Low-Cost Robots for Building and Research Activities in Schools

Martin Fislake
https://orcid.org/0000-0003-3189-8741
University of Koblenz, Germany

Laura Aileen Thran
University of Koblenz, Germany

ABSTRACT

Due to their simple construction, brushbots are popular and widely known as craft objects. Building instructions promise tinkering fun and a promoting effect but minor as low-cost research objects. They are regarded as microbots and give opportunities for discussions. One questions to answer is whether their chaotic motion can be influenced. Therefore, this contribution deals with an examination of factors influencing their movement behaviour. The aim was to show experimental work and to discuss the potential of brushbots in three directions: first, to propose a cheap platform to evaluate the motion capabilities of different designs; second, to offer an approach to improve vibrobots concerning experimentation results in the design process; and third, to provide a sample that can be transferred to schools. For this purpose, the basics of brushbots are presented, and the experiments and their results are described. The results offer inspiration for teaching applications that go beyond mere tinkering and provide approaches for further research.

INTRODUCTION

In the education sector, the call for technology education is increasing. Because in a constantly evolving digitalized world, young people should be able to familiarize themselves with technical areas and be able

DOI: 10.4018/978-1-7998-7443-0.ch012

to act in them (ITEEA, 2007, 2020). Therefore, pupils are introduced to machine and robot technology at an early age, while educational robots are appreciated as one of the tools to technology education.

Robots in general are ubiquitous in today's world, so it is almost unimaginable to live life without them. They are used in all areas of life and the world. They are not only considered a core area of physics, technology and industry, but also play an interesting, important role in natural disasters, bomb disposal, medicine, education, in the home as a supportive cleaner or as toys, and have long since become indispensable.

Brushbots in particular are popular in schools and the maker scene. They offer options to propose a cheap research platform for design and research activities in schools. To enhance this, all experiments presented later in this chapter may be used as examples to show that research is an immanent part of the design process and how a design process can be supported through research activities. In addition a few examples of school activities will illustrate the use of brushbots in schools not only as tinker stuff.

Brushbots are also a frequent cause for discussion and debate. The questions that come to the fore are how is a brushbot constructed, why does it move through the vibration motor? Are there similar bots that work according to the same principle and is it possible to influence the movement behaviour of brushbots, as they have so far been characterized by their chaotic movements in space. These questions have received little attention in the scientific context so far. In order to make a contribution to this topic, this work is concerned with the practical investigation of factors influencing the (un)directed movement behaviour of brushbots. The aim is to develop a solution approach in which a brushbot can be steered.

Subsequently, the focus is on experiments about locomotion through vibration. Due to a lack of relevant literature and insufficient research about the movement of classic brushbots, information on the state of research is given but largely based on internet sources and forum discussions in the technical maker scene.

Equivalent to suggestions and assumptions from children in learning settings those from the maker scene will be taken into account for the experiments. The aim is a development of scientifically based findings on the influencing factors of brushbots in the form of tests and studies. All intermediate results will be recorded in reports and pictures.

With this background knowledge, the paper introduces the practical investigation and the function of brushbots is examined in more detail and various assumptions about movement behaviour will be presented. Different assertions and ideas are worked out that can influence movement behaviour, in order to subsequently substantiate or resolve them through various experimental set-ups. The factors influencing the movement behaviour will be analyzed in the form of experimental set-ups in order to be able to verify or falsify them. To begin with, a classic brushbot is built in order to look at and document the undirected movement behaviour more specifically. The classic brushbot is then modified by various changes in order to be able to examine the influencing factors of its movement more closely. Different brushes, mounting positions, bases and additional elements are considered in order to approach conscious control and most successful approaches were varied and investigated in further runs.

These experiments can be carried out in a didactically reduced manner with pupils of primary and lower secondary schools in the area of technology teaching in school lessons. The focus is on the practical implementation of different experimental set-ups which also may be applicated in schools. Different prototypes have been used for the experiments and should be able to provide information about which changes in the structure, the elements, the brushes or other factors influence the movement behaviour of brushbots.

For that purpose, the Tracker free video analysis and modeling tool was selected. It is designed to be used in learning settings and was used to show how it can be applicated to reproduce or continue the presented experiments. It also offers the possibility to transfer the recorded movement behaviour of the individual prototypes in a diagram to validate it. Thus, meets the requirements of education in the digital world to promote technological understanding and interest in scientific-technical occupational fields.

Defining and Explaining Brushbots

Brushbots are considered to be easy-to-construct mini-motion devices that can move by vibration (AXL et al., 2012). The term is derived from the English word bristlebot (bristle - brush, -bot - short form of robot - robot). Since they cannot be steered in a controlled manner, the term "robot" or "bot" is commonly used in everyday language, but does not correspond to the robot definitions such as those of RIA (Robot Institute of America), VDI (Association of German Engineers) and JIRA (Japan Industrial Robot Association). The classic brushbot consists of a toothbrush or equivalent, a vibration motor as drive unit and an energy unit in form of a button cell. The various components are attached and fixed to the brush with the help of adhesive tape, glue or holders (Aschoff, 2015; Cowen 2013)

An uneven rotating mass creates free centrifugal forces, which in turn generate vibrations and are transmitted to the brush. The brushbot moves due to the repetitive forces, the bending of the bristles and the release of the load, which causes the bristles to pull back to their original position (Giomi et al., 2013).

Due to their function, layout and size, they belong to so-called vibrobots, microscale legged robots or inch-scale robots. Since their structure does not follow any fixed rules, they differ in their speed, stability, appearance and strength (Finio, 2020).

In addition to the DIY versions with vary instructions from the maker scene, there is a wide range of assembly kits for schools and also of ready-made models (AXL et al., 2012), as exemplified by the HEXBUG nano from Innovation Labs or the I-Insect beetle bug from Doridia Imports. They are similarly constructed and show a comparable behaviour described by Innovation First Labs (2018) as a movement that is driven by a vibration motor and randomly changes its directional behavior as soon as it encounters another object or an obstacle.

Brief History of Bristlebots

The term bristlebot was probably largely coined by Oskay and the Evil Mad Scientist Laboratories (Oskay, 2007). He reported about bristlebots and provided instructions for their reproduction. He used a toothbrush head and a vibration motor, double-sided adhesive tape to attach the components (Figure 1) and presented an 1,5 and a 3,0 Volt version.

Corresponding to Innovation First Labs (2018) the idea of locomotion via vibration transmission on brushes seems to be older. However, mobile micro-robot using centrifugal forces like those of Ioi (1999) are guiding to their roots, while Downie had already presented building instructions in 2001 and named his brushbots Dougalls (Figure 2) derived from a long-haired dog in a French cartoon.

Murphy (Murphy, 2009) continued the idea of Downie conceptually. In addition to various ideas and suggestions, he presented two maker-kits for building brushbots out of toothbrush heads and partitions to set up a maze for the roaming brushbots.

Figure 1. Bristlebot made by Oskay
(Oskay 2008)

Figure 2. Sketch of a Dougall
(Downie 2001)

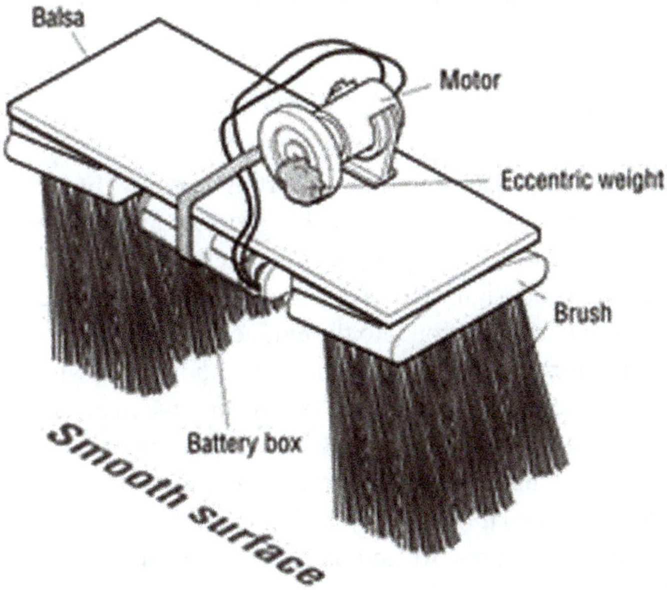

RELATED WORK

Research Papers

Papers regarding vibration driven microbots as research subjects may be divided into two threats. There are at one hand social scientists and others like (Mammes et al., 2013; Underling, 2010, Ross et al., 2017) who uses brushbots as miniature-robots in general to investigate their effects promoting interests in STEM subjects or for example to compare different educational approaches to maker concepts like Kimberly & Konopasky (2015). At the other hand there are engineers and natural scientists like (Becker et al., 2011; Ioi, 1999; Vartholomeos & Papadopoulos, 2008) who are concentrated on mathematical modeling of vibrations or on long-range motion dynamics of bristle robots, which takes into account the slip-stick motion of the bristles with the substrate like Gandra (2019) did.

In contrast Morin (2011) gained more practically interesting results from his bristlebot velocity tests. His results showed as the surface roughness was decreased, the performance of the bristlebots velocity increased. To address swarm robotics the Kilobot (Rubenstein et al., 2012) was designed. It is a low-cost robot that moves on three pins to make testing and evaluating collective algorithms and control methods for groups on hundreds or thousands of robots accessible to robotics researchers.

In addition, Notomista et al. (2019) presented a theoretical and experimental study of brushbots to develop improved robotic platforms, a fully actuated bot on six pillars and a differential-drive-like brushbot. Therefore, a series of experiments has been used to validate the derived theoretical models and to characterize their range of applicability. Furthermore, swarms of both versions have been used to showcase swarm-robotics applications.

Akbarimajd & Sotoudeh (2014) also developed a small vibrobot controlled by a microcontroller to setup a real low-cost robot. Like the larger Kilobot it used three pillars to make a triangular supporting area for the robot instead of bristles and was then evaluated by simulations and some standard experiments. Moreover, the robot was successfully tested in a line following task, as a typical task for mobile robots in educational settings.

Applications in Schools

Most applications of brushbots in learning settings reporting about STE(A)M fun activities or as promoting effect for technical interest but minor as teaching tool or low-cost research objects. Tucker & Miller (2021) an exception to this rule, as they report a concrete use of brushbots as a learning tool in a primary school during their annual design week. They used a design challenge and HexBug NanoBots as readymade brushbots to teach an introduction into the engineering design process (EDP). The challenge was to create an ultimate maze for a robotic bug to complete without any assistance (aside from turning it on) with the goal of including the most elements and completing the maze in the shortest time possible.

Different criteria were given like the minimum distance, handling rules and design conditions like an elevation change, a sound making element and a tunnel to pass. For the design and building process a few steps were explained, how to do it right. They followed the PLTW (Project Lead The Way) design process (2014) to explore the constraints, to find and sketch their ideas followed by an evaluation of built examples. While this challenge takes the brushbot as given other school activities use brushbots as concrete occasion to teach soldering and other handicraft skills.

BUILDING BRUSHBOTS AS TECHNOLOGY EDUCATION CONCEPT

While ready-made brushbots in the forementioned examples of applications in schools were taken as a given object to deal with this section takes brushbots as a cheap experimentation platform for engineering, construction, design and research activities in schools. Building own brushbots offers the opportunity to teach soldering and other handicraft skills, to validate and evaluate the motion capabilities of different

Figure 3. Engineering design process with sub circle

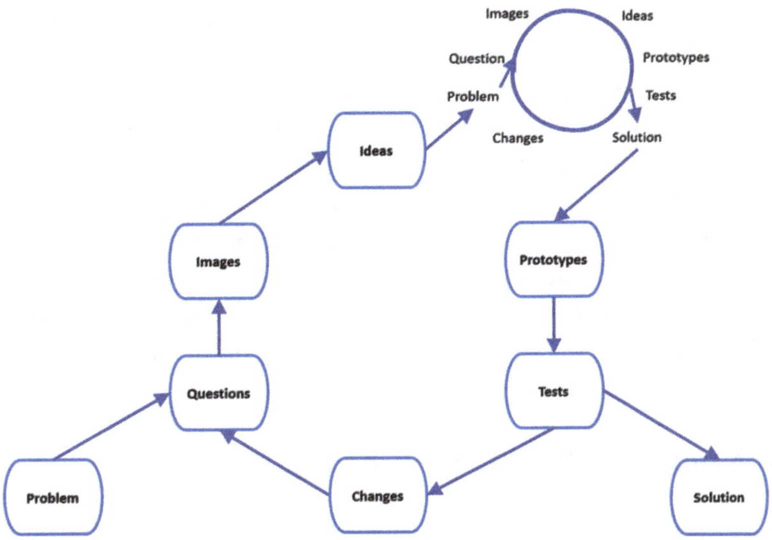

self-made designs and to teach the EDP as an approach how to improve vibrobots through experimentation results. To this, the EDP is explained in short and the research concept is presented and elaborated out as an approach whereas craft basics of how to build brushbots are unconsidered. The EDP is therefore taken in this case duality as an educational concept and as a valuable tool for problem solving that can be taught to primary and lower secondary schools.

Fislake (2021) summarizes that the history of EDP could be traced back to Leonardo da Vinci, as he is known as one of the first to use a systematic variation of possible solutions, while the valuable integration into schools started with the transformation of industrial arts to technology education in the 1980s and at the latest 1990s. However, according to (Feldhusen & Grote, 2013) the verifiable development of important contributions to the current used systematic structured design process in mechanical engineering started at the end of the nineteenth century. With a focus on computing Stager (2019) depicts alternative forms to the classic EDP as design models from the real world or design models for learning. In concrete, he names and describes the spiral design model by Boehm (1988), the common iterative development model used by programmers and designers, and the design model according to Resnick (2007). They all have an iterative, cyclical nature instead the common linear step-by step concept that is used in non-reflective waterfall design concepts and causal driven scientific methods.

The key to engineering design to facilitate learning for general education purposes is more than just applied science and also known as adapted concepts named the technology education process. The technology design process is the way to use an iterative, highly reflective and not linear thinking strategy. It was developed to solve (not only technical) problems systematically and concludes minimum three steps like there are to identify a problem, to search for solutions and to realize an appropriate one.

The basic EDP is known and used in many different versions in engineering practice as reported by ElMaraghy (2015) as well as in learning settings as can be seen exemplary by Sullivan et al. (2013) for Pre-Kindergarten, Farmer (2015) for Grades 9–12 and Kaya et al. (2017) as part of an engineering unit in an elementary science teaching methods course. It is because the EDP has to be adapted to the aspired situation in a certain way. In concrete there are usually much more steps to follow like shown in Figure 3.

Anyway, the actual construction process starts with the problem and leads through the solving process, generating ideas and prototypes that have to be evaluated. Normally the first results do not lead to satisfactory results within one run and however the whole circle or even smaller versions of it starts as sub circles for one or more of these steps if appropriate until a solution is found. Nevertheless, steps like questions and generating ideas are important and don't have to be skipped.

They are intersections where problem-solving techniques lead to inspirations and forces students to think in alternatives and innovations that do not have to favor one or another concept. The EDP helps to takes all possible solutions into account, and the decision about the most appropriate solution is one of its immanent core components. Going through this process often collections of assumptions are generated to find new ideas and approaches that have to be proved. In this chapter, equivalent to suggestions and assumptions from children in learning settings those from the maker scene will be taken into account for the experiments.

Ideas and Assumptions in the Design Process

Assumptions and ideas are an important and an immanent factor for problem solving in the design process. If there are no ideas, additional problem-solving methods like brainstorming must be used to generate new approaches and innovative solutions. In contrast assumptions about the movement behaviour of brushbots are popular, provide impulses and are discussed, but remain without practical implementation and without proof.

With a focus on experiments about locomotion through vibration of brushbots subsequently collected suggestions and assumptions from relevant forums of the maker scene are presented. Due to the lack (because of the covid19 pandemic) of suggestions and assumptions they are taken as equivalent to those from children in learning settings.

Collection of Assumptions to Be Proved

For the intended purpose many opinions, suggestions and assumptions can be collected from the discussions that had a peak around 2007 to 2010 in relevant forums of the maker scene. Oskay (2007), for example, recommends evenly angled bristles in order to be able to influence the direction of movement of the brushbot. In his opinion, toothbrushes with straight bristles can also be used if they have been curved in the corresponding direction of movement beforehand. Bristles of uneven length should be shortened and those with a different direction removed. Since the brushbot would then rest evenly on the ground, it would gain stability. According to his observations, the Brushbot steers either to the right or to the left when started. He suspects that the placement and weight distribution of the battery, motor and the direction of rotation of the motor, in addition to the bristle shape, influence the movement behaviour as well as the stability (Oskay, 2007).

An unknown user comments on these assumptions that shortening the bristles in overall height would provide more stability to the brushbot and thus protect it from falling over (Anonymous, 2007). Bristlebotics (2017) also argues similarly, bringing angled bristles for locomotion and wider brushes for more stability into the discussion. In his view, larger brushes would also travel more slowly and splayed bristles would provide more stability. AXL, Jo and Manupool also assume that the bristle bend influences the direction of movement. However, they recommend additionally removing the middle bristles of the toothbrush without going into further explanations (AXL et al., 2012).

Another unknown user, on the other hand, suggests using a microcontroller to control a weight shift on the brush and thus influence the direction of movement of the brushbot (Anonymous, 2009). In a similar article, it is suggested to control the weight distribution with a small servo motor in order to let the brushbot move mostly straight and, due to the servo motor's deceleration, in serpentine lines (Ringo, 2008). Rickharris, in contrast, suggests converting a toothpick so that it is used as a steering rod similar to a rudder on a boat (Rickharris, 2010).

The user Slurp suspects that straight bristles lead to a random direction of movement. The rotation of the motor would have an effect on the directional behaviour, but to actually influence the direction of movement, angled bristles would have to be used. By connecting two brushes with angled bristles, the shorter bristles being the front of the brush bot, to a piece of wood and controlling them with two motors simultaneously, he guarantees a stable forward movement (Slurp, 2010). Manf also sees a possibility of directional control in a two-brush-head system (Manf, 2008a).

This idea is taken up by MichaelM and extended by the use of two differently controlled motors (MichaelM, 2008). Lunarman uses the concept of tank control and proposes to couple two brushes via an elastic but firm connection so that the vibrations of one brush are not transmitted to the other. For this, he recommends a spring made from a biro (Lunarman, 2008). He also continues that a difference in movement could be in the procurement unit of the bristles, as softer bristles are more easily compressed by vibration (Manf, 2008b).

HannoHupmann then relates the movement behaviour to the choice of eccentric motor. He distinguishes between the use of a coin-operated vibration motor and an unbalanced motor with an external unbalance (HannoHupmann, 2008), while MadMage says that in order to control an optimal brushbot, more total weight and a vertically mounted vibration motor with two opposing unbalanced weights are needed to ensure uniform vibration. The weight must be on the opposite side of the direction of movement and by rotating the unbalance motor he assumes that curves can be driven (MadMage, 2008).

User Spion suggests mounting two excentre motors in parallel on the bot. The direction of rotation of the vibration motors should be able to control the movement behaviour. If both eccentric weights turn in the same direction, the brushbot will follow a curve. If the direction of rotation is different, the brushbot would move forward (Spion, 2008).

According to Szczys, brushbots with a vibration motor positioned centrally between two large brushes should be able to be steered by a long stick and additional weight. The necessary weight shift would be made possible by a servo motor positioned at the front in the direction of movement (Szczys, 2010).

Another user adds to this contribution that the idea could also be implemented with electromagnets in a smaller format by pulling the stick as a swivel arm to the respective side by magnetism, but the arm would have to be connected to a spring so that it can pull itself back to the central starting point (Underling, 2010).

Furthermore, Strider_mt2k wonders if the idea of weight shifting can also be implemented with a micro RC module, so the direction of movement of the brushbot would be remotely controllable (Strider_mt2k, 2010). Fhunter is considering the idea of steering with two differently controlled vibratory motors as proposed by MichaelM, but mounted on two connected brushes. One motor should be positioned on each brush, so he wants to avoid the weight shift with heavy components (Fhunter, 2010).

In another blog, another steering option is presented by the author Underling (2011). This version is presumably a remote-controlled brush-bot. It is to be steered by a servo-motor attached to the rear part of the brush. Each side of the motor arm is connected to a paddle. By reversing the polarity of the motor, the respective paddle should be activated and push the brushes away. This function should contribute to a change in direction. Table 1 summarizes the assumptions made.

Table 1. Assumptions on the movement behaviour of brushbots

Author/Source	Assumptions on the Movement Behaviour of Brush Bots
Oskay	- Angled bristles - Weight distribution on the brush influences the direction of movement and stability - Direction of rotation of the unbalance motor
AXL, Jo & Manupool	- Angled bristles - Bending bristles outwards (stability) - Removing the middle bristles
Anonymous, 2007	- Short bristles give more stability
Anonymous, 2009	- Weight shifting with the aid of a microcontroller
Ringo	- Control of the weight distribution by a servo motor
Bristlebotics	- Stability through wider brushes - Stability through spread bristles - Angled bristles
Rickharris	- Rudder steering (boat)
Slurp	- Straight bristles lead to random movement behaviour - Use of angled bristles - Two-brush head system (simultaneously controlled vibration motors)
Manf	- Two-brush head system (simultaneously controlled vibration motors) - Bristle texture (softer bristles can be more compressed by vibration)
Fhunter	- Two-brush head system (differently controlled vibration motors)
MichaelM	- Armour control (two brushes via an elastic connection to decouple vibration transmission).
HannoHupmann	- Choice of vibration motor (coin-operated vibration motor, vibration motor with external unbalance)
MadMage	- Attachment of additional weight on the brush (weight attachment on opposite direction of travel) - Rotatable double-headed vibration motor on the brush
Spion	- Use of two motors on one brush (steering by direction of rotation of the eccentric weights)
Mike Szczys	- Control by weight shift (bolt with additional weight) via servo motor
Underling	- Cable remote controlled brushbot with paddle control - Control by weight shift with the aid of electric motors
Strider_mt2k	- Weight transfer via RC module

From Assumptions to Results

From an educational point of view, this kind of non-proofed discussions about assumptions apply not only to brushbots, but in general. It is thus one of the crucial educational challenges to guide young people to an enlightened and knowledge-based behavior in which cause-effect relationships are not discussed on the basis of opinions but acted upon on the basis of evaluated results.

In the design process, this means that ideas and the assumptions behind them are consciously and planned tested in a sub circle for their feasibility in suitable theoretical and/or practical procedures. Only in case of success the next step in the design process can be taken as illustrated in Figure 3. Otherwise, the design loop or its sub circles start all over again with the search for new ideas and assumptions. Consequently, this means that brushbots can be used not only for tinkering and as a teaching tool for applying the EDP from ideas to improved design concepts presented in the application in school section, but also as a low-cost research platform as shown below.

As a research project they also offer research opportunities for the use in learning settings as shown below. Watching these robots move, a lot of research questions may arise that can be answered cheap and easy through a series of experiments using a camera and a free software. Following the concept of student experiments in science lessons, where they independently formulate hypotheses, test these hypotheses through experiments, and interpret the results in relation to their hypotheses, it is important to select single factors from the multitude of influencing factors in the experiments.

Subsequently it is also important to vary these experiments as systematically as possible and to observe them under appropriate conditions. As a result, conclusions can be drawn about the brushes, the selection and position of the motor, the correct speed, the center of gravity and many other factors to explain basic mechanics or physics for example. As a result, recommendations for an idealized setup can be made. The aim is a development of scientifically based findings on the influencing factors of brushbots in the form of tests and studies. All intermediate results will be recorded in reports and pictures. For this purpose, the structure and the basic function of brushbots are presented and the experiments and their results are described. The results offer inspiration for teaching applications that go beyond mere tinkering and provide approaches for further research.

To enhance this, all experiments presented in this chapter may be used as examples to show that research is an immanent part of the design process and how a design process can be supported through research activities.

RESEARCH CONCEPT

In the first step of the study, the question of how toothbrush heads differ in their movement behaviour depending on their bristle position, weight and the use of different motors is investigated. In the further investigation it is to be found out whether the use of different brushes changes the behaviour of the brushbots. For this purpose, different brushes are used that differ in their shape, size and material composition.

When selecting the materials, a conscious decision was made to use brushes with rubber, boar and plastic bristles in order to include as wide a spectrum as possible in the investigation. Brushes, e.g., wire brushes, whose bristles are made of metal or similar, are neglected in this investigation, as the bristles do not retract into their starting position due to a lack of elasticity.

In the following section, four methods are selected for further investigation that have an influence on the steering of the brushbots. For this purpose, the ideas from the users of the forum contributions are partly used and modified in their structure and their mode of operation for the corresponding experiments.

The investigation of the steering of brushbots begins with an experiment on weight shifting through the use of a servo motor. In another steering experiment, the direction of movement is to be influenced with a paddle control. In a third experiment, the steering is to be set up on one and two brushes by two unbalanced motors operated independently of each other and operated with the aid of a cable remote control.

Table 2. Parts of the classic brushbot build on a vibration motor

Brush Type	Motor	Battery
Toothbrush head (vertical bristles)	- Motor - unbalance located outside- 0.18 W - 7500 rpm- Size: Approx.: 21.7 * 7 mm- Weight: 3 g	CR2016 3V

In the final experiment, the possibility of steering by means of a self-designed rudder is to be tested. Subsequently, the results and implications will be tabulated in Table 4 for overview. The results will be used to create a brushbot whose movement behaviour can be influenced in a controlled manner.

Experimental Set-Up

The first series of experiments were conducted on a wooden smooth flat, round plate with a diameter of 60 cm and a rim of 2 cm. It later changed to larger brushes and a smooth and even PVC floor for longer distances.

In order to retrace and measure the trajectory of the evaluated brushbots the video analysis and modeling application Tracker (Brown, 2008; Wee & Lee, 2012) was selected. It was applied instead of systems based on AprilTag or Whycon fiducial markers on the robots because it is designed for education in schools and offers the possibility to transfer the recorded movements in a diagram manually in a low-cost manner. It is built on the Open-Source Physics (OSP) Java framework and was develop to combine videos with computer modeling to overlay simple dynamic particle models on a video clip.

For this purpose, the motion behavior in each experiment was recorded with a digital camera and reproduced several times to avoid special cases and to obtain typical movements of the individual brushbots. The recorded sequences were played back in steps to the application on the computer and the movements of the brushbots were manually marked in the coordinate system. These markings were then used to convert the image position into a diagram and resulted in a graphically representation of the individual movement patterns of the tested brushbots.

Test Series I: Toothbrush Heads

Initially, a classic brushbot was built to observe and document the undirected movement behaviour. It was then modified to gain knowledge about possible influencing factors in relation to the mounting setup, weight and the use of different motors.

Experiment 1: Classically Setup

To get a first impression of the movement behaviour of this type of brushes, a classic brushbot (Figure 1) was first built from the head of a toothbrush with straight bristles, following Oskay (2007). The motor was attached to the front part of the brush and the button cell to the end.

This arrangement of components (Table 2) suggested rotation around the vertical axis, but the high centre of gravity, combined with a small footprint and the vibrations, caused it to regularly tip over sideways. A graphical representation of the direction of movement could therefore not yet be made.

Figure 4. Motor between the bristles

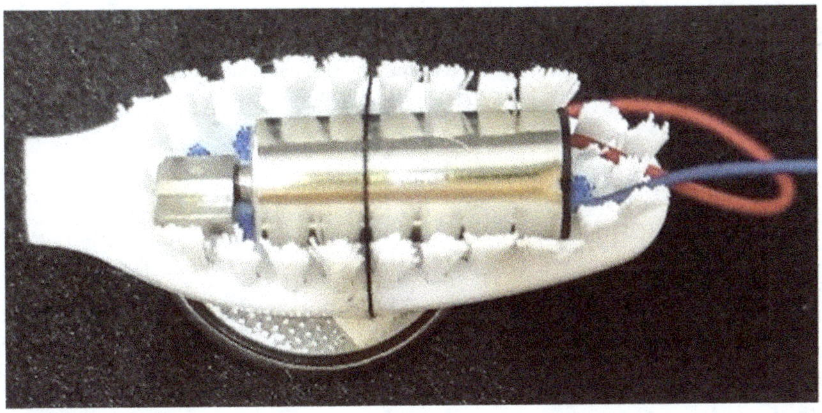

Experiment 2: Shortened Bristles

For this reason, the bristles were shortened evenly, following (Anonymous, 2007). However, this brushbot also repeatedly tipped over to the side, which again made a graphical representation impossible.

Experiment 3: Motor Underside

In order to create more stability and a lower centre of gravity, the middle bristles were cut off in experiment three and the vibration motor was installed in between the bristles (Figure 4). The brushbot then moved in fast and tight circles (Figure 5) without falling over. The direction of the circles can be influenced by shifting the weight position of the battery on the brush.

Figure 5. Recorded path

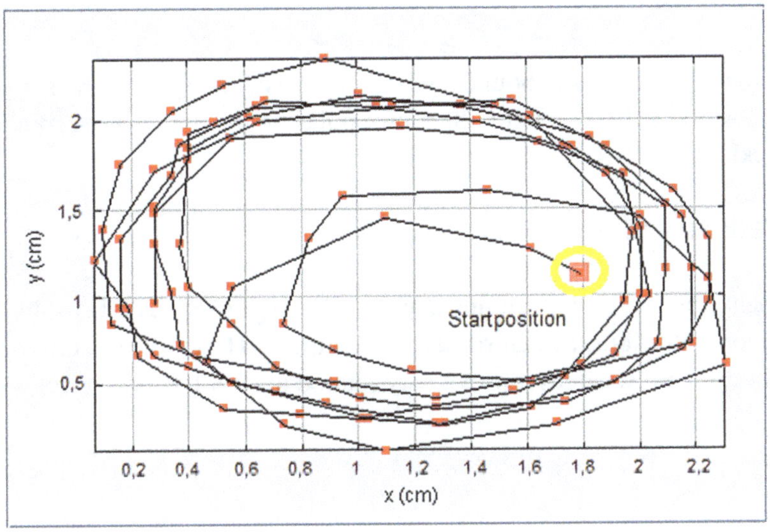

Table 3. Parts of the classic brushbot build on a coin vibration motor

Brush Type	Motor	Battery
Toothbrush head (vertical bristles)	- Coin vibration motor - internal unbalance- 0.15 W - 6000 rpm- Size: Approx. 10 * 2.7 mm- Weight: 1 g	CR2016 3V

Experiment 4: Coin Vibration Motor

In the fourth test, according to (HannoHupmann, 2008) a horizontally positioned coin vibration motor was used instead of the previously one (Table 3). It was attached to the front while the battery was fixed to the end (Figure 6). This version rotated in rapid gyrations around the vertical axis (Figure 7) and only rarely tilted over to the side.

Figure 6. Brushbot with coin vibration motor

When it came into contact with the boundary edge, it moved along then. As soon as it left it, it continued its previous movement. By reversing the motor polarity, the direction of rotation can be changed as described by (Rickharris, 2010).

Experiment 5: Vertically Motor

In the next experiment, the motor was mounted vertically (Figure 8). The brushbot then continued to show a rotating motion behaviour, as can be seen in Figure 9. However, the circling movements now showed a slight sideways shift whereas it previously rotated in very tight fast circles almost on the spot.

Figure 7. Recorded path

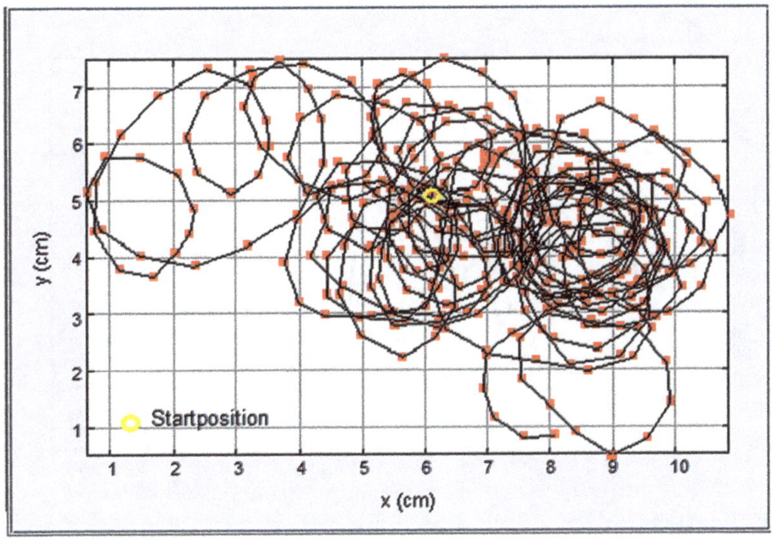

Figure 8. Brushbot, motor vertically

Experiment 6: Angled Bristles

Ross et al. (2017) recommends, without giving reasons, that for effective forward movement of brushbots, the bristles must be bent backward (usually with hot water) by about 15 degrees. But this case is following the recommendations of (AXL et al., 2012) that the movement is influenced by angled bristles and two brushes were bent by about 20 degrees one forward, one backward (Figure 10 and Figure 12).

In addition, they were pressed onto the floor to create an increased contact to the surface at the same time. According to (AXL et al., 2012), this should give the brushbots better stability against tilting.

Figure 9. Recorded path

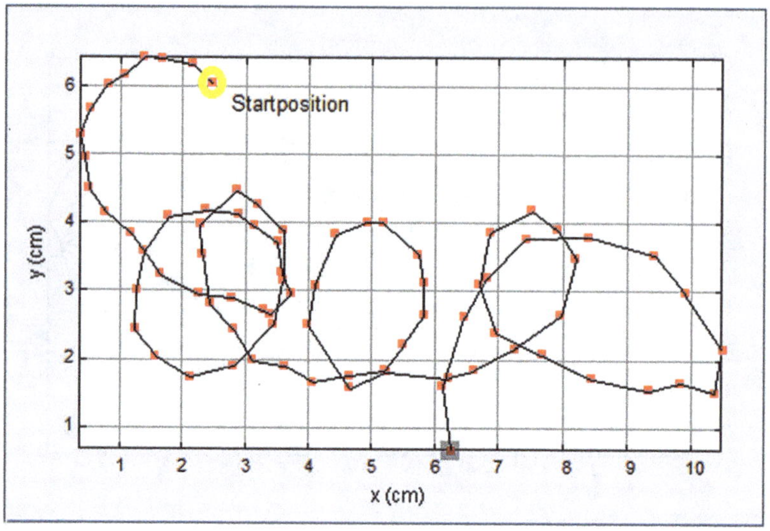

Both brushes were equipped with a vertically positioned coin vibration motor and a button cell. In both cases, the brushbot rotated around its vertical axis. The centre of rotation was the battery.

When the battery was positioned as displayed in Figure 10, the brushbots steered in the respective direction of the bristle bending. Both versions moved in short straight stretches that were interrupted by sudden changes in direction, so that no constant course of movement could be observed (Figure 11).

Figure 10. Moving direction, angled bristles

If the weight of the battery was shifted to the centre of the brush, as can be seen in Figure 12, the movement behaviour basically changed from short straight distances to a repeating steady circular movement (Figure 13). Consequently, the direction of movement in this experiment was determined by bending the bristles and shifting the weight.

Figure 11. Recorded path

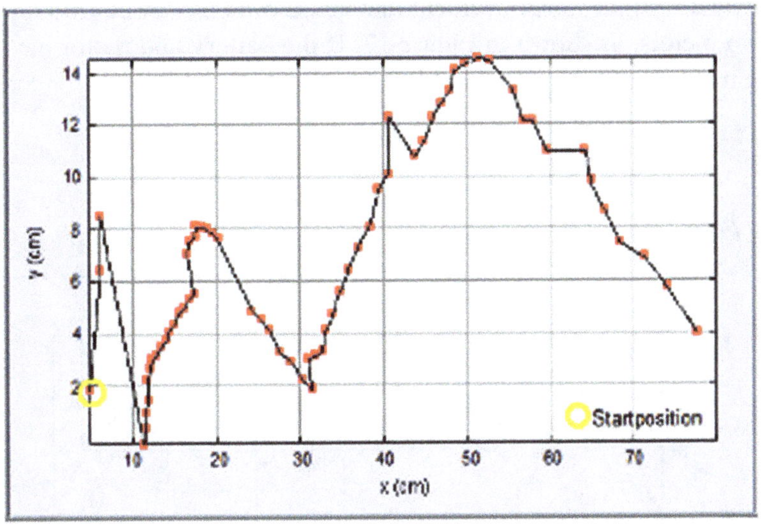

Figure 12. Battery central mounted

Experiment 7: Removed Bristles

Test seven addressed the suggestion to additionally remove the middle bristles in longitudinal direction. As a result, the brushbot (Figure 14) rotated in a radius of about 8 cm (Figure 15) in very uniform circular movements. Even a shift in weight did not influence this behaviour.

Experiment 8: Diagonally Shortened Bristles

In order to create a pure forward movement in the following test, a recommendation from Ross et al. (2017) has been taken into account. But in contrast to their concrete recommendation of 10 degrees the bristle was shortened at an angle of about 15 degrees and a setup like displayed in Figure 16 was used. The analysis confirmed a straight-line movement that could even be reproduced on level ground and at a distance of over two meters, as shown in Figure 17. If the battery and motor have been arranged the other way round, the movement was again circular.

Figure 13. Recorded path

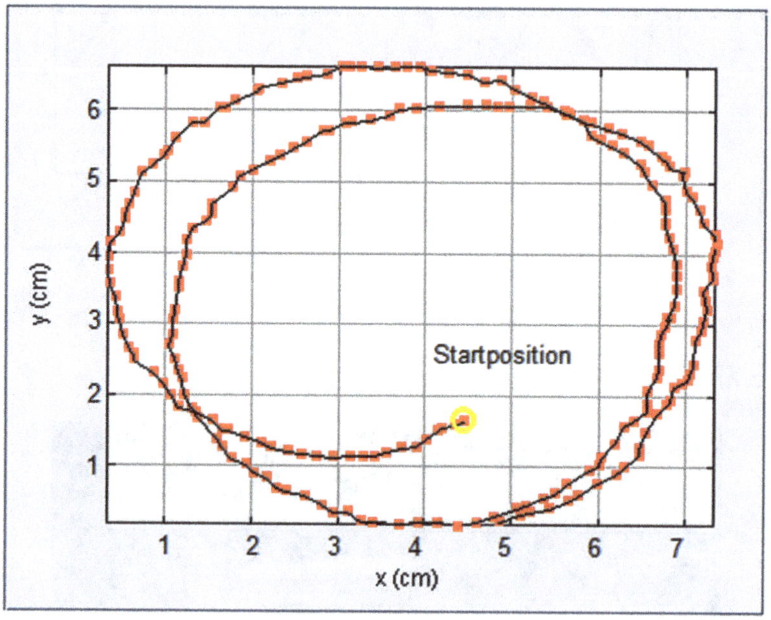

Figure 14. Without middle bristles in longitudinal direction

Figure 15. Recorded path

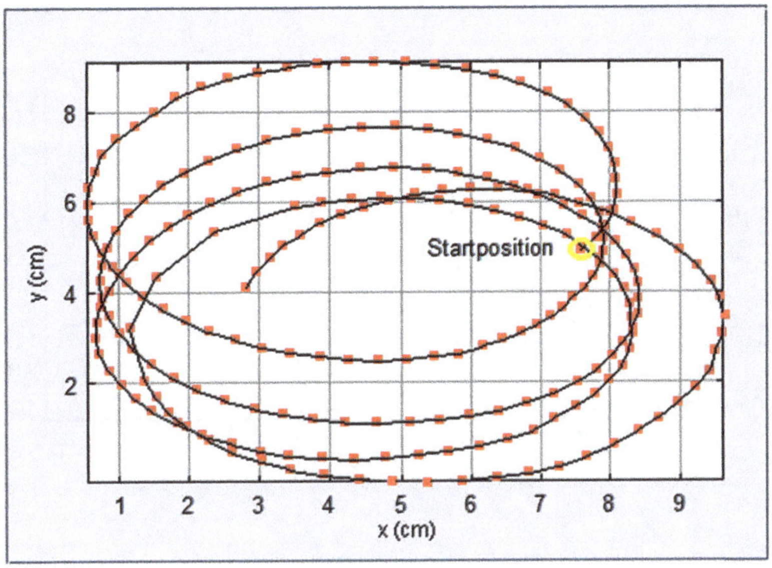

Figure 16. Bristles shortened at an angle

Test Series II: Elongated Brushes

In the second test series so-called giant bristle bots were built. Like the smaller versions they are very common as tinker and maker objects and follow alternative construction proposals and models from the internet. In addition to a selection of bigger brushes, larger motors, batteries and battery holders were used to test which brushes were overall most suitable for further experiments on directional movement.

Figure 17. Recorded path

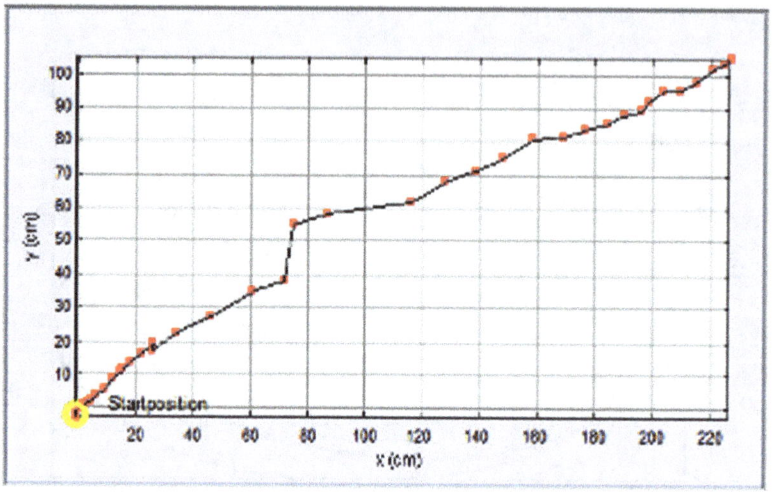

All components were firmly connected (glued) to the brush as recommended by (Aschoff, 2015; Cowen, 2013) like before.

Experiment 9: Pointed Scrub Brush

For the first four experiments in this series, a brush with one rounded end and one pointed end was used. The bristles were consisting of hard synthetic fibers which were fanned out broadly on all sides.

Figure 18. Arced moving, rotation centre

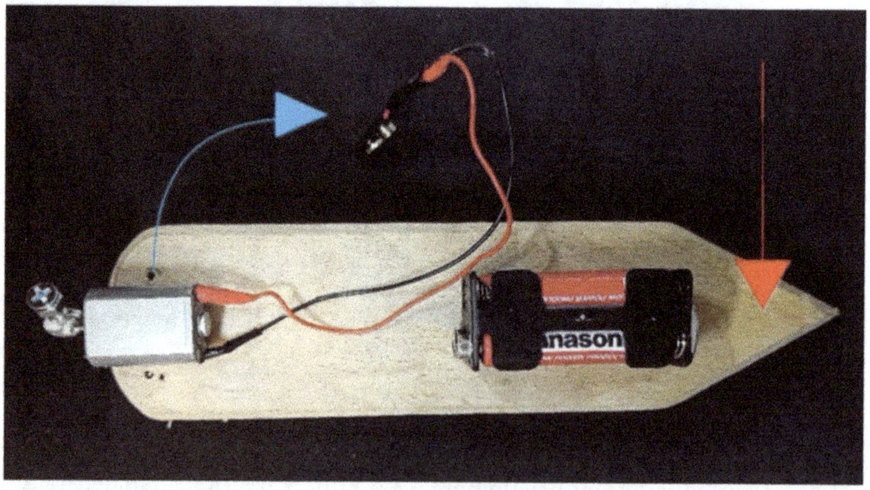

In the first version, the motor was mounted on to the rounded end. In operation, it followed a circular path, while the tip of the brush approximately represented the centre of rotation (Figure 18 and Figure 19).

Figure 19. Recorded path

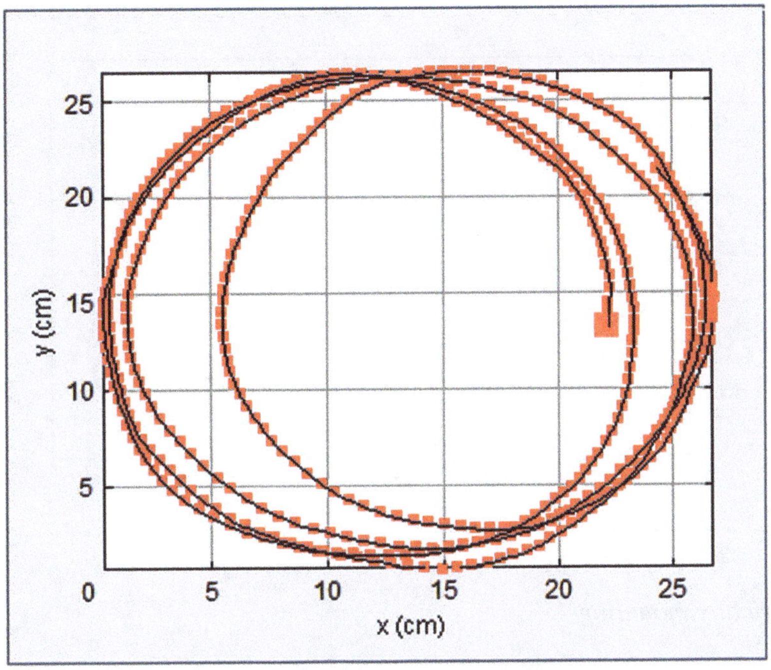

The behaviour remains the same even after the positions of the components have been swapped. The motor moved on a circular path, while the other end of the brush was approximately the centre of rotation. Then the motor was positioned still longitudinally but centrally while the centre of rotation shifted to the middle of the brush.

If the motor was placed transversely and centrally, different movements could be observed depending on the position of the battery holder. Once the battery holder was attached to the pointed end, it also represented the pivot point, while the direction of rotation could be changed with the direction of rotation of the vibration motor. When placed on the rounded side of the brush, the brushbot moved forward in a large arc against the motor (Figure 20 and Figure 21). The rounded end described the outer arc.

Figure 20. Arced lateral movement

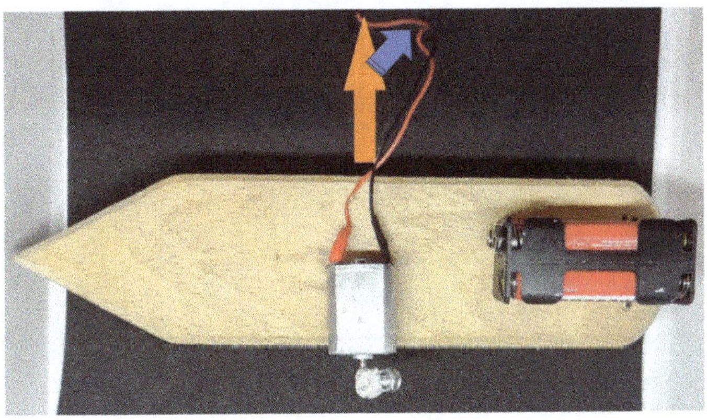

Figure 21. Recorded path

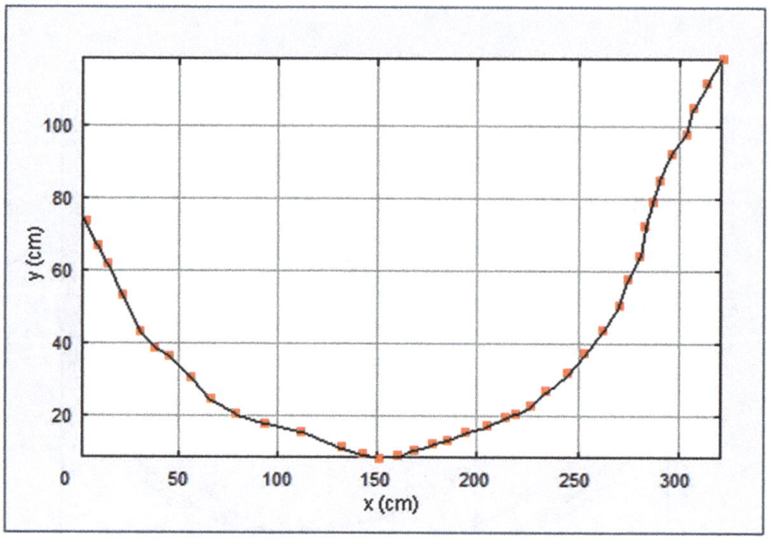

Figure 22. Symmetrically mounting

Figure 23. Recorded path

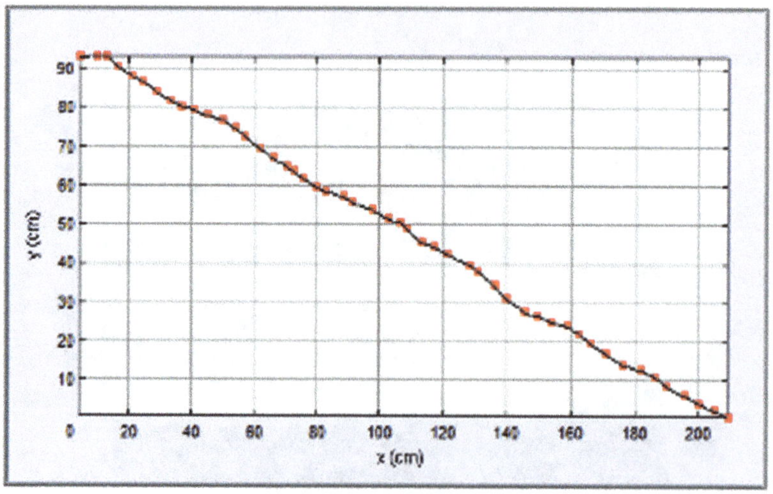

Figure 24. Bristlebot with s-shaped brush

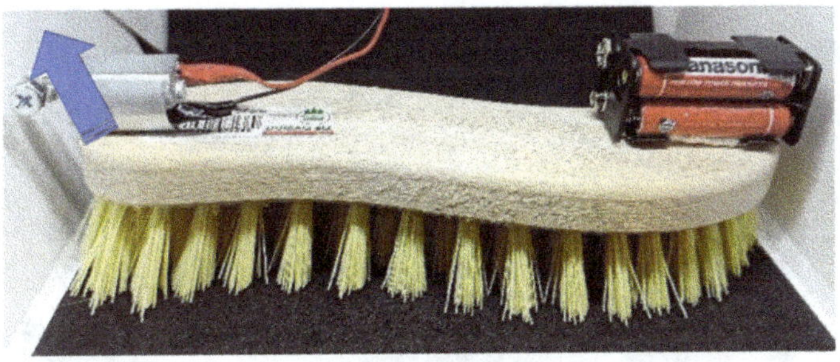

Experiment 10: Symmetrical Brush

In test 10, the results from test 9 were checked with a symmetrical brush. With the otherwise same configuration a movement behaviour was shown like before. A graphical representation was therefore omitted.

However, with a second, symmetrically mounted battery holder (Figure 22), an almost straight forward movement could be stimulated (Figure 23). The long side of the brush with the rotating eccentric weight formed the front side of the brush.

Experiment 11 and 12: Changing Material of Bristles

Brushbots with rubber bristles and a combination of soft plastic and boar bristles also behaved similarly to experiments nine and ten, each with the same structure. However, the softer the bristles, the slower the robot's movements.

Figure 25. Recorded path

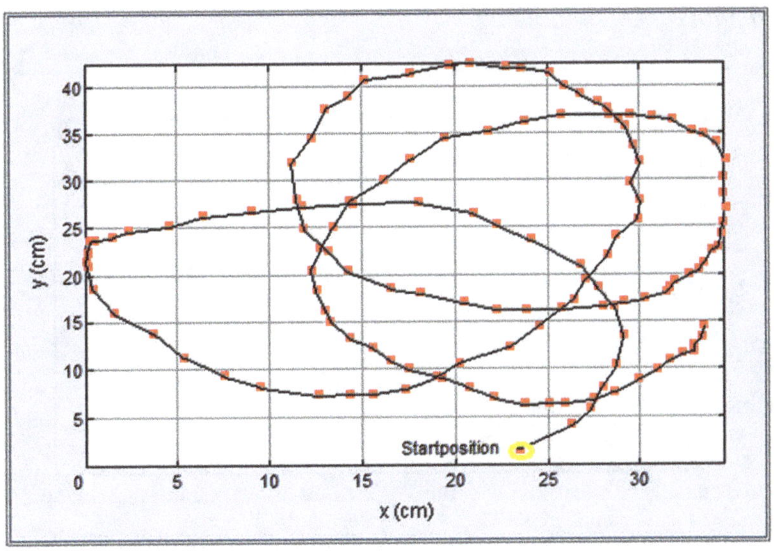

Figure 26. Setup for forward movement

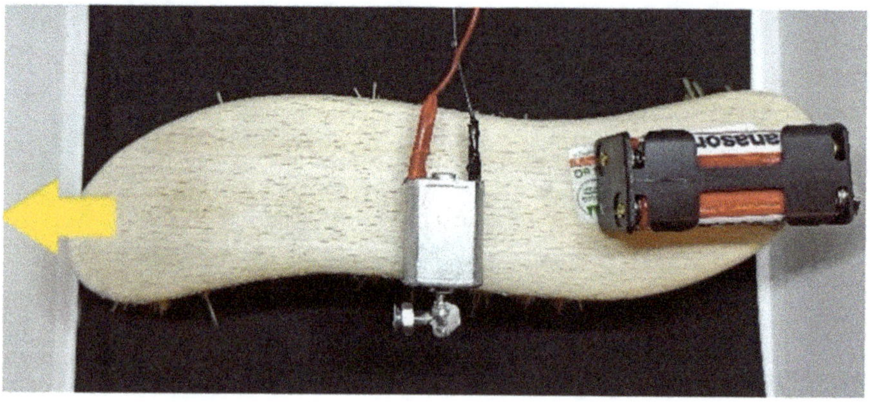

Experiment 13: S-Shaped Brush

Experiments with an S-shaped brush finally showed a deviation in the movement pattern. Apart from the shape, it corresponded to the one from experiment nine. With a similar arrangement of the components to test nine (Figure 24), this time the bot performed arc-shaped movements (Figure 25), with the arc radius decreasing the closer the battery holder was attached to the motor.

Once the motor was attached in the middle (Figure 26), a serpentine movement (Figure 27) was observed in several runs, which could be slightly influenced by shifting the weight. This means, the assumption of MadMage (2008) regarding the influence of the direction of movement when additional weight is applied can be accepted when using an S-shaped brush.

Figure 27. Recorded path

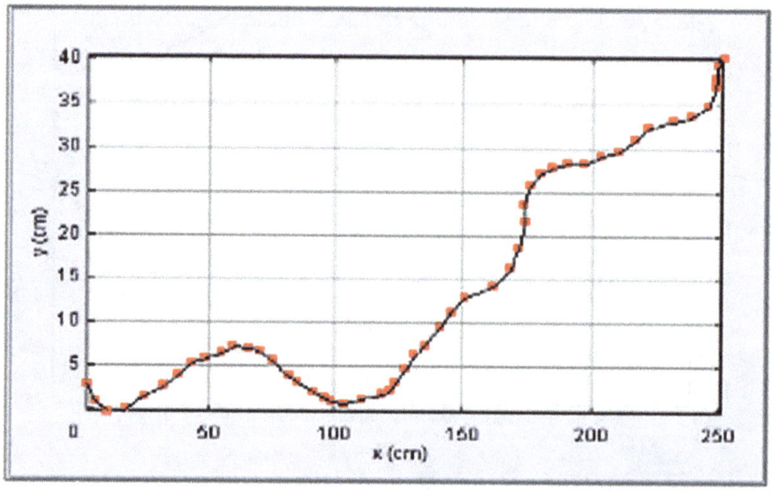

Figure 28. Bristlebot with short brush

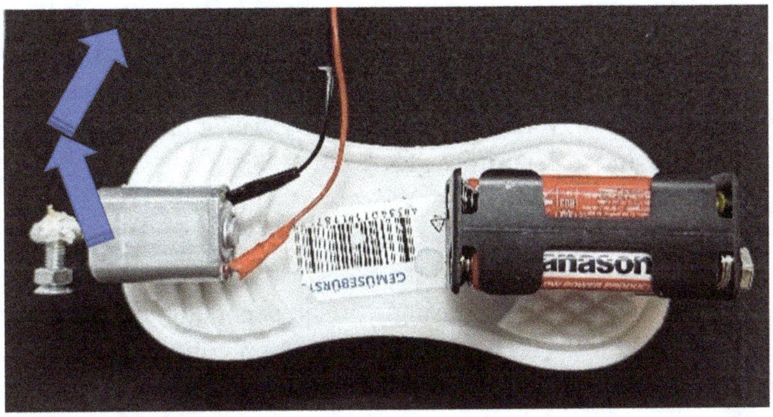

Test Series III: Short Brushes

Experiment 14: Hourglass Brush

With a shorter body, the brushbot moved on an arc-shaped path (Figure 28 and Figure 29) with the same setup as in test nine and did not change it when the unbalance motor was mounted centrally. With a second, symmetrically mounted battery holder, only the radius of the arc-shaped path was reduced.

Figure 29. Recorded path

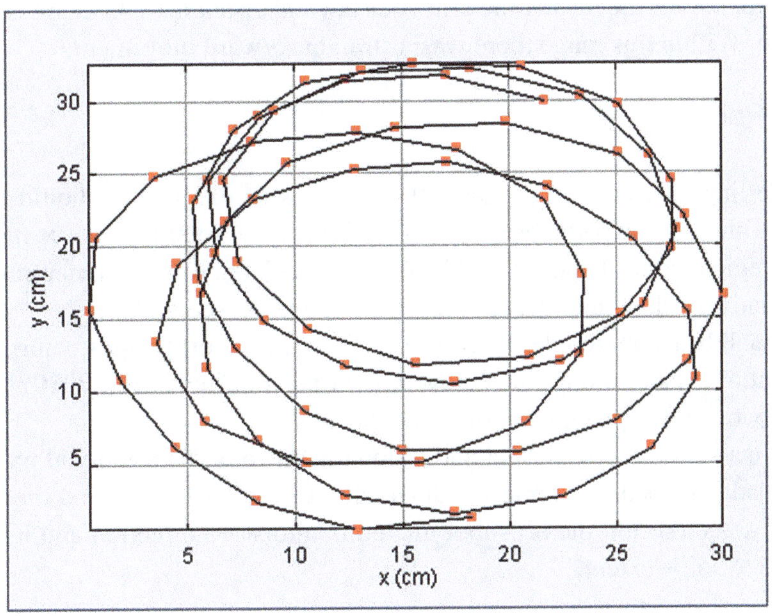

Experiment 15 and 16: Round and Very Short Brushes

Tests 15 and 16 partly confirmed the assumed movement pattern made in experiment 14 for short brushes. Despite the different positions of the components on the brush, both brushbots essentially rotated steadily around their own vertical axis. However, noticeably was observed a faster rotation speed of the smaller brush in experiment 16 compared to experiment 15.

Test Series IV: Additional Elements

In contrast to a classic brushbot set-up out of a single brush, one motor and one battery a small number of variations was also built and tested. They are also based on collected assumptions how to steer a brushbot from the forum. Therefore, different additional elements have been mounted and validated. Even the last experiments then keep in touch with successful steering experiences reported in related work.

Experiment 17: Weight Shifting I

Since the S-shaped brush was the only one that moved in a straight line, it was again used in steering experiments with a weight shifting according to (Anonymous, 2007; Ringo, 2008; Szczys, 2010). For this purpose, an extension arm was provided with additional weights of up to 24g and mounted onto a servo motor. To prevent the brushbot from tipping over the inclination was limited to 45 degrees from vertical.

Experiment 18: Weight Shifting II

Because test 17 showed that an extension arm can in principle effect a change in direction, the results have to be checked with another elongated but symmetrical brush. However, the servo motor was placed in the middle and the vibration motor as well as the battery pack have been placed at one end each.

If the extension arm was in neutral position (vertically) the bot moved similar to experiment ten. It increasingly strove for curves the further the arm goes beyond an angle of 45 degrees on both sides from the vertical position. Within this range there was a straight forward movement.

Experiment 19: Paddle

According to (Underling, 2010), paddles that press sideways on the bristles should steer have been designed and a brush with long and softer bristles was used this time. Again, the servo motor was controlled by a manual cable remote control and the paddle did not touch the bristles in neutral position. The soft bristles caused a relatively slow motion and the bot moved in curved paths.

The relatively small brushbot toppled over due to its long and soft bristles being pushed too far. For this reason, the central bristles were not removed, contrary to (Underling, 2010), but shortened in a similar way to test two and the paddle length was reduced.

It then resulted in a small steering effect and a slowing down of the rotational motion. Therefore, as in test eight, the bristles have been shortened diagonally towards the front and the paddle device was adjusted accordingly. As a result, the brushbot moved in a forward direction and its motion behaviour was also influenced to some extent.

Figure 30. Motors longitudinal

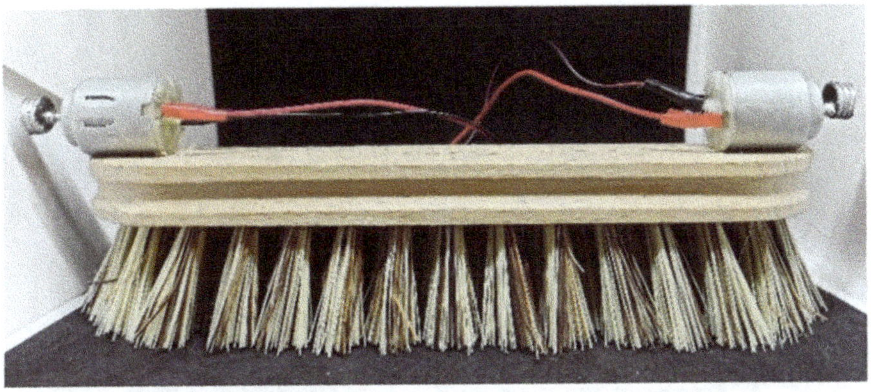

Experiment 20 and 21: Rudder

In experiment 20 and 21, a rudder steering system according to (Rickharris, 2010) was built and tested on a symmetrical brush and a brush with one rounded and one pointed end.

In both tests the steering was only possible once the motor was mounted at one tip of the brush, although the movement behaviour should be independent of the motor position according to the results from experiment nine and ten.

It was also proved to be more effective with the elongated symmetrical brush. Finally, using an RC remote control and an electronic speed controller for the vibration motor, the possibility to control the brushbot could be significantly improved again.

Experiment 22: Two Motors I

Spion (2008) recommended the use of two motors to enable control. For the experiment, an elongated symmetrical brush, two identical vibration motors and a cable remote control with two buttons have been used. As a variation of the classic layout, it is mentioned early by Ioi (1999) shown good results of steering.

For the first run, both motors were mounted longitudinally (Figure 30), offset from each other. If one of them was then activated, the brushbot rotated around a vertical axis, which was marked by the second motor. Once both motors were switched on at the same time, the brushbot remained in place.

Figure 31 shows the second setup, which brought an improvement in control because the brushbot could now be steered forward in a serpentine line.

The assumption of (Lunarman, 2008) that the removal of the middle bristles in the direction of movement (Figure 32) could result in decoupling and thus led to an improvement in the movement behaviour could not be confirmed in conclusion.

Experiment 23: Two Motors II

In test 23, the two-head control was taken up again in an analogous way and, following the contributions of (Fhunter, 2010; Manf, 2008a; Manf 2008b; Slurp, 2010), implemented with two beveled toothbrush

Figure 31. Steerable movement

heads. The motors were attached like in Figure 33 and controlled by a cable control. The shortened bristles form the front, while the eccentric weights were mounted at the rear. This also showed a good possibility for steering and confirms the design layout mentioned by Morin (2009), Tossel et al. (2010), and Ross et al. (2017).

Figure 32. Brush bot with bristles removed

RESULTS AND CONCLUSION

As part of this contribution, collected assumptions (Table 1) about the motion behaviour of brushbots were tested and validated for their feasibility and effect. The experiments also indirectly picked up research suggestions of (Ioi, 1999) and showed that the movement depends on the shape of the brush, the overall geometry, the position of the components and the condition of the bristles. It seems that control is only possible with additional components as shown summarized in Table 4.

Figure 33. Steerable setup

The directional behaviour of a brushbot, for example, became more random as the length of the brush decreases, while most of the elongated symmetrical brushes aimed for an arc-shaped path in the longitudinal direction. Here, only the S-shaped brush was able to perform a straight-line movement. Reversing the polarity of the motor only changed the direction of rotation of the brushbot.

The arrangement of the components and the changed centre of gravity had a much greater influence on the directional behaviour. A change in the radius of the arc path and a shift of the centre of the arc path was observed several times for example. On the other hand, a change between horizontal and vertical mounting of the vibration motor, only resulted in a shifted and changed arc radius. With an even weight distribution in the longitudinal axis and a central arrangement of the vibration motor, a transversely directed but straight movement could be recorded.

It could also be seen that bots with softer bristles moved more slowly than bots with stiffer bristles. The tests also showed that curved and beveled bristles can influence the direction. If the centre of gravity is then on the long bristles, the brushbot strives to move in the direction of the short bristles. In contrast, shortening the bristles evenly and removing the angled middle bristles showed no change in the directional behaviour.

In the penultimate series of experiments, a small number of variations was also built and tested in contrast to a classic brushbot set-up out of a single brush, one motor and one battery. Therefore, a servo motor, a manual cable remote control and other components have been used to test the steering effect of weight shifts, paddle controls and rudder steering. These experiments all showed a steering effect, but the rudder steering turned out to be the most effective.

Finally, taking into account the knowledge gained, the use of two independently controllable vibration motors resulted in good steerability and smooth forward motion. It confirms the design concepts used by Ioi (1999) Morin (2009), Tossel et al. (2010), and Ross et al. (2017).

Table 4. Results of the experiments listed by numbers

Assumptions Movement Behavior/Steering	#	Brush Type	Result/Impact
Short bristles give more stability	1-2	Toothbrush	Does not lead to any improvement in stability
Choice of vibration motor influences movement behavior	3-4		Choice of motor influences motion behavior and stability
			Direction of rotation of the brush bot depends on the direction of rotation of the unbalance
			Alignment of the vibration motor influences the direction of movement
Weight distribution on the brush influences stability	5		Leads to improve stability
Angled and outward bent bristles increase stability	6		Angled bristles, in addition to bristle bending and weight shift on the brush bot, have a decisive influence on the movement behavior
Removal of the middle bristles leads to a change in behavior	7		Does not result in any directional movement of the brush bot
Direction of movement is influenced by the use of angled bristles	8		Leads to a directed forward movement when weight is shifted to the longer bristles
Stability due to wider brushes Different brush shapes	9-16	Sink brush (S-shape) Sink brush (pointed) Rinsing brush (symmetrical) Rubber brush Boar brush	Movement behavior changes with the positioning of the motor and the weight by the battery holder. Forward movement in the direction of the long side of the brush is possible
		Hourglass brush	Steady circular motion regardless of the orientation of the components
		Butterfly brush Sink brush (round)	Steady circular motion around its own axis regardless of the orientation of the components
Separate weight shifting via servo motor	17-18	Sink brush (S-shape) Sink brush (symmetrical)	Steering is possible with adapted additional weight
Paddle steering	19	Hand brush	Vague steering possible, high precision of servo motor operation necessary
Rudder steering	20-21	Sink brush (pointed) Sink brush (symmetrical)	Conscious control possible
Use of two motors on one brush	22	Sink brush (symmetrical)	Steering is possible
Two-brush head system	23	Toothbrushes	Steering is possible

FUTURE RESEARCH DIRECTIONS

To continue the presented work, it could be investigated whether an RC-control can in principle compensate for the disadvantages of a cable control and whether an even better steering result can be achieved with a two speed controller for both motors. The modification of this control by a microcontroller could then be the subject of a further investigation and turn the simple tinkering object into a real (educational) robot, while an educational approach might evaluate different scenarios of using the robot to explain basic mechanics or physics

SUMMARY

At the beginning of this chapter, some theoretical basics are clarified in order to create an approach to classify brushbots as robots as well as educational robots for design and research activities in learning settings. It starts with the current state of knowledge and related work on brushbots will be summarized in order to obtain an overview on research on brushbots and their use in schools.

It follows an overview about the experimental work done. It is presented in detail to validate and discuss the potential of brushbots in three directions. First a cheap platform was proposed to evaluate the motion capabilities of different designs. Doing this it provides a sample how to improve vibrobots concerning experimentation results in the design process. It was done in a manner that it can be transferred to primary and lower secondary schools in the area of technology teaching in school lessons

Initially, a classic brushbot was first built to observe and document the undirected movement behaviour. It was then modified to gain knowledge about possible influencing factors in relation to the bristle position, weight and the use of different motors.

After that, the motions of different types and shapes of brushes was investigated. This is followed by the experimental setup of different theories on steering approaches of brushbots, as proposed and discussed in relevant internet-forums of the maker-scene as a substitute of ideas generated from children. Finally, the results obtained will be used to propose a movement control system.

In order to graphically represent the movement patterns of the tested brushbots, the movement behaviour was tracked, recorded and evaluated with Tracker, a video analysis software. For this purpose, the sequences were marked each within the software and their movement radius was transferred to the coordinate system and finally displayed in a diagram. This section is followed by displaying the obtained results and a conclusion section, before it closes with some future research directions.

Consequently, this means that brushbots can be used not only for tinkering and as a teaching tool for applying the engineering design process from ideas to improved design concepts presented in the application in school section, but also as a low-cost research platform as shown below.

REFERENCES

Akbarimajd, A., & Sotoudeh, N. (2014). Design and motion analysis of vibration-driven small robot Rizeh. *Advanced Robotics*, *28*(2), 105–117. doi:10.1080/01691864.2013.843788

Anonymous. (2007, December 20). *Bristlebot: A tiny directional vibrobots*. https://www.evilmadscientist.com/2007/bristlebot-a-tiny-directional-vibrobot/comment-page-1/#comment-2521

Anonymous. (2009, January 17). *Bristlebot: A tiny directional vibrobot*. https://www.evilmadscientist.com/2007/bristlebot-a-tiny-directional-vibrobot/comment-page-1/#comment-2533

Aschoff, C. (2015, March 20). *Aufbauanleitung für einen Bristle-Bot*. http://www.christians-bastel-laden.de/BRISTLE_BOT_MANUAL/index.html

Axl, J., & Manupool. (2012, January 12). *Bristlebots*. https://www.hackerspace-ffm.de/wiki/index.php?title=Bristlebots

Becker, F., Minchenya, V., Zeidis, I., & Zimmermann, K. (2011). Modelling and dynamical simulation of vibration-driven robots. In *Proceeding of the 56th International scientific colloquium*. Ilmenau University of Technology.

Boehm, B. W. (1988). A spiral model of software development and enhancement. *Computer*, *21*(5), 61–72. doi:10.1109/2.59

Bristlebotics. (2017). *Light Sensing Bristlebot*. https://www.instructables.com/Light-Sensing-Bristlebot

Brown, D. (2008). *Video Modeling: Combining Dynamic Model Simulations with Traditional Video Analysis*. AAPT Summer Meeting.

Cowen, A. (2013, August 30). *Building Bristlebots: Basic Toothbrush Robotics*. https://www.sciencebuddies.org/blog/building-bristlebots-basic-toothbrush-robotics?from=Blog

Downie, N. A. (2001). *Vacuum Bazookas, Electric Rainbow Jelly and 27 Other Saturday Science Projects*. Princeton University Press.

ElMaraghy, W. H. (2019). Engineering Design. In CIRP Encyclopedia of Production Engineering. Springer. (pp. 608-612). doi:10.1007/978-3-662-53120-4_16781

Farmer, C. (2015). Engineer your World Engineering Design and Problem Solving. In C. L. Sneider (Ed.), *The Go-To Guide for Engineering Curricula, Grades 9–12: Choosing and Using the Best Instructional Materials for Your Students Engineer your World Engineering Design and Problem Solving* (pp. 73–89). SAGE Publications. doi:10.4135/9781483388373.n8

Feldhusen, J., & Grote, K. H. (2013). Der Produktentstehungsprozess (PEP). In J. Feldhusen & K. H. Grote (Eds.), *Pahl/Beitz Konstruktionslehre* (pp. 11–14). Springer Vieweg. doi:10.1007/978-3-642-29569-0_2

Fhunter. (2010, November 4). *RC Bristlebots shifts weight for steering*. https://hackaday.com/2010/11/03/rc-bristlebot-shifts-weight-for-steering/#comment-207574

Finio, B. (2020, June 23). *Design Your Own Vibrobots*. https://www.sciencebuddies.org/science-fair-projects/project-ideas/Robotics_p030/robotics/vibrobots

Fislake, M. (2021). Educational Robotics Between Coding and Engineering Education. In S. Papadakis & M. Kalogiannakis (Eds.), *Handbook of Research on Using Educational Robotics to Facilitate Student Learning* (pp. 107–139). IGI Global. doi:10.4018/978-1-7998-6717-3.ch004

Gandra, C. (2019). Dynamics of a Vibration Driven Bristle Bot. *All Theses*. https://tigerprints.clemson.edu/all_theses/3149

Giomi, L., Hawley-Weld, N., & Mahadevan, L. (2013). Swarming, swirling and stasis in sequestered bristle-bots. *Proc R Soc A*, *469*(2151), 20120637. doi:10.1098/rspa.2012.0637

HannoHupmann. (2008, January 1). *Extrem cool -> "BürstenBot"*. https://www.roboternetz.de/community/threads/33177-Extrem-cool-B%C3%BCrstenBot?p=314629&viewfull=1#post314629

Innovation First Labs, Inc. (2018, June 13). *HEXBUG nano*. https://www.hexbug.com/hexbug-nano-single.html

Ioi, K. (1999). Mobile micro-robot using centrifugal forces. *Journal of the Robotics Society of Japan, 17*(3), 736–741. doi:10.7210/jrsj.17.396

ITEEA, International Technology and Engineering Educators Association. (2007). Standards for Technological Literacy: Content for the Study of Technology (3rd ed.). ITEEA.

ITEEA, International Technology and Engineering Educators Association. (2020). *Standards for Technological and Engineering Literacy: Defining the Role of Technology and Engineering in STEM Education*. ITEEA.

Kaya, E., Yesilyurt, E., Deniz, H., Newley, A., & Newley, P. (2017). Research and teaching: Introducing engineering design to a science teaching methods course through educational robotics and exploring changes in views of preservice elementary teachers. *Journal of College Science Teaching, 047*(02), 66–75. doi:10.2505/4/jcst17_047_02_66

Lunarman. (2011, January 11). *Extrem cool -> "BürstenBot"*. https://www.roboternetz.de/community/threads/33177-Extrem-cool-B%C3%BCrstenBot?p=314868&viewfull=1#post314868

MadMage. (2008, January 11). *Extrem cool -> "BürstenBot"*. https://www.roboternetz.de/community/threads/33177-Extrem-cool-B%C3%BCrstenBot?p=314697&viewfull=1#post314697

Mammes, I., Schäffer, K. & Tuncsoy, M. (2013). „Robotik" in der Grundschule. Beschreibung des Baus des „Miniatur-Roboters" Bürstenfuzzi. *Tu - Zeitschrift für Technik im Unterricht, 147*(13), 41-45.

Manf. (2008a, January 9). *Extrem cool -> "BürstenBot"*. https://www.roboternetz.de/community/threads/33177-Extrem-cool-B%C3%BCrstenBot?p=314134&viewfull=1#post314134

Manf. (2008b, January 10). *Extrem cool -> "BürstenBot"*. https://www.roboternetz.de/community/threads/33177-Extrem-cool-B%C3%BCrstenBot?p=314648&viewfull=1#post314648

Martinez, S. L., & Stager, G. S. (2019). Invent to Learn Making, Tinkering, and Engineering in the Classroom (2nd ed.). Constructing Modern Knowledge Press.

Michael, M. (2008, January 11). *Extrem cool -> "BürstenBot"*. https://www.roboternetz.de/community/threads/33177-Extrem-cool-B%C3%BCrstenBot?p=314857&viewfull=1#post314857

Morin, J. W. (2011). *Design, fabrication and mechanical optimization of multi-scale anisotropic feet for terrestrial locomotion*. Massachusetts Institute of Technology. Dept. of Mechanical Engineering. https://dspace.mit.edu/handle/1721.1/65314

Notomista, G., Mayya, S., Mazumdar, A., Hutchinson, S., & Egerstedt, M. (2019). A Study of a Class of Vibration-Driven Robots: Modeling, Analysis, Control and Design of the Brushbot. *2019 IEEE/RSJ International Conference on Intelligent Robots and Systems (IROS)*, 5101-5106. 10.1109/IROS40897.2019.8968490

Oskay, W. (2007, December 19). *Bristlebot: A tiny directional vibrobot.* https://www.evilmadscientist.com/2007/bristlebot-a-tiny-directional-vibrobot

Resnick, M. (2007). All I really need to know (about creative thinking) I learned (by studying how children learn) in kindergarten. In B. Shneiderman, G. Fischer, E. Giaccardi, & M. Eisenberg (Eds.), *Proceedings of the 6th acm sigchi conference on creativity & cognition - c&c '07* (pp. 1–6). ACM Press. 10.1145/1254960.1254961

Rickharris. (2010, January 3). *Bristlebots?* https://picaxeforum.co.uk/threads/bristlebots.14219/post-121433

Ringo. (2008, March 26). *Mein 10min. IC-Bot (BristleBot/Vibrobot).* https://www.roboternetz.de/community/threads/35068-mein-10min-IC-Bot-%28BristleBot-Vibrobot%29?p=333234&viewfull=1#post333234

Ross, R., Stanger, J., & Console, A. (2017). Phototropic BristleBot activity for robotics and STEM engagement. In *2017 IEEE International Conference on Mechatronics (ICM)* (pp. 425-430). 10.1109/ICMECH.2017.7921144

Rubenstein, M., Ahler, C., & Nagpal, R. (2012) Kilobot: A low cost scalable robot system for collective behaviors. *2012 IEEE International Conference on Robotics and Automation*, 3293-3298. 10.1109/ICRA.2012.6224638

Sheridan, K., & Konopasky, A. (2015). *An Experimental Study Comparing Two Educational Approaches to Making With Simple Circuits* [Paper presentation]. Conference American Educational Research Association.

Slurp. (2010, January 3). *Bristlebots?* https://picaxeforum.co.uk/threads/bristlebots.14219/post-121456

Spion. (2008, March 26). *Mein 10min. IC-Bot (BristleBot/Vibrobot).* https://www.roboternetz.de/community/threads/35068-mein-10min-IC-Bot-%28BristleBot-Vibrobot%29?p=333265&viewfull=1#post333265

Strider_mt2k. (2010, November 4). *RC bristlebot shifts weight for steering.* https://hackaday.com/2010/11/03/rc-bristlebot-shifts-weight-for-steering/#comment-207571

Sullivan, A., Kazakoff, E. R., & Bers, M. U. (2013). The Wheels on the Bot go Round and Round: Robotics Curriculum in Pre-Kindergarten. *Journal of Information Technology Education: Innovations in Practice*, *12*, 203–219. http://www.jite.org/documents/Vol12/JITEv12IIPp203-219Sullivan1257.pdf

Szczys, M. (2010, November 3). *RC Bristlebots shifts weight for steering.* https://hackaday.com/2010/11/03/rc-bristlebot-shifts-weight-for-steering

Tossel, K., Hammond, A., Arvelo, E., & Martins, N. (2010). Vision-based Micro-robot Control [poster presentation]. University of Maryland.

Tucker, M. & Miller, T. (2021, March 24). *Navigate with nanobots - (steam "212" design challenge).* ITEEA conference 2021.

Underling. (2010, November 3). *RC Bristlebots shifts weight for steering.* https://hackaday.com/2010/11/03/rc-bristlebot-shifts-weight-for-steering/#comment-207526

Underling. (2011, March 3). *Bristlebot.* http://underlinglabs.blogspot.com/2011/03/bristle-bot.html

Vartholomeos, P., & Papadopoulos, E. (2008). Dynamics, design and simulation of a novel microrobotic platform employing vibration microactuators. *Journal of Dynamic Systems, Measurement, and Control, 128*, 122–133.

Wee, L.K. & Lee, T. (2012). *Video Analysis and Modeling Tool for Physics Education: A workshop for Redesigning Pedagogy.* Academic Press.

ADDITIONAL READING

ABET - Accreditation Board for Engineering and Technology. (2017). *Criteria for accrediting engineering programs*. Engineering Accrediting Commission.

Becker, F., Boerner, S., Lysenko, V., Zeidis, I., & Zimmermann, K. (2014) On the Mechanics of Bristle-Bots - Modeling, Simulation and Experiments. *ISR/Robotik 2014; 41st International Symposium on Robotics*, 1-6.

Biochemtronics. (2008, July 8). *SOCBOT - the next generation vibrobot*. https://www.instructables.com/SOCBOT-the-next-generation-vibrobot

Blikstein, P. (2015). Computationally Enhanced Toolkits for Children: Historical Review and a Framework for Future Design. Foundations and Trends. *Human-Computer Interaction*, *9*(1), 1–68. doi:10.1561/1100000057

Cicconofri, G., & DeSimone, A. (2015). Motility of a model bristle-bot: A theoretical analysis. *International Journal of Non-linear Mechanics*, *76*, 233–239. doi:10.1016/j.ijnonlinmec.2014.12.010

Eguchi, A. (2014). Educational Robotics for Promoting 21st Century Skills. *Journal of Automation. Mobile Robotics & Intelligent Systems*, *8*(1), 5–11. doi:10.14313/JAMRIS_1-2014/1

Kühnel, D. T., Helps, T., & Rossiter, J. (2016). Kinematic Analysis of VibroBot: A Soft, Hopping Robot with Stiffness and Shape-Changing Abilities. *Frontiers in Robotics and AI*, *3*, 60. doi:10.3389/frobt.2016.00060

Robomaniac. (2008, May 8). *How to Build the One Motor Walker!* https://www.instructables.com/How-to-build-the-one-motor-walker/

KEY TERMS AND DEFINITIONS

Bristlebot: A bristlebot, also known as bristle-bots, is one of the simplest of all mobile robots, both in its function and in its construction. The term is synonymous with brushbots, but is based on the English term bristle. They are popular and widespread as craft objects in the maker scene and in schools.

Brushbot: Brushbots, also known as brush-bots, are considered as a play type of vibrobots. They consist of a toothbrush or equivalent, a vibration motor as drive unit and an energy storage unit in the form of a battery. The various components are attached and fixed to the brush using adhesive tape, glue, or holders.

Giant Bristlebot: See giant brushbot.

Giant Brushbot: Brushbot consisting of a larger brush, a larger motor, batteries and battery holder. Known as craft objects in the maker scene, they promise tinkering fun and a promotional effect for technical interest when building such a bot.

Inch-Scale Robot: Designation for robots, up to a size of one inch. Used in certain professional circles and as a classification. Sometimes used similar for brushbots.

Microbot: Very small robot made possible by the progressive miniaturization of mechanical and electronic systems in the last decade of the 20th century. They can be programmed and controlled by a microcontroller. It is larger than a nanobot, potentially very cheap and is often used in large numbers as swarm robotics.

Microscale Legged Robot: Characterization of very small robots with legs. Sometimes used synonymously to brushbots in certain communities.

Vibrobot: A special kind of walking robots that is driven by any kind of actuation that causes vibrations like clockworks, rubber bands, motors, and others. They have a long history, pre-dating the electric bristlebots and a broader diversity. They are known as toys, artwork, and research objects.

Chapter 13
Educational Robotic Competitions:
Methodology, Practical Appliance, and Experience

Diego E. Galvez-Aranda
Texas A&M University, USA

Mauricio P. Galvez-Legua
https://orcid.org/0000-0002-4845-4218
Universidad Nacional de Ingeniería, Peru

ABSTRACT

In this chapter, the authors present a methodology for the formation and preparation of teams in order to participate in educational robotics competitions composed of five stages: recruitment, training, team conformation and preparation, prototyping, and competition day. The chapter begins with an introduction to educational robotics applied in schools and its use to implement design-based learning, followed by a description of the characteristics of the educational robotics kits and a brief description of the 5PER methodology used in the fourth stage, prototyping, with a description of the main robotics competencies worldwide. The five stages methodology is presented in detailed so it can be replicated by other educators around the world. The exposed methodology is based on the experience of the authors in the preparation of equipment to participate in various robotics competitions at the Faculty of Electrical and Electronic Engineering of the Universidad Nacional de Ingenieria Lima, Peru.

INTRODUCTION

Research is one of the great challenges faced by today's students in order to grow in their field of action. Research gives us freedom of knowledge, the power to develop our own industry. Technological independence is possible if each professional develops research processes in their own areas. Two key aspects

DOI: 10.4018/978-1-7998-7443-0.ch013

in research are the access of the information and the equipment. Nowadays any student independently of university or school, can access information, which is currently available in great volume thanks to internet. Regarding equipment, several schools are implementing laboratories where the student can experiment and test their ideas.

In this scenario, Educational Robotics (ER) have been growing in the last years until becoming one of the most popular activities in schools from K-12 to graduate level.(Arís & Orcos, 2019; Eguchi, 2010) The development of the ER have been influenced by several authors and theories, starting from the **constructivism** theory of Jean Piaget in the 1960s,(Wadsworth, 1996) passing through the **constructionism** of Seymour Pappert in the 1990s,(Harel & Papert, 1991) reaching the **connectivism** of George Siemens in 2004, which is a pedagogical perspective that explains the effect that technology has had on the way we currently live, communicate and learn.

In this chapter, the authors are not going to explain in detail the development of ER because it is not the goal of the chapter. However, the definition of ER must be clear. Different definitions can be found in the literature;(Angel-Fernandez & Vincze, 2018; Eguchi, 2014; Mikropoulos & Bellou, 2013) and the authors have synthesized them as: ER is a learning environment that allows the understanding of abstract concepts (mathematics, geometry, programming, etc.), as well as physical concepts of our environment (structures, forces, speed, center of gravity, etc.) through the construction of prototypes. ER stimulates imagination, creativity, motivation, concentration, and manual skills in students when solving challenges. ER has the main virtue of serving as an interface between the logical world and the physical world and it does so in a playful way. The main goal of ER is not to make the students learn about robotics, that is one only of the benefits. The ER must be understood as a learning environment. (Alimisis, 2013) Nowadays most of school programs includes ER lessons indicating the growth of this field.(Eguchi, 2007)

ER kits are currently a powerful piece of equipment that allows students to design and implement robotic models that behave in a predefined way. ER allows students to "build their ideas." In the words of Mitchel Resnick, the goal of an ER kit is to design things that will allow kids to design things. (Resnick & Silverman, 2005) An ER kit carry out programming that interacts with the physical world, so that the programs "move things" and do not remain only in the sphere of the abstract or virtual environment. This interaction with the real world allows the student to approach research and take an interest in the career courses. The control system based on a microcontroller is a key element in an ER kit, thanks to it, students can build prototypes that go from static structures to dynamic machines, exhibiting a certain behavior and interacting with the real world. The control system is a programmable device, that allows to "embed" formulas and algorithms that "define" the prototype behavior. In general, an ER kit, is a tool that encouraged innovation, creativity, and imagination.

Besides the application of ER in formal learning environments, classroom, another great application of ER that are being widely used are the robotic competitions. Thanks to the increasing influence of ER in school around the world, robotics competitions and events have also increased locally as well as internationally. Different kind of competitions are hosted all around the world, where students develop different kind of robots to solve a series of challenges.

In this chapter the authors present a methodology for the selection, training, and preparation of teams with a view on participation in robotics championships through the use of educational robotics kits. The methodology could help teachers from different school and universities around the world to elaborate a good program in the preparation of student teams for a robotic competition. The practical application of this methodology occurred for 3 years, from 2012 to 2014, with students from the electrical engineering department at Universidad Nacional de Ingenieria, Lima-Peru. As a result, the university obtained the first places in the WRO 2012, 2013 and 2014 national edition, as well as a fourth place in the Latin American Robotics Counsil (LARC) 2013.

The environment a student faces in a robotic competition allow them to develop capabilities such as: emotional intelligence, work under pressure, teamwork, communication, and problem-solving skills. Even though the practical application was performed in first year undergraduate student, the methodology can be also applied to other educational levels such as elementary and high schools.

BACKGROUND

ER Application in Schools

ER has generated great interest from teachers as a medium to develop cognitive, procedural and aptitude skills in students from preschool to university, supporting learning in mathematics and science. ER emerged as learning medium that can offer fun and practical activities in an engaging learning environment increasing students' interest and curiosity.(Eguchi, 2010). Later as a learning environment, where skills for research, design and implementation are promoted.

In many schools, ER is used as an end and not as a means or environment, which promotes learning. Robot construction and programming courses have been included in the curriculum, which in some cases are not developed through educational robotics kits, but with electronic platforms such as SBC (System Board Computer) and pre-manufactured chassis. It is true that in many cases, this type of infrastructure has a lower cost than educational robotics kits, but it is not possible to mechanically design things since they lack construction parts with which the development of psychomotor skills does not occur to a great extent.

Depending on the way the construction pieces are connected, you can have a great variety of combinations that practically allow you to build "anything we can imagine". That is why, in part, most international competitions are based on the use of educational robotics kits because they favor constructive creativity, which is very necessary in the design of solutions for the real world.

ER-Based DBL

Design Based Learning (DBL) is a form of project-based learning to foster creative, problem-solving skills and support students in learning and applying knowledge through participation in cross-cutting challenges related to the real world. Currently in schools there is a growing emphasis on the importance of fostering creativity, critical thinking, collaboration and communication, all inherent design skills. DBL fosters these competencies while supporting the learning of knowledge. The complexity of a design often requires collaboration and role assignment, giving students the opportunity to begin to specialize in a particular area and thereby discover their "areas of interest" (vocation). Design projects require students to set goals and constraints, come up with ideas, build prototypes, and test them until requirements are met.

One way to implement DBL is through the ER, because ER provides a powerful infrastructure that allows creating an environment where creativity is fostered and allows facing motivating challenges as well as promoting teamwork. An example of ER-based DBL are robotics competitions, which are design-based learning activities, where each member of a participating team is assigned a role to: design, build, program, test and document the solution to the challenge posed.

"The educational robotics competitions are essentially spaces of creation where the human intellect is put to test."

ER Kit Features

There are several ER kits available in the market. The authors have identified five features that a good educational robotic kit should have:

1. It should be easy to assemble and disassemble the prototype during its design.

Some ER kits provide snap-together assembly method, in which there is no need of an additional tool such as screwdriver to assembly the pieces.

For example, LEGO education bricks can be connected through their patented stud-and-tube coupling which consists of studs on top of the bricks and tubes underneath the bricks making a rigid, precise, and easy to assemble junction.(Lauwaert, 2008)

For young students, the pieces of the ER kit should not have sharp points so they can be manipulated without any problem.

2. It should be modular, quickly modifiable.

A modular construction system must contain different pieces in terms of size and types, such as: bricks, rods, beams, axles, gears, plates, hinges, etc) and these pieces can be assembled in many different ways.

Modularity allows the development of different robot models using the same number of pieces, simply assembling them in a different way. For example, using the same set of pieces you can build different robotics models such as bipeds, quadrupeds, humanoids, with different motions such as walking, crawling and jumping.(Galvez-Aranda & Galvez Legua, 2021a, 2021b, 2021c, 2021d, 2021e)

Another example of modularity in a model would be the 6 LEGO bricks combination problem. The number of different ways to combine six 2 × 4 LEGO bricks of the same color is 915 103 765. (Abrahamsen & Eilers, 2011)

3. It should allow the implementation of intelligent prototypes by employing a CPU-based control system. Nowadays every electronic equipment has a CPU, which provides:
    a. Computational intelligence: based on a microprocessor or microcontroller.
    b. Programmable: the "behavior" of the prototype depends on a program. This gives the prototype more flexibility.
    c. Communication capacity: the prototype can communicate with other equipment to transmit or receive information.
4. You must have a powerful programming language that allows you to program relatively complex algorithms.

There are two types of programming languages: textual and iconographic. Ichnographic languages have the advantage of using icons as representation of instructions and thus eliminating the complexity of the syntax that all textual programming languages have, in addition to that their learning curve is faster. The advantage of textual languages is that they generate smaller programs than iconographic languages when compiled, meaning less consumption of RAM memory space on the CPU.

Examples of iconographic programming languages,(Plaza et al., 2020) we have: Scratch, LabView and NTX-G. Examples of textual programming languages we have: Java, C, python, etc..(Conversy, 2014)

Programming languages are integrated into a development environment that allows the creation of programs, called the Integrated Development Environment (IDE).

5. It should be easy to document:

Documentation of the robotic model is important, to replicate it can be improved or used to the design another robotic model. Several 3D software can be found online to create a virtual version of your robotic model, such as MLCAD,(Clague & Agullo, 2003) Sketchup,(Murdock, 2009) Lego Digital Designer (LDD),(Borrego & Gómez) Fritzing,(Knörig et al., 2009) etc.

Table 1 lists some of the different ER kits along with platforms such as Arduino, Raspberry PI, and CompactRIO, which do not meet all the features but can eventually be used in conjunction with another educational robotic kit to meet all 5 features. For example, the Arduino board only meets 3 out of 5 features: control system, programming, and documentation. However, by adding building materials to the Arduino board, you can get a complete kit that meets all 5 characteristics. Building materials can come from another ER kit or can even be generated using manufacturing techniques like 3D printing.

5 Phases of ER (5PER)

5PER is a methodology created by the authors for the design and implementation of prototypes through the use of ER and is based on the traditional scientific method. In this section authors explain the most relevant information about the 5PER methodology, however, more information about the development and the theory behind this methodology and their phases can be found in the corresponding chapter: 5 Phases of ER (5PER) of the present book.

In the 5PER methodology, the design and implementation of a prototype is divided into 5 phases, which are connected in a closed-loop structure. In Figure 1, each of these 5 phases are represented in a diagram showing the closed-loop connection between phase, each phase is an activity that define a manual or cognitive action in the execution of ER.

1. **Design Phase:** The idea and its representation based on the need to solve a problem. It will lead to the design and development of a prototype, mockup or model. The initial design could be inspired by an example from reality (imitation) or using the student own imagination to create something new (creation).
2. **Build Phase:** Based on the design developed in the previous phase, now is the moment to bring that idea to the real world by building it, using parts, connectors, sensors, actuators, and connections. At the end of this phase, students should have a prototype based on their idea elaborated in the design phase.

Table 1. Feature analysis of different educational robotic kits

	Easy to Assemble	Modular	CPU-based Control System	Programming Language	Documentation Software
LEGO Mindstorms NXT	Yes	Yes	Yes	Yes	Yes
LEGO Mindstorms EV3	Yes	Yes	Yes	Yes	Yes
LEGO SPIKE PRIME	Yes	Yes	Yes	Yes	Yes
VEX IQ	Yes	Yes	Yes	Yes	Yes
TETRIX	Yes	Yes	No	Yes	Yes
mBOT Makeblock	Yes	No	Yes	Yes	No
LEGO WeDo	Yes	Yes	requires to be connected to a PC	Yes	Yes
CompactRIO	No (Does not come with any mechanical part)	No	Yes	Yes	Yes
Arduino	No (Does not come with any mechanical part)	No	Yes	Yes	Yes
Raspberry PI	No (Does not come with any mechanical part)	No	Yes	Yes	Yes

3. **Program Phase:** using a programming language, students can, for example, activate motors and generate movements in the prototype. Creativity is a key characteristic in this phase, as students think of a possible solution to solve a problem, and translate that solution into an unambiguous, finite and ordered sequence of tasks (instructions) that the prototype will perform.
4. **Test Phase:** The behavior of the prototype to perform the scheduled task is verified by visual inspections. Any potential problem is identified in this phase. Depending on the problem nature, student must go back to the corresponding phase to fix it. For example, if there is a mechanical problem with your prototype while performing the tasks, such as some pieces drop or disassembly, that is an issue that needs to be checked in the Build Phase. If the issue is about the programmed tasks, which are not behaving as expected, student must go back to the Program Phase to solve it.

Figure 1. The Five Phases of ER (5PER)

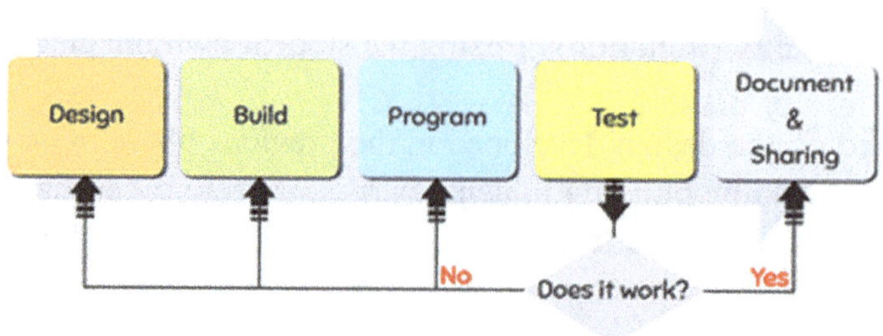

Finally, the prototype can be mechanically stable and the scheduled task performs as expected, however, the problem is not solved. In this scenario, the prototype design is not adequate to solve the problem and another idea must be implemented.

In this phase the prototype must be tested for its mechanical strength and compliance with the design criteria. All possible failure cases must be anticipated, otherwise we expose our prototype in the event of an unforeseen event.

5. **Document and Share Phase:** Once the prototype has been tested and works accordingly to the design criteria, the documentation of the entire process is carried out. This can be done in a number of ways: freehand drawing, word processor, graphics editor, specialized software to document a robotic model, such as CAD or 3D software. Documentation is important because by sharing our work the knowledge is spread.

Robotic Competitions

There are several robotics competitions internationally. Most of the competitions are associated with the use of a certain brand of robotics kit. Among the most recognized competitions we can mention the World Robot Olympiad (WRO), which is an international robotics competition organized in more than 85 countries around the world and teams can participate in four different categories with different age groups from 8 to 19 years. Similarly, FIRST is another international team competition with 3 categories: FIRST LEGO League, FIRST Tech Challenge and FIRST robotics competitions, involving students from 4 to18 years old. VEX Robotics Competition is a world robot championship in which team of students develop a robot to play against other teams in a game-based engineering challenge.

There are also regional competitions that involve several countries in the same region, such as the LARC, in which students from universities throughout the latin american region participate. All these competitions have one feature in common: the challenges. Participants must develop a working robot which can perform a series of task in order to solve the proposed challenges. The competition winner is usually determined based on the number of challenges a robot can successfully solve, in case of a tie, the time will be the tie breaker. Other competition mode is based on a direct match, in which two teams compete in the same arena at the same time and based on a scoring system a winner is declared after an amount of time. Several case studies were reported in high school and university students that have participated in robotic competitions, showing a positive impact, improving student capabilities and interest in science.(Chiang, Liu, Feng, Zhuang, & Sun, 2020; Pisciotta, Vello, Bordo, & Morgavi, 2010; Sigüenza, Guillen, Arroyo, & Cuellar, 2017; Welch, 2010)

Several robotic competitions are listed below:

1. **World Robot Olympiad (WRO):** It is a worldwide competition where Lego robotics kits are used. Currently, it has four categories. For example, the first category is about a challenge that the robot must solve. Additionally, a surprise rule is announced the day of the competition, which requires to perform modifications to the robot or the program.(Chiang et al., 2020)
2. **FIRST Challenges:** It is a worldwide competition where the participating teams represent their countries. Participants are between 14 and 18 years old. Competition consists of solving a challenge based on a real-world scientific theme. It is closely related to STEM.(Chris, 2013)

3. **VEX Robotics Competition:** Two teams compete in a arena simultaneously. Robots form both teams are controlled remotely. It is related to STEM. VEX robotics kits should be used.(Caro, 2011)
4. **Latin American Robotic Competition (LARC):** It is a robotics event sponsored by the IEEE, RAS section (Society for Robotics and Automation) at the Latin American level. You have different types of competitions that take place in several days in a certain Latin American country. It is not subject to the use of a specific robotics kit.(Norouzi, Fattahi, & Massah)
5. **Local Robotics Competition:** These are competitions that take place within a country. Usually, a university or a school host this kind of competition and they are open to the community.

Robotic competitions can be classified in various ways. For example:

- Due to how the robots compete:
    - Direct match mode (Figure 2): In this type of competition the teams face each other to determine a winner. Example: Sumo Robots, two robots face each other within a defined circular arena to try to get each other out of that arena.
    - Time-Score match mode (Figure 3): In this type of competition, each team competes individually and the main criterion to define the winner is the time the robot took to complete the challenge. Example: Line follower, the robot must follow a line drawn on a track and travel it in the shortest time possible.
- Due to how the robot is controlled:
    - Autonomous: No participant in the group can directly or remotely control the robot. The robot must have been programmed all the algorithms that allow it to carry out the challenge without human intervention.
    - Guided: One or more participants control the robot either directly or remotely.

Figure 2. Direct Match challenge. Two or more team robots participated simultaneously in the competition arena. From left to right the photos correspond to teams participating in LARC 2013, a Local ZUMOBOT competition in 2013 and LARC soccer-bot 2013.

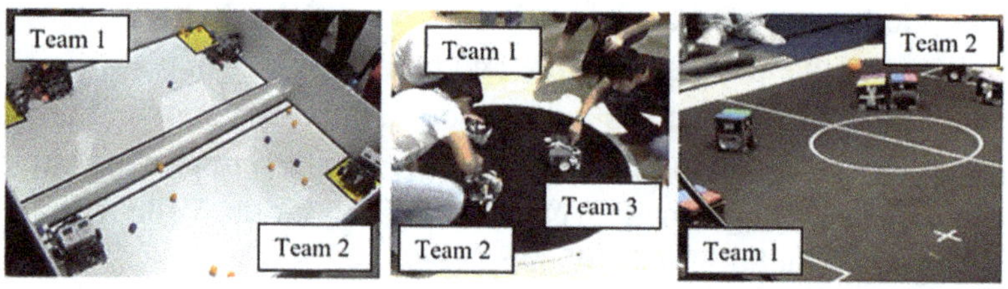

ROBOTIC COMPETITON: THE METHODOLOGY

In this section the authors explain the methodology developed and applied by them in the formation and preparation of teams to participate in educational robotics competitions. The methodology covers the stage of recruitment, training, team conformation and preparation, prototyping and competition day (Figure 4). For each stage, its requirements, the evaluation criteria, and the variables that will be considered to measure its success are specified.

Figure 3. Time-Score match mode. Only one team robot participates in the competition arena. The first goal is to accomplish the requested challenges and do it in the least time possible. From left to right the photos correspond to teams participating in WRO 2012, WRO 2013 and WRO 2015.

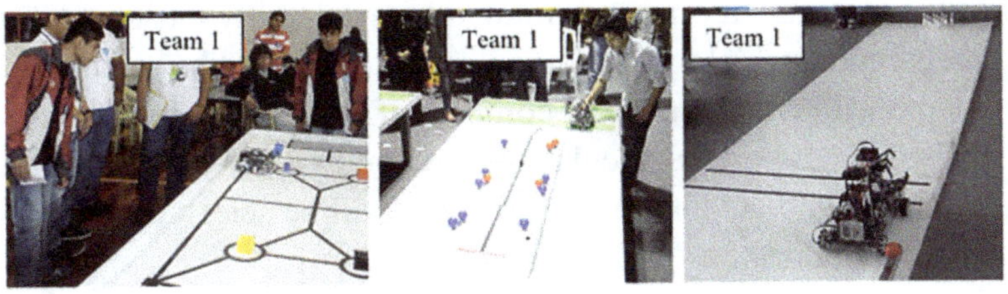

Figure 4. Methodology for the formation and preparation of teams for RE competencies

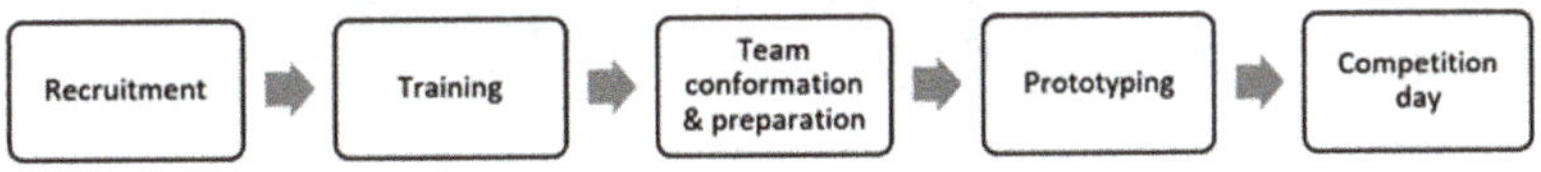

FIRST STAGE: RECRUITMENT

The first stage is a student selection process, which based on certain previously defined selection criteria, will be selected to be trained in competitive robotics.

Goals

- Select students who have a certain profile to participate as members of a competition team in educational robotics.

Selection Criteria

The selection criteria are based on the student profile that is required for an educational robotics competition. Attitudinal competencies are privileged, which, according to the authors' experience, is the most valued in competitive environments. The criteria are:

- Time availability: Regardless of the educational robotics competition in which one is going to participate, time is an important factor, since in many cases the design and testing activities require a lot of time for their development, in some cases arriving at dawn.
- Sports practice: Students who play sports usually develop a competitive character, discipline, perseverance and a great self-esteem. They know that to win you have to prepare and that all success includes sacrifice. It also helps to develop a leader profile.
- Art practice: Art is synonymous of creativity, it allows to develop imagination, increases concentration, improves motor coordination, and self-esteem.

- Robotics experience: Includes past experience handling robotics kits and participation in competitions.

Depending on the type of competition, a minimum time availability must be defined that the student must assign to their preparation, which includes hours of training and hours for the design and implementation of the prototype with which they will compete. If the student does not have that availability, he/she will not be selected. The score is calculated based on two values: the geometric mean and the arithmetic sum. The students who obtain the highest score will be selected. Table 2 shows an example of the scoring of 4 different students.

Table 2. Evaluation of the availability of time, skills and experience of each student

Student	Time Availability - TA (Hours)	Sports Practice - SP (Years)	Art Practice - AP (Years)	Robotics Experience -RE (Years)	Score (Geometric Mean + Arithmetic Sum)
Student 1	8	4	3	2	30.85=13.85+17
Student 2	6	4	4	1	24.79=9.79+15
Student 3	7	6	1	2	25.16=9.16+16
Student 4	5	0	3	0	8=0+8

SECOND STAGE: TRAINING

The second stage is the training process in robotics issues. A workshop is developed where knowledge of robotics and the evaluation of each student regarding their abilities to integrate a competitive team are taught.

Goals

- Discover each student's skills in creativity (design), construction, programming, testing, leadership, and teamwork.
- Train the students in the development of prototype based on the use of ER kits.
- Understand how a robot works and describe its main parts.
- Develop psychomotor skills in handling the building materials.
- Know how to use the electronic components such as sensors and actuators to provide perception (sensors) and action (motors) functions to the prototype.
- Develop programs that implement various actions (tasks) performed by the prototype.
- Differentiate the types of mechanisms and use them in the construction of prototype.
- Check the performance of the prototype.

Teacher Tasks

- Structure the content of the training session, using the robotics kits.

- Support students in their learning process, through the support of material (manuals, guides, graphics, and animations).
- Encourage high and respectful participation among the students of the workshop.
- Evaluate student learning during all workshop sessions.

Recommendations Before Each Session: Employment and Safety Rules in the Use of ER Kits

These rules must be noticeably clear for everybody prior the beginning of the training sessions, following them will allow a better development of the activities involving the use of an educational robotic kit during the initial training session.

1. Regarding the building materials
   a. The use of trays to store the building materials is highly recommendable. Always keep the building material in their trays while they are not being used. If a tray does not come with the building material, please provide one at the beginning of the session. A tray is useful to store the materials decreasing the chances to a potential loss of pieces.
   b. The work session must begin by assigning a kit to a group of students, and the teacher designate a team member as the person responsible for the ER kit.
   c. Each session begins with a physical inventory of all the components of the kit. This inventory must be also performed at the end of the session.
   d. The control system (intelligent brick or other microcontroller) is the main component of the ER kit; therefore, its handling must be careful and only take it out from the tray at the time of use.
2. Regarding the electronic components: motors, sensors, cables, and NXT programmable brick
   a. These components can be cleaned with a dry cloth.
   b. They should be prevented from coming into contact with liquid material.
   c. They must be operated with care, preventing them to suffer strong blows (falling for example), as this could cause damage. Unused components must be kept on the tray.
   d. Be careful while managing the electronic connections, cables must not be pulled drastically.
   e. If the ER kit uses a rechargeable battery to operate, at the beginning of the session, always check that the battery is fully charged. If the device uses regular batteries such as AA, just have extra batteries ready to go if during the session the ones being used get discharged.

Workshop Features

The course should emphasize:

- Creativity: The challenges that are presented in the educational robotics competitions require that students have a lot of creativity to generate ideas that are then "translated" into designs to solve the proposed challenges.
- Construction: Due to the experience of the authors, it is one of the weak points in the students, many of them have no experience handling an ER kit, nor in the assembly of parts. Concepts of

simple machines and mechanisms are very important to provide the prototype with movement and displacement, required in educational robotics competitions.

- Programming: It is another of the weak points in the students and deserves more time in the training. Students are required to master "hardware-oriented programming", that is, make programs that allow a certain hardware (prototype) to display a desired behavior (depending on the challenge). Example: A bit is simply a one element of a byte, which can represent a quantity (from 0 to 255) or a character (A, a). In a ER environment, a bit can mean the activation of a motor (ON) or its shutdown (OFF).

Workshop Development

The workshop is divided into two parts:

First part: Concepts of simple machines, mechanisms, and assembly criteria. Activities where the student tests their creativity and constructive skills are emphasized. This first part is done individually (one ER kit per student) and attitudinal skills such as leadership are explored. All activities must be completed by each of the students, this is done because one of the fundamental requirements in every member of an educational robotics team is perseverance, not giving up in the face of adversity.

In the examples shown, the LEGO WeDo 1.0 ER kit (158 pieces) is used, showing some basic activities that seek to measure the level of creativity that each of the students have. The time it takes to perform each of the activities is measured.

- Example 1: build the decimal digits of the same size: length and width (Figure 5a).
- Example 2: build geometric figures such as triangles, quadrilaterals, trapezoids, rhombuses, etc (Figure 5b).
- Example 3: build a cube whose sides should try to be the same size (Figure 5c).

Guided construction activities and basic programming are then carried out, where students:

- Students learn various mechanisms to generate movement such as: linkages, motion and energy transformation.
- Students compete by demonstrating leadership skills, continuous improvement, and respect for their peers.

The duration of the first part depends on the design activities that the teacher has developed. An average of 10 activities is recommended, which can be divided into two sessions of 3 hours each.

At the end of the first part of the workshop, considering that 10 design activities were carried out, the Table 3 shows and example of the scoring assigned to each student. It is recommended that all students complete all activities, regardless of the time they take.

"As important as winning, is to finish what you do, this allows later to make an analysis of continuous improvement"

Second part: Student teams are formed (the number of members depends on the rules of the ER competition to which they are going to participate). The members of each team are selected based on

Figure 5. (a) The decimal digits from 0 to 9, (b) geometric figures and (c) Building a cube.

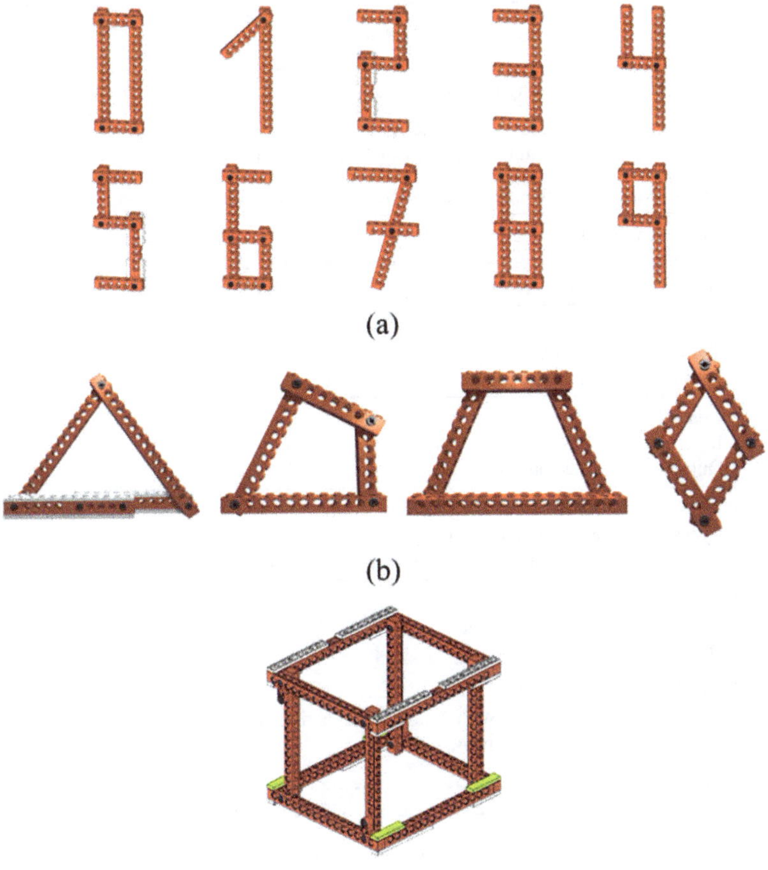

(a) (b) (c)

the score obtained in the first part of the training. A balance must be found and combining high-scoring students with low-scoring students. We must be clear that the main reason for participating in a ER competition is not to win an award, but to strengthen the teamwork skills, perseverance, instill values such as friendship, tolerance, in short, to train better people. This second part of the workshop focuses on developing skills in the programming and construction of prototypes, as well as teamwork. Although it is true that the teacher forms the teams, the assignment of roles to be fulfilled by each of the members is recommended that they decide themselves. They must be given a certain degree of autonomy so that the members of each team have a space to develop their leadership, negotiation skills and tolerance (Figure 6).

Table 3. Evaluation of the number of activities solved and time of implementation of each student.

Student	Number of Activities Solved - N	Time to Solve the Activities – t (Hours)	Score (N/t)
Student 1	10	1.8	10/1.8 =5.55
Student 2	10	2.1	10/2.1 = 4.76
Student 3	10	2.4	10/2.4 = 4.16

Table 4. Example of a training session table of content

Main Topic	Specific Topics	Theory (Hours)	Lab (Hours)
Mechanic principles	Recognizing the building materials: beams, bricks, plates, axles, gears, etc.	0.5	0.5
	Transforming the motion: mechanisms (Chebyshev, Slider-crank)	0.5	1.0
	Transferring the motion: train gear, worm gear, bevel gears	0.5	1.0
Programming principles	Understanding the programing software Defining concepts: algorithm, flowchart, pseudocode, and program. Basic programming examples: logic gates (AND and OR)	1.0	1.0
	Programming loops Conditional Programming Parallel programming	1.0	1.5
Electronic principles	Control System Input devices (sensor) Output devices (actuators)	0.5	
	Working with sensors Types of sensors: digital and analog Leading with noise while reading the sensors signals	0.5	2.0
	Working with actuators Motor specifications: torque and rotation speed ON/OFF motor control Motor ramp starting/stopping	0.5	2.0
Application: Rover	Building of a simple rover with 2 motors for motion and the use of sensor to interact with the environment. Flowchart design		2.0
	Avoiding obstacles using distance sensor, touch sensor or light sensors.		2.0
	Development and programming of a basic line follower robot.		2.0
	Development and programming of a basic sumo robot		2.0
Application: Grabbing objects	Test of different robot arm mechanisms to grab objects such as cans		2.0
	Mounting the robot arm into a rover, and grab a can located in a particular distance/position from the robot using the sensors.		2.0
	Move a can from one initial location to other using the robot model.		2.0
Application: Sensor Calibration	Change the environment conditions such as the indoor lights. Calibrate the sensors accordingly. Test the performance of the sensors and overall performance of the rover at the different environment conditions.		2.0

Table 4 shows an example of a training session table of content. In the example, 6 big topics have been defined. The first three: mechanical, programming, and electronic principles are focused on defining the basic theory of the different components of a robot model. Theory and lab session are divided in each of these three initial main topics. The last three main topic are focused on application, putting in practice the acquired knowledge from the first three main topics. The applications main topics are oriented to prepare the students to a robotic competition environment. The total amount of hours in this example is 30 hours, and these can be divided in sessions of 3 hours. During each session, the teacher is evaluating cognitive, procedural, and fundamentally attitudinal skills of each student. Each session can be divided by theory and Laboratory. During the theory, the teacher explains the concepts corresponding

Figure 6. Students of the FIEE-UNI during the workshop activities

to an specific topic. During the laboratory, students through hands-on session experiment and apply the learned concepts during the theory session.

At this stage is where the dropouts of some students usually occur, who for study reasons (exams, work) or personal issues, no longer continue. It is especially important to take attendance at each session. During the training sessions, the teacher should not only focus on fulfilling the activities assigned in each session, but also on how each of the team members behave and interact during the development of the activities and, above all, determine the skills of each of them based on the roles that have been assigned. At the end of the training, the teacher has a table with the skills that he has measured during the training of each student. This information will serve for the formation of teams that will participate in the RE competition (Table 5).

Table 5. Assessment of specific skills of students.

Group	Student	Hard Skills			Soft Skills		Score
		Design Skills	Construction Skills	Programming Skills	Leader Skills	Teamwork Skills	
I	Student 1	2	1	1	0	1	5
	Student 2	1	2	2	1	2	8
	Student 3	1	1	2	2	2	8
II	Student 4						
	Student 5						
	Student 6						

The values in each column range from 0 (no skill), 1 (low skill), 1 (intermediate skill), and 2 (very good skill)

THIRD STAGE: TEAM CONFORMATION AND PREPARATION

From the skills measured to the students in the previous stage, the teacher forms homogeneous teams, where each of its members stands out in some of the competencies required for the competition, for example in programming, construction, etc.

"A good team is one that exhibits a combination of specific and soft skills"

Goals

- Formation of solid and balanced teams in terms of the abilities of each member.

Team Conformation

With the information obtained in the previous stages, the teacher forms the teams with the number of members that is specified in the rules of the competition in which they will participate. Ideally, a team should consist of student with different but complementary roles. For example:

- The designer: person with skills to generate ideas contextualized to the problem to be solved.
- The builder: skills for the construction of mechanisms and knowledge about the various building materials the robotics kit has.
- The electronic: skills in the operation of sensors, actuators. For example: for certain applications, a push-button sensor is better than infrared because the latter depends on the type of surface and the light in the environment.
- The programmer: skills for the development of algorithms and programming.
- The tester: skills in recognizing what is malfunctioning. Together with the programmer they can detect if the problem is hardware or software.

The teacher must ensure that each team is made up of students who have, together, the roles mentioned above. Depending on the complexity of the robotic competition challenge, a team member can have more than one of the roles listed above. Even in the case that one student has one role, that does not mean that a student from a different role cannot provide ideas or solutions into a field out from his role.

Team Preparation

During the teams' preparation, the evaluation is constant. Evaluation of team performance should not be performed exclusively regarding the ability in the development of a robot to solve the specific challenge, but also about how the team work together solving the different problems that could appear while developing the robot.

The teacher poses a series of challenges to measure how individual skills complement each other to face a problem. It is about evaluating the strength of the team in activities under pressure. The teacher must try to take the challenge requirement to the limit so that it can be seen if communication problems, stress or lack of leadership are generated to face the problems.

If problems are detected regarding the lack of some of the required skills, the teacher can make changes of team members, seeking balance.

Team Evaluation

The evaluation must be focused on the team, for this it is important to be clear about the differences between team and group showed in Table 6.

Table 6. Differences between team and group.

Group	Team
Tasks are shared equally.	Tasks are divided according to skills and abilities.
You compete, you seek to stand out. Individual commitment.	You cooperate, you seek to belong. Collective commitment
A single leader defines the actions to be carried out. Passive attitude.	Diverse leaders who suggest, debate and look for the best alternatives. Proactive attitude.
It seeks not to lose. Evaluation focused on the result.	It seeks to learn and improve. Evaluation focused on the process and the result.
It is discussed, it is imposed.	It is debated, it is convinced.

Based on the Group-Team differences, Table 7 is prepared to evaluate the members of each team.

Table 7. Evaluation of the teamwork of each member.

<table>
<tr><th>Team</th><th>Member</th><th>Collaborate With Your Colleagues</th><th>Debate With Respect</th><th>Exhibits Leadership</th><th>Score</th></tr>
<tr><td rowspan="3">I</td><td>Student 1</td><td>1</td><td>1</td><td>1</td><td rowspan="3">5</td></tr>
<tr><td>Student 2</td><td>1</td><td>0</td><td>0</td></tr>
<tr><td>Student 3</td><td>0</td><td>1</td><td>0</td></tr>
<tr><td rowspan="3">II</td><td>Student 4</td><td>1</td><td>1</td><td>1</td><td rowspan="3">8</td></tr>
<tr><td>Student 5</td><td>1</td><td>1</td><td>0</td></tr>
<tr><td>Student 6</td><td>1</td><td>1</td><td>1</td></tr>
</table>

The values entered in each column must be binary values.

In Table 7, it is appreciated that Team II obtains a higher score than Team I, this is due to the strengths of each member and their contribution to teamwork. The authors call this, degree of cohesion, and it is of vital importance in work situations under pressure that are characteristic of ER competencies.

Figure 7. Results of the team cohesion test.

Team	1	1	1	2	2	2	3	3	3	4	4	4	5	5	5	6	6	6
1		1	-1		1			-1										
1	1					1		-1	-1									
1								-1	1									
2			-1		1			-1								1		
2				1		1		-1	-1									
2				1	1			-1								-1		
3																		
3			-1			-1			1			1						
3				1	-1		1									-1		
4											1	1			-1	-1		
4			-1	-1						1		1						
4										1			-1		-1		1	
5						-1						-1		1	1			
5													1		1	-1		
5						-1							1	1			-1	
6						-1			-1						1			1
6											-1		-1			1		1
6						-1						-1				1	1	
	1	1	-4	2	2	-3	1	-6	-1	2	0	1	0	2	1	-1	1	2

"A team may have technical weaknesses, but it should NEVER stop being a team."

To corroborate the results obtained in Table 7, authors present an example of a quick test to evaluate how well the team members get along with which other: team cohesion test. This test can be done even in the second stage. Provides valuable information to verify the strength of the group or take actions that can lead to the reassignment of team members. In authors' example, there are 6 teams, and each team is composed by 3 students. For the test, the teacher gives a small piece of paper to each student. In that small piece of paper, the student should write:

- Two names of students with whom they would like to make a new team.
- Two names of students with whom you would not like to make a new team.

Figure 7 show the results of the test. Each student that appears in the paper containing with whom they would like to make a new team receive a 1 point. Each student that appears in the paper containing with whom they would not like to make a new team receive a -1 point.

Table 8. Evaluation of the degree of cohesion of each team.

Team	Score	Characteristic	Action to Execute
5	6	Strong cohesion	None
2	5	cohesion	None
4	5	cohesion	None
6	5	cohesion	None
1	2	No cohesion	Rearrange the team
3	2	No cohesion	Rearrange the team

The total score obtained by each student (see the last row in Figure 7) indicates the level of acceptance that student has in front of the others.

Regarding the authors example, it is worrying that a student gets a total score of -6, meaning that there are 6 people who would not like to have him as a member of their team. There is also a student with a score of - 4 and another with - 3. These 3 students must be observed in a special way to see their development in future activities. We can see that the most "solid" group is group 5, where each of its members has chosen their two teammates to form a "new team", showing that there is a strong relationship between them and that they appreciate each other. In Table 8, are the results of the test and the corresponding action to execute.

The teacher must ensure that all teams have an acceptable degree of cohesion to start the next state. The following evaluation is about the results obtained by each team in the process of solving the challenges posed. It is based on the number of challenges solved and the time spent on them. In Table 9, 5 challenges are posed to the teams.

Table 9. Evaluation in the resolution of challenges by team

Team	Number of Challenges - N	Time to Solve the Challenges – t (Hours)	Score (N/t)
I	5	4.2	5/4.2 =1.19
II	5	3.8	5/3.8 = 1.32
III	5	5.2	5/5.2 = 0.96

FOURTH STAGE: PROTOTYPING

This is one of the most complicated stages, since a prototype that meets the requirements of the challenge must be designed. The requirements must be clearly understood and there should not be any doubt regarding them. At this stage, there is used a constant trial-and-error methodology, that is why the motivation that the teacher must give his students is especially important, since there will be groups that will go faster than others and that is good because it will serve so all the other teams will be dealing with the pressure, that they will live when competing for real: emotional intelligence.

Figure 8. Different types of robotic competition arena. From left to right: Flat arena used in line follower competitions, Sand surface arena used to emulate trash cans collection in a beach, and a Embossed arena containing different elements such as ladders.

Goals

- Discover the skills of each student in the design, construction, programming, testing, and documenting of a prototype (5PER).
- Support students in the process of designing and implementing a prototype to solve the specific robotic competition challenge.
- Support in the definition of strategies to solve the corresponding challenge of the robotic competition.
- Promote in students systemic thinking, teamwork, tolerance, public speaking, etc.
- Enhance in students the skills for the design and construction of prototypes by understanding various mechanisms.
- Promote research and problem-solving skills.
- Document the final prototype design so it can be published and shared among peers.

Robotic Competition Arena

Emulate the condition and the environment that the robots will face the day of the competition. If possible, build an arena similar to the one in which the teams will compete.

- The arena can be flat such as the one used in a regular line follower.
- The arena can be a single flat piece such as the arena used in sumo robot competitions.
- The arena can be a more complex surface such as: a sand arena, or an arena that involves embossed elements such as ladders.

Figure 8 show some examples of the different types of robotic competition arenas.

5 Phases of ER Applied in Robotic Competition

For the design of the prototype, we are going to use the methodology called: The Five Phases of ER (5PER), which is based on the traditional scientific method. 5PER are a set of sequential and relatively independent activities that define a manual and/or intellectual action in the use of ER.

We use the word prototype to refer to any build that may or may not have a CPU. Additionally, because it refers to something that is not finished yet, that is being improved until reaching the final product.

Figure 9. Teams during (a) Phase I, (b) Phase II and (c) Phase III and Phase IV.

(a) (b) (c)

- Phase I - Design: The requirements of the challenge must be considered to define a strategy to compete. For example, if on the day of the competition the robot must be assembled from "scratch" then it is advisable that the prototype be modular, so that each member of the team can assemble a part and thereby gain time or ensure that the assembly does not have errors because it is easier to remember (Figure 9a).
- Phase II - Build: The prototype construction should be as simple as possible. The greater the number of pieces, the greater the number of possible errors that can occur. The prototype must facilitate redesign. The prototype must be mechanically reliable (Figure 9b).
- Phase III - Program: The programming must make use of constants and variables that facilitate the change of the behavior of the prototype. It is also recommended to use functions that implement functionalities of the prototype and that can be quickly improved or modified according to the tests that are being carried out.
- Phase IV - Test: The prototype must be tested, and the team must observe if the robot meets all the requirements of the challenge. It is recommended to verify that mechanically everything works well, this will be achieved by testing the prototype several times. For example, if the prototype has wheels, they start to come off after various tests. That means that part of the construction is not reliable and must be corrected (Figure 9c).
- Phase V – Documents & Sharing: Once the prototype passes all the tests, it is documented using a CAD program. This is necessary in some educational robotics competitions, where, as part of the competition, the team must make a presentation of their prototype.

As an example of the application of the 5PER methodology, authors show the evolution of the prototype used in the WRO 2012 competition (Table 10). The goal was to reduce the travel time of the prototype by meeting the requirements of the challenge but keeping its mechanical reliability. Table 10 shows an example of continuous improvement evaluation of the prototype designed by a team during the prototyping stage, showing how the time to solve the challenge is improved.

FIFTH STAGE: COMPETITION DAY

General Recommendations

Table 10. Evaluation in continuous improvement by each team.

Team I	Prototype I	Prototype II	Prototype III	Prototype IV (Final version)
Model				
Time to solve the challenge	Did not work due to mechanical issues	72 seconds	52 seconds	37 seconds
Reliability	Low	High	Average	High
Design	Monolithic	Modular	Modular	Modular
Number of sensors	1	4	4	4
Number of motors	3	3	3	3

- All the members of the teams concentrate in a specific place and together they move to the place of the competition; arriving together at the place of competition, provides confidence, helps to minimize the natural fear of competing.
- The presence of the teacher is essential on the day of the competition: he is the person who has accompanied each team from the beginning and has guided them until that day. The teacher is the mentor.
- It is important that the teams have the presence of fans in the venue where the competition will take place; Educational robotics competencies are technical and psychological competencies.
- In many cases, educational robotics competitions are direct elimination. If you are participating with several teams, the teams that are eliminated must remain until the end, supporting the teams that continue to compete; This allows to strengthen the bonds of friendship and respect between peers.

Recommendations for the Teams

- Bring additional mechanical and electronic parts in case of failure.
- If your robot uses rechargeable batteries, bring at least one extra, and check that both batteries are fully charged at the beginning of the competition. If your robot uses non-rechargeable batteries, just bring extra batteries.
- Laptops used to program robots must have their batteries 100% charged. You must bring your battery chargers and extensions.
- Have backup copies of robot programs.

On the day of competition, it is recommended to evaluate the performance of each team, its strengths and weaknesses. This allows to propose improvement actions for future competences. (Table 11)

Table 11. Team evaluation on the day of the competition

Team	Member	Leadership	Teamwork	Communication	Strengths	Weakness	Final Ranking
I	Student 1	2	2	2	Communication	They argued a lot	
	Student 2	1	2	2			
	Student 3	1	1	1			
II	Student 4						
	Student 5						
	Student 6						

EXPERIENCE IN UNIVERSITY STUDENTS

The methodology presented was tested at the Faculty of Electrical and Electronic Engineering (FIEE) of the National University of Engineering (UNI), Lima-Peru.

First year undergraduate students from that faculty participated in three editions of the World Robotics Olympiad (WRO) national version 2012, 2013, 2014 and Latin American Robotics Competition (LARC) 2013.

Programming courses are taught in the second year of studies, so at the time of the competition the students had no previous programming experience. As a result, the students obtained the first place in the WRO (national version) 2012, 2013 and 2014. In WRO 2012 the university presented 3 teams for the competition obtaining the first, second and sixth place. In WRO 2013 the university presented 6 teams, obtaining the first, third, fourth, fifth, sixth and tenth place. In WRO 2014 the university presented 3 teams, obtaining the first, third and sixth place. In addition, the university submitted a team for the 2013 LARC competition and placed fourth (Figure 10).

Figure 10. (a) FIEE-UNI students participating in LARC (b) FIEE-UNI students and Prof. Mauricio P. Galvez-Legua with 4 of the 5 trophies obtained in WRO 2012 and 2013.

(a)

(b)

The participation of students in the Educational Robotics competitions has allowed to promote a greater interest in engineering careers, showing the relationship that exists between robotics and the subjects they study such as: digital electronics, microprocessors, microcontrollers, control systems and programming (cognitive skills). It has also allowed them to potentiate their manual and attitudinal skills such as teamwork, work under pressure, tolerance, public speaking, etc.

CONCLUSION

Authors present a methodology composed of 5 stages to prepare a team of student that will participate in any robotic competition. The 5 stages are: (1) Recruitment, (2) Training, (3) Team conformation & preparation, (4) Prototyping, and (5) Competition Day. The methodology in the preparation of teams for robotic competitions is based on the selection of the members of each team based on their specific abilities (design, construction and programming), their general abilities (respect, camaraderie, leadership, teamwork) and the degree of cohesion that is formed during the preparation stage. The latter is, in the authors' opinion, the most important factor in the formation of a competitive team.

The use of the 5PER methodology helps the members of a team to clearly define the phases that must be developed in the creation of a robotic prototype (Stage 4), it helps them to assign tasks to each member based on their specific abilities. Evaluations throughout the team preparation process is extremely important to ensure a solid and balanced team formation in terms of the skills of each member. Authors have recommended a series of evaluations with certain characteristics that are summarized in Figure 11.

In the first stage, recruitment, the workshop has an eminently practical development, which arouses great interest in the students. Normally in the first sessions, a lack of psychomotor skills in the handling of construction materials can be noticed, which is largely overcome with the following training sessions. Most students who participate in the preparation of teams for robotics competitions have not worked with educational robotics kits, however they quickly come to master their use, thereby demonstrating the great potential that robotics has as a teaching medium.

As a result of the application of this methodology, the educational institution obtained the first places in the national edition WRO 2012, 2013 and 2014, as well as a fourth place in the Latin American Robotics Council (LARC) 2013. The achievements obtained were the result of work planned by the educational institution, teachers and the enthusiasm of the students who participated. All the students were from the first year of engineering baccalaureate, which reaffirmed themselves in the choice of their respective careers.

RECOMMENDATION

For a successful implementation of ER in an educational institution and future participation in a robotic competition, the authors recommend:

- Mechanical ability is one of the most difficult to find in students, going from design to implementation presupposes knowledge of simple machines and relatively complex mechanisms that most students do not know. Training should emphasize this point to ensure that: "good ideas get to be implemented."

Figure 11. Methodology in the preparation of teams to participate in robotic competitions

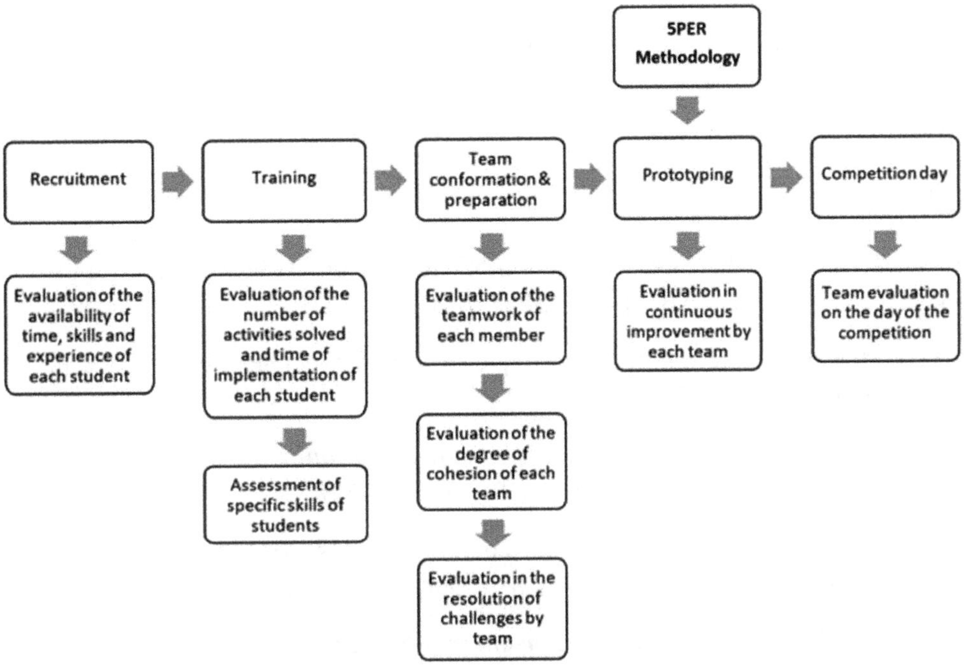

- The prototype was programmed using an iconographic programming language, however certain shortcomings were noted when developing the programs. It is necessary that the programming skill be intensified through the development of specific activities.
- Complement the training of students with recreational and artistic activities that help improve their procedural and attitudinal skills.
- Los talleres de ER deben desarrollarse en las instituciones educativa para que los estudiantes puedan:
  - Apply the knowledge learned in such as: programming, electronic devices, control, sensors, etc.
  - Acquire procedural skills (assembling things) that enhance the management of equipment in the industry and activities such as welding, screwing, etc.
  - Learn how to work in teams, which allows fostering communication with respect among peers and tolerance.
- Implement robotics laboratories with educational robotics kits in educational institutions to provide students with a space for research, design and implementation of solutions. An educational robotics kit allows:
  - Program them with a variety of languages.
  - Command them with different additional software program, for the:
    - Data acquisition and mathematical analysis.
    - Communication and control of instruments of any manufacturer.
    - Industrial automation.

ACKNOWLEDGMENT

This research received no specific grant from any funding agency in the public, commercial, or not-for-profit sectors.

REFERENCES

Abrahamsen, M., & Eilers, S. (2011). On the Asymptotic Enumeration of LEGO Structures. *Experimental Mathematics*, *20*(2), 145–152. doi:10.1080/10586458.2011.564539

Alimisis, D. (2013). Educational robotics: Open questions and new challenges. *Themes in Science and Technology Education*, *6*(1), 63–71.

Angel-Fernandez, J. M., & Vincze, M. (2018). *Towards a definition of educational robotics.* Paper presented at the Austrian Robotics Workshop 2018. 10.15203/3187-22-1-08

Arís, N., & Orcos, L. (2019). Educational robotics in the stage of secondary education: Empirical study on motivation and STEM skills. *Education Sciences*, *9*(2), 73. doi:10.3390/educsci9020073

Caro, I. A. (2011). *VEX robotics: STEM program and robotics competition expansion into Europe.* Paper presented at the International Conference on Research and Education in Robotics.

Chiang, F.-K., Liu, Y., Feng, X., Zhuang, Y., & Sun, Y. (2020). Effects of the world robot Olympiad on the students who participate: A qualitative study. *Interactive Learning Environments*, 1–12. doi:10.1080/10494820.2020.1775097

Chris, C. (2013). *Learning with FIRST LEGO league.* Paper presented at the Society for Information Technology & Teacher Education International Conference.

Conversy, S. (2014). Unifying textual and visual: A theoretical account of the visual perception of programming languages. *Proceedings of the 2014 ACM International Symposium on New Ideas, New Paradigms, and Reflections on Programming & Software.* 10.1145/2661136.2661138

Eguchi, A. (2007). *Educational robotics for elementary school classroom.* Paper presented at the Society for Information Technology & Teacher Education International Conference.

Eguchi, A. (2010). *What is educational robotics? Theories behind it and practical implementation.* Paper presented at the Society for information technology & teacher education international conference.

Eguchi, A. (2014). *Robotics as a learning tool for educational transformation.* Paper presented at the Proceeding of 4th international workshop teaching robotics, teaching with robotics & 5th international conference robotics in education, Padova, Italy.

Galvez-Aranda, D., & Galvez Legua, M. (2021a). Biomimetic Robots. In D. Galvez-Aranda & M. Galvez Legua (Eds.), *Robotics Models Using LEGO WeDo 2.0: Design, Build, Program, Test, Document and Share* (pp. 281–334). Apress. doi:10.1007/978-1-4842-6846-9_7

Galvez-Aranda, D., & Galvez Legua, M. (2021b). Biped Robots. In D. Galvez-Aranda & M. Galvez Legua (Eds.), *Robotics Models Using LEGO WeDo 2.0: Design, Build, Program, Test, Document and Share* (pp. 79–128). Apress. doi:10.1007/978-1-4842-6846-9_3

Galvez-Aranda, D., & Galvez Legua, M. (2021c). Crawling Robots. In D. Galvez-Aranda & M. Galvez Legua (Eds.), *Robotics Models Using LEGO WeDo 2.0: Design, Build, Program, Test, Document and Share* (pp. 129–176). Apress. doi:10.1007/978-1-4842-6846-9_4

Galvez-Aranda, D., & Galvez Legua, M. (2021d). Humanoids. In D. Galvez-Aranda & M. Galvez Legua (Eds.), *Robotics Models Using LEGO WeDo 2.0: Design, Build, Program, Test, Document and Share* (pp. 229–280). Apress. doi:10.1007/978-1-4842-6846-9_6

Galvez-Aranda, D., & Galvez Legua, M. (2021e). Quadruped Robots. In D. Galvez-Aranda & M. Galvez Legua (Eds.), *Robotics Models Using LEGO WeDo 2.0: Design, Build, Program, Test, Document and Share* (pp. 177–228). Apress. doi:10.1007/978-1-4842-6846-9_5

Harel, I. E., & Papert, S. E. (1991). *Constructionism*. Ablex Publishing.

Lauwaert, M. (2008). Playing outside the box–on LEGO toys and the changing world of construction play. *History and Technology*, *24*(3), 221–237. doi:10.1080/07341510801900300

Mikropoulos, T. A., & Bellou, I. (2013). Educational robotics as mindtools. *Themes in Science and Technology Education*, *6*(1), 5–14.

Norouzi, S., Fattahi, A., & Massah, A. (2010). *LARC*. Competition–Mixed Reality League S&TRN.

Pisciotta, M., Vello, B., Bordo, C., & Morgavi, G. (2010). *Robotic competition: A classroom experience in a vocational school.* Paper presented at the Actas del 6th International Conference on Educational Technologies.

Plaza, P., Peixoto, A., Sancristobal, E., Castro, M., Blazquez, M., Menacho, A., . . . Muñoz, M. (2020). *Visual block programming languages and their use in educational robotics.* Paper presented at the 2020 IEEE Global Engineering Education Conference (EDUCON). 10.1109/EDUCON45650.2020.9125219

Resnick, M., & Silverman, B. (2005). Some reflections on designing construction kits for kids. *Proceedings of the 2005 conference on Interaction design and children*. 10.1145/1109540.1109556

Sigüenza, M., Guillen, D., Arroyo, D., & Cuellar, F. (2017). *Mobile robots development: A case study from robotics competitions and course projects.* Paper presented at the 2017 IEEE XXIV International Conference on Electronics, Electrical Engineering and Computing (INTERCON). 10.1109/INTERCON.2017.8079703

Wadsworth, B. J. (1996). *Piaget's theory of cognitive and affective development: Foundations of constructivism*. Longman Publishing.

Welch, A. G. (2010). Using the TOSRA to assess high school students' attitudes toward science after competing in the FIRST robotics competition: An exploratory study. *Eurasia Journal of Mathematics, Science and Technology Education*, *6*(3), 187–197. doi:10.12973/ejmste/75239

KEY TERMS AND DEFINITIONS

Algorithm: Algorithm is a finite and ordered sequence of tasks to follow in order to solve a problem.

Electrical Motor: An actuator that transform electrical energy into mechanical energy.

Electronics: Electronics allows the addition of senses to the prototype, making it able to detect and recognized its environment.

Flowchart: A flowchart is a diagram that represent a sequence of tasks (algorithm).

Program: A program is a set of instructions that a computer understands and execute.

Robotics: Is the design and construction of machines with a certain degree of intelligence, capable of replacing human beings in certain activities.

Technology: Is the set of skills, methods or processes used for the design and construction of machines or services to satisfy human needs.

Compilation of References

Abrahamsen, M., & Eilers, S. (2011). On the Asymptotic Enumeration of LEGO Structures. *Experimental Mathematics, 20*(2), 145–152. doi:10.1080/10586458.2011.564539

Abt, C. C. (1970). *Serious Games*. Viking Press.

Abuhmaid, A. (2011). ICT training courses for teacher professional development in Jordan. *The Turkish Online Journal of Educational Technology, 10*(4).

acatech und VDI. (2009). *Nachwuchsbarometer Technikwissenschaften*. Acatech und VDI.

acatech. (2011). *Monitoring von Motivationskonzepten für den Techniknachwuchs (MoMoTech)*. Springer.

Adler, C. G., & Coulter, B. L. (1978). Galileo and the Tower of Pisa experiment. *American Journal of Physics, 46*(3), 199–201. doi:10.1119/1.11165

Afrotechmodes. (2010). *Comparator tutorial & clapper circuit* [Videotutorial]. YouTube. https://www.youtube.com/watch?v=y0Q0ERSP24A

Agatolio, F., Pivetti, M., Di Battista, S., Menegatti, E., & Moro, M. (2017). A training course in educational robotics for learning support teachers. In *International Conference EduRobotics 2016* (pp. 43-57). Springer. 10.1007/978-3-319-55553-9_4

Agyei, D. D., & Voogt, J. (2012). Developing technological pedagogical content knowledge in pre-service mathematics teachers through collaborative design. *Australasian Journal of Educational Technology, 28*(4). Advance online publication. doi:10.14742/ajet.827

Agyei, D., & Voogt, J. (2014). Examining factors affecting beginning teachers' transfer of learning of ICT-enhanced learning activities in their teaching practice. *Australasian Journal of Educational Technology, 30*(1). Advance online publication. doi:10.14742/ajet.499

Ahmed, H., & La, H. M. (2019). *Education-Robotics Symbiosis: An Evaluation of Challenges and Proposed Recommendations*. Academic Press.

Akbarimajd, A., & Sotoudeh, N. (2014). Design and motion analysis of vibration-driven small robot Rizeh. *Advanced Robotics, 28*(2), 105–117. doi:10.1080/01691864.2013.843788

Akca, O. F. (2020). *The effect of inquiry based robotic activities on students' conceptual success, logical thinking and science process skills in science centers* [Unpublished master's dissertation]. Gazi University, Ankara, Turkey.

Akman-Selcuk, N. (2019). *The investigation of educational robotics activities on middle school students' course motivation, robotic attitudes and success* [Unpublished doctoral dissertation]. Istanbul University, Istanbul, Turkey.

Aksu, F. N. (2019). *Robotic coding and robotic competitions from the perspective of the information and communication technology teachers* [Unpublished master's dissertation]. Balıkesir University, Balıkesir, Turkey.

Alammary, J., Al-haiki, F., & Al-muqahwi, K. (2017). The impact of assistive technology on down syndrome students in Kingdom of Bahrain. *The Turkish Online Journal of Educational Technology, 16*, 103–119.

Alayyar, G. M., Fisser, P., & Voogt, J. M. (2011). ICT integration through design teams in science teacher preparation. *International Journal of Learning Technology, 6*(2), 125–145. doi:10.1504/IJLT.2011.042645

Alimisis, D. (2013). Educational robotics: Open questions and new challenges. *Themes in Science and Technology Education, 6*(1), 63–71. http://earthlab.uoi.gr/theste/index.php/theste/article/view/119

Alimisis, D. (2014). Educational robotics in teacher education: An innovative tool for promoting quality education. *The Teacher of the 21st Century: Quality Education for Quality Teaching*, 14.

Alimisis, D., & Boulougaris, G. (2014). Robotics in physics education: Fostering graphing abilities in kinematics. In D. Alimisis, G. Granosik, & M. Moro (Eds.), *Proceedings from the 4th international workshop teaching robotics: Teaching with robotics & 5th international conference robotics in education* (pp. 2–10). Padova, Italy: Academic Press.

Alimisis, D. (2009). *Teacher education on robotics-enhanced constructivist pedagogical methods*. Athens: School of Pedagogical and Technological Education.

Alimisis, D. (2012). Robotics in Education & Education in Robotics: Shifting Focus from Technology to Pedagogy. *Proceedings of the 3rd International Conference on Robotics in Education*, 7–14.

Alimisis, D. (2013). Educational robotics: Open questions and new challenges. *Themes in Science and Technology Education, 6*(1), 63–71.

Alimisis, D. (2013). Educational robotics: Open questions and new challenges. *Themes Science and Technology Education, 6*(1), 63–71.

Alimisis, D. (2019). Teacher Training in Educational Robotics: The ROBOESL Project Paradigm. Technology. *Knowledge and Learning, 24*(2), 279–290. doi:10.100710758-018-9357-0

Alimisis, D. (Ed.). (2009). *Teacher Education on Robotics-Enhanced Constructivist Pedagogical Methods. School of Pedagogical and Technological Education*. ASPETE.

Alimisis, D., & Plessas, A. (2011). Teacher distance training in educational robotics through synchronous audiographic conferencing: towards a learner-centered approach. In *Proceedings of EdMedia+ Innovate Learning* (pp. 901–906). Association for the Advancement of Computing in Education AACE.

Alkahtani, K. D. F. (2013). Teachers' knowledge and use of assistive technology for students with special educational needs. *Journal of Studies in Education, 3*(2), 65–86. doi:10.5296/jse.v3i2.3424

Allen, R., Davis, R., & Hill, E. (2013). The effects of autism and alexithymia on physiological and verbal responsiveness to music. *Journal of Autism and Developmental Disorders, 43*(2), 432–444. doi:10.100710803-012-1587-8 PMID:22752845

Almeida, T. O., Netto, J. F. d M., & Rios, M.L. (2017). Remote robotics laboratory as support to teaching programming. In *Proceedings of IEEE Frontiers in Education Conference* (pp. 1–6). Indianapolis, IN: IEEE. 10.1109/FIE.2017.8190472

Altin, H., & Pedaste, M. (2013). Learning approaches to applying robotics in science education. *Journal of Baltic Science Education, 12*(3), 365.

American Psychiatric Association. (2013). *Diagnostic and statistical manual of mental disorders* (5th ed.). Author.

Ames, C. W. (2016). *Teacher perceptions of factors influencing technology integration in k-12 schools*. Utah State University.

Amsters, R., & Slaets, P. (2020). Turtlebot 3 as a Robotics Education Platform. In M. Merdan, W. Lepuschitz, G. Koppensteiner, R. Balogh & D. Obdržálek (Eds.), *Robotics in Education. RiE 2019. Advances in Intelligent Systems and Computing* (Vol. 1023). Springer. doi:10.1007/978-3-030-26945-6_16

Anastasiadis, T., Lampropoulos, G., & Siakas, K. (2018). Digital game-based learning serious games in education. *International Journal of Advances in Scientific Research and Engineering*, *4*(12), 139–144. doi:10.31695/IJASRE.2018.33016

Anewalt, K. (2002). Experiences teaching writing in a computer science course for the first time. *Journal of Computing Sciences in Colleges*, *18*(2), 346–355.

Angel-Fernandez, J. M., & Vincze, M. (2018). *Towards a definition of educational robotics.* Paper presented at the Austrian Robotics Workshop 2018. 10.15203/3187-22-1-08

Angeli, C., & Makridou, E. (2018). Developing Third-Grade Students' Computational Thinking Skills with Educational Robotics. In E. Langran & J. Borup (Eds.), *Proceedings of Society for Information Technology & Teacher Education International Conference* (pp. 1-8). Association for the Advancement of Computing in Education (AACE). Retrieved January 20, 2021 from https://www.learntechlib.org/primary/p/182493/

Angeli, C., & Giannakos, M. (2020). Computational thinking education: Issues and challenges. *Computers in Human Behavior*, *105*, 106185. doi:10.1016/j.chb.2019.106185

Angeli, C., & Valanides, N. (2020). Developing young children's computational thinking with educational robotics: An interaction effect between gender and scaffolding strategy. *Computers in Human Behavior*, *105*, 105954. doi:10.1016/j.chb.2019.03.018

Angeli, C., Voogt, J., Fluck, A., Webb, M., Cox, M., Malyn-Smith, J., & Zagami, J. (2016). A K-6 Computational Thinking Curriculum Framework- Implications for Teacher Knowledge. *Journal of Educational Technology & Society*, *19*(3), 47–57.

Anonymous. (2007, December 20). *Bristlebot: A tiny directional vibrobots.* https://www.evilmadscientist.com/2007/bristlebot-a-tiny-directional-vibrobot/comment-page-1/#comment-2521

Anonymous. (2009, January 17). *Bristlebot: A tiny directional vibrobot.* https://www.evilmadscientist.com/2007/bristlebot-a-tiny-directional-vibrobot/comment-page-1/#comment-2533

Anthony, L., Brown, Q., Nias, J., & Tate, B. (2013). Examining the need for visual feedback during gesture interaction on mobile touchscreen devices for kids. *Proceedings of the 12th International Conference on Interaction Design and Children -IDC '13*, 157–164. 10.1145/2485760.2485775

Antonenko, P. D., Dawson, K., & Sahay, S. (2017). A framework for aligning needs, abilities and affordances to inform design and practice of educational technologies. *British Journal of Educational Technology*, *48*(4), 916–927. doi:10.1111/bjet.12466

Anwar, S., Bascou, N. A., Menekse, M., & Kardgar, A. (2019). A systematic review of studies on educational robotics. *Journal of Pre-College Engineering Education Research*, *9*(2), 19–42. doi:10.7771/2157-9288.1223

Ardito, G., Mosley, P., & Scollins, L. (2014). We, robot: Using robotics to promote collaborative and mathematics learning in a middle school classroom. *Middle Grades Research Journal*, *93*(3), 73–88.

Arís, N., & Orcos, L. (2019). Educational robotics in the stage of secondary education: Empirical study on motivation and STEM skills. *Education Sciences*, *9*(2), 73. doi:10.3390/educsci9020073

Arnab, S., Lim, T., Carvalho, M. B., Bellotti, F., De Freitas, S., Louchart, S., Suttie, N., Berta, R., & De Gloria, A. (2015). Mapping learning and game mechanics for serious games analysis. *British Journal of Educational Technology*, *46*(2), 391–411. doi:10.1111/bjet.12113

Aschoff, C. (2015, March 20). *Aufbauanleitung für einen Bristle-Bot.* http://www.christians-bastel-laden.de/BRISTLE_BOT_MANUAL/index.html

Ashton, T. S. (1997). *The industrial revolution 1760-1830.* OUP Catalogue.

Astrachan, O., Barnes, T., Garcia, D., Paul, J., Simon, B., & Snyder, L. (2011). CS principles: piloting a new course at national scale. In *Proceedings of the 42nd ACM technical symposium on Computer science education* (pp. 397-398). 10.1145/1953163.1953281

Ater-Kranov, A., Bryant, R., Orr, G., Wallace, S., & Zhang, M. (2010). Developing a community definition and teaching modules for computational thinking: accomplishments and challenges. *Proceedings of the 2010 ACM conference on Information technology education.* 10.1145/1867651.1867689

Atmatzidou, S., & Demetriadis, S. (2012). Evaluating the role of collaboration scripts as group guiding tools in activities of educational robotics. *IEEE International Conference on Advanced Learning Technologies (ICALT 2012).* 10.1109/ICALT.2012.111

Atmatzidou, S., & Demetriadis, S. (2015). Advancing students' computational thinking skills through educational robotics: A study on age and gender relevant differences. *Robotics and Autonomous Systems*, *75*, 661–670. doi:10.1016/j.robot.2015.10.008

Atmatzidou, S., & Demetriadis, S. (2017). A Didactical Model for Educational Robotics Activities: A Study on Improving Skills through Strong or Minimal Guidance. In D. Alimisis, M. Moro, & E. Menegatti (Eds.), *Educational Robotics in the Makers Era. Edurobotics 2016. Advances in Intelligent Systems and Computing* (Vol. 560, pp. 58–72). Springer. doi:10.1007/978-3-319-55553-9_5

Atmatzidou, S., Demetriadis, S., & Nika, P. (2018). How Does the Degree of Guidance Support Students' Metacognitive and Problem Solving Skills in Educational Robotics? *Journal of Science Education and Technology*, *27*(1), 70–85. doi:10.100710956-017-9709-x

Axl, J., & Manupool. (2012, January 12). *Bristlebots.* https://www.hackerspace-ffm.de/wiki/index.php?title=Bristlebots

Baek, Y., Jung, J., & Kim, B. (2008). What makes teachers use technology in the classroom? Exploring the factors affecting facilitation of technology with a Korean sample. *Computers & Education*, *50*(1), 224–234. doi:10.1016/j.compedu.2006.05.002

Baek, Y., Wang, S., Yang, D., Ching, Y.-H., Swanson, S., & Chittoori, B. (2019). Revisiting second graders' robotics with an Understand/Use-Modify-Create (U2MC) strategy. *European Journal of STEM Education*, *4*(1), 1–12. doi:10.20897/ejsteme/5772

Baharuddin, M. Z., Abidin, I. Z., Mohideen, S. S. K., Siah, Y. K., & Chuan, J. T. T. (2005). *Analysis of line sensor configuration for the advanced line follower robot.* University Tenaga Nasional.

Bahceci, B. (2019). *To examine attitudes of teachers working in the field of special education towards assistive technologies and individual innovation levels* [Unpublished master's dissertation]. Konya Necmettin Erbakan University, Konya, Turkey.

Baker, D. R., Wood, L., Corkins, J., & Krause, S. (2015). Tinkering and Technical Self-Efficacy of Engineering Students at the Community College. *Community College Journal of Research and Practice*, *39*(6), 555–567. doi:10.1080/10668926.2014.902780

Baker, R. (2017). Problem-solving. In *Agile UX Storytelling* (pp. 107–116). Springer. doi:10.1007/978-1-4842-2997-2_13

Bakker, A. (2018). *Design research in education: A practical guide for early career researchers*. Routledge. doi:10.4324/9780203701010

Balch, T., Summet, J., Blank, D., Kumar, D., Guzdial, M., O'hara, K., ... Tansley, S. (2008). Designing personal robots for education: Hardware, software, and curriculum. *IEEE Pervasive Computing*, *7*(2), 5–9. doi:10.1109/MPRV.2008.29

Banzi, M., & Shiloh, M. (2014). *Getting started with Arduino: the open source electronics prototyping platform*. Maker Media, Inc.

Barak, M., & Assal, M. (2018). Robotics and STEM learning: Students' achievements in assignments according to the P3 Task Taxonomy—practice, problem solving, and projects. *International Journal of Technology and Design Education*, *28*(1), 121–144. doi:10.100710798-016-9385-9

Barbosa, M. B., Rêgo, A. B., & de Medeiros, I. (2015). *HEEG: Heuristic Evaluation for Educational Games*. Academic Press.

Barr, V., & Stephenson, C. (2011). Bringing computational thinking to K-12: What is Involved and what is the role of the computer science education community? *ACM Inroads*, *2*(1), 48–54. doi:10.1145/1929887.1929905

Basawapatna, A. R., Koh, K. H., & Repenning, A. (2010). Using scalable game design to teach computer science from middle school to graduate school. *Proceedings of the Fifteenth Annual Conference on Innovation and Technology in Computer Science Education*, 224–228. 10.1145/1822090.1822154

Beard, L., Carpenter, L., & Johnston, L. (2011). *Assistive technology: Access for all students* (2nd ed.). Merrill Prentice-Hall.

Becker, F., Minchenya, V., Zeidis, I., & Zimmermann, K. (2011). Modelling and dynamical simulation of vibration-driven robots. In *Proceeding of the 56th International scientific colloquium*. Ilmenau University of Technology.

Beede, D. N., Julian, T. A., Langdon, D., McKittrick, G., Khan, B., & Doms, M. E. (2011). *Women in STEM: A gender gap to innovation*. Economics and Statistics Administration Issue Brief (04-11).

Begum, M., Serna, R. W., & Yanco, H. A. (2016). Are Robots Ready to Deliver Autism Interventions? A Comprehensive Review. *International Journal of Social Robotics*, *8*(2), 157–181. doi:10.100712369-016-0346-y

Belpaeme, T., Kennedy, J., Ramachandran, A., Scassellati, B., & Tanaka, F. (2018). Social robots for education: A review. *Science Robotics*, *3*(21), 1–10. doi:10.1126cirobotics.aat5954 PMID:33141719

Benitti, F. B. V. (2012). Exploring the educational potential of robotics in schools: A systematic review. *Computers & Education*, *58*(3), 978–988. doi:10.1016/j.compedu.2011.10.006

Benjes, H., Zuschlag, F., & Welschehold, E. (1992). *Fertigen, Konstruieren und Steuern im Technikunterricht: Lehrerhandbuch für die Arbeit mit dem Halbzeugsystem UMT*. NLI.

Bergner, N., & Müller, K. (2018). Fachempfehlung Informatiksysteme. In N. Bergner, H. Köster, J. Magenheim, K. Müller, R. Romeike, U. Schroeder, & C. Schulte (Eds.), *Frühe informatische Bildung – Ziele und Gelingensbedingungen für den Elementar- und Primarbereich* (pp. 268–301). Barbara Budrich.

Bers, M. U., Rogers, C., Beals, L., Portsmore, M., Staszowski, K., Cejka, E., . . . Barnett, M. (2006, June). Innovative session: Early childhood robotics for learning. In *Proceedings of the 7th International Conference on Learning Sciences, International Society of the Learning Sciences*, (pp. 1036-1042). Academic Press.

Bers, M. U. (2010). The TangibleK robotics program: Applied computational thinking for young children. *Early Childhood Research & Practice, 12*(2), 2.

Bers, M. U. (2018). Coding, playgrounds and literacy in early childhood education: The development of KIBO robotics and ScratchJr. In *2018 IEEE Global Engineering Education Conference (EDUCON)* (pp. 2094-2102). IEEE. 10.1109/EDUCON.2018.8363498

Bers, M. U., Flannery, L., Kazakoff, E. R., & Sullivan, A. (2014). Computational thinking and tinkering: Exploration of an early childhood robotics curriculum. *Computers & Education, 72*, 145–157. doi:10.1016/j.compedu.2013.10.020

Bers, M., Flannery, L., Kazakoff, E., & Crouser, J. (2010). *A Curriculum Unit on Programming and Robotics. Computer Science*. Faculty Publications, Smith College. https://scholarworks.smith.edu/csc_facpubs/9

Bevan, B. (2017). The promise and the promises of Making in science education. *Studies in Science Education, 53*(1), 75–103. doi:10.1080/03057267.2016.1275380

Bevan, B., Petrich, M., & Wilkinson, K. (2014). Tinkering is serious play. *Educational Leadership, 72*(4), 28–33.

Biggs, J. (1996). Enhancing teaching through constructive alignment. *Higher Education, 32*(3), 347–364. doi:10.1007/BF00138871

Biggs, J. (2003). Aligning teaching and assessing to course objectives. Teaching and Learning in Higher Education. *New Trends and Innovations, 2*(April), 13–17.

Biggs, J. (2003). Aligning teaching and assessing to course objectives. *Teaching and Learning in Higher Education: New Trends and Innovations, 2*(April), 13–17.

Billard, A., Robins, B., Nadel, J., Dautenhahn, K. (2007). Building Robota, a Mini Humanoid Robot for the Rehabilitation of Children With Autism. *Assistive Technology, 19*, 37-49. . doi:10.1080/10400435.2007.10131864

Bingimlas, K. A. (2009). Barriers to the successful integration of ICT in teaching and learning environments: A review of the literature. *Eurasia Journal of Mathematics, Science and Technology Education, 5*(3). Advance online publication. doi:10.12973/ejmste/75275

Birk, A. (2011). What is robotics? an interdisciplinary field is getting even more diverse. *IEEE Robotics & Automation Magazine, 18*(4), 94–95. doi:10.1109/MRA.2011.943235

Bitesize, B. B. C. (2017). *Introduction to computational thinking*. Online verfügbar unter: http://www. bbc. co. uk/education/guides/zp92mp3/revision

Blackhurst, A. (1997). Perspectives on technology in special education. *Teaching Exceptional Children, 29*(5), 41–48. doi:10.1177/004005999702900509

Blanchard, S., Freiman, V., & Lirrete-Pitre, N. (2010). Strategies used by elementary schoolchildren solving robotics-based complex tasks: Innovative potential of technology. *Procedia: Social and Behavioral Sciences, 2*(2), 2851–2857. doi:10.1016/j.sbspro.2010.03.427

Blikstein, P. (2013). Digital fabrication and "making" in education: The democratization of invention. In J. Walter-Herrmann & C. Büching (Eds.), FabLabs: Of machines, makers and inventors. Bielefeld: Transcript Publishers.

Blikstein, P., & Krannich, D. (2013). The Makers' Movement and FabLabs in Education: Experiences, Technologies, and Research. *IDC '13 Proceedings of the 12th International Conference on Interaction Design and Children.* 10.1145/2485760.2485884

BMBWF. (2021). *Digitale Kompetenzen-Informatische Bildung.* digikomp4. https://digikomp.at/index.php?id=530

Boccanfuso, L., & O'Kane, J. M. (2011). CHARLIE: An adaptive robot design with hand and face tracking for use in autism therapy. *International Journal of Social Robotics, 3*(4), 337–347. doi:10.100712369-011-0110-2

Bocconi, S., Chioccariello, A., Dettori, G., Ferrari, A., & Engelhardt, K. (2016). *Developing computational thinking in compulsory education-Implications for policy and practice* (No. JRC104188). Joint Research Centre (Seville site).

Boehm, B. W. (1988). A spiral model of software development and enhancement. *Computer, 21*(5), 61–72. https://doi.org/10.1109/2.59

Bonar, J., & Soloway, E. (1983, January). Uncovering principles of novice programming. In *Proceedings of the 10th ACM SIGACT-SIGPLAN symposium on Principles of programming languages* (pp. 10-13). ACM.

Borg, J., Lindstrom, A., & Larsson, S. (2009). Assistive technology in developing countries: National and international responsibilities to implement the convention on the rights of persons with disabilities. *Lancet, 374*(28), 1863–1865. doi:10.1016/S0140-6736(09)61872-9 PMID:19944867

Boscariol, P., & Richiedei, D. (2019). Trajectory design for energy savings in redundant robotic cells. *Robotics, 8*(1), 15. Advance online publication. doi:10.3390/robotics8010015

Bostock, M. (1983). BBC Buggy. *Your Computer, 3*(4), 51–53.

Boticki, I., Pivalica, D., & Seow, P. (2018). The Use of Computational Thinking Concepts in Early Primary School. *International Conference on Computational Thinking Education 2018.*

Bottino, R., & Chioccariello, A. (2015). Computational Thinking: videogames, educational robotics, and other powerful ideas to think with. *KEYCIT 2014: Key competencies in informatics and ICT, 7,* 301.

Boud, D., & Prosser, M. (2002). Appraising new technologies for learning: A framework for development. *Educational Media International, 39*(3-4), 237–245. doi:10.1080/09523980210166026

Bower, M. (2008). Affordance analysis – Matching learning tasks with learning technologies. *Educational Media International, 45*(1), 3–15. doi:10.1080/09523980701847115

Brandhofer, G. (2017). Coding und Robotik im Unterricht. *Erziehung und Unterricht, 167*(7–8), 630–637.

Brandhofer, G., Baumgartner, P., Ebner, M., Köberer, N., Trültzsch-Wijnen, C., & Wiesner, C. (2018). Bildung im Zeitalter der Digitalisierung. *Nationaler Bildungsbericht Österreich, 2,* 307–362.

Braun, V., & Clarke, V. (2006). Using thematic analysis in psychology. *Qualitative Research in Psychology, 3*(2), 77–101. doi:10.1191/1478088706qp063oa

Breazeal, C., & Aryananda, L. (2002). Recognition of affective communicative intent in robot-directed speech. *Autonomous Robots, 12*(1), 83–104. doi:10.1023/A:1013215010749

Breiner, J. M., Harkness, S. S., Johnson, C. C., & Koehler, C. M. (2012). What is STEM? A discussion about conceptions of STEM in education and partnerships. *School Science and Mathematics, 112*(1), 3–11. doi:10.1111/j.1949-8594.2011.00109.x

Bristlebotics. (2017). *Light Sensing Bristlebot.* https://www.instructables.com/Light-Sensing-Bristlebot

Brown, D. (2008). *Video Modeling: Combining Dynamic Model Simulations with Traditional Video Analysis*. AAPT Summer Meeting.

Brown, J. S., Collins, A., & Duguid, P. (1989). Situated cognition and the culture of learning. *Educational Researcher, 18*(1), 32–42. doi:10.3102/0013189X018001032

Bryant, D. P., & Bryant, B. R. (2003). *Assistive technology for people with disabilities*. Allyn and Bacon.

Bryant, D. P., Smith, D. D., & Bryant, B. R. (2008). *Teaching students with special needs in inclusive settings*. Allyn & Bacon.

Buabeng-Andoh, C. (2012). Factors influencing teachers' adoption and integration of information and communication technology into teaching: A review of the literature. *International Journal of Education and Development Using Information and Communication Technology, 8*(1), 136.

Burdick, A. (2009). *Design without designers*. Paper presented at the conference on the future of art and design education in the 21st century, Brighton, UK.

Burke, Q., & Kafai, Y. B. (2012). The writers' workshop for youth programmers: digital storytelling with scratch in middle school classrooms. In *Proceedings of the 43rd ACM technical symposium on Computer Science Education* (pp. 433-438). Academic Press.

Burke, Q., & Kafai, Y. B. (2010). Programming & storytelling: opportunities for learning about coding & composition. In *Proceedings of the 9th international conference on interaction design and children* (pp. 348-351). 10.1145/1810543.1810611

Bybee, R. W. (2013). *The case for STEM education: Challenges and opportunities*. NSTA press.

Cabibihan, J. J., Javed, H., Ang, M., & Aljunied, S. M. (2013). Why Robots? A Survey on the Roles and Benefits of Social Robots in the Therapy of Children with Autism. In International Journal of Social Robotics (Vol. 5, Issue 4, pp. 593–618). doi:10.100712369-013-0202-2

Cagiltay, N. E., Ozcelik, E., & Ozcelik, N. S. (2015). The effect of competition on learning in games. *Computers & Education, 87*, 35–41. doi:10.1016/j.compedu.2015.04.001

Cakmak, S. (Ed.). (2016). *Özel eğitim ve yardımcı teknolojiler* [Lexical characteristics of Turkish language]. Vize Publishing.

Campeau-Lecours, A., Maheu, V., Lepage, S., Lamontagne, H., Latour, S., Paquet, L., & Hardie, N. (2016, July). JACO assistive robotic device: Empowering people with disabilities through innovative algorithms. *Rehabilitation Engineering and Assistive Technology Society of North America (RESNA) Annual Conference.*

Campigotto, R., McEwan, R., & Demmans Epp, C. (2013). Especially social: Exploring the use of an iOS application in special needs classrooms. *Computers & Education, 60*(1), 74–86. doi:10.1016/j.compedu.2012.08.002

Cano, S., Arteaga, J. M., Collazos, C. A., Gonzalez, C. S., & Zapata, S. (2016). Toward methodology for serious games design for children with auditory impairments. *IEEE Latin America Transactions, 14*(5), 2511–2521. doi:10.1109/TLA.2016.7530453

Caro, I. A. (2011). *VEX robotics: STEM program and robotics competition expansion into Europe*. Paper presented at the International Conference on Research and Education in Robotics.

Cascalho, J., Mendes, A., Ramos, A., Pedro, F., Bonito, N., Almeida, D., Augusto, P., Leite, P., Funk, M., & Garcia, A. (2020). Azoresbot. *An Arduino Based Robot for Robocup Competitions., 6*, 542–552. doi:10.1007/978-3-030-35990-4_44

Castledine, A. R., & Chalmers, C. (2011). LEGO Robotics: An authentic problem solving tool? *Design and Technology Education: An International Journal, 16*(3).

Castro, E., Cecchi, F., Salvini, P., Valente, M., Buselli, E., Menichetti, L., Calvani, A., & Dario, P. (2018). Design and impact of a teacher training course, and attitude change concerning educational robotics. *International Journal of Social Robotics, 10*(5), 669–685. doi:10.100712369-018-0475-6

Castro, M. D. B., & Tumibay, G. M. (2021). A literature review: Efficacy of online learning courses for higher education institution using meta-analysis. *Education and Information Technologies, 26*(2), 1367–1385. doi:10.100710639-019-10027-z

Catlin, D. (2016). 29 Effective ways you can use robots in the classroom. In *International Conference EduRobotics* (pp. 135-148). Springer.

Çavdar, G., & Doe, S. (2012). Learning through writing: Teaching critical thinking skills in writing assignments. *PS, Political Science & Politics, 45*(2), 298–306. doi:10.1017/S1049096511002137

Cay, E., Yikmis, A., & Sola-Ozguc, C. (2020). Experiences and opinions of special education teachers regarding the use of technology. *Journal of Qualitative Research in Education, 8*(2), 629–648. doi:10.14689/issn.2148-624.1.8c.2s.9m

Cecchi, R., Verotti, M., Capata, R., Dochshanov, A. M., Broggiato, G. B., Crescenzi, R., Balucani, M., Natali, S., Razzano, G., Lucchese, F., Bagolini, A., Bellutti, P., Sciubba, E., & Belfiore, N. P. (2015). Development of micro-grippers for tissue and cell manipulation with direct morphological comparison. *Micromachines, 6*(11), 1710–1728. doi:10.3390/mi6111451

Chalmers, C., & Nason, R. (2017). Systems thinking approach to robotics curriculum in schools. In *Robotics in STEM Education* (pp. 33–57). Springer. doi:10.1007/978-3-319-57786-9_2

Chan Lye, N., Wonga, K. W., & Chiou, A. (2013). Framework for educational robotics: A multiphase approach to enhance user learning in a competitive arena. *Interactive Learning Environments, 21*(2), 142–155. doi:10.1080/10494820.2012.705853

Chan, N., Wai, K., & Chiou, A. (2013). *Framework for educational robotics : a multiphase approach to enhance user learning in a competitive arena*. Academic Press.

Chandra, V. (2014). Developing students' technological literacy through robotics activities. *Literacy Learning: The Middle Years, 22*(3), 24–29.

Chang, A., Tilahun, L., & Breazeal, C. (2014). Visualisations of Data from the Literacy Tablet Reading Project in Rural Ethiopia. In *Proceedings of the Electronic Visualisation and the Arts (EVA)*, 142–149. 10.14236/ewic/EVA2014.35

Chang, C. W., Lee, J. H., Chao, P. Y., Wang, C. Y., & Chen, G. D. (2010). Exploring the possibility of using humanoid robots as instructional tools for teaching a second language in primary school. *Journal of Educational Technology & Society, 13*(2), 13–24.

Chassignol, M., Khoroshavin, A., Klimova, A., & Bilyatdinova, A. (2018). Artificial Intelligence trends in education: A narrative overview. *Procedia Computer Science, 136*, 16–24. doi:10.1016/j.procs.2018.08.233

Chaudhary, V., Agrawal, V., Sureka, P., & Sureka, A. (2016, December). An experience report on teaching programming and computational thinking to elementary level children using lego robotics education kit. In *IEEE Eighth International Conference on Technology for Education (T4E)* (pp. 38-41). IEEE. 10.1109/T4E.2016.016

Chen, G. D., & Wang, C. Y. (2011). A survey on storytelling with robots. In *International Conference on Technologies for E-Learning and Digital Entertainment* (pp. 450-456). Springer.

Chen, G., Shen, J., Barth-Cohen, L., Jiang, S., Huang, X., & Eltoukhy, M. (2017). Assessing elementary students' computational thinking in everyday reasoning and robotics programming. *Computers & Education, 109*, 162–175. doi:10.1016/j.compedu.2017.03.001

Chen, T. L., Ciocarlie, M., Cousins, S., Grice, P. M., Hawkins, K., Hsiao, K., ... Takayama, L. (2013). Robots for humanity: Using assistive robotics to empower people with disabilities. *IEEE Robotics & Automation Magazine*, *20*(1), 30–39. doi:10.1109/MRA.2012.2229950

Chen, X. (2013). *STEM Attrition: College Students' Paths into and out of STEM Fields. Statistical Analysis Report. NCES 2014-001*. National Center for Education Statistics.

Cheryan, S., Ziegler, S. A., Montoya, A. K., & Jiang, L. (2017). Why are some STEM fields more gender balanced than others? *Psychological Bulletin*, *143*(1), 1–35. doi:10.1037/bul0000052 PMID:27732018

Chevalier, M., Giang, C., Piatti, A., & Mondada, F. (2020). Fostering computational thinking through educational robotics: A model for creative computational problem solving. *International Journal of STEM Education*, *7*(1), 39. doi:10.118640594-020-00238-z

Chevalier, M., Riedo, F., & Mondada, F. (2016). Pedagogical Uses of Thymio II. *IEEE Robotics & Automation Magazine*, *23*(2), 16–23. doi:10.1109/MRA.2016.2535080

Chia, K. S., Siew, K., & Chee, E. N. G. (2014). *Build A Line Follower Robot: A User-Friendly Guide*. Academic Press.

Chiang, F.-K., Liu, Y., Feng, X., Zhuang, Y., & Sun, Y. (2020). Effects of the world robot Olympiad on the students who participate: A qualitative study. *Interactive Learning Environments*, 1–12. doi:10.1080/10494820.2020.1775097

Ching, Y. H., Yang, D., Wang, S., Baek, Y., Swanson, S., & Chittoori, B. (2019). Elementary school student development of STEM attitudes and perceived learning in a STEM integrated robotics curriculum. *TechTrends*, *63*(5), 1–12. doi:10.100711528-019-00388-0

Chookaew, S., Howimanporn, S., Pratumsuwan, P., Hutamarn, S., Sootkaneung, W., & Wongwatkit, C. (2018). Enhancing High-School Students' Computational Thinking with Educational Robotics Learning. In *Proceedings - 2018 7th International Congress on Advanced Applied Informatics, IIAI-AAI 2018* (pp. 204–208). Institute of Electrical and Electronics Engineers Inc. 10.1109/IIAI-AAI.2018.00047

Chris, C. (2013). *Learning with FIRST LEGO league*. Paper presented at the Society for Information Technology & Teacher Education International Conference.

Chung, E. Y. (2019). Robotic Intervention Program for Enhancement of Social Engagement among Children with Autism Spectrum Disorder. *Journal of Developmental and Physical Disabilities*, *31*(4), 419–434. doi:10.100710882-018-9651-8

Cingolani, R., & Metta, G. (2015). *Nanotechnology for Humans and Humanoids A vision of the use of nanotechnology in future robotics*. Academic Press.

Citil, M. (2016). Bilişsel yetersizliği olan çocuklar için yardımcı teknolojiler. In Özel eğitimde yardımcı teknolojiler (pp. 152-180). Ankara: Vize Publishing.

Clark, K. R. (2018). Learning theories: Behaviorism. *Am Soc Radiol Tech.*

Clark, R., Kirschner, P., & Sweller, J. (2012). Putting students on the path to learning: The case for fully guided instruction. *American Educator*, *36*(1), 6–11.

Clarkson, E., & Arkin, R. C. (2007). Applying heuristic evaluation to human-robot interaction systems. *Proceedings of the Twentieth International Florida Artificial Intelligence Research Society Conference, FLAIRS 2007*, 44–49.

Coeckelbergh, M., Pop, C., Simut, R., Peca, A., Pintea, S., David, D., & Vanderborght, B. (2016). A survey of expectations about the role of robots in robot-assisted therapy for children with ASD: Ethical acceptability, trust, sociability, appearance, and attachment. *Science and Engineering Ethics*, *22*(1), 47–65. doi:10.100711948-015-9649-x PMID:25894654

Cohen, S. A. (1987). Instructional Alignment: Searching for a Magic Bullet. *Educational Researcher, 16*(8), 16–20. doi:10.3102/0013189X016008016

Collins, A. (2006). Cognitive apprenticeship. In R. K. Sawyer (Ed.), *The Cambridge handbook of the learning sciences* (pp. 47–60). Cambridge University Press.

Combefis, S., Beresnevičius, G., & Dagienė, V. (2016). Learning programming through games and con-tests: Overview, characterisation and discussion. *Olympiads in Informatics, 10*(1), 39–60. doi:10.15388/ioi.2016.03

Conchinha, C., Osório, P., & de Freitas, J. C. (2015). Playful learning: Educational robotics applied to students with learning disabilities. In *Proceedings of International symposium on computers in education (SIIE)* (pp. 167-171). Setubal, Portugal: IEEE. 10.1109/SIIE.2015.7451669

Conde, M. Á., Rodríguez-Sedano, F. J., Fernández-Llamas, C., Gonçalves, J., Lima, J., & García-Peñalvo, F. J. (2021). Fostering STEAM through challenge-based learning, robotics, and physical devices: A systematic mapping literature review. *Computer Applications in Engineering Education, 29*(1), 46–65. doi:10.1002/cae.22354

Constantinou, V., & Ioannou, A. (2018). *Development of computational thinking skills through educational robotics* [Paper presentation]. *13th European Conference On Technology Enhanced Learning*, Leeds, UK.

Constantinou, V., & Ioannou, A. (2018). Development of Computational Thinking Skills through Educational Robotics. In *EC-TEL*. Practitioner Proceedings.

Conti, D., Commodari, E., & Buono, S. (2017). Personality factors and acceptability of socially assistive robotics in teachers with and without specialized training for children with disability. *Life Span and Disability, 20*(2), 251–272.

Conversy, S. (2014). Unifying textual and visual: A theoretical account of the visual perception of programming languages. *Proceedings of the 2014 ACM International Symposium on New Ideas, New Paradigms, and Reflections on Programming & Software.* 10.1145/2661136.2661138

Copley, J., & Ziviani, J. (2004). Barriers to the use of assistive technology for children with multiple disabilities. *Occupational Therapy International, 11*(4), 229–243. doi:10.1002/oti.213 PMID:15771212

Costa, A., Steffgen, G., & Ziafati, P. (2017). *Socially assistive robots for teaching emotional abilities to children with autism spectrum disorder.* CIFIL: Characterization of Railway Infrastructure using Lidar View project Quality of work View project. www.luxai.eu

Costa, M. F., & Fernandes, J. (2004). Growing up with robots. *Proceedings of Hsci2004.* http://www.hsci.info/hsci2004/PROCEEDINGS/FinalPapers/E00461377837.pdf

Coull, N. J., & Duncan, I. M. (2011). Emergent requirements for supporting introductory programming. *Innovation in Teaching and Learning in Information and Computer Sciences, 10*(1), 78–85. doi:10.11120/ital.2011.10010078

Cowen, A. (2013, August 30). *Building Bristlebots: Basic Toothbrush Robotics.* https://www.sciencebuddies.org/blog/building-bristlebots-basic-toothbrush-robotics?from=Blog

Cox, M. J., Cox, K., & Preston, C. (2000). *What factors support or prevent teachers from using ICT in their classrooms?* Academic Press.

Crnokić, B., Grubišić, M., & Volarić, T. (2017). Different applications of mobile robots in education. *International Journal on Integrating Technology in Education, 6*(3), 15–28. doi:10.5121/ijite.2017.6302

Crowder, R. M., & Zauner, K. P. (2013). A project-based biologically-inspired robotics module. *IEEE Transactions on Education, 56*(1), 82-87. Advance online publication. doi:10.1109/TE.2012.2215862

Csizmadia, A., Curzon, P., Dorling, M., Humphreys, S., Ng, T., Selby, C., & Woollard, J. (2015). *Computational thinking-A guide for teachers*. Computing at School. https://eprints.soton.ac.uk/424545/1/150818_Computational_Thinking_1_.pdf

Cuban, L. (2001). *Oversold and Underused: Reforming Schools Through Technology, 1980- 2000*. Harvard University Press.

Cullinan, M. A., Panas, R. M., Dibiasio, C. M., & Culpepper, M. L. (2012). Scaling electromechanical sensors down to the nanoscale [Review]. *Sensors and Actuators. A, Physical*, *187*, 162–173. doi:10.1016/j.sna.2012.08.035

Curto, B., & Moreno, V. (2016). Robotics in education. *Journal of Intelligent & Robotic Systems*, *81*(1), 3–4. doi:10.100710846-015-0314-z

Cutumisu, M., Adams, C., & Lu, C. A. (2019). Scoping Review of Empirical Research on Recent Computational Thinking Assessments. *Journal of Science Education and Technology*, *28*(6), 651–676. doi:10.100710956-019-09799-3

D'Agostino, L. (2013). *Dalle competenze agite al "Campo delle competenze trasversali"*. Academic Press.

Damianidou, D., Eidels, A., & Arthur-Kelly, M. (2020). The Use of Robots in Social Communications and Interactions for Individuals with ASD: A Systematic Review. *Advances in Neurodevelopmental Disorders*, *4*(4), 357–388. doi:10.100741252-020-00184-5

Dardani, C., Kaburlasos, V., & Amanatidis, A. (2020). *Enhancing the Application of Interactive Social Robots in Educational Interventions for Autism Spectrum Disorders: A Research Initiative in Greece*. E-Poster Presented in INSAR 2020. https://insar.confex.com/insar/2020/techapp/eposterview.cgi?eposterid=278

Dautenhahn, K., Nehaniv, C. L., Walters, M. L., Robins, B., Kose-Bagci, H., Mirza, N. A., & Blow, M. (2009). KASPAR – A Minimally Expressive Humanoid Robot for Human–Robot Interaction Research. *Applied Bionics and Biomechanics*, *6*(3–4), 369–397. . doi:10.1155/2009/708594

Dautenhahn, K., & Billard, A. (2002). Games children with autism can play with Robota, a humanoid robotic doll. In *Universal access and assistive technology* (pp. 179–190). Springer London. doi:10.1007/978-1-4471-3719-1_18

David, D. O., Costescu, C. A., Matu, S., Szentagotai, A., & Dobrean, A. (2020). Effects of a Robot-Enhanced Intervention for Children With ASD on Teaching Turn-Taking Skills. *Journal of Educational Computing Research*, *58*(1), 29–62. doi:10.1177/0735633119830344

De Vries, M. (2005). Teaching about Technology: An Introduction to the Philosophy of Technology for Non-philosophers. In *Science & Technology Education Library* (Vol. 27). Springer. doi:10.1007/1-4020-3410-5

De Vries, M., & Tamir, A. (1997). Shaping Concepts of Technology: What Concepts and How to Shape Them. *International Journal of Technology and Design Education*, *7*, 3–10.

Demetriou, G. A., Lambrou, A., Eteokleous, N., & Sisamos, C. (2013). The Engino Robotics Platform (ERP) controller for education. In *Control & Automation (MED), 2013 21st Mediterranean Conference on* (pp. 567-572). IEEE.

Deng, W., Pi, Z., Lei, W., Zhou, Q., & Zhang, W. (2019). Pencil Code improves learners' computational thinking and computer learning attitude. *Computer Applications in Engineering Education*, *2020*(28), 90–104. doi:10.1002/cae.22177

Després, C., & George, S. (1999). *Computer-supported distance learning-an example in educational robotics*. Université du Maine.

Desurvire, H., Caplan, M., & Toth, J. A. (2004). Using heuristics to evaluate the playability of games. *Extended Abstracts of the 2004 Conference on Human Factors and Computing Systems - CHI '04*, 1509. 10.1145/985921.986102

Dewey, J. (1997). *How we think*. Courier Corporation.

Di Battista, S., Pivetti, M., Moro, M., & Menegatti, E. (2020). Teachers' opinions towards educational robotics for special needs students: An exploratory Italian study. *Robotics*, *9*(3), 72. doi:10.3390/robotics9030072

DiGiacomo, D. K., & Gutiérrez, K. D. (2014). *Learning and becoming in an after school program: The relationship as a tool for equity within the practices of making and tinkering* [Conference Paper]. Academic Press.

Dimitriadou, A., Djafarova, N., Turetken, O., Verkuyl, M., & Ferworn, A. (2021). Challenges in serious game design and development: Educators' experiences. *Simulation & Gaming*, *52*(2), 132–152. doi:10.1177/1046878120944197

Dionne, L., Lemyre, F., & Savoie-Zajc, L. (2010). Vers une définition englobante de la communauté d'apprentissage (CA) comme dispositif de développement professionnel. *Revue des Sciences de l'Education*, *36*(1), 25–43. doi:10.7202/043985ar

Dittert, N., Robinson, S., Thestrup, K., & Knudsen, J. (2019). Supporting creativity and collaboration: Considerations for the development of a technologically enhanced toolkit for kindergartens. *Proceedings of the FabLearn Europe 2019 Conference*, 1–3. 10.1145/3335055.3335064

Dochshanov, A., & Lapina, M. (2019). Robotics in STEM education: A multiperspective strategy case study. *CEUR Workshop Proceedings*.

Dochshanov, A., & Tramonti, M. (2018). Re-making Classroom Borders with Tinkering Approach. *Conference Proceedings*.

Dochshanov, A., & Tramonti, M. (2020). Computational Design Thinking and Physical Computing: Preliminary Observations of a Pilot Study. *Robotics*, *9*(3), 71. https://www.mdpi.com/2218-6581/9/3/71

Dochshanov, A., Verotti, M., & Belfiore, N. P. (2017). A Comprehensive Survey on Microgrippers Design: Operational Strategy [Review]. *Journal of Mechanical Design*, *139*(7), 070801. Advance online publication. doi:10.1115/1.4036352

Dong, C., Cao, S., & Li, H. (2020). Young children's online learning during covid-19 pandemic: Chinese parents' beliefs and attitudes. *Children and Youth Services Review*, *118*, 105440. doi:10.1016/j.childyouth.2020.105440 PMID:32921857

Dong, L., Subramanian, A., & Nelson, B. J. (2007). Carbon nanotubes for nanorobotics. *Nano Today*, *2*(6), 12–21. doi:10.1016/S1748-0132(07)70169-X

Doppelt, Y. (2009). Assessing creative thinking in design-based learning. *International Journal of Technology and Design Education*, *19*(1), 55–65. doi:10.100710798-006-9008-y

Dos Santos Lopes, M. S., Gomes, I. P., Trindade, R. M., da Silva, A. F., & Lima, A. C. (2016). Web environment for programming and control of a mobile robot in a remote laboratory. *IEEE Transactions on Learning Technologies*, *10*(4), 526–531. doi:10.1109/TLT.2016.2627565

Downie, N. A. (2001). *Vacuum Bazookas, Electric Rainbow Jelly and 27 Other Saturday Science Projects*. Princeton University Press.

Drent, M., & Meelissen, M. (2008). Which factors obstruct or stimulate teacher educators to use ICT innovatively? *Computers & Education*, *51*(1), 187–199. doi:10.1016/j.compedu.2007.05.001

Duncan, C., Bell, T., & Atlas, J. (2017). What Do the Teachers Think?: Introducing Computational Thinking in the Primary School Curriculum. In *Proceedings of the Nineteenth Australasian Computing Education Conference* (pp. 65-74). ACM. 10.1145/3013499.3013506

Duquette, A., Michaud, F., & Mercier, H. (2008). Exploring the use of a mobile robot as an imitation agent with children with low-functioning autism. *Autonomous Robots*, *24*(2), 147–157. doi:10.100710514-007-9056-5

Durak, H. Y., & Saritepeci, M. (2018). Analysis of the relation between computational thinking skills and various variables with the structural equation model. *Computers & Education*, *116*, 191–202. doi:10.1016/j.compedu.2017.09.004

Durak, H. Y., Yilmaz, F. G. K., & Yilmaz, R. (2019). Computational thinking, programming self-efficacy, problem solving and experiences in the programming process conducted with robotic activities. *Contemporary Educational Technology, 10*(2), 173–197. doi:10.30935/cet.554493

Ebrahimipour, V., & Suzuki, K. (2005). US Department of Labor Bureau of Labor Statistics; Report 948 US Department of Labor Bureau of Labor Statistics; Report 948, 2000. *Journal of Chemical Engineering of Japan, 38*(6), 436–445. doi:10.1252/jcej.38.436

educ8s.tv. (2015, 26 September). *Line Follower Robot without Arduino or Raspberry Pi - A DIY project* [Video]. YouTube. https://www.youtube.com/watch?v=BcUCDADzndw

Eguchi, A. (2007). *Educational robotics for elementary school classroom.* Paper presented at the Society for Information Technology & Teacher Education International Conference.

Eguchi, A. (2010). What is educational robotics? Theories behind it and practical implementation. In *Proceedings of Society for Information Technology & Teacher Education International Conference* (pp. 4006-4014). Academic Press.

Eguchi, A. (2010). What is Educational Robotics? Theories behind it and practical implementation. In Society for information technology & teacher education international conference (pp. 4006-4014). Association for the Advancement of Computing in Education (AACE).

Eguchi, A. (2010). *What is educational robotics? Theories behind it and practical implementation.* Paper presented at the Society for information technology & teacher education international conference.

Eguchi, A. (2010, 03). What is Educational Robotics? Theories behind it and practical implementation.

Eguchi, A. (2014). Educational robotics for promoting 21st century skills. *Journal of Automation, Mobile Robotics and Intelligent Systems*, 5-11.

Eguchi, A. (2014). *Robotics as a learning tool for educational transformation.* Paper presented at the Proceeding of 4th international workshop teaching robotics, teaching with robotics & 5th international conference robotics in education, Padova, Italy.

Eguchi, A. (2014, July). Robotics as a learning tool for educational transformation. *4th international workshop teaching robotics, teaching with robotics & 5th international conference robotics in education,* 27–34.

Eguchi, A. (2014). Robotics as a Learning Tool for Educational Transformation. *International Workshop Teaching Robotics, Teaching with Robotics & International Conference Robotics in Education*, 27–34. 10.4018/978-1-4666-8363-1.ch002

Eguchi, A. (2017). Bringing robotics in classrooms. In M. Khine (Ed.), *Robotics in STEM education* (pp. 3–31). Springer. doi:10.1007/978-3-319-57786-9_1

El-Hamamsy, L., Bruno, B., Chessel-Lazzarotto, F., Chevalier, M., Roy, D., Zufferey, J. D., & Mondada, F. (2021). The symbiotic relationship between educational robotics and computer science in formal education. *Education and Information Technologies*, 1–31. PMID:33841027

El-Hamamsy, L., Chessel-Lazzarotto, F., Bruno, B., Roy, D., Cahlikova, T., Chevalier, M., Parriaux, G., Pellet, J.-P., Lanarès, J., & Zufferey, J. D. (2020). A computer science and robotics integration model for primary school: Evaluation of a large-scale in-service K-4 teacher-training program. *Education and Information Technologies*, 1–31. PMID:33162777

Elicin, Ö. (2016). *Özel eğitimde teknoloji destekli öğretim.* Vize Publishing.

ElMaraghy, W. H. (2019). Engineering Design. In CIRP Encyclopedia of Production Engineering. Springer. (pp. 608-612). doi:10.1007/978-3-662-53120-4_16781

Eordanidis, S., Gee, E., & Carmichael, G. (2017). The effectiveness of pairing analog and digital games to teach computer science principles to female youth. *Journal of Computing Sciences in Colleges*, *32*(3), 12–19.

Erlam, G. D., Smythe, L., & Clair, W.-S. (2017). Simulation is not a pedagogy. *Open Journal of Nursing*, *7*, 779–787. doi:10.4236/ojn.2017.77059

Eroglu, S., & Bektas, O. (2016). Ideas of science teachers took STEM Education about STEM based activities. *Journal of Qualitative Research in Education - JOQRE*, *4*(3), 43-67.

Ersoy, H., Madran, R. O., & Gulbahar, Y. (2011, February). *Programlama dilleri öğretimine bir model önerisi: Robot programlama.* XIII. Akademik Bilişim Konferansı, Malatya, Turkey.

Erten, E. (2019). *A case study on coding and robotic teaching* [Unpublished master's dissertation]. Balıkesir University, Balıkesir, Turkey.

Esteve-Mon, F. M., Adell-Segura, J., Llopis Nebot, M. Á., Valdeolivas Novella, M. G., & Pacheco Aparicio, J. (2019). *The Development of Computational Thinking in Student Teachers through an Intervention with Educational Robotics.* Academic Press.

Estivill-Castro, V. (2019). Inviting teachers to use educational robotics to foster mathematical problem-solving. *International Conference on Robotics and Education RiE 2017*, 248–261.

Eteokleous-Grigoriou, N., & Psomas, C. (2013). Integrating Robotics as an Interdiscipli-nary-Educational Tool in Primary Education. In R. McBride & M. Searson (Eds.), *Proceedings of Society for Information Technology & Teacher Education International Conference 2013* (pp. 3877-3881). Chesapeake, VA: AACE.

Eteokleous, N. (2008). Evaluating computer technology integration in a centralized educational system. *Computers in Education Journal*, *51*(2), 669–686. doi:10.1016/j.compedu.2007.07.004

Eteokleous, N. (2018). Employing educational robotics for the development of problem-based learning skills. In *Encyclopedia of Information Science and Technology* (4th ed., pp. 2492–2502). IGI Global.

Eteokleous, N., & Ktoridou, D. (2014). Educational robotics as learning tools within the teaching and learning practice. In *2014 IEEE Global Engineering Education Conference (EDUCON)* (pp. 1055-1058). IEEE. 10.1109/EDUCON.2014.6826237

Eteokleous, N., Neophytou, R., Kolani, E., & Christodoulou, C. (2020). The case of the Robotics Academy@ Frederick University: 21st Century Skills Developed through a Non-formal Educational Setting. *International Conference in Open & Distance Learning, 10*(1B), 162-170. 10.12681/icodl.2434

Euler, M. (2009). Schülerlabore. Lernen; forschen und kreative Potenziale entfalten. In Kursbuch 2010. Schülerlabore in Deutschland. Tectum.

Fachantidis, N., Syriopoulou-Delli, C. K., Vezyrtzis, I., & Zygopoulou, M. (2020). Beneficial effects of robot-mediated class activities on a child with ASD and his typical classmates. *International Journal of Developmental Disabilities*, *66*(3), 245–253. doi:10.1080/20473869.2019.1565725 PMID:34141387

Fachantidis, N., Syriopoulou-Delli, C. K., & Zygopoulou, M. (2018). The effectiveness of socially assistive robotics in children with ASD. *International Journal of Developmental Disabilities.* Advance online publication. doi:10.1080/20473869.2018.1495391 PMID:34141373

Fadel, C., Bialik, M., & Trilling, B. (2015). *Four-dimensional education.* Academic Press.

Fagerlund, J., Häkkinen, P., Vesisenaho, M., & Viiri, J. (2021). Computational thinking in programming with scratch in primary schools: A systematic review. *Computer Applications in Engineering Education, 29*(1), 12–28. doi:10.1002/cae.22255

Fan, J., Yu, W., & Hunter, L. (2004). *Clothing appearance and fit: Science and technology*. Woodhead Publishing.

Fanny, B., Julie, H., & Anne-Sophie, C. (2020). Developing a critical robot literacy for young people from conceptual metaphors analysis. *2020 IEEE Frontiers in Education Conference (FIE),* 1–7.

Farias, G., Fabregas, E., Peralta, E., Vargas, H., Dormido-Canto, S., & Dormido, S. (2019). Development of an easy-to-use multi-agent platform for teaching mobile robotics. *IEEE Access: Practical Innovations, Open Solutions, 7*, 55885–55897. doi:10.1109/ACCESS.2019.2913916

Farmer, C. (2015). Engineer your World Engineering Design and Problem Solving. In C. L. Sneider (Ed.), *The Go-To Guide for Engineering Curricula, Grades 9–12: Choosing and Using the Best Instructional Materials for Your Students Engineer your World Engineering Design and Problem Solving* (pp. 73–89). SAGE Publications. doi:10.4135/9781483388373.n8

Fast, L. (2006). Zur Genese von Baukästen im Technikunterricht. *Unterricht - Arbeit + Technik, 8*(29), 50-53.

Fearing, R. S. (1995). Survey of sticking effects for micro parts handling. *Proceedings 1995 IEEE/RSJ International Conference on Intelligent Robots and Systems. Human Robot Interaction and Cooperative Robots.*

Feldhusen, J., & Grote, K. H. (2013). Der Produktentstehungsprozess (PEP). In J. Feldhusen & K. H. Grote (Eds.), Pahl/Beitz Konstruktionslehre (pp. 11–14). Springer Vieweg. https://doi.org/10.1007/978-3-642-29569-0_2.

Feldhusen, J., & Grote, K. H. (2013). Der Produktentstehungsprozess (PEP). In J. Feldhusen & K. H. Grote (Eds.), *Pahl/Beitz Konstruktionslehre* (pp. 11–14). Springer Vieweg. doi:10.1007/978-3-642-29569-0_2

Ferrari, E., Robins, B., & Dautenhahn, K. (2009). Therapeutic and educational objectives in robot assisted play for children with autism. *Proceedings - IEEE International Workshop on Robot and Human Interactive Communication,* 108–114. 10.1109/ROMAN.2009.5326251

Fessakis, G., Gouli, E., & Mavroudi, E. (2013). Problem solving by 5–6 years old kindergarten children in a computer programming environment: A case study. *Computers & Education, 63*, 87–97. doi:10.1016/j.compedu.2012.11.016

Fessakis, G., Komis, V., Mavroudi, E., & Prantsoudi, S. (2018). Exploring the scope and the conceptualization of computational thinking at the K-12 classroom level curriculum. In M. S. Khine (Ed.), *Computational thinking in the STEM disciplines: Foundations and research highlights* (pp. 181–212). Springer. doi:10.1007/978-3-319-93566-9_10

Fhunter. (2010, November 4). *RC Bristlebots shifts weight for steering*. https://hackaday.com/2010/11/03/rc-bristlebot-shifts-weight-for-steering/#comment-207574

Filippov, S., Ten, N., Fradkov, A., & Shirokolobov, I. (2017). Robotics education in Saint Petersburg secondary school. *International Conference on Robotics and Education RiE 2017.*

Finio, B. (2020, June 23). *Design Your Own Vibrobots.* https://www.sciencebuddies.org/science-fair-projects/project-ideas/Robotics_p030/robotics/vibrobots

Fischer, E. (2000). *Die andere Bildung: Was man von den Naturwissenschaften wissen sollte*. Ullstein.

Fislake, M. (2018). Robotics in Technology Education. In M. de Vries (Ed.), Handbook of Technology Education (pp. 361–384). Springer. https://doi.org/10.1007/978-3-319-44687-5_28.

Fislake, M. (2021). Educational Robotics Between Coding and Engineering Education. In S. Papadakis & M. Kalogiannakis (Eds.), Handbook of Research on Using Educational Robotics to Facilitate Student Learning (pp. 107–139). IGI Global. https://doi.org/10.4018/978-1-7998-6717-3.ch004.

Fislake, M. (2001). *Determinanten und Interdependenzen Technischer Bildung aus didaktischer und bildungsökonomischer Sicht*. Peter Lang.

Fislake, M. (2010). Kinder Technik Ferien Camps in Rheinland-Pfalz - Bildungsarbeit zwischen Markt und Mission. Bildungsprozesse im MINT. In C. Quaiser-Pohl & M. Endepohls-Ulpe (Eds.), *Bildungsprozesse im MINT-Bereich. Interesse, Partizipation und Leistungen von Mädchen und Jungen* (pp. 127–141). Waxmann.

Fislake, M. (2021). Educational Robotics Between Coding and Engineering Education. In S. Papadakis & M. Kalogiannakis (Eds.), *Handbook of Research on Using Educational Robotics to Facilitate Student Learning* (pp. 107–139). IGI Global. doi:10.4018/978-1-7998-6717-3.ch004

Fojtik, R. (2017). The Ozobot and education of programming. *New Trends and Issues Proceedings on Humanities and Social Sciences, 4*(5).

Fok, D., Polgar, J. M., Shaw, L., & Jutai, J. W. (2011). Low vision assistive technology device usage and importance in daily occupations. *Work (Reading, Mass.), 39*(1), 37–48. doi:10.3233/WOR-2011-1149 PMID:21815311

Forum, W. E. (2020). *The Future of Jobs Report 2020*. Academic Press.

Foster, I. (2006). 2020 Computing: A two-way street to science's future. *Nature, 440*(7083), 419–419. doi:10.1038/440419a PMID:16554785

Foster, J. M. Jr, I. R. C., & Romesberg, F. E. (2014). The Clinical use of robots for individuals with Autism Spectrum disorder. *Research in Autism Spectrum Disorders, 509*(7500), 385–388. doi:10.1016/j.rasd.2011.05.006.The

Frazer, A., Recio, A., Gilbert, L., & Wills, G. (2014). Profiling the educational value of computer games. *Interaction Design & Architecture (s) Journal-IxD&A, 19*, 1–19.

Freire, P. (2018). *Pedagogy of the oppressed*. Bloomsbury Publishing USA.

Fromm, E. (1973). *Review of Skinner's theory*. Revista de Psicoanálisis Psiquiatrîa y Psicologîa.

Fujimoto, I., Matsumoto, T., de Silva, P. R. S., Kobayashi, M., & Higashi, M. (2011). Mimicking and evaluating human motion to improve the imitation skill of children with autism through a robot. *International Journal of Social Robotics, 3*(4), 349–357. doi:10.100712369-011-0116-9

Fukuda, T., Nakajima, M., Liu, P., & ElShimy, H. (2009). Nanofabrication, nanoinstrumentation and nanoassembly by nanorobotic manipulation. *The International Journal of Robotics Research, 28*(4), 537–547.

Fullan, M. (2007). *The new meaning of educational change*. Routledge.

Futschek, G. (2016). Computational Thinking im Unterricht. *Schule Aktiv*, 4–5.

Galvez-Aranda, D., & Galvez Legua, M. (2021). What to Know Before You Start? In D. Galvez-Aranda & M. Galvez Legua (Eds.), *Robotics Models Using LEGO WeDo 2.0: Design, Build, Program, Test, Document and Share* (pp. 1–25). Apress. doi:10.1007/978-1-4842-6846-9_1

Galvez-Aranda, D., & Galvez Legua, M. (2021a). Biomimetic Robots. In D. Galvez-Aranda & M. Galvez Legua (Eds.), *Robotics Models Using LEGO WeDo 2.0: Design, Build, Program, Test, Document and Share* (pp. 281–334). Apress. doi:10.1007/978-1-4842-6846-9_7

Galvez-Aranda, D., & Galvez Legua, M. (2021b). Biped Robots. In D. Galvez-Aranda & M. Galvez Legua (Eds.), *Robotics Models Using LEGO WeDo 2.0: Design, Build, Program, Test, Document and Share* (pp. 79–128). Apress. doi:10.1007/978-1-4842-6846-9_3

Galvez-Aranda, D., & Galvez Legua, M. (2021c). Crawling Robots. In D. Galvez-Aranda & M. Galvez Legua (Eds.), *Robotics Models Using LEGO WeDo 2.0: Design, Build, Program, Test, Document and Share* (pp. 129–176). Apress. doi:10.1007/978-1-4842-6846-9_4

Galvez-Aranda, D., & Galvez Legua, M. (2021d). Humanoids. In D. Galvez-Aranda & M. Galvez Legua (Eds.), *Robotics Models Using LEGO WeDo 2.0: Design, Build, Program, Test, Document and Share* (pp. 229–280). Apress. doi:10.1007/978-1-4842-6846-9_6

Galvez-Aranda, D., & Galvez Legua, M. (2021e). Quadruped Robots. In D. Galvez-Aranda & M. Galvez Legua (Eds.), *Robotics Models Using LEGO WeDo 2.0: Design, Build, Program, Test, Document and Share* (pp. 177–228). Apress. doi:10.1007/978-1-4842-6846-9_5

Gandra, C. (2019). Dynamics of a Vibration Driven Bristle Bot. *All Theses*. https://tigerprints.clemson.edu/all_theses/3149

García-Valcárcel-Muñoz-Repiso, A., & Caballero-González, Y. (2019). Robotics to develop computational thinking in early Childhood Education. *Comunicar. Media Education Research Journal, 27*(1).

Garofalo, L., Sandler, S., & Seth, D. (2020). *Evaluation of a Snake Jaw Robot to Teach Integrated Biology*. Mathematics, and Engineering.

Garratt, G. R. M. (1994). *The early history of radio: from Faraday to Marconi*. Iet.

Gauch, H. G. Jr, & Gauch, H. G. (2003). *Scientific method in practice*. Cambridge University Press.

Gaudiello, I., & Zibetti, E. (2016). *Learning robotics, with robotics, by robotics: Educational robotics*. Wiley-ISTE. doi:10.1002/9781119335740

Gaycken, S. (2005). *Was ist technisches Wissen?: Philosophische Grundlagen technischer Wissenschaften*. https://events.ccc.de/congress/2005/fahrplan/attachments/575-paper_technischesWissen.pdf

Gee, J. P. (2003). *What Video Games Have to Teach Us about Learning and Literacy*. Palgrave Macmillan. doi:10.1145/950566.950595

Gee, J. P. (2007). *Good video games + good learning: Collected essays on video games, learning and literacy*. Peter Lang. doi:10.3726/978-1-4539-1162-4

Gee, J. P. (2008). *What video games have to teach us about learning and literacy, revised and updated*. Palgrave Macmillan.

Ge, X., & Land, S. (2003). Scaffolding students' problem-solving processes in an ill-structured task using question prompts and peer interactions. *ETR&D, 51*(1), 21–38. doi:10.1007/BF02504515

Gezici, H., Kocaoglu, S., Cosgun, E., Yilmazlar, E., & Tuna, M. (2017). Planning the Arduino embedded programming course with robot project application for mechatronics programs. *Electronic Journal of Vocational Colleges, 7*(1), 1–7.

Giang, C. (2020). *Towards the alignment of educational robotics learning systems with classroom activities* (Ph.D. dissertation). EPFL.

Giang, C., Chevalier, M., Negrini, L., Peleg, R., Bonnet, E., Piatti, A., & Mondada, F. (2018). Exploring Escape Games as a Teaching Tool in Educational Robotics. *Proceedings of Educational Robotics 2018 (EDUROBOTICS)*.

Giang, C. (2020). *Towards the alignment of educational robotics learning systems with classroom activities*. EPFL.

Giang, C., & Negrini, L. (2021). Educational Robotics in Online Distance Learning: An Experience from Primary School. In M. Merdan, W. Lepuschitz, G. Koppensteiner, R. Balogh, & D. Obdržálek (Eds.), *Robotics in Education. RiE 2021. Advances in Intelligent Systems and Computing* (Vol. 1359). Springer.

Giang, C., Piatti, A., & Mondada, F. (2019). Heuristics for the Development and Evaluation of Educational Robotics Systems. *IEEE Transactions on Education*, *62*(4), 1–10. doi:10.1109/TE.2019.2912351

Gil-Flores, J., Rodríguez-Santero, J., & Torres-Gordillo, J. J. (2017). Factors that explain the use of ICT in secondary-education classrooms: The role of teacher characteristics and school infrastructure. *Computers in Human Behavior*, *68*, 441–449. doi:10.1016/j.chb.2016.11.057

Giomi, L., Hawley-Weld, N., & Mahadevan, L. (2013). Swarming, swirling and stasis in sequestered bristle-bots. *Proc R Soc A*, *469*(2151), 20120637. doi:10.1098/rspa.2012.0637

Giullian, N., Ricks, D., Atherton, A., Colton, M., Goodrich, M., & Brinton, B. (2010). Detailed requirements for robots in autism therapy. *Conference Proceedings - IEEE International Conference on Systems, Man and Cybernetics*, 2595–2602. 10.1109/ICSMC.2010.5641908

Glasser, O. (1993). *Wilhelm Conrad Röntgen and the early history of the Roentgen rays.* Norman Publishing.

goBilda. (2021). *What is goBILDA?* https://www.gobilda.com

Gouin-Vallerand, C., Ferreira, S. M., & Hotte, R. (2018, November). Towards a mobile serious game environment for children self-learning. In *Proceedings of the 4th EAI International Conference on Smart Objects and Technologies for Social Good* (pp. 100-105). 10.1145/3284869.3284901

Greene, B., & Land, S. (2000). A qualitative analysis of scaffolding use in a resource-based learning environment involving the World Wide Web. *Journal of Educational Computing Research*, *23*(2), 151–179. doi:10.2190/1GUB-8UE9-NW80-CQAD

Green, J. L. (2011). *The ultimate guide to assistive technology in special education: Resources for education, intervention, and rehabilitation.* Prufrock Press.

Grover, S., & Pea, R. (2018). Computational Thinking: A competency whose time has come. *Computer science education: Perspectives on teaching and learning in school*, 19.

Grover, S., & Pea, R. (2013). Computational Thinking in K–12 A Review of the State of the Field. *Educational Researcher*, *42*(1), 38–43. doi:10.3102/0013189X12463051

Grover, S., & Pea, R. (2018). Computational thinking: A competency whose time has come. In S. Sentance, E. Barendsen, & C. Schulte (Eds.), *Computer science education: Perspectives on teaching and learning* (pp. 19–38). Bloomsbury Academic. doi:10.5040/9781350057142.ch-003

Guest, G., MacQueen, K. M., & Namey, E. E. (2012). *Validity and reliability (credibility and dependability) in qualitative research and data analysis. In Applied thematic analysis.* Sage Publications.

Gulbahar, Y. (2018). *Bilgi işlemsel düşünmeden programlamaya.* Pegem Akademi. doi:10.14527/9786052411117

Gundogdu, B. (2020). *The effect of algorithm teaching on vocational high school students' computational thinking, cognitive load and achievement* [Unpublished master's dissertation]. Atatürk University, Erzurum, Turkey.

Gutwill, J. P., Hido, N., & Sindorf, L. (2015). Research to Practice: Observing Learning in Tinkering Activities. *Curator*, *58*(2), 151–168. https://doi.org/10.1111/cura.12105

Hacker, L. (2003). *Robotics in education: ROBOLAB and robotic technology as tools for learning science and engineering* [Unpublished Senior Thesis]. Department of Child Development, Tufts University.

Hadjithoma, C., & Eteokleous, N. (2007). ICT in primary schools: Explaining the integration in relation to the context. *Mediterranean Journal of Educational Studies, 12*(1), 1–25.

Hammer, D. (1997). Discovery learning and discovery teaching. *Cognition and Instruction, 15*(4), 485–529.

Hamner, E., Cross, J., & Zito, L. (2016). *Training teachers to integrate engineering into non-technical middle school curriculum. In 2016 IEEE Frontiers in education conference*. FIE.

HannoHupmann. (2008, January 1). *Extrem cool -> "BürstenBot"*. https://www.roboternetz.de/community/threads/33177-Extrem-cool-B%C3%BCrstenBot?p=314629&viewfull=1#post314629

Harel, I. E., & Papert, S. E. (1991). *Constructionism*. Ablex Publishing.

Harel, I., & Papert, S. (Eds.). (1991). *Constructionsim*. Ablex Publishing Corporation.

Harris, E. E. (2014). *Hypothesis and perception: The roots of scientific method.* Routledge. doi:10.4324/9781315830018

Hart, S., Margheri, A., Paci, F., & Sassone, V. (2020). Riskio: A serious game for cyber security awareness and education. *Computers & Security, 95*, 101–827. doi:10.1016/j.cose.2020.101827

Hattie, J. (2009). *Visible Learning - A synthesis of over 800 meta-analyses relating to achievement*. Routledge.

Hattie, J. (2012). *Visible learning for teachers: maximizing impact on learning*. Routledge. doi:10.4324/9780203181522

Haydn, T. A., & Barton, R. (2007). Common needs and different agendas: How trainee teachers make progress in their ability to use ICT in subject teaching. Some lessons from the UK. *Computers & Education, 49*(4), 1018–1036. doi:10.1016/j.compedu.2005.12.006

Hedges, H. (2010). Blurring the boundaries: Connecting research, practice and professional learning. *Cambridge Journal of Education, 40*(3), 299–314. doi:10.1080/0305764X.2010.502884

Heersmink, R. (2013). A Taxonomy of Cognitive Artifacts: Function, Information, and Categories. *Review of Philosophy and Psychology, 4*(3), 465–481. doi:10.100713164-013-0148-1

Heidegger, M. (1977). Science and reflection. *The question concerning technology and other essays, 173*.

Heinssen, R. K. Jr, Glass, C. R., & Knight, L. A. (1987). Assessing computer anxiety: Development and validation of the computer anxiety rating scale. *Computers in Human Behavior, 3*(1), 49–59. doi:10.1016/0747-5632(87)90010-0

Hemal, A. K., & Menon, M. (2018). *Robotics in genitourinary surgery*. Springer. doi:10.1007/978-3-319-20645-5

Henderson, P. B., Cortina, T. J., & Wing, J. M. (2007). Computational thinking. In *Proceedings of the 38th SIGCSE Technical Symposium on Computer Science Education-SIGCSE* '07, 195–19. 10.1145/1227310.1227378

Henseler, K., & Höpken, G. (1996). *Methodik des Technikunterrichts*. Klinkhardt.

Henshilwood, C. S., & Dubreuil, B. (2009). Reading the artefacts: gleaning language skills from the Middle Stone Age in southern Africa. *The Cradle of Language, 2*, 61-92.

Hersh, M. A., & Johnson, M. A. (2008). On modelling assistive technology systems. Part I: Modelling framework. *Technology and Disability, 20*(3), 193–215. doi:10.3233/TAD-2008-20303

Hiltunen, J., & Jarvinen, E. M. (2000). Automation Technology in Elementary Technology Education. *Journal of Industrial Teacher Education, 37*(4), 51–76.

Himpsl-Gutermann, K., Brandhofer, G., Frick, K., Fikisz, W., Steiner, M., Bachinger, A., Gawin, A., Gawin, P., Szepannek, P., & Lechner, I. (2017). *Denken lernen–Probleme lösen (DLPL).* Primarstufe.

Hixon, R. (2007). Teaching software engineering principles using Robolab and Lego Mindstorms. *International Journal of Engineering Education, 23*(5), 868–873.

Hobart, C. G., Mazumdar, A., Spencer, S. J., Quigley, M., Smith, J. P., Smith, J. P., . . . Buerger, S. P. (2020). Achieving Versatile Energy Efficiency with the WANDERER Biped Robot. *IEEE Transactions on Robotics, 36*(3), 959-966. doi:10.1109/TRO.2020.2969017

Hodges, C., Moore, S., Lockee, B., Trust, T., & Bond, A. (2020). *The difference between emergency remote teaching and online learning.* Retrieved August 2021 from https://er.educause.edu/articles/2020/3/the-difference-between-emergency-remote-teaching-and-online-learning

Holeva, V., Nikopoulou, V. A., Papadopoulou, M., Vrochidou, E., Papakostas, G. A., & Kaburlasos, V. G. (2019). Toward Robot-Assisted Psychosocial Intervention for Children with Autism Spectrum Disorder (ASD). In Lecture Notes in Computer Science: Vol. 11876. *Social Robotics. ICSR 2019.* Springer. doi:10.1007/978-3-030-35888-4_45

Holland, J. M. (2004). *Designing autonomous mobile robots: Inside the mind of an intelligent machine.* Elsevier.

Honey, M., Pearson, G., & Schweingruber, H. A. (2014). *STEM integration in K-12 education: Status, prospects, and an agenda for research* (Vol. 500). National Academies Press Washington.

Hsu, T. C., Chang, S. C., & Hung, Y. T. (2018). How to learn and how to teach computational thinking: Suggestions based on a review of the literature. *Computers & Education, 126,* 296–310. doi:10.1016/j.compedu.2018.07.004

Huijnen, C. A. G. J., Lexis, M. A. S., Jansens, R., & de Witte, L. P. (2019). Roles, Strengths and Challenges of Using Robots in Interventions for Children with Autism Spectrum Disorder (ASD). *Journal of Autism and Developmental Disorders, 49*(1), 11–21. doi:10.100710803-018-3683-x PMID:30019273

Huijnen, C., Lexis, M., & Witte, L. (2017). Robots as New Tools in Therapy and Education for Children with Autism. *Robots as New Tools in Therapy and Education for Children with Autism., 4*(4), 1–4. doi:10.4172/2376-0281.1000278

Humbert, L., Herper, H., Best, A., Borowski, C., Freudenberg, R., Fricke, M., Haselmeier, K., Hinz, V., Müller, D., & Schwill, A. (2019). *Empfehlungen der GI–Kompetenzen für informatische Bildung im Primarbereich.* Informatik für alle.

Humphries, L., & McDonald, S. (2011). Emotion faces: the design and evaluation of a game for preschool children. *Proceedings of the 2011 annual conference extended abstracts on Human factors in computing systems (CHIEA '11),* 1453. 10.1145/1979742.1979790

Hußmann, S., Thiele, J., Hinz, R., Prediger, S., & Ralle, B. (2013). Gegenstandsorientierte Unterrichtsdesigns entwickeln und erforschen. In Der lange Weg zum Unterrichtsdesign. Zur Begründung und Umsetzung fachdidaktischer Forschungs- und Entwicklungsprogramme. Münster: Waxmann.

Hutchison, A. (2012). Literacy teachers' perceptions of professional development that increases integration of technology into literacy instruction. *Technology, Pedagogy and Education, 21*(1), 37–56. doi:10.1080/1475939X.2012.659894

Hymavathi, V., & Kumar, V. (2011). Design and implementation of double line follower robot. *International Journal of Engineering Science and Technology,* 3.

Innovation First Labs, Inc. (2018, June 13). *HEXBUG nano.* https://www.hexbug.com/hexbug-nano-single.html

Ioannou, I., & Angeli, C. (2016). A Framework and an Instructional Design Model for the Development of Students' Computational and Algorithmic Thinking. In MCIS (p. 19). Academic Press.

Ioannou, A., & Makridou, E. (2018). Exploring the potentials of educational robotics in the development of computational thinking: A summary of current research and practical proposal for future work. *Education and Information Technologies, 23*(6), 2531–2544. doi:10.100710639-018-9729-z

Ioi, K. (1999). Mobile micro-robot using centrifugal forces. *Journal of the Robotics Society of Japan, 17*(3), 736–741. doi:10.7210/jrsj.17.396

ISTE & CSTA. (2011). *Operational Definition of Computational Thinking*. https://cdn.iste.org/www-root/ct-documents/computational-thinking-operational-definition-flyer.pdf?sfvrsn=2

ITEEA, International Technology and Engineering Educators Association. (2007). Standards for Technological Literacy: Content for the Study of Technology (3rd ed.). ITEEA.

ITEEA, International Technology and Engineering Educators Association. (2020). *Standards for Technological and Engineering Literacy: Defining the Role of Technology and Engineering in STEM Education*. ITEEA.

ITEEA, International Technology and Engineering Educators Association. (2020). *Standards for Technological and Engineering Literacy: Defining the Role of Technology and Engineering in STEM Education*. Pre-Publication Copy. Author.

Jablonski, J., & Posey, J. (1985). Robotics terminology. In *Handbook of industrial robotics*. Wiley.

Jacobson, M. J., & Wilensky, U. (2006). Complex Systems in Education: Scientific and Educational Importance and Implications for the Learning Sciences. *Journal of the Learning Sciences, 15*(1), 11–34. https://doi.org/10.1207/s15327809jls1501_4

Janka, P. (2008). Using a programmable toy at preschool age: Why and how. In *Teaching with robotics: didactic approaches and experiences. Workshop of International Conference on Simulation, Modeling and Programming Autonomous Robots* (pp. 112-121). Academic Press.

Jeretin-Kopf, M., Kosack, M., & Wiesmüller, M. (2015). Technikdidaktische Medien – Einfluss verschiedener technikdidaktischer Medien auf die kindliche Motivation, problemlösendes Denken und technische Kreativität. In G. Graube, M. Jeretin-Kopf, W. Kosack, I. Mammes, O. Renn & C. Wiesmüller (Eds.), Wissenschaftliche Untersuchungen zur Arbeit der Stiftung "Haus der kleinen Forscher". Stiftung Haus der kleinen Forscher.

Jim, C. K. W. (2010). *Teaching with LEGO mindstorms robots: Effects on learning environment and attitudes toward science* [Unpublished master's dissertation]. The University of Texas at Dallas, United States. Available from ProQuest Dissertations and Theses database. (UMI No. 1489918)

Jimoyiannis, A. (2010). Designing and implementing an integrated technological pedagogical science knowledge framework for science teachers professional development. *Computers & Education, 55*(3), 1259–1269. doi:10.1016/j.compedu.2010.05.022

Johnston, C., Thurlow, M., Altman, J., Timmons, J., & Kato, K. (2009). Assistive technology approaches for large-scale assessment: Perceptions of teachers of students with visual. *Impairments Exceptionality, 17*(2), 66–75. doi:10.1080/09362830902805756

Jona, K., Wilensky, U., Trouille, L., Horn, M. S., Orton, K., Weintrop, D., & Beheshti, E. (2014, January). Embedding computational thinking in science, technology, engineering, and math (CT-STEM). In Future directions in computer science education summit meeting, Orlando, FL.

Jonassen, D. H., & Hung, W. (2015). All problems are not equal: Implications for problem-based learning. *Essential readings in problem-based learning*, 7-41.

Ju, A., Chen, D., & Tang, J. (2020). Design of Educational Robot Platform Based on Graphic Programming. In M. Atiquzzaman, N. Yen, & Z. Xu (Eds.), *Big Data Analytics for Cyber-Physical System in Smart City* (pp. 382–391). Springer Singapore. doi:10.1007/978-981-15-2568-1_53

Jung, S. E., & Won, E. S. (2018). Systematic review of research trends in robotics education for young children. *Sustainability (Switzerland), 10*(4), 1–24. doi:10.3390u10040905

Junior, L. A., Neto, O. T., Hernandez, M. F., Martins, P. S., Roger, L. L., & Guerra, F. A. (2013). A low-cost and simple arduino-based educational robotics kit. *Cyber Journals: Multidisciplinary Journals in Science and Technology Journal of Selected Areas in Robotics and Control, 3*(12), 1–7.

Kalogiannakis, M., Nirgianaki, G. M., & Papadakis, S. (2018). Teaching magnetism to preschool children: The effectiveness of picture story reading. *Early Childhood Education Journal, 46*(5), 535–546. doi:10.100710643-017-0884-4

Kalogiannakis, M., & Papadakis, S. (2017). Combining mobile technologies in environmental education: A Greek case study. *International Journal of Mobile Learning and Organisation, 11*(2), 108–130. doi:10.1504/IJMLO.2017.084272

Kandlhofer, M., & Steinbauer, G. (2016). Evaluating the impact of educational robotics on pupils' technical-and social-skills and science related attitudes. *Robotics and Autonomous Systems, 75*, 679–685. doi:10.1016/j.robot.2015.09.007

Karaahmetoğlu, K. (2019). *The effect of project-based Arduino educational robot applications on students' computational thinking skills and their perception of basic STEM skill levels* [Unpublished master's dissertation]. Amasya University, Amasya, Turkey.

Karakosta, E., Dautenhahn, K., Syrdal, D. S., Wood, L. J., & Robins, B. (2019). *Using the humanoid robot Kaspar in a Greek school environment to support children with Autism Spectrum Condition.* De Gruyter.

Karakoyun, F., & Lindberg, O. J. (2020). Preservice teachers' views about the twenty-first century skills: A qualitative survey study in Turkey and Sweden. *Education and Information Technologies, 25*(4), 2353–2369. doi:10.100710639-020-10148-w

Karthik, K. P., & Chandra, R. P. (2014). An overview of agricultural robots. *International Journal & Magazine of Engineering, Technology, Management and Research.* http://www.yuvaengineers.com/an-overview-of-agricultural-robots-p-koteswara-karthik-p-ravi-chandra

Kasari, C., Freeman, S., & Paparella, T. (2006). Joint attention and symbolic play in young children with autism: A randomized controlled intervention study. *Journal of Child Psychology and Psychiatry, and Allied Disciplines, 47*(6), 611–620. doi:10.1111/j.1469-7610.2005.01567.x PMID:16712638

Kaya, E., Yesilyurt, E., Deniz, H., Newley, A., & Newley, P. (2017). Research and teaching: Introducing engineering design to a science teaching methods course through educational robotics and exploring changes in views of preservice elementary teachers. *Journal of College Science Teaching, 047*(02), 66–75. doi:10.2505/4/jcst17_047_02_66

Kaygisiz, G. M., Uzumcu, O., & Ucar, F. M. (2020). The case of prospective teachers' integration of coding-robotics practices into science teaching with STEM approach. *Elementary Education Online, 19*(3), 1200–1213. doi:10.17051/ilkonline.2020.728020

Kazakoff, E. R., Sullivan, A., & Bers, M. U. (2013). The effect of a classroom-based intensive robotics and programming workshop on sequencing ability in early childhood. *Early Childhood Education Journal, 41*(4), 245–255. doi:10.100710643-012-0554-5

Kazakoff, E., & Bers, M. (2012). Programming in a robotics context in the kindergarten classroom: The impact on sequencing skills. *Journal of Educational Multimedia and Hypermedia*, *21*(4), 371–391.

Kazimoglu, C., Kiernan, M., Bacon, L., & Mackinnon, L. (2012). A Serious Game for Developing Computational Thinking and Learning Introductory Computer Programming. *Procedia: Social and Behavioral Sciences*, *47*, 1991–1999. doi:10.1016/j.sbspro.2012.06.938

Kearns, L. R. (2012). Student Assessment in Online Learning: Challenges and Effective Practices. *Journal of Online Learning and Teaching*, *8*(3), 198–208.

Keen, D., & Halle, J. W. (2016). *Prelinguistic and Minimally Verbal Communicators on the Autism Spectrum* (H. Meadan & N. C. Brady, Eds.). Springer Verlag. doi:10.1007/978-981-10-0713-2

Khanlari, A., & Mansourkiaie, F. (2015, July). Using robotics for STEM education in primary/elementary schools: Teachers' perceptions. *2015 10th International Conference on Computer Science Education (ICCSE).*

Khanlari, A. (2013). Effects of robotics on 21st century skills. *European Scientific Journal*, *9*(27), 26–36.

Khanlari, A. (2013). Effects of robotics on 21st century skills. *European Scientific Journal, ESJ*, *9*(27), 26–36.

Khasawneh, O. Y. (2018). Technophobia without boarders: The influence of technophobia and emotional intelligence on technology acceptance and the moderating influence of organizational climate. *Computers in Human Behavior*, *88*, 210–218. doi:10.1016/j.chb.2018.07.007

Kids, B. (2019). *Framework for 21st century learning definitions*. Battelle for Kids.

Kilinc, A. (2014). *The using of robotic technology in teaching of light unit 7th grade* [Unpublished master's dissertation]. Erciyes University, Turkey.

Kilpatrick, W. H. (1918). *The Project Method: The Use of the Purposeful Act in the Educative Process*. Teachers College, Columbia University.

Kim, E.S., Berkovits, L.D., Bernier, E.P., Leyzberg, D., Shic, F., Paul, R., & Scassellati, B. (2013). Social robots as embedded reinforcers of social behavior in children with autism. *Journal of Autism Development Disorder*, *43*, 1038–1049. . doi:10.1007/s10803-012-1645-2

Kim, C. J., & Santiago, R. (2005). Construction of e-learning environments in Korea. *Educational Technology Research and Development*, *53*(4), 108–114. doi:10.1007/BF02504690

Kim, C., Kim, D., Yuan, J., Hill, R. B., Doshi, P., & Thai, C. N. (2015). Robotics to promote elementary education pre-service teachers' STEM engagement, learning, and teaching. *Computers & Education*, *91*, 14–31. doi:10.1016/j.compedu.2015.08.005

Kim, C., Kim, M. K., Lee, C., Spector, J. M., & DeMeester, K. (2013). Teacher beliefs and technology integration. *Teaching and Teacher Education*, *29*, 76–85. doi:10.1016/j.tate.2012.08.005

Kim, C., Yuan, J., Kim, D., Doshi, P., Thai, C., Hill, R., & Melias, E. (2019). Studying the usability of an intervention to promote teachers' use of robotics in STEM education. *Journal of Educational Computing Research*, *56*(8), 1179–1212. doi:10.1177/0735633117738537

Kim, S., Yang, J. W., Lim, J., Lee, S., Ihm, J., & Park, J. (2021). The impact of writing on academic performance for medical students. *BMC Medical Education*, *21*(1), 1–8. doi:10.118612909-021-02485-2 PMID:33468118

Kim, Y. W., Kim, Y. G., Kim, N. J., & Woo, Y. G. (2003). A study on the Korea teacher's perception in using assistive technology. *Journal of Asia-Pacific Special Education*, *3*(1), 35–48.

Kirschner, P., Strijbos, J. W., Kreijns, K., & Beers, P. J. (2004). Designing electronic collaborative learning environments. *Educational Technology Research and Development, 52*(3), 47–66. doi:10.1007/BF02504675

Kirschner, P., Sweller, J., & Richard, E. C. (2006). Why minimal guidance during instruction does not work: An analysis of the failure of constructivist, discovery, problem-based, experiential and inquiry-based teaching. *Educational Psychologist, 41*(2), 75–86. doi:10.120715326985ep4102_1

Klopfer, E., Osterweil, S., & Salen, K. (2009). *Moving learning games forward*. The Education Arcade.

Knight, V. F., Wright, J., & DeFreese, A. (2019). Teaching Robotics Coding to a Student with ASD and Severe Problem Behavior. *Journal of Autism and Developmental Disorders, 49*(6), 2632–2636. doi:10.100710803-019-03888-3 PMID:30734176

Knight, V. F., Wright, J., Wilson, K., & Hooper, A. (2019). Teaching Digital, Block-Based Coding of Robots to High School Students with Autism Spectrum Disorder and Challenging Behavior. *Journal of Autism and Developmental Disorders, 49*(8), 3113–3126. doi:10.100710803-019-04033-w PMID:31055684

Knud, I. (2004). *The three dimensions of learning*. Krieger Pub. Co.

Koc, A. (2019). *The comparison of stem implementations with robotic-assisted and simple materials in preschool and basic science education* [Unpublished doctoral dissertation]. Erciyes University, Kayseri, Turkey.

Koehler, M. J., Mishra, P., & Cain, W. (2013). What is Technological Pedagogical Content Knowledge (TPACK)? *Journal of Education, 193*(3), 13–19. doi:10.1177/002205741319300303

Koh, J. H. L., & Chai, C. S. (2014). Teacher clusters and their perceptions of technological pedagogical content knowledge (TPACK) development through ICT lesson design. *Computers & Education, 70*, 222–232. doi:10.1016/j.compedu.2013.08.017

Kölling, M., & McKay, F. (2016). Heuristic Evaluation for Novice Programming Systems. *Trans. Comput. Educ., 16*(3), 12:1--12:30. doi:10.1145/2872521

Komis, Romero, Depover, & Karsenti, (2019). Robotics in Primary Education–Robotique en Éducation Primaire: Introduction. Review of Science. *Mathematics and ICT Education, 13*(1), 3-6.

Kong, S.-C., Abelson, H., & Lai, M. (2019). Introduction to computational thinking education. In *Computational thinking education* (pp. 1–10). Springer. doi:10.1007/978-981-13-6528-7_1

Kordaki, M., & Kakavas, P. (2017). Digital storytelling as an effective framework for the development of computational thinking skills. *EDULEARN2017*, 3–5.

Kosmas, P., & Zaphiris, P. (2019, September). Embodied interaction in language learning: Enhancing students' collaboration and emotional engagement. In *IFIP Conference on Human-Computer Interaction* (pp. 179-196). Springer. 10.1007/978-3-030-29384-0_11

Kotsopoulos, D., Floyd, L., Khan, S., Namukasa, I. K., Somanath, S., Weber, J., & Yiu, C. (2017). A pedagogical framework for computational thinking. *Digital Experiences in Mathematics Education, 3*(2), 154–171. doi:10.100740751-017-0031-2

Kozima, H., Nakagawa, C., Kawai, N., Kosugi, D., & Yano, Y. (2004). A humanoid in company with children. *2004 4th IEEE-RAS International Conference on Humanoid Robots, 1*, 470–477. 10.1109/ICHR.2004.1442138

Kozima, H., Michalowski, M., & Nakagawa, C. (2009). Keepon:A Playful Robot for Research, Therapy, and Entertainment. *International Journal of Social Robotics, 3*(18), 3–18. Advance online publication. doi:10.100712369-008-0009-8

Kraetzschmar, G. K. (2009). Educational robotics: On the role of robotics in learning and education. In C. Zacharias, K. W. ter Horst, K.-U. Witt, V. Sommer, M. Ant, U. Essmann, & L. Mülheims (Eds.), *Forschungsspitzen und Spitzenforschung* (pp. 83–90). Springer. doi:10.1007/978-3-7908-2127-7_9

Kramer, R. K. (2015). Soft electronics for soft robotics. *Micro-and Nanotechnology Sensors, Systems, and Applications VII.*

Kroiss, A., Sonego, D., & Rollins, P. (2014). *Using a Humanoid Robot as a Co-therapist with Children with ASD Mean Percent Engaged Child instructs Milo.* . doi:10.1109/TNSRE.2012.2230188.Diehl

Kucuk, S., & Sisman, B. (2017). Instructor experiences in one-to-one robotics instruction. *Elementary Education Online, 16*(1), 312–325. doi:10.17051/io.2017.12092

Kucuk, S., & Sisman, B. (2018). Pre-service teachers' experiences in learning robotics design and programming. *Informatics in Education, 17*(2), 301–320. doi:10.15388/infedu.2018.16

Kules, B. (2016). Computational thinking is critical thinking: Connecting to university discourse, goals, and learning outcomes. *Proceedings of the Association for Information Science and Technology, 53*(1), 1–6. doi:10.1002/pra2.2016.14505301092

Kulich, M., Chudoba, J., Kosnar, K., Krajnik, T., Faigl, J., & Preucil, L. (2012). Syrotek—Distance teaching of mobile robotics. *IEEE Transactions on Education, 56*(1), 18–23. doi:10.1109/TE.2012.2224867

Kumazaki, H., Muramatsu, T., Yoshikawa, Y., Matsumoto, Y., Ishiguro, H., Kikuchi, M., Sumiyoshi, T., & Mimura, M. (2020). Optimal robot for intervention for individuals with autism spectrum disorders. *Psychiatry and Clinical Neurosciences, 74*(11), 581–586. doi:10.1111/pcn.13132 PMID:32827328

Kumazaki, H., Warren, Z., Swanson, A., Yoshikawa, Y., Matsumoto, Y., Yoshimura, Y., Shimaya, J., Ishiguro, H., Sarkar, N., Wade, J., Mimura, M., Minabe, Y., & Kikuchi, M. (2019). Brief Report: Evaluating the Utility of Varied Technological Agents to Elicit Social Attention from Children with Autism Spectrum Disorders. *Journal of Autism and Developmental Disorders, 49*(4), 1700–1708. doi:10.100710803-018-3841-1 PMID:30511126

Kutlu, M., Schreglmann, S., & Cinisli, N. A. (2018). The opinions of special education teachers on the use of assistive technologies in special education. *YYU Journal of Education Faculty, 15*(1), 1540–1569. doi:10.23891/efdyyu.2018.115

Lammer, L., Lepuschitz, W., Kynigos, C., Giuliano, A., & Girvan, C. (2017). ER4STEM educational robotics for science, technology, engineering and mathematics. In *Robotics in Education* (pp. 95–101). Springer.

Lancioni, G. E., Sigafoos, J., O' Reilly, M. F., & Singh, N. N. (2013). Instructional technology for promoting writing, work, and leisure skills. In *Assistive Technology* (pp. 73–105). Springer. doi:10.1007/978-1-4614-4229-5_4

Lassahn, R. (1979). Technik und Bildung. In G. P. Bunk & R. Lassahn (Eds.), *Technik und Bildung*. Quelle & Meyer.

Lauwaert, M. (2008). Playing outside the box–on LEGO toys and the changing world of construction play. *History and Technology, 24*(3), 221–237. doi:10.1080/07341510801900300

Lauwers, T. (2010). *Aligning capabilities of interactive educational tools to learner goals.* ProQuest Dissertations and Theses, 3470163, 366. http://ezproxy.net.ucf.edu/login?url=http://search.proquest.com/docview/742526809?accountid=10003%5Cnhttp://sfx.fcla.edu/ucf?url_ver=Z39.88-2004&rft_val_fmt=info:ofi/fmt:kev:mtx:dissertation&genre=dissertations+&+theses&sid=ProQ:ProQuest+Dissertations+&+T

Lauxmann, F. (1999). *Mit Hegel auf der Datenautobahn. Über die Freude an geistigen Zusammenhängen.* DTV.

Lee, I., Martin, F., Denner, J., Coulter, B., Allan, W., Erickson, J., Malyn-Smith, J., & Werner, L. (2011). Computational thinking for youth in practice. *Acfm Inroads, 2*(1), 32–37. doi:10.1145/1929887.1929902

Leekam, S. R., & Ramsden, C. A. (2006). Dyadic orienting and joint attention in preschool children with autism. *Journal of Autism and Developmental Disorders, 36*(2), 185–197. doi:10.100710803-005-0054-1 PMID:16502142

Lee, T. Y., Mauriello, M. L., Ahn, J., & Bederson, B. B. (2014). CTArcade: Computational thinking with games in school age children. *International Journal of Child-Computer Interaction, 2*(1), 26–33. doi:10.1016/j.ijcci.2014.06.003

Lee, Y., & Lee, J. (2014). Enhancing pre-service teachers' self-efficacy beliefs for technology integration through lesson planning practice. *Computers & Education, 73*, 121–128. doi:10.1016/j.compedu.2014.01.001

Leonard, J., Mitchell, M., Barnes-Johnson, J., Unertl, A., Outka-Hill, J., Robinson, R., & Hester-Croff, C. (2018). Preparing teachers to engage rural students in computational thinking through robotics, game design, and culturally responsive teaching. *Journal of Teacher Education, 69*(4), 386–407. doi:10.1177/0022487117732317

Leoste, J., Pastor, L., López, J. S. M., Garre, C., Seitlinger, P., Martino, P., & Peribáñez, E. (2020). Using Robots for Digital Storytelling. A Game Design Framework for Teaching Human Rights to Primary School Students. *International Conference on Robotics in Education (RiE)*, 26–37.

Letherby, G. (2013). Theorised subjectivity. *Objectivity and subjectivity in social research*, 59-78.

Liesa-Orús, M., Latorre-Cosculluela, C., Vázquez-Toledo, S., & Sierra-Sánchez, V. (2020). The technological challenge facing higher education professors: Perceptions of ICT tools for developing 21st century skills. *Sustainability, 12*(13), 5339. doi:10.3390u12135339

Lindh, J., & Holgersson, T. (2007). Does Lego training stimulate pupils' ability to solve logical problems? *Computers & Education, 49*(4), 1097–1111. doi:10.1016/j.compedu.2005.12.008

Liu, E. Z.-F., Lin, C.-H., Liou, P.-Y., Feng, H.-C., & Hou, H.-T. (2013). An Analysis of Teacher-Student Interaction Patterns in a Robotics Course for Kindergarten Children: A Pilot Study. *Turkish Online Journal of Educational Technology-TOJET, 12*(1), 9–18.

Lloyd, G. E. R. (2012). *Early greek science: Thales to Aristotle.* Random House.

Lorenzo, G., Lledó, A., Arráez-Vera, G., & Lorenzo-Lledó, A. (2018). The application of immersive virtual reality for students with ASD: A review between 1990–2017. *Education and Information Technologies, 24*(1), 1–25. doi:10.100710639-018-9766-7

Loughlin, C., Lygo-Baker, S., & Lindberg-Sand, Å. (2021). Reclaiming constructive alignment. *European Journal of Higher Education, 11*(2), 119–136. doi:10.1080/21568235.2020.1816197

Lu, Y. (2017). Industry 4.0: A survey on technologies, applications and open research issues. *Journal of Industrial Information Integration, 6*, 1-10.

Lu, J., & Fletcher, G. (2009). Thinking about computational thinking. *ACM SIGCSE Bulletin, 41*(1), 260–264. doi:10.1145/1539024.1508959

Lunarman. (2011, January 11). *Extrem cool -> "BürstenBot".* https://www.roboternetz.de/community/threads/33177-Extrem-cool-B%C3%BCrstenBot?p=314868&viewfull=1#post314868

Lye, S. Y., & Koh, J. H. L. (2014). Review on teaching and learning of computational thinking through programming: What is next for K-12? *Computers in Human Behavior, 41*, 51–61. doi:10.1016/j.chb.2014.09.012

Lykke, M., Coto, M., Mora, S., Vandel, N., & Jantzen, C. (2014, April). Motivating programming students by problem based learning and LEGO robots. In *2014 IEEE Global Engineering Education Conference (EDUCON)* (pp. 544-555). 10.1109/EDUCON.2014.6826146

Lytridis, C., Bazinas, C., Sidiropoulos, G., Papakostas, G. A., Kaburlasos, V. G., Nikopoulou, V. A., Holeva, V., & Evangeliou, A. (2020). Distance special education delivery by social robots. *Electronics (Basel)*, *9*(6), 1–13. doi:10.3390/electronics9061034

MacPhail, A., Tannehill, D., Leirhaug, P. E., & Borghouts, L. (2021). Promoting instructional alignment in physical education teacher education. *Physical Education and Sport Pedagogy*, 1–12. doi:10.1080/17408989.2021.1958177

Mader, A., & Dertien, E. (2016). Tinkering as method in academic teaching. *DS 83: Proceedings of the 18th International Conference on Engineering and Product Design Education (E&PDE16), Design Education: Collaboration and Cross-Disciplinarity.*

MadMage. (2008, January 11). *Extrem cool -> "BürstenBot"*. https://www.roboternetz.de/community/threads/33177-Extrem-cool-B%C3%BCrstenBot?p=314697&viewfull=1#post314697

Malec, J. (2001). Some thoughts on robotics for education. *Proceedings of the AAAI Spring Symposium on Robotics and Education.*

Malliarakis, C., Satratzemi, M., & Xinogalos, S. (2014). Educational games for teaching computer programming. In *Research on e-Learning and ICT in Education* (pp. 87–98). Springer.

Mammes, I., Schäffer, K. & Tuncsoy, M. (2013). „Robotik" in der Grundschule. Beschreibung des Baus des „Miniatur-Roboters" Bürstenfuzzi. *Tu - Zeitschrift für Technik im Unterricht, 147*(13), 41-45.

Mandrikas, A., Michailidi, E., & Stavrou, D. (2020). Teaching nanotechnology in primary education. *Research in Science & Technological Education*, *38*(4), 377–395.

Manf. (2008a, January 9). *Extrem cool -> "BürstenBot"*. https://www.roboternetz.de/community/threads/33177-Extrem-cool-B%C3%BCrstenBot?p=314134&viewfull=1#post314134

Manf. (2008b, January 10). *Extrem cool -> "BürstenBot"*. https://www.roboternetz.de/community/threads/33177-Extrem-cool-B%C3%BCrstenBot?p=314648&viewfull=1#post314648

Marino, F., Chilà, P., Sfrazzetto, S. T., Carrozza, C., Crimi, I., Failla, C., Busà, M., Bernava, G., Tartarisco, G., Vagni, D., Ruta, L., & Pioggia, G. (2020). Outcomes of a Robot-Assisted Social-Emotional Understanding Intervention for Young Children with Autism Spectrum Disorders. *Journal of Autism and Developmental Disorders*, *50*(6), 1973–1987. doi:10.100710803-019-03953-x PMID:30852783

Martinez, S. L., & Stager, G. (2019). *Invent to learn: Making, tinkering, and engineering in the classroom*. Academic Press.

Martinez, S. L., & Stager, G. S. (2019). Invent to Learn Making, Tinkering, and Engineering in the Classroom (2nd ed.). Constructing Modern Knowledge Press.

Martínez-Córcoles, M., Teichmann, M., & Murdvee, M. (2017). Assessing technophobia and technophilia: Development and validation of a questionnaire. *Technology in Society*, *51*, 183–188. doi:10.1016/j.techsoc.2017.09.007

Martinez, S. L., & Stager, G. S. (2019). *Invent to learn: Making, tinkering, and engineering in the classroom*. Constructing Modern Knowledge Press.

Martin, F. (1988). *Children, cybernetics, and programmable turtles*. Massachusetts Institute of Technology.

Martin, F. G. (1994). *Circuits to control—learning engineering by designing LEGO robots*. Massachusetts Institute of Technology.

Maryuningsih, Y., Hidayat, T., Riandi, R., & Rustaman, N. Y. (2020). Profile of information and communication technologies (ICT) skills of prospective teachers. *Journal of Physics: Conference Series*, *1521*(4), 042009. doi:10.1088/1742-6596/1521/4/042009

Mataric, M. J. (2004, March). Robotics education for all ages. In *Proceeding of the American Association for Artificial Intelligence (AAAI) Spring Symposium on Accessible, Hands-on AI and Robotics Education* (pp. 22-24). Retrieved from https://www.aaai.org/Papers/Symposia/Spring/2004/SS-04-01/SS04-01-004.pdf

Mataric, M. J., Koenig, N. P., & Feil-Seifer, D. (2007, March). Materials for enabling hands-on robotics and STEM education. *AAAI Spring Symposium: Semantic Scientific Knowledge Integration* (pp. 99-102). AAAI.

Mathevula, M. D., & Uwizeyimana, D. E. (2014). The challenges facing the integration of ICT in teaching and learning activities in South African rural secondary schools. *Mediterranean Journal of Social Sciences*, *5*(20), 1087. doi:10.5901/mjss.2014.v5n20p1087

Ma, Y., & Williams, D. C. (2013). The Potential of a First LEGO League Robotics Program in Teaching 21 st Century Skills. *An Exploratory Study*, *6*, 13–28.

Mayring, P. (2015). Qualitative Inhaltsanalyse. Grundlagen und Techniken, 4, 58.

McCormick, M. P., Weiland, C., Hsueh, J., Maier, M., Hagos, R., Snow, C., Leacock, N., & Schick, L. (2020). Promoting content-enriched alignment across the early grades: A study of policies & practices in the Boston Public Schools. *Early Childhood Research Quarterly*, *52*, 57–73. doi:10.1016/j.ecresq.2019.06.012

McKenney, S., & Reeves, T. C. (2018). *Conducting educational design research*. Routledge. doi:10.4324/9781315105642

McKnight, L., & Fitton, D. (2010). Touch-screen Technology for Children: Giving the Right Instructions and Getting the Right Responses. *Proceedings of the 9th International Conference on Interaction Design and Children -IDC '10*, 238. 10.1145/1810543.1810580

McNeill, K., Lizotte, D., Krajcik, J., & Marx, R. (2006). Supporting students' construction of scientific explanations by fading scaffolds in instructional materials. *Journal of the Learning Sciences*, *15*(2), 153–191. doi:10.120715327809jls1502_1

Mead, R. A., Thomas, S. L., & Weinberg, J. B. (2012). From grade school to grad school: An integrated STEM pipeline model through robotics. In *Robots in K-12 Education: A New Technology for Learning* (pp. 302–325). IGI Global. doi:10.4018/978-1-4666-0182-6.ch015

Mehalik, M. M., Doppelt, Y., & Schuun, C. D. (2008). Middle-school science through design-based learning versus scripted inquiry: Better overall science concept learning and equity gap reduction. *Journal of Engineering Education*, *97*(1), 71–85. doi:10.1002/j.2168-9830.2008.tb00955.x

Mehrotra, A., Giang, C., Duruz, N., Dedelley, J., Mussati, A., Skweres, M., & Mondada, F. (2020). Introducing a Paper-Based Programming Language for Computing Education in Classrooms. *Proceedings of the 2020 ACM Conference on Innovation and Technology in Computer Science Education*, 180–186. 10.1145/3341525.3387402

Mehrotra, A., Giang, C., El-Hamamsy, L., Guinchard, A., Dame, A., Zahnd, G., & Mondada, F. (2021). Accessible Maker-Based Approaches to Educational Robotics in Online Learning. *IEEE Access: Practical Innovations, Open Solutions*, *9*, 96877–96889. doi:10.1109/ACCESS.2021.3094158

Menary, R. (2007). Writing as thinking. *Language Sciences*, *29*(5), 621–632. doi:10.1016/j.langsci.2007.01.005

Mertala, P. (2019). Teachers' beliefs about technology integration in early childhood education: A meta-ethnographical synthesis of qualitative research. *Computers in Human Behavior*, *101*, 334–349. doi:10.1016/j.chb.2019.08.003

Mester, G. (2006). Distance learning in robotics. In *Proceedings of The Third International Conference on Informatics, Educational Technology and New Media in Education* (pp. 239–245). Academic Press.

Michael, M. (2008, January 11). *Extrem cool -> “BürstenBot”*. https://www.roboternetz.de/community/threads/33177-Extrem-cool-B%C3%BCrstenBot?p=314857&viewfull=1#post314857

Michael, D., & Chen, S. (2006). *Serious games: Games that educate, train, and inform*. Thomson Course Technology.

Michaud, F., Duquette, A., & Nadeau, I. (2003). Characteristics of mobile robotics toys for children with pervasive developmental disorders. *Proceedings of the IEEE International Conference on Systems, Man and Cybernetics*, *3*(May), 2938–2943. 10.1109/ICSMC.2003.1244338

Michaud, F., Salter, T., Duquette, A., Mercier, H., Lauria, M., Larouche, H., & Larosse, F. (2007). Mobile robots engaging children in learning. *CMBES Proceedings*, 30.

Mikropoulos, T. A., & Bellou, I. (2013). Educational robotics as mindtools. *Themes in Science and Technology Education*, *6*(1), 5–14.

Mitchell, F. M. (1999). *All Students Can Learn: Effects of Curriculum Alignment on the Mathematics Achievement of Third-Grade Students*. Academic Press.

Mitchell, F. M. (1999, April). *All Students Can Learn: Effects of Curriculum Alignment on the Mathematics Achievement of Third-Grade Students*. Paper presented at the Annual Meeting of the American Educational Research Association, Montreal, Quebec, Canada.

Mitnik, R., Nussbaum, M., & Recabarren, M. (2009). Developing cognition with collaborative robotic activities. *Journal of Educational Technology & Society*, *12*(4), 317–330.

Moeller, P. (2012). *Where the jobs will be in 2020*. U.S. News and World Report. http://money.usnews.com/money/careers/articles/2012/09/10/where–the–jobs–will–be–in–2020

Mohamed, H., & Jaafar, A. (2010). Development and potential analysis of Heuristic Evaluation for Educational Computer Game (PHEG). *Computer Sciences and Convergence Information Technology (ICCIT), 2010 5th International Conference On*, 222–227. doi:10.1109/ICCIT.2010.5711061

Mondada, F., Bonani, M., Riedo, F., Briod, M., Pereyre, L., Retornaz, P., & Magnenat, S. (2017). Bringing Robotics to Formal Education: The Thymio Open-Source Hardware Robot. *IEEE Robotics & Automation Magazine*, *24*(1), 77–85. doi:10.1109/MRA.2016.2636372

Mondada, F., Bonnet, E., Davrajh, S., Johal, W., & Stopforth, R. (2016). R2T2: Robotics to integrate educational efforts in South Africa and Europe. *International Journal of Advanced Robotic Systems*, *13*(5). doi:10.1177/1729881416658165

Morin, J. W. (2011). *Design, fabrication and mechanical optimization of multi-scale anisotropic feet for terrestrial locomotion*. Massachusetts Institute of Technology. Dept. of Mechanical Engineering. https://dspace.mit.edu/handle/1721.1/65314

Morrison, G. R., Ross, S. J., Morrison, J. R., & Kalman, H. K. (2019). *Designing effective instruction*. John Wiley & Sons.

Mostajo, S. T., Legaspi, O. M., Camarse, M. G., & Salva, R. A. (2021). Exploring the potentials of robotics in supporting children with autism spectrum disorder. *IAFOR Journal of Education*, *9*(1), 77–93. doi:10.22492/ije.9.1.05

Moulton, B., & Johnson, D. (2010). Robotics education: A review of graduate profiles and research pathways. *World Transactions on Engineering and Technology Education*, *8*(1), 26–31.

Mubin, O., Stevens, C. J., Shahid, S., Al Mahmud, A., & Dong, J.-J. (2013). a Review of the Applicability of Robots in Education. *Technology for Education and Learning*, *1*(1). Advance online publication. doi:10.2316/Journal.209.2013.1.209-0015

Mundy, P., & Neal, R. (2001). Neural plasticity, joint attention and autistic developmental pathology. *International Review of Research in Mental Retardation, 23*, 139–168. doi:10.1016/S0074-7750(00)80009-9

Mussati, A., Giang, C., Piatti, A., & Mondada, F. (2019). A Tangible Programming Language for the Educational Robot Thymio. *2019 10th International Conference on Information, Intelligence, Systems and Applications (IISA)*, 1–4.

Mussati, A., Giang, C., Piatti, A., & Mondada, F. (2019). A tangible programming language for the educational robot Thymio. In *Proceedings of 10th International Conference on Information, Intelligence, Systems and Applications (IISA)* (pp. 1-4). IEEE. 10.1109/IISA.2019.8900743

National Research Council. (2010). *Committee for the workshops on computational thinking: Report of a workshop on the scope and nature of computational thinking.* National Academy Press. doi:10.17226/12840

Negrini, L., & Giang, C. (2019). How do pupils perceive educational robotics as a tool to improve their 21st century skills? *Journal of e-Learning and Knowledge Society, 15*(2), 77-87.

Negrini, L. (2019). Teacher Training in Educational Robotics. An Experience in Southern Switzerland: The PReSO Project. In W. Lepuschitz, M. Merdan, G. Koppensteiner, R. Balogh, & D. Obdrzalek (Eds.), *Robotics in Education: Methods and Applications for Teaching and Learning* (pp. 92–97). Springer. doi:10.1007/978-3-319-97085-1_10

Negrini, L. (2019). Teacher training in educational robotics. In *International Conference on Robotics and Education RiE 2017* (pp. 92-97). Springer.

Negrini, L. (2020). Teachers' attitudes towards educational robotics in compulsory school. *Italian Journal of Educational Technology, 28*(1), 77–90.

Negrini, L., & Giang, C. (2019). How do pupils perceive educational robotics as a tool to improve their 21st century skills? *Journal of E-Learning and Knowledge Society, 15*(2).

Negrini, L., Reyes-Mury, S., & Moonnee, D. (2021). A robotic teacher community to foster the integration of educational robotics in school. In W. Lepuschitz, M. Merdan, G. Koppensteiner, R. Balogh, & D. Obdrzalek (Eds.), *Robotics in Education: Advances in Intelligent Systems and Computing* (pp. 354–359). Springer. doi:10.1007/978-3-030-67411-3_32

NGSS Lead States. (2013). *Next Generation Science Standards: For States, by States.* The National Academies Press.

NGSS Lead States. (2013). *Next Generation Science Standards: For States, By States.* The National Academies Press.

Nickell, G. S., & Pinto, J. N. (1986). The computer attitude scale. *Computers in Human Behavior, 2*(4), 301–306. doi:10.1016/0747-5632(86)90010-5

Nielsen, J., & Molich, R. (1990). Heuristic Evaluation of User Interfaces. *Proceedings of the SIGCHI Conference on Human Factors in Computing Systems*, 249–256. 10.1145/97243.97281

Norouzi, S., Fattahi, A., & Massah, A. (2010). *LARC.* Competition–Mixed Reality League S&TRN.

Notomista, G., Mayya, S., Mazumdar, A., Hutchinson, S., & Egerstedt, M. (2019). A Study of a Class of Vibration-Driven Robots: Modeling, Analysis, Control and Design of the Brushbot. *2019 IEEE/RSJ International Conference on Intelligent Robots and Systems (IROS)*, 5101-5106. 10.1109/IROS40897.2019.8968490

Nourbakhsh, I. R., Crowley, K., Bhave, A., Hsium, T., Hammer, E., & Perez Bergquist, A. (2005). The robotic autonomy mobile robotics course: Robot design, curriculum design and educational assessment. *Autonomous Robots, 18*(1), 103–127. doi:10.1023/B:AURO.0000047303.20624.02

Nückles, M., Hübner, S., Dümer, S., & Renkl, A. (2010). Expertise reversal effects in writing-to-learn. *Instructional Science, 38*(3), 237–258. doi:10.100711251-009-9106-9

Nugent, G. C., Barker, B. S., & Grandgenett, N. (2014). The impact of educational robotics on student STEM learning, attitudes, and workplace skills. In *Robotics: Concepts, methodologies, tools, and applications* (pp. 1442–1459). IGI Global.

Nugent, G., Barker, B., Grandgenett, N., & Adamchuk, V. I. (2010). Impact of robotics and geospatial technology interventions on youth STEM learning and attitudes. *Journal of Research on Technology in Education*, *42*(4), 391–408. doi:10.1080/15391523.2010.10782557

Osborne, R. (2014). *An ecological approach to educational technology : affordance as a design tool for aligning pedagogy and technology*. Academic Press.

Oskay, W. (2007, December 19). *Bristlebot: A tiny directional vibrobot*. https://www.evilmadscientist.com/2007/bristlebot-a-tiny-directional-vibrobot

Otto, D. (2020). Hochschullehre und Digitalisierung: Digital Storytelling als Lehr-Lernmethode für Kompetenzen in der digitalen Welt. *Bildung und Digitalisierung*, 135–152.

Ozdamar, O. (2016). *The views of special education teachers about the use of assistive technology in lessons* [Unpublished master's dissertation]. Anadolu University, Eskisehir, Turkey.

Ozer, F. (2019). *Effect of using robotics in teaching coding on achievement, motivation and problem solving skills of middle school students* [Unpublished master's dissertation]. Hacettepe University, Ankara, Turkey.

Pachidis, T. P., Macedonia, E., Vrochidou, E., Kaburlasos, V., & Macedonia, E. (2018). *Social Robotics in Education : State-of-the-Art and Directions Social Robotics in Education : State-of-the-Art and Directions In recent years, robots attract considerable attention and become ever more popular in*. Academic Press.

Pahl, G., Beitz, W., Feldhusen, J., & Grote, K. H. (2007). Introduction. In G. Pahl, W. Beitz, J. Feldhusen & K.H. Grote (Eds.), *Engineering Design. A Systematic Approach* (pp. 1-25). Springer. doi:10.1007/978-1-84628-319-2_1

Pakdaman, M., Sanaatiyan, M. M., & Rezaei, M. (2010). *A line follower robot from design to implementation: Technical issues and problems*. Academic Press.

Pane, J. F., & Myers, B. a. (1996). Usability Issues in the Design of Novice Programming Systems. *Education*, *78*(August). http://repository.cmu.edu/isr/820/

Papadakis, S., Vaiopoulou, J., Sifaki, E., Stamovlasis, D., Kalogiannakis, M., & Vassilakis, K. (2021). Factors That Hinder in-Service Teachers from Incorporating Educational Robotics into Their Daily or Future Teaching Practice. In CSEDU (2) (pp. 55-63). Academic Press.

Papadopoulos, P., Demetriadis, S., Stamelos, I., & Tsoukalas, I. (2011). The value of writing-to-learn when using question prompts to support web-based learning in ill-structured domains. *Educational Technology Research and Development*, *59*(1), 71–90. doi:10.100711423-010-9167-0

Papakostas, G. A., Sidiropoulos, G. K., Papadopoulou, C. I., Vrochidou, E., Kaburlasos, V. G., Papadopoulou, M. T., Holeva, V., Nikopoulou, V.-A., & Dalivigkas, N. (2021). Social robots in special education: A systematic review. *Electronics (Basel)*, *10*(12), 1398. doi:10.3390/electronics10121398

Papert, S. (1980). *"Mindstorms" Children. Computers and powerful ideas*. Academic Press.

Papert, S. (1980). Mindstorms: Children, computers and powerful ideas. In New Ideas in Psychology (Vol. 1). doi:10.1016/0732-118X(83)90034-X

Papert, S. A. (1973). *Uses of technology to enhance education*. Academic Press.

Papert, S. (1987). Information technology and education: Computer criticism vs. technocentric thinking. *Educational Researcher, 16*(1), 22–30. doi:10.3102/0013189X016001022

Papert, S. (1991). Situating constructionism. In S. Papert & I. Harel (Eds.), *Constructionism*. Ablex Publishing Corporation.

Papert, S. (1993). *Mindstorms: children, computers, and powerful ideas* (2nd ed.). Basic Books.

Papert, S. (1993). *The children's machine: Rethinking school in the age of the computer*. ERIC.

Papert, S. A. (2020). *Mindstorms: Children, computers, and powerful ideas*. Basic books.

Papert, S., & Harel, I. (1991). Situating constructionism. *Constructionism, 36*(2), 1–11.

Parsazadeh, N., Cheng, P.-Y., Wu, T.-T., & Huang, Y.-M. (2020). Integrating Computational Thinking Concept Into Digital Storytelling to Improve Learners' Motivation and Performance. *Journal of Educational Computing Research*. doi:10.1177/0735633120967315

Paull, L., Tani, J., Ahn, H., Alonso-Mora, J., Carlone, L., Cap, M., Chen, Y. F., Choi, Ch., Dusek, J., Fang, Y., Hoehener, D., Liu, S.-Y., Novitzky, M., Okuyama, I. F., Pazis, J., Rosman, G., Varricchio, V., Wang, H.-Ch., Yershov, D., . . . Censi, A. (2017). Duckietown: An open, inexpensive and flexible platform for autonomy education and research. In 2017 IEEE International Conference on Robotics and Automation. ICRA. https://doi.org/10.1109/ICRA.2017.7989179.

Pelikan, E. R., Lüftenegger, M., Holzer, J., Korlat, S., Spiel, C., & Schober, B. (2021). Learning during COVID-19: The role of self-regulated learning, motivation, and procrastination for perceived competence. *Zeitschrift für Erziehungswissenschaft, 24*(2), 393–418. doi:10.100711618-021-01002-x PMID:33686344

Peng, Y. H., Lin, C. W., Mayer, N. M., & Wang, M. L. (2014, November). Using a humanoid robot for music therapy with autistic children. In *2014 CACS International Automatic Control Conference (CACS 2014)* (pp. 156-160). IEEE. 10.1109/CACS.2014.7097180

Pennisi, P., Tonacci, A., Tartarisco, G., Billeci, L., Ruta, L., Gangemi, S., & Pioggia, G. (2016). Autism and social robotics: A systematic review. *Autism Research, 9*(2), 165–183. doi:10.1002/aur.1527 PMID:26483270

Pérez-Vázquez, E., Lorenzo, G., Lledó, A., & Lorenzo-Lledó, A. (2020). Evolution and Identification from a Bibliometric Perspective of the Use of Robots in the Intervention of Children with ASD. Technology. *Knowledge and Learning, 25*(1), 83–114. doi:10.100710758-019-09415-8

Petre, M., & Price, B. (2004). Using robotics to motivate "back door" learning. *Education and Information Technologies, 9*(2), 147–158. doi:10.1023/B:EAIT.0000027927.78380.60

Petscher, Y. M., Schatschneider, C., & Compton, D. L. (Eds.). (2013). *Applied quantitative analysis in education and the social sciences*. Routledge. doi:10.4324/9780203108550

Pettersson, I., & Fahlström, G. (2010). Roles of assistive devices for home care staff in Sweden: A qualitative study. *Disability and Rehabilitation. Assistive Technology, 5*(5), 295–304. doi:10.3109/17483100903100285 PMID:19968568

Phal, D. D., Phal, K. D., & Jacob, S. (2014). Design, implementation and reliability estimation of speech-controlled mobile robot. *IEEE, 2014 2nd International Conference on Emerging Technology Trends in Electronics, Communication and Networking*.

Piaget, J. (1973b). *To understand is to invent: The future of education*. Academic Press.

Piaget, J. (1964). Part I: Cognitive development in children: Piaget development and learning. *Journal of Research in Science Teaching, 2*(3), 176–186.

Piaget, J. (1973a). *The child and reality: Problems of genetic psychology* (A. Rosin, Trans.). Grossman.

Picka, K., Dosedla, M., & Stuchlikova, L. (2020). Robotic didactic aid Ozobot in Czech schools. *2020 18th International Conference on Emerging eLearning Technologies and Applications (ICETA)*, 525–533.

Pickover, C. (2008). *Archimedes to Hawking: Laws of science and the great minds behind them.* Oxford University Press.

Piedade, J. M. N. (2021). Pre-service and in-service teachers' interest, knowledge, and self-confidence in using educational robotics in learning activities. *The Educational Forum*, *6*(1), 2.

Pioggia, G., Sica, M. L., Ferro, M., Igliozzi, R., Muratori, F., Ahluwalia, A., & De Rossi, D. (2007). Human-robot interaction in autism: FACE, an android-based social therapy. *Proceedings - IEEE International Workshop on Robot and Human Interactive Communication*, 605–612. 10.1109/ROMAN.2007.4415156

Pisciotta, M., Vello, B., Bordo, C., & Morgavi, G. (2010). *Robotic competition: A classroom experience in a vocational school.* Paper presented at the Actas del 6th International Conference on Educational Technologies.

Player-Koro, C. (2012). Factors influencing teachers' use of ICT in education. *Education Inquiry*, *3*(1), 93–108. doi:10.3402/edui.v3i1.22015

Plaza, P., Peixoto, A., Sancristobal, E., Castro, M., Blazquez, M., Menacho, A., . . . Muñoz, M. (2020). *Visual block programming languages and their use in educational robotics.* Paper presented at the 2020 IEEE Global Engineering Education Conference (EDUCON). 10.1109/EDUCON45650.2020.9125219

Plexousakis, S., Georgiadi, M., Kourkoutas, E., Gkintoni, E., Halkiopoulos, K., & Roumeliotou, V. (2020). Enhancing sexual awareness in children with Autism Spectrum Disorder: A case study report. In *Cases on Teaching Sexuality Education to Individuals with Autism* (pp. 79–98). IGI Global. doi:10.4018/978-1-7998-2987-4.ch005

Pliasa, S., & Fachantidis, N. (2019). Using Daisy Robot as a Motive for Children with ASD to Participate in Triadic Activities. *Themes in ELearning*, *12*(12), 35–50.

Plomp, T., & Nieveen, N. (2007). An introduction to educational design research. *Proceedings of the seminar conducted at the East China Normal University*, 23.

Plomp, T., & Nieveen, N. (2013). *Educational design research part A: An introduction.* SLO, Enschede.

Ponticorvo, M., Rubinacci, F., Marocco, D., Truglio, F., & Miglino, O. (2020). Educational Robotics to Foster and Assess Social Relations in Students' Groups. *Frontiers in Robotics and AI*, *7*, 78. doi:10.3389/frobt.2020.00078 PMID:33501245

Pop, C. A., Simut, R., Pintea, S., Saldien, J., Rusu, A., David, D., Vanderfaeillie, J., Lefeber, D., & Vanderborght, B. (2013). Can the social robot Probo help children with autism to identify situation-based emotions? A series of single case experiments. *International Journal of HR; Humanoid Robotics*, *10*(03), 1350025. doi:10.1142/S0219843613500254

Popper, S. K. (1999). *Falsificationism.* Academic Press.

Poulakis, E., & Politis, P. (2021). *Computational Thinking Assessment: Literature Review. In Research on E-Learning and ICT in Education: Technological, Pedagogical and Instructional Perspectives.* Springer International Publishing., doi:10.1007/978-3-030-64363-8_7

Prensky, M. (2001). *Digital game-based learning.* McGraw-Hill.

Prensky, M. (2006). *Don't Bother Me, Mom, I'm Learning! How Computer and Video Games Are Preparing Your Kids for 21st Century Success and How You Can Help!* Paragon House.

Rahman, S. M. M., Krishnan, V. J., & Kapila, V. (2017, June), Exploring the dynamic nature of TPACK framework in teaching STEM using robotics in middle school classrooms. *2017 ASEE Annual Conference & Exposition.* 10.18260/1-2--28336

Rahwan, I., Cebrian, M., Obradovich, N., Bongard, J., Bonnefon, J.-F., Breazeal, C., Crandall, J. W., Christakis, N. A., Couzin, I. D., Jackson, M. O., Jennings, N. R., Kamar, E., Kloumann, I. M., Larochelle, H., Lazer, D., McElreath, R., Mislove, A., Parkes, D. C., Pentland, A., Roberts, M. E., Tenenbaum, J. B., & Wellman, M. (2019). Machine behaviour. *Nature, 568*(7753), 477-486. doi:10.1038/s41586-019-1138-y

Rainie, L., & Anderson, J. (2017). *The Future of Jobs and Jobs Training.* Pew Research Center.

Rapanta, C., Botturi, L., Goodyear, P., Guardia, L., & Koole, M. (2020). Online university teaching during and after the covid-19 crisis: Refocusing teacher presence and learning activity. *Postdigital Science and Education, 2*(3), 923–945. doi:10.100742438-020-00155-y

Raskind, M. H., & Higgins, E. L. (1999). Speaking to read: The effects of speech recognition technology on the reading and spelling performance of children with learning disabilities. *Annals of Dyslexia, 49*(1), 251–281. doi:10.100711881-999-0026-9

Reed, P., & Bowser, G. (2005). Assistive technology and the IEP. In Handbook of special education technology research and practice (pp. 61–77). Knowledge by Design.

Requicha, A. A. (2003). Nanorobots, NEMS, and nanoassembly. *Proceedings of the IEEE, 91*(11), 1922–1933.

Resnick, M. (2007). All I really need to know (about creative thinking) I learned (by studying how children learn) in kindergarten. In Proceedings of the 6th ACM SIGCHI conference on Creativity & cognition (C&C '07) (pp. 1–6). Association for Computing Machinery. https://doi.org/10.1145/1254960.1254961.

Resnick, M. (2014). Give P's a chance: Projects, peers, passion, play. In *Constructionism and creativity: Proceedings of the third international constructionism conference.* Austrian Computer Society.

Resnick, M. (1993). Behavior construction kits. *Communications of the ACM, 36*(7), 64–71. https://doi.org/10.1145/159544.159593

Resnick, M. (2007). All I really need to know (about creative thinking) I learned (by studying how children learn) in kindergarten. In B. Shneiderman, G. Fischer, E. Giaccardi, & M. Eisenberg (Eds.), *Proceedings of the 6th acm sigchi conference on creativity & cognition - c&c '07* (pp. 1–6). ACM Press. 10.1145/1254960.1254961

Resnick, M., Berg, R., & Eisenberg, M. (2000). Beyond Black Boxes: Bringing Transparency and Aesthetics Back to Scientific Investigation. *Journal of the Learning Sciences, 9*(1), 7–30. https://doi.org/10.1207/s15327809jls0901_3

Resnick, M., Martin, F., Berg, R., Borovoy, R., Colella, V., Kramer, K., & Silverman, B. (1998). Digital manipulatives: new toys to think with. In *Proceedings of the SIGCHI Conference on Human Factors in Computing Systems* (pp. 281-287). 10.1145/274644.274684

Resnick, M., & Silverman, B. (2005). Some reflections on designing construction kits for kids. *Proceedings of the 2005 conference on Interaction design and children.* 10.1145/1109540.1109556

Richmond, G., Bartell, T., & Dunn, A. H. (2016). Beyond "Tinkering": Enacting the Imperative for Change in Teacher Education in a Climate of Standards and Accountability [Editorial]. *Journal of Teacher Education, 67*(2), 102–104. https://doi.org/10.1177/0022487116628837

Rich, P. J., Browning, S. F., Perkins, M., Shoop, T., Yoshikawa, E., & Belikov, O. M. (2019). Coding in K-8: International trends in teaching elementary/primary computing. *TechTrends, 63*(3), 311–329. doi:10.100711528-018-0295-4

Rickharris. (2010, January 3). *Bristlebots?* https://picaxeforum.co.uk/threads/bristlebots.14219/post-121433

Riedler, A. (1919). *Wirklichkeitsblinde in Wissenschaft und Technik*. Springer.

Riedo, F., Chevalier, M., Magnenat, S., & Mondada, F. (2013). Thymio II, a robot that grows wiser with children. *Proceedings of IEEE Workshop on Advanced Robotics and Its Social Impacts, ARSO*, 187–193. 10.1109/ARSO.2013.6705527

Ringo. (2008, March 26). *Mein 10min. IC-Bot (BristleBot/Vibrobot)*. https://www.roboternetz.de/community/threads/35068-mein-10min-IC-Bot-%28BristleBot-Vibrobot%29?p=333234&viewfull=1#post333234

Roberts-Yates, C., & Silvera-Tawil, D. (2019). Better education opportunities for students with autism and intellectual disabilities through digital technology. *International Journal of Special Education*, *34*(1), 197–210.

Robin, B. (2006). The educational uses of digital storytelling. *Society for Information Technology & Teacher Education International Conference*, 709–716.

Robin, B. R. (2016). The power of digital storytelling to support teaching and learning. *Digital Education Review*, *30*, 17–29.

Robins, B., Dautenhahn, K., & Dubowski, J. (2006). Does appearance matter in the interaction of children with autism with a humanoid robot? *Interaction Studies: Social Behaviour and Communication in Biological and Artificial Systems*, *7*(3), 479–512. doi:10.1075/is.7.3.16rob

Robinson, M. (2005). Robotics-driven activities: Can they improve middle school science learning? *Bulletin of Science, Technology & Society*, *25*(1), 73–84. doi:10.1177/0270467604271244

Rockland, R., Bloom, D. S., Carpinelli, J., Burr-Alexander, L., Hirsch, L. S., & Kimmel, H. (2010). Advancing the. *The Journal of Technology Studies*, *36*(1), 53–64.

Rodriguez del Rey, Y. A., Cawanga Cambinda, I. N., Deco, C., Bender, C., Avello-Martínez, R., & Villalba-Condori, K. O. (2020). Developing computational thinking with a module of solved problems. *Computer Applications in Engineering Education*, 1–11. doi:10.1002/cae.22214

Rogers, C., & Portsmore, M. (2004). Bringing engineering to elementary school. *Journal of STEM Education: Innovations and Research*, *5*(3&4), 17–28.

Rogers, E. M. (2003). *Diffusion of innovations* (5th ed.). Free Press.

Roni, S. M., Merga, M. K., & Morris, J. E. (2020). *Conducting quantitative research in education*. Springer. doi:10.1007/978-981-13-9132-3

Ronsivalle, G. B., Boldi, A., Gusella, V., Inama, C., & Carta, S. (2019). How to Implement Educational Robotics' Programs in Italian Schools: A Brief Guideline According to an Instructional Design Point of View. Technology. *Knowledge and Learning*, *24*(2), 227–245. doi:10.100710758-018-9389-5

Rosenblatt, M., & Choset, H. (2000). Designing and implementing hands-on robotics labs. *IEEE Intelligent Systems & their Applications*, *15*(6), 32–39. doi:10.1109/5254.895856

Rosenshine, B., Meister, C., & Chapman, S. (1996). Teaching students to generate questions: A review of the intervention studies. *Review of Educational Research*, *66*(2), 181–221. doi:10.3102/00346543066002181

Ross, R., Stanger, J., & Console, A. (2017). Phototropic BristleBot activity for robotics and STEM engagement. In *2017 IEEE International Conference on Mechatronics (ICM)* (pp. 425-430). 10.1109/ICMECH.2017.7921144

Rubenstein, M., Ahler, C., & Nagpal, R. (2012) Kilobot: A low cost scalable robot system for collective behaviors. *2012 IEEE International Conference on Robotics and Automation*, 3293-3298. 10.1109/ICRA.2012.6224638

Saidin, D., Khalid, F., Martin, R., Kuppusamy, Y., & Nalini, A. P. (2021). Benefits and Challenges of Applying Computational Thinking in Education. *International Journal of Information and Education Technology. 11*(5), 248–254. .11.5.1519 doi:10.18178/ijiet.2021

Samson, A. C., Huber, O., & Gross, J. J. (2012). Emotion regulation in Asperger's syndrome and high-functioning autism. *Emotion (Washington, D.C.), 12*(4), 659–665. doi:10.1037/a0027975 PMID:22642342

Sang, G., Valcke, M., van Braak, J., & Tondeur, J. (2009). Investigating teachers' educational beliefs in Chinese primary schools: Socioeconomic and geographical perspectives. *Asia-Pacific Journal of Teacher Education, 37*(4), 363–377. doi:10.1080/13598660903250399

Sani-Bozkurt, S. (2017). Özel eğitimde dijital destek: Yardımcı teknolojiler. *Açıköğretim Uygulamaları ve Araştırmaları Dergisi, 3*(2), 37–60.

Sapounidis, T., & Alimisis, D. (2021). Educational Robotics Curricula: Current Trends and Shortcomings. In M. Malvezzi, D. Alimisis, & M. Moro (Eds.), *Education in & with Robotics to Foster 21st-Century Skills. Studies in Computational Intelligence* (Vol. 982). Springer. doi:10.1007/978-3-030-77022-8_12

Sawyer, B., & Rejeski, D. (2002). *Serious Game Initiative.* Academic Press.

Scaradozzi, D., Screpanti, L., Cesaretti, L., Storti, M., & Mazzieri, E. (2019). Implementation and assessment methodologies of Teachers' training courses for STEM activities. Technology. *Knowledge and Learning, 24*(2), 247–268. doi:10.100710758-018-9356-1

Scassellati, B., Admoni, H., & Matari, M. (2012). Robots for Use in Autism Research. *Annual Review of Biomedical Engineering, 14*(1), 275–294. doi:10.1146/annurev-bioeng-071811-150036 PMID:22577778

Scheffler, I. (1982). *Science and subjectivity.* Hackett Publishing.

Schenk, J. (1920). *Zur Reform des Unterrichtes des Maschinenbauwesens an den Technischen Hochschulen.* Dülfer.

Schina, D., Esteve-González, V., & Usart, M. (2021). An overview of teacher training programs in educational robotics: Characteristics, best practices and recommendations. *Education and Information Technologies, 26*(3), 2831–2852. doi:10.100710639-020-10377-z

Schlegel, C., Steck, A., Brugali, D., & Knoll, A. (2010). Design abstraction and processes in robotics: From code-driven to model-driven engineering. *International Conference on Simulation, Modeling, and Programming for Autonomous Robots*.

Schmayl, W. (1981). *Das Experiment im Technikunterricht - methodologische und didaktische Studien zur Grundlegung einer Unterrichtsmethode*. Franzbecker.

Schmidt, H., Loyens, S., Van Gog, T., & Paas, F. (2007). Problem-based learning is compatible with human cognitive architecture: Commentary on Kirschner, Sweller, and Clark (2006). *Educational Psychologist, 42*(2), 91–97. doi:10.1080/00461520701263350

Schwanitz, D. (1999). *Bildung. Alles was man wissen muß.* Eichborn.

Seau Yoon, F., Ho, J., & Hedberg, J. G. (2005). Teacher understandings of technology affordances and their impact on the design of engaging learning experiences. *Educational Media International, 42*(4), 297–316. doi:10.1080/09523980500237716

Seeman, N. C. (2005). Structural DNA nanotechnology. *NanoBiotechnology Protocols*, 143-166.

Selby, C. C. (2014). *How can the teaching of programming be used to enhance computational thinking skills?* (Unpublished doctoral dissertation). University of Southampton, Southampton, UK.

Seng, J. (2021). Constructing Robots for Undergraduate Projects Using Commodity Aluminum Build Systems. In M. Merdan, W. Lepuschitz, G. Koppensteiner, R. Balogh, D. Obdržálek (Eds.), Robotics in Education. RiE 2021. Advances in Intelligent Systems and Computing. Academic Press.

Sengupta, D. L., & Sarkar, T. K. (2003). Maxwell, Hertz, the Maxwellians, and the early history of electromagnetic waves. *IEEE Antennas & Propagation Magazine*, *45*(2), 13–19. doi:10.1109/MAP.2003.1203114

Sengupta, P., Kinnebrew, J. S., Basu, S., Biswas, G., & Clark, D. (2013). Integrating computational thinking with K-12 science education using agent-based computation: A theoretical framework. *Education and Information Technologies*, *18*(2), 351–380. doi:10.100710639-012-9240-x

Sentence, S., Waite, J., Hodges, S., MacLeod, E., & Yeomans, L. (2017). "Creating cool stuff" pupils' experience of the bbc micro:bit. In *Proceedings of the 2017 ACM SIGCSE technical symposium on computer science education* (pp. 531–536). ACM. 10.1145/3017680.3017749

Sergis & Koutromanos. (2013). Effects of an INSET in ICT in education to the educators. *Journal of Issues of Science and Technology in Education*, *6*, 1–2.

Setiawan, A. R., & Koimah, S. (2019). *Effective learning and teaching*. Academic Press.

Settle, A., & Perkovic, L. (2010). *Computational Thinking across the Curriculum: A Conceptual Framework*. Technical Reports.

Shamsuddin, S., Yussof, H., Ismail, L. I., Mohamed, S., Hanapiah, F. A., & Zahari, N. I. (2012). Humanoid robot NAO interacting with autistic children of moderately impaired intelligence to augment communication skills. *Procedia Engineering*, *41*(Iris), 1533–1538. . doi:10.1016/j.proeng.2012.07.346

Shanahan, M.-C., Burke, L. E. C.-A., & Francis, K. (2016). Using a boundary object perspective to reconsider the meaning of STEM in a Canadian context. *Canadian Journal of Science, Mathematics and Technology Education*, *16*(2), 129–139.

Shanmugam, L., Yassin, F., & Khalid, F. (2019). Enhancing students' motivation to learn computational thinking through mobile development module (M-CT). *International Journal of Engineering and Advanced Technology*, *8*(5), 1293–1303.

Sheridan, K., & Konopasky, A. (2015). *An Experimental Study Comparing Two Educational Approaches to Making With Simple Circuits* [Paper presentation]. Conference American Educational Research Association.

Shin, J., Siegwart, R., & Magnenat, S. (2014). *Visual Programming Language for Thymio II Robot. Interaction Design and Children*. IDC. doi:10.3929/ethz-a-010144554

Shin, W. S. (2020). An Analysis of Teachers' TPACK on Robotics in Education. *Journal of the Korea Academia-Industrial Cooperation Society*, *21*(6), 196–203.

Shire, S. Y., Gulsrud, A., & Kasari, C. (2016). Increasing responsive parent–child interactions and joint engagement: Comparing the influence of parent-mediated intervention and parent psychoeducation. *Journal of Autism and Developmental Disorders*, *46*(5), 1737–1747. doi:10.100710803-016-2702-z PMID:26797940

Shute, V. J., Sun, C., & Asbell-Clarke, J. (2017). Demystifying computational thinking. *Educational Research Review*, *22*, 142–158. doi:10.1016/j.edurev.2017.09.003

Siciliano, B., & Khatib, O. (2016). Springer handbook of robotics. Springer.

Siemens, G. (2017). Connectivism. *Foundations of Learning and Instructional Design Technology*.

Sigüenza, M., Guillen, D., Arroyo, D., & Cuellar, F. (2017). *Mobile robots development: A case study from robotics competitions and course projects.* Paper presented at the 2017 IEEE XXIV International Conference on Electronics, Electrical Engineering and Computing (INTERCON). 10.1109/INTERCON.2017.8079703

Sisman, B., & Kucuk, S. (2019). An Educational Robotics Course: Examination of Educational Potentials and Pre-Service Teachers' Experiences. *International Journal of Research in Education and Science, 5*(2), 510–531.

Sisman, B., & Kucuk, S. (2019). Educational robotics course: Examination of educational potentials and preservice teachers' experiences. *International Journal of Research in Education and Science, 5*(2), 510–531.

Skinner, B. (1986). Programmed instruction revisited. *Phi Delta Kappan, 68*(2), 103–110.

Skinner, B. F. (1960). Teaching Machines. *The Review of Economics and Statistics, 42*(3), 189–191. doi:10.2307/1926170

Slangen, L., Keulen, H., & Gravemeijer, K. (2011). Preparing teachers to teach robotics in primary schools. In Professional Development for Primary Teachers in Science and Technology (Vol. 9, pp. 181-198). doi:10.1007/978-94-6091-713-4_14

Slangen, L., Keulen, H., & Gravemeijer, K. (2010). What pupils can learn from working with robotic direct manipulation environments. *International Journal of Technology and Design Education, 21*(4), 449–469. doi:10.100710798-010-9130-8

Slurp. (2010, January 3). *Bristlebots?* https://picaxeforum.co.uk/threads/bristlebots.14219/post-121456

Smith, D. W. (2008). *Assistive technology competencies for teachers of students with visual impairments: A delphi study* [Unpublished doctoral dissertation]. Texas Tech University.

Smyrnova-Trybulska, E., Staniek, D., & Zegzuła, D. (2020). Robotics in education a survey report: Case study. *International Journal of Research in E-learning, 6*(1), 1–18. doi:10.31261/IJREL.2020.6.2.01

Snow, C. P. (1967). *Die zwei Kulturen. Literarische und naturwissenschaftliche Intelligenz.* Klett.

Sola-Ozguc, S. (2017). Özel eğitimde yardımcı teknolojiler. In Özel Eğitim içinde (pp. 391-404). Vize Akademik Publishing.

Solomon, C., Harvey, B., Kahn, K., Lieberman, H., Miller, M. L., Minsky, M., . . . Silverman, B. (2020). History of Logo. *Proc. ACM Program. Lang., 4*(HOPL), Article 79. 10.1145/3386329

So, W. C., Cheng, C. H., Lam, W. Y., Huang, Y., Ng, K. C., Tung, H. C., & Wong, W. (2020). A Robot-Based Play-Drama Intervention May Improve the Joint Attention and Functional Play Behaviors of Chinese-Speaking Preschoolers with Autism Spectrum Disorder: A Pilot Study. *Journal of Autism and Developmental Disorders, 50*(2), 467–481. doi:10.100710803-019-04270-z PMID:31655965

Spion. (2008, March 26). *Mein 10min. IC-Bot (BristleBot/Vibrobot).* https://www.roboternetz.de/community/threads/35068-mein-10min-IC-Bot-%28BristleBot-Vibrobot%29?p=333265&viewfull=1#post333265

Stanger, C. A., & Cawley, M. F. (1996). Demographics of rehabilitation robotics users. *Technology and Disability, 5*(2), 125–137. doi:10.3233/TAD-1996-5202

Stangl, W. (2021). *Stichwort: "Scaffolding".* Online Lexikon für Psychologie und Pädagogik: https://lexikon.stangl.eu/13399/scaffolding/

Starr-Glass, D. (2020). Encouraging engagement: Video-conference augmentation of online distance learning environments. *On the Horizon, 28*(3), 125–132. doi:10.1108/OTH-06-2020-0020

Steen-Utheim, A., & Wittek, A. L. (2017). Dialogic feedback and potentialities for student learning. *Learning, Culture and Social Interaction, 15*, 18–30.

Stewart, T., Myers, A., & Culley, M. (2010). Enhanced learning and retention through "writing to learn" in the psychology classroom. *Teaching of Psychology, 37*(1), 46–49. doi:10.1080/00986280903425813

Stewart, W., Baek, Y., Kwid, G., & Taylor, K. (2021). Exploring Factors That Influence Computational Thinking Skills in Elementary Students' Collaborative Robotics. *Journal of Educational Computing Research, 59*(6), 1208–1239. Advance online publication. doi:10.1177/0735633121992479

Stoeckelmayr, K., Tesar, M., & Hofmann, A. (2011). Kindergarten children programming robots: A first attempt. *Proceedings of 2nd International Conference on Robotics in Education (RIE).*

Stork, M. G. (2020). Supporting Twenty-First Century Competencies Using Robots and Digital Storytelling. *Journal of Formative Design in Learning, 4*(1), 1–8. doi:10.100741686-019-00039-w

Strider_mt2k. (2010, November 4). *RC bristlebot shifts weight for steering.* https://hackaday.com/2010/11/03/rc-bristlebot-shifts-weight-for-steering/#comment-207571

Stripling, T., & Simmons, B. (2016). Get students revved up! Robotics brings excitement to STEM. *Tech Directions, 75*(7), 13–17.

Sullivan, A., Kazakoff, E. R., & Bers, M. U. (2013). The wheels on the bot go round and round: Robotics curriculum in pre-kindergarten. *Journal of Information Technology Education, 12*, 203–219.

Sullivan, A., Kazakoff, E. R., & Bers, M. U. (2013). The Wheels on the Bot go Round and Round: Robotics Curriculum in Pre-Kindergarten. *Journal of Information Technology Education: Innovations in Practice, 12*, 203–219. http://www.jite.org/documents/Vol12/JITEv12IIPp203-219Sullivan1257.pdf

Sullivan, F. R. (2008). Robotics and Science Literacy : Thinking Skills. *Science Process Skills and Systems Understanding, 45*(3), 373–394.

Sullivan, F. R., & Heffernan, J. (2016). Robotic construction kits as computational manipulatives for learning in the STEM disciplines. *Journal of Research on Technology in Education, 48*(2), 105–128. doi:10.1080/15391523.2016.1146563

Swaid, S. I. (2015). Bringing computational thinking to STEM education. *Procedia Manufacturing, 3*, 3657–3662. doi:10.1016/j.promfg.2015.07.761

Sweller, J., Kirschner, P., & Clark, R. E. (2007). Why minimally guided teaching techniques do not work: A reply to commentaries. *Educational Psychologist, 42*(2), 115–121. doi:10.1080/00461520701263426

Szczys, M. (2010, November 3). *RC Bristlebots shifts weight for steering.* https://hackaday.com/2010/11/03/rc-bristlebot-shifts-weight-for-steering

Takeuchi, M. A., Sengupta, P., Shanahan, M. C., Adams, J. D., & Hachem, M. (2020). Transdisciplinarity in STEM education: A critical review [Review]. *Studies in Science Education, 56*(2), 213–253. https://doi.org/10.1080/03057267.2020.1755802

Talaiver, M., & Bowen, R. (2010). Developing 21st-Century Skills: Game Design and Robotics Exploration. In *Society for Information Technology & Teacher Education International Conference* (pp. 2089-2090). Association for the Advancement of Computing in Education (AACE).

Talan, T. (2020). Eğitsel robotik uygulamaları üzerine yapılan çalışmaların incelenmesi. *Yaşadıkça Eğitim Dergisi, 34*(2), 503–522. doi:10.33308/26674874.2020342177

Tang, X., Yin, Y., Lin, Q., Hadad, R., & Zhai, X. (2020). Assessing computational thinking: A systematic review of empirical studies. *Computers & Education, 148*, 103798. doi:10.1016/j.compedu.2019.103798

Tan, Y. J., Susanto, G. J., Anwar Ali, H. P., & Tee, B. C. (2021). Progress and Roadmap for Intelligent Self-Healing Materials in Autonomous Robotics. *Advanced Materials, 33*(19), 2002800.

Tapus, A., Peca, A., Aly, A., Pop, C., Jisa, L., Pintea, S., Rusu, A., & David, D. (2012). Children with Autism Social Engagement in Interaction with Nao, an Imitative Robot-A Series of Single Case Experiments. *Interaction Studies: Social Behaviour and Communication in Biological and Artificial Systems, 13*(3). Advance online publication. doi:10.1075/is.13.3.01tap

Tatlisu, M. (2020). *The effect on problem solving skills of elementary schools students of problem based learning in educational robotic* [Unpublished master's dissertation]. Bursa Uludağ University.

Tauber, A. I. (2016). *Science and the Quest for Reality*. Springer.

Taylor, P., & C. (2016). Why is a STEAM curriculum perspective crucial to the 21st century? In *14th Annual conference of the Australian Council for Educational Research*. Brisbane: Murdoch University Research Repository.

Taylor, A. T., Berrueta, T. A., & Murphey, T. D. (2021). Active learning in robotics: A review of control principles. *Mechatronics, 77*, 102576.

Tekinarslan, E., & Yikmis, A. (2005). Özel eğitim kurumlarında çalışan öğretmenlerin teknoloji kullanımına yönelik görüşleri ve beklentileri. *Journal of Abant Social Sciences, 2*, 211–220.

Tengler, K. (2021). *Prototyping a robotics-based learning environment supporting computational thinking skills*. Submitted.

Tengler, K., Kastner-Hauler, O., & Sabitzer, B. (2021a). A Robotics-based Learning Environment Supporting Computational Thinking Skills – Design and Development. In *2021 IEEE Frontiers in Education Conference (FIE)*. IEEE.

Tengler, K., Kastner-Hauler, O., & Sabitzer, B. (2021c). Identifying preliminary design principles. *ICCSE 2021. The 4th International Conference on Computer Science & Education*, 733–738.

Tengler, K., Sabitzer, B., & Rottenhofer, M. (2019). "Fairy tale computer science"—Creative approaches for early computer science in primary schools. *ICERI Proceedings*, 8968-8974.

Tengler, K. (2020). Klein, kreativ, Ozobot: Förderung von Kreativität und informatischem Denken durch spielerisches Programmieren. *Resource (Saint Joseph, Mich.)*.

Tengler, K., Kastner-Hauler, O., & Sabitzer, B. (2021b). Enhancing Computational Thinking Skills using Robots and Digital Storytelling. *CSEDU*, (1), 157–164.

Thacker, C., McCreight, E., Lampson, B., Sproull, R., & Boggs, D. (1981). *Alto: A personal computer*. Academic Press.

Theodoropoulos, A., Antoniou, A., & Lepouras, G. (2017). How do different cognitive styles affect learning programming? Insights from a game-based approach in Greek schools. *ACM Transactions on Computing Education, 17*(1), 3. doi:10.1145/2940330

Toh, L. P. E., Causo, A., Tzuo, P.-W., Chen, I.-M., & Yeo, S. H. (2016). A Review on the Use of Robots in Education and Young Children. *Journal of Educational Technology & Society, 19*(2), 148–163. http://www.jstor.org/stable/jeductechsoci.19.2.148

Tondeur, J., Valcke, M., & Van Braak, J. (2008). A multidimensional approach to determinants of computer use in primary education: Teacher and school characteristics. *Journal of Computer Assisted Learning, 24*(6), 494–506. doi:10.1111/j.1365-2729.2008.00285.x

Tossel, K., Hammond, A., Arvelo, E., & Martins, N. (2010). Vision-based Micro-robot Control [poster presentation]. University of Maryland.

Tramonti, M., & Dochshanov, A. (2018). *Students' engagement through computational thinking and robotics*. Digital Presentation and Preservation of Cultural and Scientific Heritage.

Tremblay, M. C., Hevner, A. R., & Berndt, D. J. (2010). The Use of Focus Groups in Design Science Research. In Design Research in Information Systems: Theory and Practice (pp. 121–143). Springer US. doi:10.1007/978-1-4419-5653-8_10

Tucker, M. & Miller, T. (2021, March 24). *Navigate with nanobots - (steam "212" design challenge)*. ITEEA conference 2021.

Ucgul, M., & Cagiltay, K. (2014). Design and development issues for educational robotics training camps. *International Journal of Technology and Design Education*, *24*(2), 203–222. doi:10.100710798-013-9253-9

Uluer, P., Akalın, N., & Köse, H. (2015). A new robotic platform for sign language tutoring. *International Journal of Social Robotics*, *7*(5), 571–585. doi:10.100712369-015-0307-x

UN. (2021). *Article 24 – Education.* https://www.un.org/development/desa/disabilities/convention-on-the-rights-of-persons-with-disabilities/article-24-education.html

Underling. (2010, November 3). *RC Bristlebots shifts weight for steering*. https://hackaday.com/2010/11/03/rc-bristlebot-shifts-weight-for-steering/#comment-207526

Underling. (2011, March 3). *Bristlebot*. http://underlinglabs.blogspot.com/2011/03/bristle-bot.html

Unglaube, H. (1995). Experimente im Sachunterricht. In R. Meier, H. Unglaube & G. Faust-Siehl (Eds.), Sachunterricht in der Grundschule (pp. 224-236). Frankfurt a. M.

Vahldick, A., Mendes, A. J., & Marcelino, M. J. (2014). A review of games designed to improve introductory computer programming competencies. *2014 IEEE Frontiers in Education Conference (FIE) Proceedings*, 1–7.

Valadao, C. T., Loterio, F., Cardoso, V., Bastos, T., Frizera-Neto, A., & Carelli, R. (2015). Robotics as a tool for physiotherapy and rehabilitation sessions. *IFAC-PapersOnLine*, *48*(19), 148–153. doi:10.1016/j.ifacol.2015.12.025

Välimaa, J., & Hoffman, D. (2008). Knowledge society discourse and higher education. *Higher Education*, *56*(3), 265–285. doi:10.100710734-008-9123-7

Valtonen, T., Kukkonen, J., Kontkanen, S., Sormunen, K., Dillon, P., & Sointu, E. (2015). The impact of authentic learning experiences with ICT on pre-service teachers' intentions to use ICT for teaching and learning. *Computers & Education*, *81*, 49–58. doi:10.1016/j.compedu.2014.09.008

Van den Akker, J., Gravemeijer, K., McKenney, S., & Nieveen, N. (2006). Introducing educational design research. *Educational Design Research*, *1*(3).

Van der Loos, H. M., Reinkensmeyer, D. J., & Guglielmelli, E. (2016). Rehabilitation and health care robotics. In B. Siciliano & O. Khatib (Eds.), *Springer Handbook of Robotics*. doi:10.1007/978-3-319-32552-1_64

Van Lith, P., & Caska, J. (2007). *RoboDidactics Manual*. Academic Press.

Vanderborght, B., Simut, R., Saldien, J., Pop, C., Rusu, A. S., Pintea, S., Lefeber, D., & David, D. O. (2012). Using the social robot Probo as a social story telling agent for children with ASD. *Interaction Studies: Social Behaviour and Communication in Biological and Artificial Systems*, *13*(3), 348–372. doi:10.1075/is.13.3.02van

Vandevelde, C., Ciocci, M.-C., Saldien, J., & Vanderborght, B. (2013). Overview of Technologies for Building Robots in the Classroom. *4th International Conference on Robots in Education*. Lodz, Poland.

Vartholomeos, P., & Papadopoulos, E. (2008). Dynamics, design and simulation of a novel microrobotic platform employing vibration microactuators. *Journal of Dynamic Systems, Measurement, and Control*, *128*, 122–133.

Vega, J., & Cañas, J. (2018). PiBot: An Open Low-Cost Robotic Platform with Camera for STEM Education. *Electronics (Basel)*, *7*(12), 430. https://doi.org/10.3390/electronics7120430

Venker, C. E., McDuffie, A., Ellis Weismer, S., & Abbeduto, L. (2012). Increasing verbal responsiveness in parents of children with autism: A pilot study. *Autism*, *16*(6), 568–585. doi:10.1177/1362361311413396 PMID:21846665

Verner, I. M., & Revzin, L. B. (2017). *Robotics in school chemistry laboratories*. Academic Press.

Verotti, M., Dochshanov, A., & Belfiore, N. P. (2017a). Compliance Synthesis of CSFH MEMS-Based Microgrippers. *Journal of Mechanical Design*, *139*(2), 022301. https://doi.org/10.1115/1.4035053

Verotti, M., Dochshanov, A., & Belfiore, N. P. (2017b). A Comprehensive Survey on Microgrippers Design: Mechanical Structure [Review]. *Journal of Mechanical Design*, *139*(6), 060801. https://doi.org/10.1115/1.4036351

Vescio, V., Ross, D., & Adams, A. (2008). A review of research on the impact of professional learning communities on teaching practice and student learning. *Teaching and Teacher Education*, *24*(1), 80–91. doi:10.1016/j.tate.2007.01.004

Vidal-Hall, C., Flewitt, R., & Wyse, D. (2020). Early childhood practitioner beliefs about digital media: Integrating technology into a child-centred classroom environment. *European Early Childhood Education Research Journal*, *28*(2), 167–181. doi:10.1080/1350293X.2020.1735727

Vieyra, M., Gilmore, J., & Timmerman, B. (2011). Requiring research may improve retention in STEM fields for underrepresented women. *Council on Undergraduate Research Quarterly*, *32*(1), 13–20.

Voit, E. O. (2019). Perspective: Dimensions of the scientific method. *PLoS Computational Biology*, *15*(9), e1007279. doi:10.1371/journal.pcbi.1007279 PMID:31513575

Vossoughi, S., & Bevan, B. (2014). Making and tinkering: A review of the literature. *National Research Council Committee on Out of School Time STEM*, *67*, 1–55.

Vrasidas, C., Pattis, I., Panaou, P., Antonaki, M., Aravi, C., Avraamidou, L., & Zembylas, M. (2010). Teacher use of ICT: Challenges and opportunities. *Proceedings of the 7th International Conference on Networked Learning*.

Vygotsky, L. S. (1978). Mind in Society: The Development of Higher Psychological Processes. doi:10.1007/978-3-540-92784-6

Vygotsky, L. S. (1978). Mind in Society: The Development of Higher Psychological Processes. Harvard University Press.

Wadsworth, B. J. (1996). *Piaget's theory of cognitive and affective development: Foundations of constructivism*. Longman Publishing.

Wainer, J., Dautenhahn, K., Robins, B., & Amirabdollahian, F. (2010). Collaborating with Kaspar: Using an autonomous humanoid robot to foster cooperative dyadic play among children with autism. *2010 10th IEEE-RAS International Conference on Humanoid Robots*.10.1109/ICHR.2010.5686346

Watson, J. B. (2017). *Behaviorism*. Routledge. doi:10.4324/9781351314329

Wecker, C., & Fischer, F. (2011). From guided to self-regulated performance of domain general skills: The role of peer monitoring during the fading of instructional scripts. *Learning and Instruction*, *21*(6), 746–756. doi:10.1016/j.learninstruc.2011.05.001

Wee, L.K. & Lee, T. (2012). *Video Analysis and Modeling Tool for Physics Education: A workshop for Redesigning Pedagogy*. Academic Press.

Weinberg, J. B., Engel, G. L., Gu, K., Karacal, C. S., Smith, S. R., White, W. W., & Yu, X. W. (2001). A multidisciplinary model for using robotics in engineering education. *Proceedings of the 2001 ASEE annual conference and exposition.*

Weinberg, J. B., White, W. W., Karacal, C., Engel, G., & Hu, A.-P. (2005). Multidisciplinary teamwork in a robotics course. *ACM SIGCSE Bulletin.*

Weinberg, J. B., & Yu, X. (2008). Low-cost platforms for teaching integrated systems. *IEEE Robotics & Automation Magazine*, 4–6.

Weintrop, D., Beheshti, E., Horn, M. S., Orton, K., Trouille, L., & Jona, K. (2014). *Interactive assessment tools for computational thinking in High School STEM classrooms.* Paper presented at the international conference on intelligent Technologies for interactive entertainment. 10.1007/978-3-319-08189-2_3

Welch, A. G. (2010). Using the TOSRA to assess high school students' attitudes toward science after competing in the FIRST robotics competition: An exploratory study. *Eurasia Journal of Mathematics, Science and Technology Education, 6*(3), 187–197. doi:10.12973/ejmste/75239

Wilhelm, J. (2007). *Engaging readers and writers with inquiry: Promoting deep understanding in language arts and the content areas with guiding questions*. Scholastic Teaching Resources.

Wilkening, F. (1982). *Unterrichtsverfahren. Lernbereich Arbeit und Technik*. Neckar.

Williams, D. C., Ma, Y., Prejean, L., Ford, M. J., & Lai, G. (2007). Acquisition of physics content knowledge and scientific inquiry skills in a robotics summer camp. *Journal of Research on Technology in Education, 40*(2), 201–216.

Williams, E. (2014). *AVR Programming: Learning to Write Software for Hardware*. Make Community, LLC.

Williams, M. D., & Williams, J. (2007). A change management approach to evaluating ICT investment initiatives. *Journal of Enterprise Information Management, 20*(1), 32–50. doi:10.1108/17410390710717129

Wing, J. M. (2011, March). *Computational thinking.* Retrieved from https://csta.acm.org/Curriculum/sub/CurrFiles/WingCTPrez.pdf

Wing, J.M. (2011). Research notebook: Computational thinking-What and why. *The Link Magazine, 6*, 20-23.

Wing, J. (2008). Computational thinking and thinking about computing. *Proceedings of the IPDPS 2008: 22nd IEEE International Parallel and Distributed Processing Symposium*, 14–18.

Wing, J. M. (2006). Computational thinking. *Communications of the ACM, 49*(3), 33–35. doi:10.1145/1118178.1118215

Wing, J. M. (2008). Computational thinking and thinking about computing. *Philosophical Transactions - Royal Society. Mathematical, Physical, and Engineering Sciences, 366*(1881), 3717–3725. doi:10.1098/rsta.2008.0118 PMID:18672462

Wing, J. M. (2017). Computational thinking's influence on research and education for all. *Italian Journal of Educational Technology, 25*(2), 7–14. doi:10.17471/2499-4324/922

Wittel, H., Muhs, D., Jannasch, D., & Voßiek, J. (2015). Allgemeine und konstruktive Grundlagen. In H. Wittel, D. Muhs, D. Jannasch, & J. Voßiek (Eds.), Roloff/Matek Maschinenelemente (pp. 1–26). Springer Vieweg. https://doi.org/10.1007/978-3-658-09082-1_1.

Wonder-McDowell, C., Reutzel, D. R., & Smith, J. A. (2011). Does instructional alignment matter? Effects on struggling second graders' reading achievement. *The Elementary School Journal, 112*(2), 259–279. doi:10.1086/661524

Wood, S. (2003). Robotics ın the classroom: a teaching tool for K-12 educators. *Symposium of Growing up with Science and Technology in the 21st Century.*

World Health Organization & World Bank. (2011). *World report on disability 2011*. World Health Organization. https://apps.who.int/iris/handle/10665/44575

Xie, Y., Fang, M., & Shauman, K. (2015). STEM education. *Annual Review of Sociology, 41*(1), 331–357. doi:10.1146/annurev-soc-071312-145659 PMID:26778893

Yadav, A., Zhou, N., Mayfield, C., Hambrusch, S., & Korb, J. T. (2011). Introducing computational thinking in education courses. *Proceedings of the 42nd ACM technical symposium on Computer science education-SIGCSE0000'11*, 465–470. 10.1145/1953163.1953297

Yang, D., Swanson, S., Chittoori, B., & Baek, Y. (2018). Work-in-Progress: Integrating computational thinking in STEM education through a project-based learning approach. *Proceedings of 2018 American Society for Engineering Education (ASEE) Annual Conference and Exposition.* 10.18260/1-2--30091

Yang, X., Zhao, Y., Wu, W., & Wang, H. (2008, September). Virtual reality based robotics learning system. In *IEEE International Conference on Automation and Logistics* (pp. 859-864). IEEE. 10.1109/ICAL.2008.4636270

Yang, Y., Long, Y., Sun, D., Van Aalst, J., & Cheng, S. (2020). Fostering students' creativity via educational robotics: An investigation of teachers' pedagogical practices based on teacher interviews. *British Journal of Educational Technology, 51*(5), 1826–1842. doi:10.1111/bjet.12985

Yang, Y.-T. C., Chen, Y.-C., & Hung, H.-T. (2020). Digital storytelling as an interdisciplinary project to improve students' English speaking and creative thinking. *Computer Assisted Language Learning*, 1–23. doi:10.1080/09588221.2020.1750431

Yavuz-Konokman, G., & Cukurbasi, B. (2019). Effects of designing LEGO robotics instructional practices on the prospective science teachers' resistive behaviors towards technology supported instruction. *Malaysian Online Journal of Educational Technology, 7*(3), 57–71. doi:10.17220/mojet.2019.03.005

Yolcu, V., & Demirer, V. (2017). A review on the studies about the use of robotic technologies in education. *SDU International Journal of Educational Studies, 4*(2), 127–139.

Younie, S., & Leask, M. (2013). *Teaching With Technologies: The Essential Guide: The Essential Guide*. McGraw-Hill Education.

Yousif, J. H., Kazem, H. A., & Chaichan, M. T. (2019). Evaluation Implementation of Humanoid Robot for Autistic Children. International. *Journal of Computation and Applied Sciences, 6*(1), 412–420.

Zhou, H., Yuen, T. T., Popescu, C., Guillen, A., & Daviset, D. G. (2015). *Designing teacher professional development workshops for robotics integration across elementary and secondary school curriculum.* Academic Press.

Zhou, L., Parker, A. T., Smith, D. W., & Griffin-Shirley, N. (2011). Assistive technology for students with visual impairments: Challenges and needs in teachers' preparation programs and practice. *Journal of Visual Impairment & Blindness, 105*(4), 197–210. doi:10.1177/0145482X1110500402

Ziaeefard, S., Mahmoudian, N., Miller, M., & Rastgaar, M. (2016). *Engaging Students in STEM Learning through Co-Robotic Hands-On Activities (Evaluation).* Paper presented at 2016 ASEE Annual Conference & Exposition. doi:10.18260/p.26975

Zyda, M. (2005). From visual simulation to virtual reality to games. *Computer, 38*(9), 25–32. doi:10.1109/MC.2005.297

About the Contributors

Nikleia Eteokleous (female) holds a PhD in Educational Administration with emphasis in Educational Technology (confirmed 2004), M. Ed in Instructional Systems (2003), and M. Ed in Educational Administration (2002) from the Pennsylvania State University and BA in Public and Business Administration (Specialization: Finance) (2001) from the University of Cyprus. She works as an Associate Professor in Educational Technology at the Educational Sciences Department, School of Educational and Social Sciences at Frederick University. She is also the Head of the Distance Learning Committee (responsible for policy development, strategic planning, and professional development training). She is also the co-founder of Robotics Academy. Her research interests focus on technology integration in the educational practice, teacher education and technology, designing learning environments enhanced with technology and online learning environments. She is also involved in research activities related to open and online distance learning, and the integration of robotics in the educational practice.

Efi Nisiforou is a Lecturer in Distance Education, at the Department of Education, University of Nicosia. Dr. Nisiforou holds a Ph.D. in Educational Technology with an emphasis in Distance Education from the Cyprus University of Technology (CY). Her research interests include implications of Human-Computer Interaction (HCI) and Educational Neuroscience for the design and development of online learning environments in all levels of education. Dr. Efi has published papers in international conferences and in refereed journals and has received EU research funding. She also serves as an external evaluator in various e-learning research projects in the EU and beyond.

* * *

Soumela Atmatzidou is Informatics teacher in Secondary schools since 1989. She received Ph.D. in Educational Robotics as a means of Developing students' Computational Thinking and Metacognition, from the Informatics Department of Aristotle University of Thessaloniki (AUTh) (2018). She holds a Bachelor Degree from the Physics Department (1983), Master Diploma (MSc) in Electronic Physics (1998) and an MSc in ICT in Education from the Informatics Department (2008), of AUTh. Her research interests focus on educational robotics activities and explore how supportive interventions during an ER activity may help students develop and improve high order skills.

Evgeniia Bonnet works at the center LEARN (Center for Learning Sciences) of the EPFL and is in charge of developing the R2T2 project and the MOOCs on Thymio. She is also the president of the non-profit association R2T2 Collaboration. Evgeniia is the creator and co-creator of various educational activities related to the Thymio robot for all school levels (from 4 to 18 years old). She has two years of experience working at secondary school where she taught computer science courses and robotics. Evgeniia graduated from the Bauman Moscow State Technical University (BMSTU) and has a master's degree in Flight Dynamics and Control of Flying Vehicles. She has 4 year-experience of working in space industry at a ballistic office of the International Space Station (RSC Energia, Russia).

Stavros Demetriadis is Full Professor at the School of Informatics, Aristotle University of Thessaloniki (AUTh), Greece. He holds a BSc degree in Physics, MSc in Electronic Physics, and PhD in Multimedia educational technology from the Aristotle University of Thessaloniki. He teaches courses and conducts research in the broader area of Technology-enhanced learning with an emphasis on Computer-supported collaborative learning (CSCL), Adaptive hypermedia systems for learning, Educational robotics and Tangible interfaces for programming, Conversational agents, Multimedia learning, Cognitive training technologies. He has published more than 150 research papers in international scientific journals and international/national conference proceedings with more than 3500 third party citations and Google reported h-index 29 (as of July 2021). He has received 3 "Best paper" awards in various international conferences and the "Winners of Open Education Challenge 2014" and "NUMA Competition 2014" awards for the "Cubes Coding" project. Lately (ended 2021) he was Principal Investigator in the EU funded "colMOOC" project exploring the use of conversational agents to support learning with MOOCs. Stavros' personal webpage is: http://mlab.csd.auth.gr/sdemetri/.

Muhammet Demirbilek is an Associate Professor of Educational Technology in the Faculty of Education at Suleyman Demirel University in Isparta /Turkey. He earned his doctoral and master's degree in Educational Technology from the University of Florida. He also holds B.S. and M.S. degrees in electronics engineering from Istanbul University. He worked as a graduate faculty and visiting Assistant Professor in Digital Worlds Institute at University of Florida (2013-2014) and Post-Doctoral Researcher at Games, Learning, and Society (GLS) at the University of Wisconsin-Madison (2008-2009). His research interests include the impact of digital media and computer games and simulations on teaching and learning. How new media (e.g., Web 2.0, online social networks, mobile media). His recent research interests are the effects of online bullying, learning analytic, big data analysis, using mobile games to improve social skills of kids with autism. How augmented reality games on handheld computers designed and used for formal and informal education. How social media, online social networks, and mobile media restructure thinking, values, actions, education, community, and culture.

Alden Dochshanov is the scientific coordinator at EU-Track (www.eu-track.eu) Association (Italy). After pursuing a PhD taken from the University of Naples "Federico II" with an emphasis on materials engineering applied to biological objects, he has collaborated with the Department of Mechanical and Aerospace Engineering of the La Sapienza University of Rome in the area of microelectromechanical systems engineering. His primary experience is in nanomaterials synthesis and characterization. Currently, his professional research interests and activities are in Applied Electronics and Educational Robotics.

Martin Fislake was trained as a mechanic during an apprenticeship in Germany before he got his first and second teacher exam as a vocational teacher for mechanical engineering in 1993. Following he was hired 1993 at the University of Koblenz and is since then continuously working as a full-time senior lecturer and head of the technology education department where he teaches primary and lower secondary student teachers. 2001 he received his doctoral in engineering education, higher education didactics and educational economics at the University of Koblenz. In addition, he operates a technology-oriented summer school comprising about one hundred week-long courses a year for children and youth as an education and research project. Currently he writes and presents widely on issues of technology education with a focus on educational robotics and his out of school project technikcamps.

Diego Galvez Aranda holds a Bachelor of Science degree in Electrical Engineering from the Universidad Nacional de Ingenieria in Lima, Peru. During his undergrad days, he started a blog named "Not Just Bricks" in which he published projects and building instructions using the LEGO WeDo and LEGO Mindstorms set. That led to a job working closely with the Ministry of Education in Peru on their "Robotics in Schools" project. He has worked on robust and applicable robotics curriculum for the Peru school systems involving LEGO WeDO and Mindstorms. Additionally, he was part of the team that won first place in the WRO 2012 - National. Currently, he is pursuing a PhD degree in electrical engineering at Texas A&M University. His thesis work involves machine learning and molecular simulation of rechargeable batteries. The programming skills he's developed thanks to building LEGO Mindstorms sets during his childhood have helped lead to this career and educational path.

Mauricio Galvez-Legua is an Electronic Engineer graduated from the National University of Engineering (UNI) in Lima Peru. Postgraduate in "Evaluation and Accreditation of Educational Quality" at the National University of Education. Postgraduate in "Specialization in Virtual Learning Environments" from the Organization of American States. Specialization in "ISO 9000 Quality Management", "Industrial Automation" and "Industrial Robotics" at SENA Colombia. Specialization in "Numerical Control, Robotics and Computer Aided Manufacturing" at ALECOP Spain. Specialization in Applied Electronics at SENAI JICA Brazil. Currently Director of Social Projection and University Extension of the Faculty of Electrical and Electronic Engineering (FIEE) of the UNI. Head of the Computing Laboratory of the FIEE. Accreditation Advisor for the FIEE Quality Office. Professor at the National University of Engineering and at the Peruvian University of Applied Sciences (UPC). Co-author of the book Robotics Models Using LEGO WeDo 2.0 Design, Build, Program, Test, Document and Share, publisher Apress and Programming Structure with C, publisher UNI.

Christian Giang received the B.Sc. and M.Sc. degrees in electrical engineering and information technology from ETH Zürich, and the Ph.D. degree in educational robotics from the École Polytechnique Fédérale de Lausanne (EPFL), in 2020. He is currently a Researcher with EPFL and the Department of Education and Learning, University of Applied Sciences and Arts of Southern Switzerland (SUPSI-DFA). His research interest includes the use of educational technologies in formal education. Specifically, he investigates the use of educational robotics in classroom activities as well as students' strategies in inquiry-based learning activities involving environments such as interactive simulations or digital games.

Panagiotis Kosmas is currently the Head of Education Unit at CARDET (www.cardet.org) and a part-time lecturer in Distance Education, in the School of Education, University of Nicosia and at the Cyprus University of Technology. Panagiotis holds a PhD in Educational Technology from the Cyprus University of Technology, an M.Ed in Pedagogical Sciences (Universita Degli Studi Roma III - Italy), a MA in New Technologies in Learning and Communication, and a BA in Classical Studies (National University of Athens - Greece). He has several years of experience in Higher Education and in his PhD research has investigated the implementation of embodied learning via technology in real classrooms. His research interests include teaching methodologies mediated by technologies, child-computer interaction, language teaching and learning. As a researcher, he has published papers in international conferences and journals and has participated in numerous research projects.

Francesco Mondada is professor at the École Polytechnique Fédérale de Lausanne (EPFL), Switzerland and director of the Center for Learning Sciences at EPFL. After a master and a PhD received at EPFL, he led the design of many miniature mobile robots, commercialized and used worldwide in thousands of schools and universities. He co-founded several companies selling these robots or other educational tools. He is author of more than a hundred publications in the field of robot design. He received several awards, including the Swiss Latsis University prize, as best young researcher at EPFL and the Credit Suisse Award for Best Teaching as best teacher at EPFL.

Chrysanthi Mpekiari graduated from Mathematics Department (Applied Mathematics) of Aristotle University of Thessaloniki in 1986. He continued her studies to get two Masters' Degrees: One of them in Total Quality and Innovations in Education in the Pedagogical Department of Aristotle University of Thessaloniki (2018). The second Master's Degree in Information and Communication Technology in Education from the Informatics Department of Aristotle University of Thessaloniki (2019). As from December 2019, she is a Ph.D. candidate at the same University and Department with her primary interest in Educational Robotics, Computational Thinking, Massive Open Online Courses (MOOCs) and Machine Learning. She teaches Computer Science in Secondary Education the last 28 years.

Lucio Negrini received his PhD in Educational Sciences from the University of Fribourg (CH) and from the University of Konstanz (D) in 2015. He also holds a Master's degree in Educational Sciences from the University of Fribourg and a Bachelor's degree in Communication Sciences form the Università della Svizzera italiana. Currently he is Professor at the Department of Education and Learning at the University of Applied Sciences and Arts of Southern Switzerland (SUPSI-DFA). He is the Head of the Laboratory Media and STEM. His research focus are on the use of educational technologies in education with a focus on educational robotics.

Raphaela Neophytou (female) holds BA in Primary Education (Frederick University), MA in New Technologies for Communication and Learning (Cyprus University of Technology), and PhD in Education (Frederick University). Since 2018 she has been working at the Frederick University Open and Distance Learning Center. Within the laboratory, she works closely with the Commission and the Distance Learning Unit on issues related to pedagogical design, development of educational materials as well as quality control of distance learning. For 3 academic years, she had the role of Teaching Assistant (TA) in distance learning courses offered by the Department of Education Sciences. Since 2014 she has been a part of the Frederick University Robotics Academy Teaching Team as an instructor, visiting schools across the country where she organizes robotics lab classes and teaches robotics based tutoring to elementary school pupils. She has also participated in organizing and delivering teacher training for both preschool and primary education on educational robotics and the integration of new technologies into education. She is also a member of the Pedagogy Team where she develops material for the Academy's Curriculum utilizing various robotics training packages and visual programming platforms. She also attended a variety of workshops and seminars related to educational technology. She has participated in various academic conferences, educational and research programs mainly on ERASMUS + (eg RoboScientists) in Cyprus and abroad related to educational technology and robotics, as well as social, cultural and charity events organized by the Academy Frederick University of Robotics. Her research interests focus on technology integration in the educational practice, the field of which is based on her doctoral dissertation.

Andrea Philippou is a Registered School Psychologist in Cyprus (Registration No. 604). She completed her undergraduate studies in Psychology, in the United Kingdom, at the University of Surrey (BSc Psychology-3years with 1 year internship) and her postgraduate studies in School Psychology at the University College of London (UCL-MSc Psychology of Education-1 year) and the University of Cyprus (MA School Psychology-3 years). During her studies, she has conducted research projects in the field of Psychology. She has worked at the Educational Psychology Service of the Ministry of Education, Culture, Sports and Youth, completing 1500 hours of supervised internship. She has collaborated with public and private schools in the district of Larnaca, where she provided student assessment and guidance and offered supporting and counseling services to parents and teachers. She has dealt with a wide range of individual cases of children and adolescents experiencing learning, emotional and social difficulties and behavioral problems. At the same time, she developed preventive intervention programs and gave lectures to teachers and parents at school. At the present time, she is employed by CARDET and she is participating in projects that are related with Computation Thinking (CT).

Alberto Piatti is director of the Department of Education and Learning (DFA) at the University of Applied Sciences and Arts of Southern Switzerland (SUPSI) and senior teacher-researcher in Mathematics and Mathematics education at SUPSI. He holds a master's degree in Mathematics from ETH Zürich and a PhD in economics from the University of Lugano. After a Postdoc and advanced research activities in the field of human knowledge modelling at the Dalle Molle Institute for Artificial Intelligence (IDSIA) and teaching activities at SUPSI and at the University of Lugano, he joined SUPSI-DFA in 2010 as a member of the Executive Board and teacher-researcher in Mathematics Education.

Stefanos Plexousakis is a clinical Psychologist Lecturer and Researcher. He studied psychology in the University of Crete and after that he completed his postgraduate studies in the University of Greenwich, London in the field of Counselling and Psychotherapy (Dipl.,MSc, BACP ACCREDITED). He worked in different contexts of the British National Health System for a period of eight years. He is teaching in programs for special education and same time he has private practice working with children with emotional-behavioural problems, complex trauma as well as developmental disorders. Also, adults with anxiety, depression, and trauma. He completed his PhD thesis on Trauma from Bullying experiences in the university of Crete. He has published articles in scientific journals on Trauma, Bullying, developmental disorders also in the field of Counselling and Psychotherapy.

Panagiotis Psomos holds a PhD in Digital Educational Technology from the University of the Aegean, a M.Sc. in Information Systems (with full scholarship) from the University of Macedonia (2007) and a Diploma (5-year Engineer Degree) in Electrical & Computer Engineering from Aristotle University of Thessaloniki (2003). His main research interests include educational technology, educational neuroscience & computer science education.

Tarik Talan is an Assistant Professor in the department of computer engineering at Gaziantep Islam Science and Technology University. He has received the BSc degree in Department of Computer Education and Instructional Technology from Inonu University in 2005; MSc degree in Educational Technology from Suleyman Demirel University in 2014 and PhD degree from Istanbul University Informatics Department in 2018. His research areas include meta-analysis, meta-thematic analysis, bibliometric analysis, blended learning, distance learning and use of social media in education. He has published his academic works in national and international journals.

Karin Tengler is a lecturer and researcher at the University of Teacher Education Lower Austria in the Department of Media Pedagogy. She holds a Bachelor's degree in Education and a Master's in Applied Knowledge Management. Currently, she is conducting a design-based research project that focuses on promoting computational thinking skills of primary school students in robotics-based learning environments.

Laura-Aileen Thran studied German, philosophy and technology education at the University of Koblenz and finished her master studies at the end of 2020 and has started her traineeship for the teaching profession at lower secondary schools in January 2021. During her studies she was involved in the technikcamps out of school project at the University of Koblenz. Initially employed as a course instructor in 2014, she moved to support the management team as a student assistant in 2018 and was responsible for content, new course concepts and organizational issues.

Michela Tramonti is currently Vice-President of the EU-Track Association, a multidisciplinary centre that promotes research and development in different areas at the international level. PhD in Methodology of Education in Mathematics, Informatics and Information Technologies at the Institute of Mathematics and Informatics - Bulgarian Academy of Science. She collaborated as researcher and as assistant lecturer with the Sociology chair at the Università degli Studi di Perugia, as researcher at the Università degli Studi del Molise and Università degli Studi di Teramo. Expert in project development and management at the international level in e-learning, innovative technologies applied to learning and teaching, multimedia didactic environments.

Potheini Vaiouli is a Special Scientist for research (postdoctoral researcher) at the Center for Applied Neuroscience (CAN), University of Cyprus. Potheini holds an MA in Music Therapy (New York University) and a PhD in Special Education (Indiana University, Bloomington). Her research interests converge around social, emotional, and academic growth of young children with autism and other neurodevelopmental disabilities. Her research has focused primarily on the role of music in social communication skills (preverbal and verbal communication) and emotional development of young children with autism and their families. Potheini is experienced in mixed methods research design, she has collaborated with interdisciplinary teams and has extensive experience teaching and supervising undergraduate and graduate students.

Index

www.ingramcontent.com/pod-product-compliance
Lightning Source LLC
LaVergne TN
LVHW060301011026
849161LV00011B/2754